Detlev Möller
Chemistry for Environmental Scientists

Also of Interest

Atmospheric Chemistry. A Critical Voyage Through the History
Detlev Möller, 2022
ISBN 978-3-11-073739-4, e-ISBN (PDF) 978-3-11-073246-7,
e-ISBN (EPUB) 978-3-11-073251-1

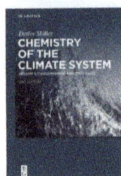

Chemistry of the Climate System. Volume 1: Fundamentals and Processes
Detlev Möller, 2019
ISBN 978-3-11-055975-0, e-ISBN (PDF) 978-3-11-056126-5,
e-ISBN (EPUB) 978-3-11-055992-7

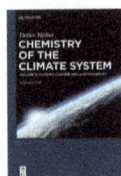

Chemistry of the Climate System. Volume 2: History, Change and Sustainability
Detlev Möller, 2020
ISBN 978-3-11-055985-9, e-ISBN (PDF) 978-3-11-056134-0,
e-ISBN (EPUB) 978-3-11-055996-5

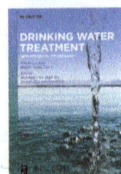

Drinking Water Treatment. New Membrane Technology
Bingzhi Dong, Tian Li, Huaqiang Chu, Huan He, Shumin Zhu, Junxia Liu
(Eds.), 2021
ISBN 978-3-11-059559-8, e-ISBN (PDF) 978-3-11-059684-7,
e-ISBN (EPUB) 978-3-11-059315-0

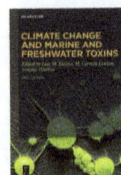

Climate Change and Marine and Freshwater Toxins
Luis M. Botana, M. Carmen Louzao, Natalia Vilarino, 2020
ISBN 978-3-11-062292-8, e-ISBN (PDF) 978-3-11-062573-8,
e-ISBN (EPUB) 978-3-11-062302-4

Detlev Möller

Chemistry for Environmental Scientists

2nd edition

DE GRUYTER

Author
Prof. Dr. Detlev Möller
Brandenburgische Technische Universität
Cottbus und Senftenberg
Platz der Deutschen Einheit 1
03046 Cottbus, Germany
de-moe@t-online.de

ISBN 978-3-11-073514-7
e-ISBN (PDF) 978-3-11-073517-8
e-ISBN (EPUB) 978-3-11-073025-8

Library of Congress Control Number: 2022934602

Bibliographic information published by the Deutsche Nationalbibliothek
The Deutsche Nationalbibliothek lists this publication in the Deutsche Nationalbibliografie;
detailed bibliographic data are available on the Internet at http://dnb.dnb.de.

© 2022 Walter de Gruyter GmbH, Berlin/Boston
Cover image: Urupong / iStock / Getty Images Plus
Back cover image: The author in front of the Keeling Building at Mauna Loa Observatory, Hawaii
in 2002
Typesetting: VTeX UAB, Lithuania
Printing and binding: CPI books GmbH, Leck

www.degruyter.com

Preface

I wrote this book preferably for non-chemists. Chemists need three fundamental books, each running to more than 1,000 pages: inorganic chemistry, organic chemist and physical chemistry. The present book comprises less than 10 % of that volume; it can never compensate for chemistry textbooks for chemists. Chemists need even more books, first on analytical chemistry, then technical chemistry, and for job specialisation many more books on special topics. Hence, there are an uncountable number of textbooks and monographs on the market, targeting almost all elements, compounds and substances, methods and subdisciplines in chemistry.

However, non-chemists in environmental sciences and engineering (e. g. physicists, biologists, ecologists, geographers, soil scientists, hydrologists, meteorologists, economists, engineers) need basic chemical knowledge to understand chemical processes in the environment. This book presents not simplified but reduced chemistry. It is not a book on environmental chemistry but a book explaining the chemical fundamentals needed for application in environmental sciences and engineering. However, as an example, you will find more on oxygen chemistry in the gas and aqueous phases here than in any other standard textbook on chemistry. This is likely also valid for the other main elements. Therefore, this book is also helpful for *chemists* beginning a job in environmental issues. A chemist would know everything from Chapters 2 and 3, but the knowledge presented in Chapters 4 and 5, you will not find in any standard chemistry textbook. There are several books on environmental chemistry or chemistry of the environment on the market, but none I might recommend. All are neither fish nor fowl. The reason for writing such textbooks is that many university departments have offered courses on environmental chemistry for almost three decades. The lecturers are mostly chemists with different backgrounds. A university chemical department will never offer a lecture simply on chemistry; the discipline of chemistry is too broad. Nevertheless, the chemistry of the environment encompasses not only the principal subdisciplines of physical, inorganic, organic and analytical chemistry but also the specific subdisciplines of atmospheric chemistry, aquatic chemistry, soil chemistry, geochemistry and biochemistry. For each subdiscipline, several voluminous textbooks are available. Moreover, interdisciplinary sciences such as toxicology, biogeochemistry and chemical engineering such as waste (water and solids) treatment and air pollution control should be included in a 'chemistry of the environment'. The reader now can likely understand that *one* book on environmental chemistry is impossible or only incomplete or even bad.

Nonetheless, scientists and engineers in the field of the environment become almost job-specialised and need additional sub-disciplinary chemical textbooks and monographs, but also books on the object, such as the atmosphere, hydrosphere, pedosphere and biosphere as well as environmental technologies.

My experience is based on 25 years (1987–2013) of teaching courses in atmospheric chemistry and air pollution control at the Humboldt University in Berlin (for geogra-

https://doi.org/10.1515/9783110735178-201

phers), the Free University in Berlin (for meteorologists) and the Technical University Cottbus (for environmental engineers). I used that teaching experience and 40 years of expertise from research in atmospheric chemistry (beginning in 1975) for my book *Chemistry of the Climate System* (first edition in 2010). In my library, I have all books (besides standard books on different chemical fields) available on atmospheric chemistry, aquatic chemistry, biogeochemistry, as well as many books on environmental and analytical chemistry. However, during all my years of teaching, I felt that there was missing a book on chemistry for non-chemists for application in environmental sciences and engineering, not too extensive, not too specific and not digressing in the endless chemistry of our environment.

There is another problem with a textbook in natural sciences. We only have three fundamental natural sciences: physics, chemistry and biology. Biology deals with the living matter and both other sciences with non-living matter. So far, biology is well defined, but an understanding of living matter also needs chemistry (biochemistry) and physics (biophysics). To separate chemistry and physics is not so simple and almost senseless. Chemistry without physics does not exist (for a chemical reaction to proceed, the substances must meet each other by transport and transfer processes). Physical chemistry is the 'theory' of chemistry. Physics, however, can be managed to a greater extent without chemistry. Mathematics is the tool to quantify processes, and there is no *a priori* need for a mathematician to know something from physics, chemistry and biology. There is no doubt that the degree of mathematisation decreases from physics to chemistry and biology. On the other hand, the complexity strongly increases from physics to chemistry and finally biology.

All explanations of chemical phenomena must ultimately be found within the electronic structure of atoms and molecules. However, deeper knowledge of this part of theoretical chemistry or quantum chemistry is not necessary to understand environmental chemistry, and the interested reader should look into adequate textbooks of general or inorganic chemistry. Due to this book's purpose, knowledge of chemical synthesis (under laboratory and technical conditions) is not needed. Furthermore, we will not deal with *pure* substances but with solutions and mixtures in different environmental reservoirs, mainly the gaseous and aqueous phases. Hence, knowledge of basic properties, uses and applications of chemical compounds will not be presented here.

There is no need to further emphasise that chemical processes in the environment are always interrelated with environmental physics. Chemical elements, substances and compounds are distributed among and cross the environmental compartments (they are in permanent motion) and undergo at the same moment chemical reactions (they are in permanent transformation). Fundamental physical and chemical laws and processes are valid at all places; in air, waters, soils, organisms and technical facilities. The only differences are given by specific conditions such as composition, volume, mass, time, temperature, pressure and so on. Therefore, understanding the stripping of sulphur dioxide in flue-gas desulphurisation equipment is not far from understanding the sub-cloud scavenging of sulphur dioxide by rain in the atmosphere. Hence, this book will focus on general and fundamental chemistry (including required physics) such as properties and bonding of matter, chemical kinetics and mechanisms, phase and chemical equilibrium, the basic features of air (gases), water (liquids) and soil (solids) and the most important substances and their reactions in the environment.

Hence, you must know the main *properties* of gaseous, liquid (aqueous) and solid matter (Chapter 2) before we go to the fundamentals of *changes* (equilibrium and reactions) in Chapter 3 and to explain the phenomena of chemical bonding and reactions (Chapter 3.3). Chapter 4 presents the elements and their compounds that are important in the environment under two aspects, namely the natural cycling and functioning of the biosphere, and which are responsible for human-made environmental problems. Finally, in Chapter 5, selected key environmental chemical processes under the light of multicomponent and multiphase chemistry are very shortly characterised; for the interested reader and those who become specialists, it is obligatory to pick up books on soils and biogeochemistry, the atmospheric and aquatic environment to understand the *specific* origin, transport, transformation and fate of chemicals. I recommend for atmospheric and global chemistry: Seinfeld and Pandis (1998), Finlayson-Pitts and Pitts (2000), Brasseur et al. (2003), Möller (2019); for biogeochemistry: Schlesinger (2013); for soil and groundwater: Langmuir (1997), Drever (1997); for soil chemistry: Sparks (2003), Sposito (2016), Tan (2010); for marine biogeochemistry: Libes (2009); for aquatic chemistry: Stumm and Morgan (1996); on trace elements in the environment: Prasad et al. (2005).

Berlin, March 2015   Detlev Möller

Preface to the 2nd edition

The publishing house deGruyter asked me to prepare a second edition. There is no need to update the text concerning general and fundamental chemistry (Chapters 2 and 3) for non-chemists: *nihil novi sub sole*. Nevertheless, research on chemicals in our environment is constantly advancing; thus, the newest results have been included, namely on halogens. On the other hand, whereas air pollution slips out more and more of our focus, abatement of *climate change* – unfortunately with a delay of two decades – is today's challenge. Understanding climate change and impacts needs interfacing and research among physics, chemistry and biology. Thus, some more on clouds, condensation nuclei and secondary aerosol are included in the chapter "Dust" to provide the reader with some fundamentals on the chain aerosol – cloud – climate.

Chemistry – in the sense of chemical processing of minerals and fossil fuels – developed over the last 200 years as the source of our property and environmental pollution as "One Way Dead End". However, chemistry can also provide sustainable solutions, nowadays called *green chemistry*, briefly introduced in an additional chapter. The message of *my* green chemistry, however, is to understand that human long-term survival is guaranteed only in a *cycling* world (product = waste = resource). Contrary to the prevailing view of a CO_2-free future, I will focus it on the chemistry of carbon and strongly believe that *fossil* fuels must be replaced by *solar* fuels. Moreover, *biogeochemistry* will experience a remake in the future as the most important environmental research. *Vernadsky*'s first ideas, about a century old, are enlightened in view of an ultimate need for human survival within a global cycling economy.

Needless to say, errors and mistakes are corrected, and wherever it was offered, additional information has been added.

Berlin, May 2022   Detlev Möller

https://doi.org/10.1515/9783110735178-202

Contents

Preface —— V

Preface to the 2nd edition —— IX

List of principal symbols —— XVII

1   Introduction —— 1
1.1   What do we mean by 'environment'? —— 1
1.2   What is chemistry? —— 2

2   Chemistry under environmental conditions —— 7
2.1   General remarks —— 7
2.2   States of matter —— 12
2.2.1   Atoms, elements, molecules, compounds and substances —— 13
2.2.2   Pure substances and mixtures —— 14
2.2.3   Concentration measures —— 15
2.3   Air and gases —— 21
2.3.1   Composition of the atmosphere —— 21
2.3.2   Properties of gases: the ideal gas —— 23
2.3.2.1 Kinetic theory of gases —— 24
2.3.2.2 Gas laws —— 25
2.3.2.3 Mean free path and number of collisions —— 28
2.3.2.4 Viscosity —— 30
2.3.2.5 Diffusion —— 32
2.4   Water and waters —— 33
2.4.1   Cycling and chemical composition of waters —— 34
2.4.2   Physical and chemical properties of water —— 40
2.4.2.1 Water structure: hydrogen bond —— 40
2.4.2.2 Water as a solvent —— 43
2.4.3   Properties of aqueous solutions —— 44
2.4.3.1 Surface tension and surface-active substances —— 44
2.4.3.2 Vapour pressure lowering: Raoult's law —— 45
2.4.3.3 Freezing point depression —— 46
2.4.3.4 Diffusion in solutions —— 47
2.4.4   Water vapour —— 47
2.5   Solid matter —— 48
2.5.1   General remarks —— 48
2.5.2   Soils —— 49
2.5.3   Dust —— 51
2.5.3.1 Soil and sea salt particles —— 52

2.5.3.2   Organic matter and soot —— 53
2.5.3.3   Condensation nuclei and cloud chemistry —— 55

3   **Fundamentals of physical chemistry —— 59**
3.1   Chemical thermodynamics —— 59
3.1.1   First law of thermodynamics and its applications —— 60
3.1.1.1   Internal energy —— 60
3.1.1.2   Molar heat capacity —— 62
3.1.1.3   Thermochemistry: heat of chemical reaction —— 64
3.1.2   Second law of thermodynamics and its applications —— 65
3.1.2.1   Entropy and reversibility —— 66
3.1.2.2   Thermodynamic potential: Gibbs–Helmholtz equation —— 68
3.1.2.3   Chemical potential —— 70
3.1.2.4   Chemical potential in real mixtures: activity —— 72
3.2   Equilibrium —— 73
3.2.1   Phase equilibrium —— 74
3.2.1.1   Gas-liquid equilibrium: evaporation and condensation —— 75
3.2.1.2   Gas-liquid equilibrium: special case for droplets —— 76
3.2.1.3   Absorption of gases in water: Henry's law —— 77
3.2.1.4   Solubility equilibrium: solid-aqueous equilibrium —— 81
3.2.1.5   Adsorption and desorption —— 82
3.2.2   Chemical equilibrium —— 84
3.2.2.1   Mass action law —— 85
3.2.2.2   Electrolytic dissociation —— 86
3.2.2.3   Acids, bases and the ionic product of water —— 87
3.2.2.4   pH value —— 92
3.2.2.5   Hydrolysis of salts and oxides —— 93
3.2.2.6   Buffer solutions —— 94
3.2.2.7   Complex ions —— 95
3.2.3   Dynamic equilibrium and steady state —— 97
3.3   Theory of chemical reactions —— 99
3.3.1   Chemical bonding —— 100
3.3.2   Types of chemical reactions —— 106
3.3.3   Chemical kinetics: reaction rate constant —— 108
3.3.4   Catalysis —— 116
3.3.5   Electrochemistry —— 116
3.3.5.1   Oxidation–reduction reaction (redox process) —— 117
3.3.5.2   Hydrated electron: a fundamental species —— 122
3.3.6   Photochemistry —— 126
3.3.6.1   Solar radiation transfer to the earth's surface —— 127
3.3.6.2   Photoexcitation: electronic states —— 130
3.3.6.3   Photodissociation: photolysis rate coefficient —— 132

3.3.6.4   Photocatalysis: photosensitising and autoxidation — 135
3.3.7   Heterogeneous chemistry — 139
3.3.8   Radicals, groups and nomenclature — 140

4   Chemistry of elements and their compounds in the environment — 145
4.1   General remarks — 145
4.2   Hydrogen — 150
4.2.1   Natural occurrence — 150
4.2.2   Compounds of hydrogen — 152
4.2.3   Chemistry — 153
4.3   Oxygen — 156
4.3.1   Natural occurrence — 158
4.3.2   Gas-phase chemistry — 158
4.3.2.1   Atomic and molecular oxygen: O, O_2 and O_3 — 158
4.3.2.2   Reactive hydrogen-oxygen compounds: OH, HO_2 and H_2O_2 — 161
4.3.3   Aqueous-phase chemistry — 163
4.3.3.1   Aqueous-phase oxygen chemistry — 166
4.3.3.2   Aqueous-phase hydrogen peroxide chemistry — 168
4.3.3.3   Aqueous-phase ozone chemistry — 171
4.3.3.4   Aqueous-phase OH chemistry — 174
4.4   Nitrogen — 175
4.4.1   Natural occurrence and sources — 179
4.4.2   Thermal dissociation of dinitrogen (N_2) — 180
4.4.3   Ammonia (NH_3) — 181
4.4.4   Dinitrogen monoxide (N_2O) — 184
4.4.5   Nitrogen oxides (NO_x) and oxoacids (HNO_x) — 185
4.4.5.1   Gas-phase chemistry — 185
4.4.5.2   Aqueous-phase chemistry — 190
4.4.6   Organic nitrogen compounds — 200
4.4.6.1   Amines, nitriles and cyanides — 201
4.4.6.2   Organic NO_x compounds — 205
4.5   Sulphur — 206
4.5.1   Natural occurrence and sources — 209
4.5.2   Reduced sulphur: H_2S, COS, CS_2, and DMS — 211
4.5.3   Oxides and oxoacids: SO_2, H_2SO_3, SO_3, and H_2SO_4 — 215
4.6   Carbon — 224
4.6.1   Elemental carbon — 225
4.6.2   Inorganic C_1 chemistry: CO, CO_2, and H_2CO_3 — 228
4.6.3   Organic carbon — 235
4.6.3.1   Hydrocarbon oxidation and organic ROS — 239
4.6.3.2 C_1 chemistry: CH_4, HCHO, CH_3OH and HCOOH — 244
4.6.3.3 C_2 chemistry: C_2H_6, CH_3CHO, C_2H_5OH, CH_3COOH and $(COOH)_2$ — 248

4.6.3.4   Alkenes, alkynes and ketones —— 254
4.6.3.5   Aromatic compounds —— 258
4.7   Halogens —— 261
4.7.1   Halogens in the environment —— 263
4.7.2   Halogen chemistry —— 272
4.8   Phosphorus —— 282
4.9   Metals and half-metals —— 287
4.9.1   General remarks —— 287
4.9.2   Alkali and alkaline earth like metals: Na, K, Mg, and Ca —— 290
4.9.3   Iron: Fe —— 291
4.9.4   Mercury: Hg —— 292
4.9.5   Cadmium: Cd —— 295
4.9.6   Lead: Pb —— 296
4.9.7   Arsenic: As —— 296
4.9.8   Silicon (Si) and aluminium (Al) —— 297

5   **Chemical processes in the environment —— 299**
5.1   Chemical evolution —— 301
5.1.1   Origin of elements and molecules —— 302
5.1.2   Formation of the Earth —— 306
5.1.3   Degassing the Earth and formation of the atmosphere —— 311
5.1.4   Evolution of life and atmospheric oxygen —— 318
5.1.5   Volcanism and weathering: inorganic CO_2 cycling —— 325
5.2   Biogeochemistry —— 328
5.2.1   Biogeochemical cycling —— 329
5.2.2   Principles of photosynthesis —— 332
5.2.3   Carbon cycle —— 337
5.2.4   Nitrogen cycle —— 343
5.2.5   Sulphur cycle —— 348
5.2.6   Acidity in the environment —— 352
5.3   Atmospheric chemistry and air pollution —— 355
5.3.1   Atmospheric acidification: "acid rain" —— 358
5.3.2   Ozone —— 365
5.3.2.1   Stratospheric ozone depletion: the "ozone whole" —— 366
5.3.2.2   Tropospheric ozone formation: "photochemical smog" —— 374
5.3.3   Halogens —— 382
5.3.3.1   Tropospheric halogen chemistry and ozone removal —— 382
5.3.3.2   Rethinking halogen loss from sea salt aerosol —— 388
5.3.4   Atmospheric removal: deposition processes —— 391
5.3.5   Radioactivity —— 395

6   **Green chemistry —— 399**

6.1   The carbon problem —— 402
6.2   The carbon economy —— 403
6.2.1   Carbon capture and storage (CCS) —— 404
6.2.2   Direct air capture (DAC) —— 406
6.2.3   Carbon dioxide cycling (DACCU) —— 409
6.2.4   Solar fuels: carbon as material and energy carrier —— 413

A   List of acronyms and abbreviations in environmental sciences found in
         literature —— 417

B   Quantities, units and some useful numerical values —— 421

C   List of the elements (alphabetically) —— 427

Bibliography —— 431

Author index —— 437

Subject index —— 439

List of principal symbols

Note: Only variables are in italic.

A, B, X	symbol for a general chemical species
[A]	concentration [square bracket] of substance A
a	activity
a	acceleration
α	degree of dissociation
α	Bunsen absorption coefficient
Acy	acidity
ads	adsorption (index)
aq	aqueous (index); in solution or dissolved in water
β	Ostwald's solubility
β	transfer coefficient
c	concentration
C_p	molar heat capacity at constant pressure
C_V	molar heat capacity at constant volume
d	diameter
d	diffusion (index)
D	diffusion coefficient
diss	dissolution (index)
e	number of elementary charges
$e, e-$	electron
ε	fraction (0...1)
E	electrical potential or electromotive force
E_A	activation energy
eq	equivalent
f	free energy (Helmholtz energy)
f	force
F	flux
F	molar free energy (Helmholtz energy)
F	Faraday constant
g	free enthalpy (Gibbs energy)
g	gaseous (index)
γ	surface tension
γ	activity coefficient
G	molar free enthalpy (Gibbs energy)
h	Planck's constant
h	enthalpy
H	Henry coefficient
H	molar enthalpy
H_0	Hammet function
het	heterogeneous (index)
i, j	specific component or particle (index)
I	electrical current
j	photolysis rate
k	Boltzman constant
k	reaction rate constant

https://doi.org/10.1515/9783110735178-203

κ	coefficient for absorption (a) or scattering (s)
K	equilibrium constant
K_f	cryoscopic constant
l	liquid (index)
l	mean-free path
LWC	liquid water content
λ	wavelength
λ	(radioactive) decay constant
m	molality
m	mass
m_m	mass of molecule or atom
M	molar mass
M	third body
max	maximum (index)
μ	chemical potential
n	amount (mole number)
n_0	Loschmidt constant
η	dynamic viscosity
N	number (of objects or subjects)
N_A	Avogadro constant
0	zero – reference concerns number, time or distance (index)
$\ominus$	index for standard conditions
Q	electric charge
Q	emission (source flux)
Q	(molar) heat
ν	frequency
ν	kinematic viscosity
p	pressure
par	particulate (index)
φ	azimuth angle
φ	electrical potential
Φ	quantum yield
q	area, surface
r	radius (of particles and droplets)
ρ	density
R	rate
R	removal (sink flux)
R	gas constant
RH	relative humidity
s	solid (index)
σ	cross section (for collision)
S	salinity
S	molar entropy
S_0	solar constant
S	saturation ratio
t	time
τ	residence time
τ	shear stress
τ	characteristic time

T	temperature
Θ	solar zenith angle
Θ	surface coverage degree (0...1)
U	molar inner energy
v	stoichiometric factor
v	velocity
v	kinematic viscosity
V	volume
V_m	molar volume
ψ	fugacity
W	(molar) work
x	mixing ratio
z	charge
z	collision number
z	ordinal number

1 Introduction

> Wherever we look upon our Earth, chemical action is seen taking place, on the land, in the air, or in the depths of the sea.

Adolph Stöckhardt (1809–1886)
"The Principles of Chemistry" (1851, p. 4)

1.1 What do we mean by 'environment'?

The terms *environ* (surround, enclose, encircle) and *environment* (surrounding) come from Old French. Thomas Carlyle (1795–1881) used the environment in 1827 to render German 'Umgebung' (today environment is rendered in German as 'Umwelt'). The German biologist Jacob von Uexküll (1864–1944) used 'Umwelt' first in 1909 in biology to denote the "surrounding of a living thing, which acts on it and influences its living conditions", nowadays termed as the biophysical environment. Whereas usually in relation to humanity, the number of biophysical environments is countless, given that it is always possible to consider an additional living organism that has its environment. The natural environment (synonym for habitat) encompasses all living and non-living things occurring naturally on Earth or some region thereof, an environment that encompasses the interaction of all living species.

Today, the expression *'environment'* often refers to the global environment, the earth system. However, each system to be defined lies in another 'mother' system, which is another surrounding or environment, hierarchically structured, where an exchange of energy and material is realised via the interfaces: cosmic system → solar system → earth system → climate system (global environment) → sub-systems (e. g. atmosphere, hydrosphere, pedosphere).

Consequently, there is no fully closed system in our world. In science and engineering, especially in thermodynamics, the environment is also known as the surroundings of a reservoir. It is the remainder of the total system that lies outside the boundaries of the regarded system. Depending on the type of system, it may interact with the environment by exchanging mass, energy, momentum or other conserved properties.

We see that there are different meanings for the term 'environment'. Following increasing use of this term in the 1950s, and related terms such as environmental pollution, environmental protection and environmental research, we must state that behind 'environment' are different natural components:

–   ecological units (habitats, ecosystems) that function as natural systems but also under human modification (note: nowadays there is no absolute natural system

https://doi.org/10.1515/9783110735178-001

on Earth without civilised human intervention), including all vegetation, microorganisms, soil, rocks, and atmospheric and natural phenomena that occur within their boundaries,
– natural resources such as air, water, soils and rocks (the climate system),
– built environment, including territories (settlements, agricultural and forest landscapes) and components (infrastructure) under strong human influence, belonging to a civilised society.

Furthermore, a geographic environment, the landscape, can be defined. However, all units such as ecosystem, landscape and habitat can be reduced to air, water, soil and living organisms. Living organisms (vegetation, microorganisms and animals, including humans) are an intrinsic part of the environment and the target of environmental protection. Air pollution control, water treatment and soil decontamination are the primary measures to avoid organism diseases. Harmful impacts on organisms are manifold: direct through toxicological effects of chemical substances, radiation, noise and land-use change; and indirectly through climate change. Naturally, the non-living world (natural resources and built environment) is also subject to the impacts of pollution and mismanagement (e. g. weathering, erosion, corrosion). Hence, the target of environmental protection is to gain a sustainable environment.

Chemistry of the environment means atmospheric chemistry, aquatic chemistry, and soil chemistry (note, we exclude biological chemistry because we consider the environment of organisms but know that understanding the environment chemistry is incomplete without considering the interaction between organism and the environment). Moreover, it is simply multiphase chemistry in and between the gas phase, the aqueous phase and the solid phase.

> There is a simple definition: *Soil chemistry* is the study of the chemical characteristics of the soil. Soil chemistry is affected by mineral composition, organic matter and environmental factors. When you exchange now the word soil for air and water, you know what atmospheric and aquatic chemistry means.

1.2 What is chemistry?

The definition of chemistry has changed over time, as new discoveries and theories add to the functionality of science. Chemistry, first established as a scientific discipline around 1650 (called chemistry) by Robert Boyle (1627–1691), had been a non-scientific discipline (alchemy) until then (Boyle 1680). Alchemy never employed a systematic approach, and because of its 'secrets', no public communication existed that would have been essential for scientific progress. In contrast, physics, established as a scientific discipline even earlier, made progress, especially concerning mechanics, thanks to the improved manufacturing of instruments in the sixteenth century.

Deep respect must be given to the two individuals who initiated the scientific revolution in both the physical and chemical understanding of the environment. First, Isaac Newton (1643–1727), who founded the principles of classical mechanics in his *Philosophiæ Naturalis Principia Mathematica* (1687). One hundred years later, Antoine Laurent de Lavoisier (1743–1794), with his revolutionary treatment of chemistry (1789), which made it possible to develop tools to analyse matter (Lavoisier 1789). This is why Lavoisier is called "the father of modern chemistry". We should not forget that the estimation of volume and mass was the sole foundation of the basic understanding of chemical reactions and physical principles after Boyle. While instruments to determine the mass (respectively weight) and volume had been known for thousands of years, the first modern analytical instruments were only developed in the late nineteenth century (spectrometry) and after 1950 (chromatography).

In the *Encyclopaedia Britannica*, published in Edinburgh in 1771 (shortly before the discovery of the chemical composition of air), chemistry is defined as: "to separate the different substances that enter into the composition of bodies [analytical chemistry in modern terms]; to examine each of them apart; to discover their properties and relations [physical chemistry in modern terms]; to decompose those very substances, if possible; to compare them together, and combine them with others; to reunite them again into one body, to reproduce the original compound with all its properties; or even to produce new compounds that never existed among the works of nature, from mixtures of other matters differently combined [synthetic chemistry in modern terms]".

This definition further evolved until, in 1947, it came to mean the science of substances: their structure, their properties, and the reactions that change them into other substances. A characterisation accepted by Linus Pauling (1901–1994) in his book *General Chemistry* (Dover Publications 1947) revolutionised the teaching of chemistry by presenting it in terms of unifying principles instead of as a body of unrelated facts. However, Wilhelm Ostwald (1853–1932) had already used such generalising principles in his book *Prinzipien der Chemie* (Leipzig 1907), subdividing chemistry into chapters of states of matter and properties of bodies, phase equilibrium, solutions and ions, chemical processes and reaction rates. The current book will follow that line.

As a short definition, chemistry is the scientific study of matter, its properties and interactions with other matter and energy. Consequently, inorganic and organic chemistry is the science of matter, physical and theoretical chemistry is the science of properties and interactions, and analytical chemistry is the science that studies the composition and structure of bodies.

As a sub-discipline of chemistry, *analytical chemistry* has the broad mission of understanding the composition of all matter. Much of early chemistry was analytical chemistry since the questions about what elements and chemicals are present in the world around us (the environment) and what is their fundamental nature are largely within the realm of analytical chemistry. Before 1800, the German term for analytical chemistry was '*Scheidekunst*' ('separation craft'); in Dutch, chemistry is still generally called '*scheikunde*'. Before developing reagents to identify substances by specific reactions, simple knowledge about the features of the chemicals (odour, colour, crystalline structure, etc.) was used to 'identify' substances. With Lavoisier's modern terminology of substances (1789) and his law of the conservation of mass, chemists acquired the basis for chemical analysis and synthesis. The German chemist Carl Remigius Fresenius (1818–1897) wrote the first textbook on analytical chemistry (1846), which is still generally valid. Today, almost all chemical analyses are based on physical methods using sophisticated instruments (such as gas chromatography – GC, liquid chromatography – LC, mass spectrometry – MS, atomic absorption spectrometry – AAS, inductively coupled plasma – ICP, combinations of them and others) requiring expert knowledge. Whereas sampling in air, water and soil is very specific, and the topic of appropriate handbooks and carried out by the (non-chemist) environmental scientist, sample treatment and analysis is mostly done by chemists.

However, "to look for definitions, to separate physics and chemistry fundamentally is impossible because they deal with the very same task, the insight into matter," the German chemist Jean D'Ans (1881–1969) wrote in the preface to his *Einführung in die allgemeine und anorganische Chemie* (Berlin 1948). Julius Adolf Stöckhardt (1808–1886) wrote in his textbook, *The Principles of Chemistry*: "Wherever we look upon our Earth, chemical action [a better translation from the original German is 'chemical processing'] is seen taking place, on the land, in the air, or in the depths of the sea" (English translation, Cambridge 1850, p. 4). Thus, chemistry is *a priori* the science of mineral (inorganic), animal and vegetable (organic) matter, the substances making our environment. First, Lavoisier systematically found that vegetable matter is composed of C, H and O and that in the animal matter, N and P are also present.

Systematic classification of chemistry into mineral, vegetable and animal according to its origin was carried out by the French chemist Nicolas Lémery (1645–1715), who wrote *Cours de chymie* (1675, cited after Kopp 1931). According to Walden (1941), the first use of the term 'organic chemistry' is now attributed to the Swedish chemist Jöns Jacob Berzelius (1779–1848), who termed it '*organisk kemie*' in a book published in 1806. After Lavoisier's revolutionary book, *Traité élémentaire de chimie* (1789), Berzelius wrote the first textbook *Lärbok i kemien* (1817–1830) in six volumes. It was soon published in French (1829) still with the now traditional title, *Traité de chimie minerale, vegetale et animale* in eight volumes with the subtitle *Chimie organique* (2ème partie – three volumes). In Germany, the first handbook, subtitled *Organic compounds*, is the third volume of *Handbuch der theoretischen Chemie* by Leopold Gmelin (1819), later rearranged into separate volumes of inorganic and organic chemistry (from 1848). It was believed that organic matter (also termed 'organised') could not be synthesised from its elements and a special force, the *vital force*, was needed for its production. However, processes not involving life can produce organic molecules. Friedrich Wöhler (1800–1882) destroyed the theory of vital force synthesising urea in 1828, an event generally seen as the turning point. Justus von Liebig (1803–1873) defined the task of organic chemistry as follows (in *Organic chemistry in its application to agriculture and physiology*, 1840):

> "The object of organic chemistry is to discover the chemical conditions which are essential to the life and perfect development of animals and vegetables, and, generally, to investigate all those processes of organic nature which are due to the operation of chemical laws" (The first phrase in Part I: *Of the chemical processes in the nutrition of vegetables*).

Gmelin (1848) wrote that carbon is the only element never missing, and hence, it is the only essential constituent in an organic compound. There has been no change in the definition since that time: the lexical database WordNet (Princeton University) defines organic chemistry as: "…the chemistry of compounds containing carbon (originally defined as the chemistry of substances produced by living organisms but now extended to substances synthesised artificially)". According to this definition, it appears, however, that (for example) carbon dioxide is an organic compound. It has been convenient to distinguish between inorganic and organic carbon compounds to explain the cycles between the biosphere and the atmosphere. On the other hand, when we state that there is prime chemistry considering a limited number of elements and an unlimited (or at least immense) number of molecules in a defined numeric relationship between the elements (including carbon), there is no need to separate chemistry into inorganic and organic chemistry. In contrast to the countless number of 'organic' carbon compounds, the number of 'inorganic' carbon compounds is rather small (of course, we count here only simple compounds found in nature and not produced in the laboratory). Hence, no separation into inorganic and organic chemistry is done in this book – we consider elements and their compounds.

With the focus of many scientific and engineering disciplines on our environment, many new sub-disciplines have arisen, such as biometeorology, bioclimatology, environmental chemistry, ecological chemistry, atmospheric environmental research, environmental meteorology, environmental physics and so on. Primary research is progressing with individuals and small groups on limited topics, and hence growing understanding (learning) and specialisation results in establishing 'scientific fields'. There is no other way to proceed in fundamental science. However, the complexity of the environment calls for an interdisciplinary approach, and we have learnt that especially at the 'interfaces' between biological, physical, chemical and geological systems, crucial and key processes occur in determining the function of the whole system. Therefore, a chemist (or non-chemist) without an understanding of the fundamental physical (or chemical) processes in the environment (and vice versa) can only work on discrete research topics and will not be able to describe the environment as a whole.

What we see is that all disciplines overlap. The original focuses of *geology* as the science of the (solid) earth, *hydrology* as the science of liquid water, and *meteorology* as the science of the atmosphere are still valid and should not be diffused. However, the modern study of earth sciences looks at the planet as a large, complex physical, chemical and biological interaction networks. This is known as a systems approach to the study of the global environment. The chemistry of the earth system (global environment) – when not considering life – is *geochemistry*. The term *geochemistry*, like many other scientific terms, has variable connotations. If geochemistry means simply the chemical study of the Earth or parts of the Earth, then geochemistry must be as old as chemistry itself and dates from the attempts of Babylonian and Egyptian metalworkers and potters to understand the nature and properties of their materials. Traditionally we may also subdivide the global environment into the atmosphere, lakes and rivers, the oceans, the soils, minerals, and volcanoes, consisting of gases, liquids and solids. Organisms – plants and animals – are distributed among these compartments.

It is self-evident that biochemistry deals with chemical processes in organisms. Thus, the chemical interaction between organisms occurs via geochemical processes; consequently, biogeochemistry is *the* chemistry of the environment that we have defined as that part of the earth system affecting life. Subdividing the geogenic part of the climate system into 'other' systems, we have the atmosphere, hydrosphere, cryosphere, and lithosphere; thus, atmospheric, aquatic and soil chemistry are the sub-disciplines of geochemistry. What now is environmental chemistry? It is all chemistry outside the laboratory and industrial reactors.

Studying the chemistry of a natural system is done by quantitative chemical analysis of the different matters. An uncountable number of analyses are needed to gain a three-dimensional distribution of the concentration of specific substances or, in other words, the chemical composition of the environment. This is the static approach to obtain an averaged chemical composition over a given time period, depending on the lifetime of the substances. However, concentration changes occur with time because of transportation (motion), transformation (chemical reactions) and emission (seasonal and diurnal variation). Hence, a dynamic approach is needed too (chemical monitoring and modelling). Chemical reactions can be studied – which means establishing the kinetics and mechanisms – only in the laboratory under controlled (and hence replicable) conditions in a large variety of reaction chambers.

2 Chemistry under environmental conditions

2.1 General remarks

As we have defined the global environment, it appears as a multi-reservoir system (air, water, soil) and a multiphase system (gaseous, aqueous and solid). Throughout the entire history of our planet, chemical, physical and biological processes have changed the composition and structure of its reservoirs. Beginning with a highly dynamic inner Earth 4.6 billion years ago, geochemical and geophysical processes have created the foundation for the Earth to become a habitat. With this, the formation of the hydrosphere (the oceans and a hydrological cycle) was the most important precondition for the evolution of living matter. Despite large changes in the chemical composition of the atmosphere, hydrosphere and lithosphere (the geospheres) over the ages, these spheres or reservoirs are well defined concerning such essential parameters as interfaces, volume, mass and others.

> The fundamental law of nature is the law of the conservation of matter. Matter occurs in two states: flowing energy and cycling material. The environment, however, is an open system and is in permanent exchange with its surrounding earth system, and thereby space.　ℹ️

Therefore, we generally expect changes and variations of internal energy and mass over time. From a chemical point of view, the interest lies in quantifying the amount – in terms of the number of particles and thereby mass – of different chemical species in a volume regarded. This quantity, called *density* (an equivalent term to *concentration*), is investigated as a function of time and space. Hence, mass, time and distance are the fundamental quantities of the environment, and all their quantities can be expressed in terms of meters, kilograms and seconds (Table B.2). The amount of matter is not static in the environment – we need to quantify its change by time and distance, mostly termed *rate* and *flux*.

The *rate R* is a ratio between two measurements, normally per unit of time: $R = \Delta\varepsilon/\Delta t$. However, a rate of change can be specified per unit of length (Δl) or mass (Δm) or another quantity, for example, $\Delta\varepsilon/\Delta x$ where Δx is the *displacement* (the shortest directed distance between any two points). In chemistry and physics, the word 'rate' is often replaced by (or synonymously used with) *speed* (see below about the difference to *velocity*). It can also be the distance covered per unit of time (i. e. *acceleration*, the rate of change in speed) or the change in speed per unit of time (i. e. *reaction rate*, the speed at which chemical reactions occur). In physics, *velocity* $\vec{v}$ is the rate of change of position. This is a vector physical quantity, and both *speed* and *direction* must define it. The scalar absolute value (magnitude) of velocity is *speed*. The *average* velocity $\bar{v}$ of an object moving through a displacement (Δx) during a time interval (Δt) is described by $\bar{v} = \Delta x/\Delta t$.

https://doi.org/10.1515/9783110735178-002

A *flux F* is defined as the amount that flows through a unit area per unit of time: $F = \Delta\varepsilon/q\Delta t$ (q area). Flux in this definition is a vector. However, in general, 'flux' in environmental research relates to the movement of a substance between compartments. This looser usage is equivalent to a rate (change of mass per time); sometimes, the term *specific* rate is used in the more exact sense of time- *and* area-related flux. Generally, the terms 'rate' and 'flux' as well as 'velocity' and 'speed' are often not separated in the literature in such an exact physical sense but used synonymously.

> A *biogeochemical cycle* comprises the sum of all transport and conversion processes that an element and its compounds can undergo in nature.

Under the biogeochemical cycle, the substances pass through several reservoirs (atmosphere, hydrosphere, pedosphere, lithosphere and biosphere) where certain concentrations accumulate because of flux rates determined by transport and reaction. A global cycle may be derived from the budget of the composition of the individual reservoirs, with a (quasi) steady state being considered to exist. It shows variations on different timescales and may be disturbed by catastrophic events (e. g., volcanism or collision with other celestial bodies). The chemical composition of the biosphere and the global environment results from continuous geochemical and biological processes within natural cycles, as mentioned above. Since the appearance of humankind, a new driving force has developed; the human matter turnover because of artificial (or anthropogenic) emissions. Therefore, today's biogeochemical cycles are very clearly no longer natural but anthropogenically modified cycles. Moreover, human-induced climate change already causes permanent change to natural fluxes, such as weathering, shifting redox and phase equilibriums.

The atmospheric reservoir plays a major role because of its high dynamic in transport and reaction processes and the global linkage between biosphere and atmosphere (Figure 2.1). At the earth–air interface, exchange of matter occurs, emission as well as deposition. Hence, the atmosphere is the global chemical reactor. The typical dictionary definition of the atmosphere is "the mixture of gases surrounding the Earth and other planets" or "the whole mass of an aeriform fluid surrounding the Earth". The terms air and atmosphere are widely used as synonyms. From a chemical point of view, it is possible to say that *air* is the substrate with which the *atmosphere* is filled. This is in analogy to the *hydrosphere*, where *water* is the substance. Nonetheless, the hydrosphere is not water as the chemical compounds; it is a solution containing living organisms and non-living solid bodies and having interfaces with the atmosphere and with the sediment where extensive exchange of matter occurs. Similarly, soils provide multiphase systems containing minerals, organic matter, microorganisms, air and water. Furthermore, *air* is an atmospheric suspension containing different gaseous, liquid (water droplets) and solid (dust particles) substances; therefore, it provides a multiphase and multicomponent chemical system. Naturally, the atmosphere is an oxidis-

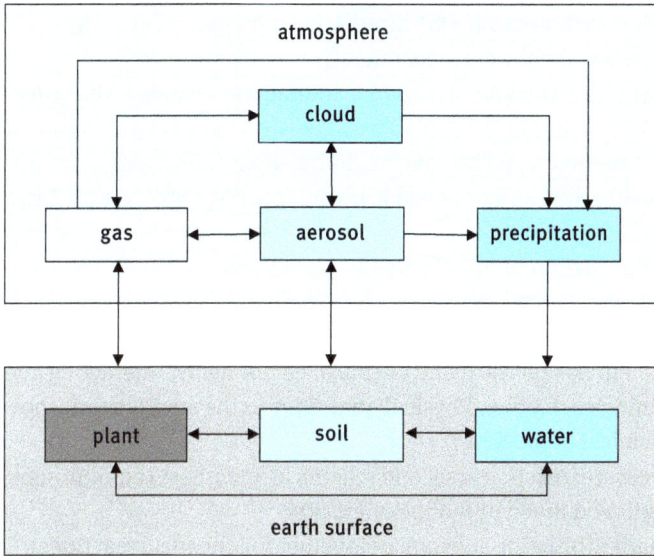

Figure 2.1: The chemical reservoirs and mass transfers in the global environment.

ing regime, and the biosphere (mostly only microorganisms and green plants) is the reducing antagonist.

Solar radiation is the primary driving force in creating gradients in pressure, temperature and concentration, resulting in transport, phase transfer and chemical processes. Considering that the incoming solar radiation do not trend over several hundred years (despite periodical variations), and accepting that natural biogenic and geogenic processes vary but do not trend over these timescales, this is only human influence on land use, contaminating soils and waters and emissions into the atmosphere that changes environmental chemical composition. Human activities affect the natural processes (biological, such as plant growth and diversity, and physical, such as radiation budget), resulting in a cascade of consequent physical and chemical developments (feedback).

Atmospheric substances possessing their physical and chemical properties will influence the environment significantly; we list the most important among them here together with impacts (there are many more impacts, parallel and synergistic effects):
- formation of cloud condensation nuclei (CCN) and subsequent cloud droplets: hydrological cycle,
- being 'greenhouse' gases (GHG): irradiative interaction (warming the atmosphere),

- being ozone-depleting substances (ODS): irradiative interaction (increasing UV radiation penetration into the lower troposphere[1]),
- formation of atmospheric aerosol: irradiative interaction (cooling the atmosphere),
- oxidation capacity: a lifetime of pollutants, oxidative stress (damages),
- acidity: chemical weathering, shifting chemical equilibriums (affecting life functions),
- toxicity: poisoning the environment (affecting life functions),
- nutrition: bioavailability of compounds essential for life.

We see that the environment has physical and chemical components, interacting and (at least partly) determining each other. Physical quantities in the environment show strong influences on chemical processes:
- temperature and pressure: reactions rate and (chemical and phase) equilibriums,
- radiation (wavelength and intensity): photochemistry,
- motion: fluxes of matter (bringing substances together for chemical reactions).

> **i** Changes in the chemical composition of air, water and soil caused by humans are termed *air, water and soil pollution*.

The terms *pollution* and *pollutant* need some comments. To begin with, the term *pollutant* should be used only for artificially (anthropogenically) released substances, even though most of them are also of natural origin. Environmental pollution represents a deviation from natural chemical composition of the environment (providing a reference level) at a given site and period (note that climate change and variation is an ongoing natural process). Therefore, environmental pollution in terms of the changing chemical composition of the environment must be identified through a *problem*, not simply by measured concentrations. The problem lies between 'dangerous' and 'acceptable' climate impact. This definition is beyond the direct role of the scientific community although scientists have many ideas about it.

Depending on the residence time of the pollutant, we can characterise the scale of pollution from local via regional to global. Air pollution nowadays is a global phenomenon because long-lived pollutants can be found to be increasing at any site on the globe. Sea pollution has also become a global issue, first through carbon dioxide acidification and shipping and river run-off. The ocean is the global pollution disposal. Atmospheric pollutants are removed relatively quickly by different deposition

---

1 The troposphere is the first lower layer of the atmosphere in contact with the earth's surface. Temperature decreases with altitude and strong mixing with clouds and precipitation characterise it. The upper limit of the troposphere is called the tropopause and lies between 8 and 17 km, depending on the latitude. Above the troposphere the stratosphere begins.

processes and transfer to soils, vegetation and waters; from soils and lakes pollutants move (much slower) via rivers to the sea where they stay for hundreds or more years before sedimentation on the sea floor and likely subduction and recycling into the deep mantle of the Earth. Within global cycling, the water cycle (or hydrological cycle) is by far the largest matter cycle and the most important for the weather and climate and the distribution of pollutants and chemical processes.

Let us now return to the atmosphere as a *multiphase system*. While gases and particles (from molecules via molecular clusters to nano- and micro-particles) are always present in the air, although with changing concentrations, condensed water (hydrometeors) is occasionally present in the air, depending on the presence of so-called cloud condensation nuclei (CCN) and water vapour supersaturation at the site of fog and cloud formation. With the formation, transportation and evaporation of clouds, huge amounts of atmospheric energy are transferred. This results in changing radiation transfer and thus 'makes' the weather and, on a long-term scale, the climate. Furthermore, clouds provide an effective 'chemical reactor' and transport medium and cause the redistribution of trace species after evaporation. When precipitating, clouds move trace substances from the air (wet deposition) to the earth's surface. Therefore, besides the continuous dry deposition process, clouds may occasionally lead to large inputs of trace substances into ecosystems. The amount of condensed water in clouds and fog is very small, with around 1 g per m^{-3} air or, in the dimensionless term, liquid water content $1 \cdot 10^{-6}$ (identical with 1 ppb). Thus, 99.99 % or more of total atmospheric water remains in the gaseous phase of an air parcel. Hydrometeors may be solid (ice crystals in different shapes and forms) or liquid (droplets ranging from a few m up to some tens of m). We distinguish the phenomenon of hydrometeors into clouds, fog and precipitation (rain, drizzle, snow, hail, etc.). This atmospheric water is always a chemical aqueous solution where the concentration of dissolved trace matter (related to the bulk quantity of water) is up to several orders of magnitude higher than in the gaseous state of the air. This analytical fact is the simple explanation of why the collection and chemical analysis of hydrometeors began much earlier than gas-phase measurements.

As chemistry is the science of matter, we consider substances moving through the environment. We have already mentioned that elementary chemical reactions are absolutely described by a mechanism and a kinetic law and, therefore, are applicable to all regimes in nature if the conditions (temperature, radiation, pressure, concentration) for this reaction correspond to those of the natural reservoir. For example, the photolytic decay of molecules is only possible in the presence of adequate radiation. Hence, in the darkness, there is no photolysis. Table 2.1 classifies how chemical regimes meet in the environment. We see that mostly 'normal' conditions occur, and extreme low and high temperatures border the environmental system in the sense of human habitat. The chemistry described in the following chapters concerns mostly these normal conditions of the environment. We focus on the earth's surface, the lower troposphere and the interfaces. For example, aqueous phase chemistry

Table 2.1: Chemical regimes in the environment, characterised by temperature T, pressure p, H_2O vapour pressure, radiation (wavelength λ) and trace concentrations.

T	p	H_2O	λ (in nm)	traces	regime
normal (~285 ± 30 K)	normal	normal	>300	normal	troposphere, plants waters, interfaces
normal (~285 ± 30 K)	normal	normal	–	normal	soils
low (<200 K)	very low	very low	<300	very low	stratosphere and upper atmosphere
very high (>1300 K)	normal-high	normal	>300	normal	lightning
high (≥1000 K)	normal	large	–	high	biomass burning
very high (≥1300 K)	very high	very low	–	–	deep in Earth
exceeded (~330 K)	normal	very high	–	very high	flue gas

in cloud droplets does not differ principally from surface water chemistry (aquatic chemistry), and much soil chemistry does not differ from aerosol chemistry (colloidal chemistry). Plant chemistry, however, is different and only by using the generic terms (see Chapter 5.2.1) of inorganic interfacial chemistry can we link it.

Pure substances (as known from the laboratory stock) are extremely rare in the natural environment. We meet solutions and mixtures, where the pollutants are mostly in low concentrations (or mixing ratios). In contrast to classical chemical textbooks, there is no need to describe here the properties of (pure) elements and substances, their synthesis and application. In the natural environment, the synthesis of (organic) compounds only occurs by green plants (assimilation), which is outside the scope of this book. Geochemical processes in the background of earth's evolution created and transformed minerals; the interested reader should look for books on geochemistry. This book will focus on life-essential elements and artificial pollutants, their transformation (where degradation is most important) and transfer processes (soil-water, water-air, soil-air), including surface chemistry (nowadays called interfacial processes).

2.2 States of matter

In physics, a *state of matter* or *phase* is one of the distinct forms that matter takes on. Four states of matter are observable in everyday life: solid, liquid, gas, and plasma. Matter in the solid-state maintains a fixed volume and shape, with component particles (atoms, molecules or ions) close together and fixed into place. Matter in the liquid state maintains a fixed volume but has a variable shape that adapts to fit its reservoir (bulk water such as river, lake, sea and droplet water). Its particles are still close together but move freely. Matter in the gaseous state has variable volume and shape, adapting both to fit its reservoir (atmosphere, soil pores). Its particles neither are close together nor fixed in place. Matter in the plasma state has variable volume and shape, but as well as neutral atoms, it contains many ions and electrons, both of which can move around freely. Plasma is the most common form of visible matter in the universe. Many compounds (matter), important in the environment, exist at the same time in different states or, in other terms, in phase equilibrium, such as solid-liquid, liquid-gaseous and solid-gaseous. The only compound existing in the environment in all three states simultaneously is water (H_2O).

2.2.1 Atoms, elements, molecules, compounds and substances

The *atom* is the smallest unit that defines the chemical elements and their isotopes; the nucleus consists of protons and neutrons. The number of protons within the atomic nucleus is called the *atomic number* equal to the number of electrons in the neutral (nonionised) atom. Recall that the number of protons in the nucleus defines an element, not the number of protons plus neutrons (which determines its weight). Elements with different numbers of neutrons are termed *isotopes*, and different elements with the same number of neutrons plus protons (nucleons) are termed *isobars*.

A *chemical element* is a pure chemical substance consisting of only one type of atom distinguished by its atomic number. **i**

Elements are divided into metals, half-metals (past name: metalloid), and non-metals. Where metals and non-metals are well defined, it is not so for half-metals (having properties between metals and non-metals). From the six half-metals (B, Si, Ge, As, Sb, Te), only silicon, boron and arsenic, as an anthropogenic pollutant, are important in environmental chemistry. Non-metals (H, C, N, O, P, S, Se, halogens, noble gases) mainly compose our atmospheric and biospheric environment and constitute evaporable molecules (from selenium, no gaseous compounds in the air are known and from phosphorus only recently).

A *molecule* is an electrically neutral group of two or more atoms held together by chemical bonds.

Only a few molecules are homonuclear, that is, consisting of atoms of a single chemical element, as with oxygen (O_2), nitrogen (N_2), hydrogen (H_2), chlorine (Cl_2) and others. Mostly they form a *chemical compound* composed of more than one element, as with water (H_2O). Chemical compounds can be molecular compounds held together by covalent bonds, salts held together by ionic bonds, intermetallic compounds held together by metallic bonds, or complexes held together by coordinate covalent bonds. Molecules as components of matter are common in organic substances (and therefore living organisms). Molecules in inorganic substances make up most of the oceans and atmosphere.

> **i** An *ion* is an atom or molecule in which the total number of electrons is not equal to the total number of protons, giving the species[2] a net positive (*anion*) or negative (*cation*) electrical charge.

Ions in their gas-like state are highly reactive and do not occur in large amounts on Earth, except in flames, lightning, electrical sparks, and other plasmas. In aqueous environments (bulk waters and atmospheric droplets), ions play a huge role, produced by chemical reactions such as electrolytic dissociation and redox processes.

However, the majority of familiar solid substances on Earth, including most of the minerals that make up the crust, mantle, and core of the Earth, contain many chemical bonds but are *not* made of identifiable molecules, called *substances*. Atoms and complexes connected by noncovalent bonds such as hydrogen or ionic bonds are generally not considered as single molecules. In addition, no typical molecule can be defined for ionic crystals (salts) and covalent crystals (network solids). However, these are often composed of repeating unit cells that extend either in a plane (such as in graphene) or three-dimensionally (such as in diamond, quartz, or sodium chloride).

Glass is an amorphous solid (non-crystalline) material that exhibits a glass transition, which is the reversible transition in amorphous materials (or in amorphous regions within semicrystalline materials) from a hard and relatively brittle state into a molten or plastic state. In glasses, atoms may also be held together by chemical bonds without any definable molecule and without any regularity of repeating units that characterise crystals. However, *natural* glasses are rare.

2.2.2 Pure substances and mixtures

With the exception of magma and crude oil, the only liquid in the natural environment is water. Pure water, however, exists only in the gaseous phase, as water vapour. Natural water is an aqueous solution of gases, ions and molecules, but it also contains undissolved, suspended and/or colloidal inorganic particles of different size and chemical composition and biogenic living and/or dead matter such as cells, plants, and so on, sometimes termed *hydrosol*. A *solution* is a homogeneous mixture composed of only one phase, such as aqueous solutions (ocean) or gases such as the atmosphere. A *mixture* is a material system made up of two or more different substances mixed but not combined chemically. Hence, natural waters and soils are always heterogeneous mixtures. It is important to note that air is not a mixture of different gases but a solution because all gases can form solutions with each other in all ratios. The specific properties of individual gases will not change. The atmosphere, however, is

---

2 The term *species* is normally only used in biology as a basic unit of biological classification. In recent years, the term *chemical species* has come into use as a common name for atoms, molecules, molecular fragments, ions, etc.

a heterogeneous mixture containing aqueous particles[3] (droplets of clouds, fog and rain) and solids (dust particles).

In physics, an *aerosol* is defined as the dispersion of solids or liquids (the dispersed phase) in a gas, specifically air (the dispersant). More generally, dispersion is a heterogeneous mixture of at least two substances that are not soluble within each other (in contrast to molecular dispersion). However, the classical scientific terms are *colloid* and *colloidal system*. Colloids exist between all gas, solid and liquid combinations except for gas-gas (all gases are mutually miscible); L – liquid, S – solid, G – gaseous:

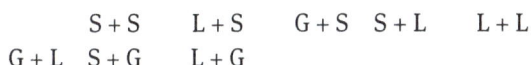

$$S + S \quad L + S \quad G + S \quad S + L \quad L + L$$
$$G + L \quad S + G \quad L + G$$

Thomas Graham (1805–1869) introduced the terms colloid and colloidal condition of matter as well as the terms *sol* and *gel* (Graham 1861). Over 100 years ago, it had already become clear that the sciences to study colloids – chemistry and physics – must deal with the colloidal state and not only with the colloid, but with the dispersed phase itself (Ostwald 1909). The colloidal system was characterised as multiphase or heterogenic. For colloids such as (G + S), cigarette smoke, atmospheric dust and for (G + L), the atmospheric fog was given as examples (Ostwald 1909).

2.2.3 Concentration measures

In chemistry, *concentration* is the measure of how much of a given substance is mixed with another substance or, in other terms, it is the abundance of a constituent divided by the total volume of a mixture.

Moreover, it is the short form for amount (of substance) concentration. It covers a group of four quantities characterising the composition of a mixtureconcerning the volume of the mixture: mass, amount, volume and number concentration. This can apply to any chemical mixture, but the concept most frequently refers to homogeneous solutions (water and air). *The amount of substance* (the SI unit *mole*) is the number of elementary entities (such as atoms and molecules) divided by the Avogadro constant. The *chemical amount* is hence the alternative name for the amount of substance. Measurable quantities are mass, volume and number; the amount of substance (mole) can be expressed by these three quantities (see below). Hence, we can express the quantity of a substance in the following terms:

---

3 Under specific conditions in the stratosphere at very low T (= −82 °C) there are liquid droplets of nitrous and sulphuric acid.

- *mass m* (related to a prototype made from iridium and stored in Paris), measured in kilograms (kg),
- *mole n* (1 mol is the *amount of substance* of a system that contains N_A elementary entities (usually atoms or molecules) where N_A is about $6.02214 \cdot 10^{23}$ mol^{-1}, named the *Avogadro constant*), dimensionless,
- *volume V* (how much three-dimensional space is occupied by a substance), measured in m^3,
- *number N* (the sum of individual particles such as atoms, molecules, droplets, dust particles), dimensionless.

The only exact measurable quantity with high accuracy is the mass of solid, liquid and gaseous bodies. Volume measurement is possible only for liquids (waters) and compact solid bodies. Soils are porous media with varying humidity; hence, mass, dry mass and pore volume must be measured to characterise the reference quantity. Measurement of a gas volume such as air is difficult (because it is highly affected by temperature and pressure, both quantities must be simultaneously measured). A standard volume of air (e. g., $1\,m^3$) is well defined but hard to measure when taking into account the volume of condensed matter (hydrometeors and particulate matter, PM). Normally, we can neglect the volume (fraction) of the condensed matter occupying a volume of air because of the small values in the order of 10^{-6} (Table 2.3), which is orders of magnitude smaller than the best gas volume measurement facilities. However, the volume of condensed matter can be well measured using optical methods. In highly polluted exhaust air, dust must be removed before the volume measurements.

The concentration c can be defined based on mass m (e. g. $g\,L^{-1}$ or $g\,m^{-3}$), number N (e. g. cm^{-3}) and mole n (molarity, e. g. $mol\,L^{-1}$ or molar or M):

$$c_i(m) = m_i/V, \quad c_i(N) = N_i/V, \quad c_i(n) = n_i/V. \tag{2.1}$$

The concentration c (more precisely, the volumetric mass concentration) of a substance is its mass per unit volume and is identical to the *density* ρ, or more precisely, the volumetric mass density.

Amount (mass, volume, number) denotes an extensive quantity, i. e. the value of an extensive quantity increases or decreases when the reference volume changes. The amount of a substance in a given gas volume is strongly determined by gas laws. By contrast, an intensive quantity (pressure, temperature) does not change with volume.

The mole together with the mass are *basic units* of the SI (Le *Système International d'Unités*) have been chosen because in 12 g of carbon-12 (^{12}C) are N_A atoms determined experimentally, each having a mass of $1.99264648 \cdot 10^{-23}$ g and expressing the *atomic mass*, the quantity of matter contained in an atom of an element.[4] The atomic mass

---

4 The atomic mass of atoms, ions, or atomic nuclei is slightly less than the sum of the masses of their constituent protons, neutrons, and electrons, due to binding energy mass loss.

may be expressed in unified *atomic mass units*; by international agreement, 1 atomic mass unit (amu) is defined as 1/12 of the mass of a single carbon-12 atom (at rest). In this scale, 1 atomic mass unit (amu) corresponds to $1.66053878 \cdot 10^{-24}$ g (one-twelfth the mass of the carbon-12 atom, $1.99264648 \cdot 10^{-23}$ g, which is assigned an atomic mass of 12 units).

When expressed in such units, the atomic mass is called the *relative isotopic mass*. Atomic mass is different from *elemental atomic weight* (also called *relative atomic mass*) and *standard atomic weight*; both refer to the averages of naturally occurring atomic mass values for samples of elements (listed in Appendix C). Most elements have more than one stable nuclide; for those elements, such an average depends on the mix of nuclides present, which may vary to some limited extent depending on the source of the sample, as each nuclide has a different mass. Atomic weights are dimensionless quantities (i. e. pure numbers), whereas molar masses shave units (in this case, g/mol).

The *molar mass M* of a compound is given by the sum of the standard atomic mass of the atoms. An average molar mass may be defined for mixtures of compounds. The molar mass (physical property) is defined as the mass m of a given substance (chemical element or chemical compound) divided by its amount of substance n (dimension mass/mole);[5] m_i is a mass of atom or molecule:

$$M = m/n \equiv m_i \cdot N_A. \tag{2.2}$$

Molar mass is closely related to the *relative molar mass* of a compound, the older term *molar weight* and to the *standard atomic masses* of its constituent elements. However, it should be distinguished from the *molecular mass* (also known as molecular weight), which is the mass of *one* molecule (of any *single* isotopic composition) and is not directly related to the atomic mass, the mass of *one* atom (of any *single* isotope). The *dalton*, symbol Da, is also sometimes used as a unit of molar mass, especially in biochemistry, with the definition $1\,\text{Da} = 1\,\text{g/mol}$, despite the fact that it is strictly a unit of molecular mass ($1\,\text{Da} = 1\,\text{amu}$).

As said, in the environment, we deal with mixtures. Hence, we define total quantities (mass, volume, mole), being the sum of partial quantities of the specific compounds and substances of the mixture:

$$m = \sum m_i, \quad V = \sum V_i, \quad \text{and} \quad n = \sum n_i. \tag{2.3}$$

We now introduce another concentration measure, the dimensionless *mixing ratio x*, defined as the abundance of one component of a mixture relative to that of all other

---

5 The base SI unit for molar mass is kg/mol. However, for historical reasons, molar masses are almost always expressed in g/mol.

components:[6]

$$x_i(m) = m_i/m, \quad x_i(V) = V_i/V, \quad \text{and} \quad x_i(n) = n_i/n. \tag{2.4}$$

These mixing ratios are also called *fractions* by the amount of mass, volume, and moles, respectively. In air and gas mixtures, mass, volume, mole and pressure (note that pressure is an intensive quantity) are linked in the general gas equation (which, however, is only valid for dilute mixtures, which is true when we consider traces in the air), as seen in equation (2.5):

$$pV = nRT, \quad p_i V = n_i RT \quad \text{as well as} \quad pV_i = n_i RT. \tag{2.5}$$

Hence, we define the total pressure to be the sum of partial pressures, $p = \sum p_i$, and it follows:

$$x_i(p) = p_i/p \equiv x_i(V) = V_i/V. \tag{2.6}$$

Using the gas law (equation (2.5)) and taking into account equation (2.2) and equation (2.4), a recalculation between mass concentration and the mixing ratio is based on

$$x_i = c(m)_i RT/pM_i, \tag{2.7}$$

where M_i is the mole mass of the substance i. Gaseous trace species were often measured in mass concentration (e. g., $\mu g\, m^{-3}$). It is obligatory to measure also pressure and temperature because otherwise, there is no way for recalculation between mass concentration and mixing ratio. For standard conditions (1 atm = $1.01325 \cdot 10^5$ Pa and 25 °C = 298.15 K) the mass concentration is listed as 1 ppb in Table 2.2.

The large advantage in its use compared with concentrations[7] (moles or mass per volume) lies in its independence from p and T. Depending on the magnitude of the mixing ratio, the most convenient units can be:
- parts per million: ppm, where 1 ppm = 10^{-6};
- parts per billion: ppb, where 1 ppb = 10^{-9};
- parts per trillion: ppt, where 1 ppt = 10^{-12}.

Normally, it must be decided on which quantity the mixing ratio is based (volume or mass) and then written as ppb(m) and ppb(V), respectively. Fractions in mass/mass

---

6 Do carefully distinguish between m_i (mass of a defined substance i), m (total mass of a mixture) and m_m (mass of the molecular entity, such as atom, molecule, ion).

7 However, whenever measuring an atmospheric substance, p and T must be co-measured to allow standard recalculations for exact averaging and intercomparisons.

Table 2.2: Recalculation between mass concentration $c(m)_i$ (in $\mu g\,m^{-3}$) and mixing ratio x_i (in ppb) under standard conditions (25 °C and 1 bar).

compound		$x(V)_i$ (in ppb)	$c(m)_i$ (in $\mu g\,m^{-3}$)
ozone	O_3	1	1.96
hydrogen peroxide	H_2O_2	1	1.39
sulphur dioxide	SO_2	1	2.63
nitrogen monoxide	NO	1	1.22
nitrogen dioxide	NO_2	1	1.89
nitrous acid	HNO_2	1	1.92
nitric acid	HNO_3	1	2.56
ammonia	NH_3	1	0.69
hydrogen chloride	HCl	1	1.63
benzene	C_6H_6	1	3.23

are often used for soil contamination; volume-based fractions are only used for air and gas mixtures. For liquids (waters), concentration is expressed in mass/volume and mole/volume only. For the main constituents in air, such as oxygen, nitrogen, water vapour; and soil, such as humidity, minerals, soil air, organic compounds; the concentration is given in percentage (%), i. e., a mixing ratio in the order of 10^{-2}.

Another concentration measure is important for ions in waters and soluble substances in solids (soils and dust): the *normality* (N) and the *equivalent* (eq). Due to the condition of electroneutrality, the sum of the equivalent concentration of all cations must be equal to the sum of the equivalent concentration of all anions in the sample. This concentration measure takes into account the *stoichiometry* of chemical reactions; one equivalent of acid reacts with one equivalent of a base, and one equivalent of a reducing species reacts with one equivalent of an oxidising species.

Normal is the one-gram equivalent of a solute per litre of solution. The definition of a gram equivalent varies depending on the type of chemical reaction being discussed – it can refer to acids, bases, redox species and ions that will precipitate. More formally, a gram equivalent of a substance taking part in a given reaction is the number of grams of the substance associated with the transfer of N_A electrons or protons or with the neutralisation of N_A negative or positive charges. The expression of concentration in equivalents per litre (or more commonly, microequivalents per litre, $\mu eq\,L^{-1}$) is based on the same principle as normality. A normal solution is one equivalent per litre of solution (eq L^{-1}), where $c_i(n)$ denotes the molar concentration (e. g., in mol L^{-1}) and e symbolises the ionic charge:

$$c_i(\text{eq}) = c_i(n) \cdot e. \tag{2.8}$$

The number concentration is only used for elementary entities (atoms, molecules, ions) and particles (solid particles in atmospheric aerosol, cloud droplets) being at very small concentrations. For example, OH radicals exist in air in the order of

10^{-6} cm^{-3}, cloud condensation nuclei's, CCN, and droplets between 20 and 300 cm^{-3} but nanoparticles in the order up to 10^5 cm^{-3}). In the air, the *liquid water content* (LWC) of clouds and fog is an important concentration measure for the condensed phase. It can be given in mass/volume (normally g/m^3) or dimensionless in volume/volume.

According to heterogeneous reactions (surface chemistry) and optical properties (reflection, scattering and absorption) in air, another quantity – the surface to volume (of air) ratio – is useful for describing the condensed phase. For droplets, it is simple to define the volume of an individual droplet based on its diameter, assuming a spherical form. The mass, number and surface quantities of particles show a very different (but characteristic) behaviour when related to the size distribution.

Concerning the chemical composition of the condensed matter in the air, there are two ways to describe the abundance. First, chemically analysing the matter (either single particle or collected mass of particles) provides the concentrations (or mixing ratios) related to the volume (or mass) of the condensed phase, for example, moles (or mass) of a substance per litre of cloud (or rain) water and mass of a substance per total mass of particulate matter (PM); see Section 2.5.3 for further details. When sampling the condensed phase, normally for hydrometeors, all droplets are collected in a bulk solution (however, multistage cloud water impactors are also available). However, for solid PM, different particle size fractions are in use. TSP means the total suspended matter (i. e. it is sampled over all particles sizes). A subscript denotes the cut-off during sampling (for example, PM$_{10}$, PM$_{2.5}$ and PM$_1$), i. e. the value denotes the aerodynamic diameter in m of (not sharp) separation between particles smaller than the given value. Second, the specific concentration of a substance within the condensed matter can be related to air volume. To do this, it is necessary to know the volume (or mass) concentration of the total condensed matter. This is given by the liquid water content (LWC) for hydrometeors and the total mass concentration of PM. The resulting value of the mass of dissolved matter or PM in a volume of air is the atmospheric abundance, and this air quality measure can be compared between different sites. However, the concentration of a substance in cloud water also depends on the cloud's physical properties. This is also valid for fog and precipitation. Hence, with a sampling of hydrometeors (cloud water, rainwater), the simultaneous registration of the LWC and rainfall amount is obligatory. The aqueous-phase concentration c_i(aq) must be multiplied by the LWC to be recalculated to the volume of air:

$$c_i(\text{air}) = c_i(\text{aq}) \cdot \text{LWC}. \tag{2.9}$$

Note that correct dimensions or factors must be used! For example:

$$\left[\frac{\text{mg}}{\text{m}^3}\right](\text{air}) = \left[\frac{\text{mg}}{\text{L}}\right](\text{droplet}) \cdot \left[\frac{\text{g}}{\text{m}^3}\right](\text{LWC}) \cdot 10^{-3}.$$

Collecting samples from soils, waters, and air for environmental analysis is very special. The reader should refer to adequate books to learn the principles for representativeness and spatial and temporal averaging.

2.3 Air and gases

Whereas air is an ancient word, the term *gas* was proposed by Jan Baptist van Helmont (1580–1644) as a new word to name and distinguish the laboratory airs (i. e. gaseous substances) from atmospheric (common) air. The gaseous substances that were observed in alchemical experiments were named fumes, vapours and airs. Atmospheric air (called common air) was still regarded as a uniform chemical substance at that time. Joseph Black (1728–1799), a Scottish chemist, carried out experiments concerning the weights of gases and other chemicals, which was the first step in quantitative chemistry. He was the first to find CO_2 in the air of Edinburgh, probably between 1752 and 1754, which he named 'fixed air'. In the second half of the eighteenth century, the air was found to consist of two different constituents, maintaining respiration and combustion (oxygen: O_2) and not maintaining it (nitrogen: N_2). The discovery of nitrogen is generally credited to Daniel Rutherford (1749–1819) in 1772. Joseph Priestley (1733–1804) wrote in 1771 about the goodness of air (air quality in modern terms) and noted that green plants could restore injured or depleted air. Priestley, starting his studies on-air, began systematically to investigate different kinds of 'air': nitrous (saltpetre) air (NO_x), acid (muriatic) air (HCl), and alkaline air (NH_3). He stated that these 'kinds of air' are not simple modifications of ordinary (atmospheric) air. He published his observations in a book titled *Observations on Different Kinds of Air*. The facts on air composition were expressed most clearly by Carl Wilhelm Scheele (1742–1786) in his booklet *Abhandlung von der Luft und dem Feuer* (*Treatise on Air and Fire*), which was published in 1777; he named the ingredients of air as '*Feuerluft*' (O_2) and '*verdorbene Luft*' (N_2). Argon was identified by William Ramsay (1852–1916) in 1894 and soon later (1898) other noble gases (neon, krypton, and xenon) by John William Strutt, Lord Rayleigh (1842–1919) and co-workers.

2.3.1 Composition of the atmosphere

Nitrogen (N_2), oxygen (O_2), water vapour (H_2O), carbon dioxide (CO_2) and rare gases are the permanent main gases in the air. Only water shows large variation in its concentration, and CO_2 is steadily increasing due to fossil fuel burning. Already in the first half of the nineteenth century, other gaseous substances had been supposed and later detected in the air. As the concentration of almost all trace gases is orders of magnitude smaller than those of the main gases (Table 2.3), it was only with the development of analytical techniques in the late nineteenth century that they were proved present

Table 2.3: Composition of the dry, remote atmosphere (global mean concentrations); note that the concentration of water vapour in the air is highly variable from less than 0.5 % in saturated polar continental air to more than 3 % in saturated (i. e. 100 % humidity) tropical air.

substance	formula	mixing ratio (in ppm)	comment
nitrogen	N_2	780830[a]	constant
oxygen	O_2	209451[a]	constant
argon	Ar	9339[a]	constant
carbon dioxide	CO_2	380	increasing
neon	Ne	18.18	constant
helium	He	5.24	constant
methane	CH_4	1.73	increasing
krypton	Kr	1.14	constant
hydrogen	H_2	0.5	constant
dinitrogen monoxide	N_2O	0.32	increasing
carbon monoxide	CO	0.120	increasing
xenon	Xe	0.087	constant
ozone	O_3	0.03	variable
particulate matter	–	0.01[d]	variable
carbonyl sulphide	COS	0.00066	increasing
nitric acid[b]	HNO_3	$\leq$0.001	variable
radicals[c]	–	<0.00001	highly variable
hydroxyl radical	OH	0.0000003	highly variable

[a] related to the "clean" atmosphere $O_2 + N_2 + Ar + CO_2$ (=100 %)
[b] and many other trace substances (NH_3, NO_x, HCl, NMVOC, H_2O_2, DMS, CFC's et al.)
[c] e. g. HO_2, NO_3, Cl
[d] corresponds to 10–20 µg m^{-3}

in the air. Table 2.3 shows the present composition of our air. It shows that the concentration range from the main constituents to the trace species covers more than ten orders of magnitude. Each component in the air has a 'function' in the environment and biogeochemical cycling (see Chapter 5.2).

Changes in the chemical composition of air caused by humans are termed *air pollution*. The terms *air pollution* and *pollutant* need some comments. To begin with, the term *pollutant* should only be used for artificial (anthropogenic) emitted substances, even though most of them are also of natural origin. Air pollution represents a deviation from natural chemical composition of air (providing a reference level) at a given site and period (note that climate change and variation is an ongoing natural process). Depending on the residence time of the pollutant, we can characterise the scale of pollution from local via regional to global. Air pollution nowadays is a global phenomenon because long-lived pollutants can be found to be increasing at any site on the globe. *Remote* air just means that the site is located far away from the sources of emissions and, consequently, this air has lower concentrations of short-lived (reactive) substances compared with sites close to sources of pollutants. Although *polluted* air is human-influenced air, *clean* air is not synonymous with *natural* air. The natural atmosphere no longer exists; it was the chemical composition of air without human influences. However, this definition is also not exact because humans are part of nature. In nature, situations may occur, such as volcanic eruptions, sand storms and biomass burning, where the air is being 'polluted' (rendered unwholesome by contaminants) or, in other words, concentrations of substances of *natural* origin are increased. Therefore, the *reference state* of natural air is a climatological figure where a mean value with its variation must be considered. The term *clean air* is also used politically in air pollution control as a target, i.e. to make our air cleaner (or less polluted) in the sense of pollutant abatement. A clean atmosphere is a political target; it represents an air chemical composition (defined in time and scale) that should permit sustainable development.

2.3.2 Properties of gases: the ideal gas

Without a doubt, one of the most important models in physical chemistry is that of an *ideal gas*, assuming the gas as the congregation of particles existing in omnidirectional stochastic motion.

The *gas laws* developed by Robert Boyle, Jacques Charles (1746–1823) and Joseph Gay-Lussac (1778–1850) is based upon empirical observations and describe the behaviour of gas in macroscopic terms, that is, in terms of properties that a person can directly observe and experience. The kinetic theory of gases describes the behaviour of molecules in a gas based on the mechanical movements of single molecules. A gas is defined as a collection of small particles (atom- and molecule-sized) with the mass m_m:
- occupying no volume (that is, they are points),
- where collisions between molecules are perfectly elastic (that is, no energy is gained or lost during the collision),
- having no attractive or repulsive forces between the molecules,
- travelling in straight-line motion independent of each other and not favouring any direction, and
- obeying Newton's Laws.

Real gases, as opposed to an ideal gas, exhibit properties that cannot be explained entirely using the ideal gas law. However, under natural environmental conditions, even for exhaust air, the ideal gas law is applicable. On the other hand, real-gas models have to be used near the condensation point of gases, near critical points, and at very high pressures, conditions found under technical applications.

2.3.2.1 Kinetic theory of gases

While colliding, gas molecules exchange energy and momentum. Collisions lead macroscopically to the *viscosity* of gas and the *diffusion* of molecules. Considering only the translation of particles (the aim of the kinetic theory of gases), there is no need for quantum mechanical description. However, quantum mechanics is essential when describing rotation and vibration in molecules (the other parts of the internal energy). The size and speed of molecules with a mass m_m can be different. Therefore, the gas is considered macroscopic by averaging individual quantities. The molecule number density $c_i(N)$ is defined as the ratio between the number of gas molecules N_i and the gas volume V (see equation (2.1)). We will later see that $c(N)$ in pure gases denotes the Loschmidt constant n_0. Please note that in the scientific literature, the number density is normally termed with n, but here that would be confused with the mole number n.

Let us now consider the motion of a molecule in x direction onto a virtual wall. Because of the three spatial directions and each positive and negative direction, the particle density in each direction amounts to $c(N)/6$. For these molecules, equality $v = v_x$ is valid because the speed in all other directions (y, z) is zero (according to the agreement, they only move in x direction). Within the time dt a molecule passes the distance vdt. At the wall with the area q, $c(N)qvdt/6$ molecules collide totally. Therefore, each molecule transfers the momentum $2m_m v$ because of the action-reaction principle (to and from the wall). All collisions, given by the collision number z (collision frequency or rate is z/dt), transfer in a given time period dt a momentum that generates the force f (equation (2.11) according to equation (2.10), Newton's equations

$$\left(\frac{d\vec{v}}{dt}\right)_m = \frac{1}{m} \cdot f, \quad \left(\frac{dm}{dt}\right)_{\vec{v}} = \frac{1}{\vec{v}} \cdot f \quad \text{and} \quad \frac{d}{dt}(m\vec{v}) = \sum_i f_i = ma, \qquad (2.10)$$

where m is mass, t is time, $\vec{v}$ is the velocity (as a vector), a is acceleration, and f is force. The last equation of motion ($m\vec{v}$ is the momentum) characterises the *fluid* atmosphere[8] (force = mass · acceleration = power/velocity and work = power · distance). *Force* is the change of energy by distance ($1\,J\,m^{-1} = 1\,N = 1\,kg\,m\,s^{-2}$). *Energy* is the capacity to do work (or produce heat); there are different kinds of energy: energy ≡ heat ≡ radiation ≡ work ($1\,J = 1\,N\,m = 1\,kg\,m^2\,s^{-2}$). *Heat* is the transfer of thermal energy from one object to another. *Power* is the rate of energy change ($1\,J\,s^{-1} = 1\,W = 1\,kg\,m^{-2}\,s^3$):

$$f = \frac{z \cdot 2m_m v}{dt} = \frac{c(N)}{6}qv \cdot 2m_m v = \frac{1}{3}c(N)q \cdot m_m v^2 \qquad (2.11)$$

where z is the number of collisions. For the *pressure p* (force/area) results:

$$p = \frac{f}{q} = \frac{1}{3}c(N) \cdot m_m v^2. \qquad (2.12)$$

---

8 It is valid for all currents of gases and liquids.

With the definition of the gas density ρ as a product of particle density $c(N)$ and molecule mass m_i, the fundamental equation of the kinetic theory of gases follows:

$$p = \frac{1}{3}\rho v^2. \tag{2.13}$$

2.3.2.2 Gas laws

Expressing the gas density ρ as the ratio between total mass m ($m = Nm_m$; see equation (2.2), taking into account that $n \cdot N_A = N_i$) and gas volume V, the Boyle–Mariotte law then follows. pV represents the quantity of energy, or more specifically the pressure-volume work of this gaseous system:

$$pV = \frac{1}{3}m \cdot v^2 = \text{constant}. \tag{2.14}$$

Empirically, it has been found that (T is the absolute temperature):

$$\frac{pV}{T} = \text{constant}. \tag{2.15}$$

This is the combined gas law that combines Charles's law ($V = \text{constant} \cdot T$ where $c(N), p = \text{constant}$), Boyle's law ($pV = \text{constant}$ where $c_N, T = \text{constant}$) and Gay-Lussac's law ($p/T = \text{constant}$ where $m, V = \text{constant}$). Avogadro's law expresses that the ratio of a given gas volume to the number of gas molecules within that volume is constant (where $T, p = \text{constant}$):

$$\frac{V}{c(N)} = \text{constant}. \tag{2.16}$$

Combining equation (2.15) and equation (2.16) leads to the *ideal gas law*, where R is the universal gas constant. The number of gas molecules we will express by the mole number $n = m/M$:

$$pV = nRT. \tag{2.17}$$

Now, we can derive an expression for the work W done by the gas volume (that is equivalent to the kinetic energy). Using the definitions $f = dW/dx$, $a = dv/dt$ and $v = dx/dt$, it follows from equation (2.10):

$$E_{kin} = \Delta W = \int_0^x f dx = \int_0^x m_m a dx = \int_0^v m_m \frac{dx}{dt} dv = \int_0^v m_m v dv = \frac{1}{2}m_m v^2. \tag{2.18}$$

Now, combining equation (2.11) and equation (2.15) and taking into account that $\rho = m/V$ and $n = m/M$, we derive an important relationship for the mean velocity of molecules:

$$v_{mean} = \sqrt{\frac{3p}{\rho}} = \sqrt{\frac{3RT}{M}} = \sqrt{\frac{3kt}{m_m}}. \tag{2.19}$$

More exactly, the velocity derived in equation (2.19) is the *root mean square velocity* v_{rms} (see equation (2.38)). Remember that m_m is the molecule mass, k is the Boltzmann constant, which is in the following relationship to the gas constant (hence, the equations with k are related to averaged single-molecule properties and those with R to the gas being the collection of molecules):

$$\frac{k}{R} = \frac{m_m}{M} = \frac{1}{N_A}. \tag{2.20}$$

From equation (2.18) and equation (2.19), another expression for the kinetic energy of molecules follows (remember that $M = m_m N_A$):

$$E_{kin} = \frac{1}{2}m_m v^2 = \frac{1}{2}\frac{M}{N_A}v^2 = \frac{3}{2}\frac{RT}{N_A} = \frac{3}{2}kT. \tag{2.21}$$

The average kinetic energy of a molecule is $3kT/2$, and the molar kinetic energy amounts to $3RT/2$. It is important to distinguish between molecular (denoted by the subscript m here) and molar quantities. Furthermore, it is valid (V_m is the molar volume):

$$\frac{p}{RT} = \frac{n}{V} = \frac{1}{V_m}. \tag{2.22}$$

Another (original) expression for the gas law then follows:

$$pV_m = RT = \frac{1}{N_A}kt. \tag{2.23}$$

The Loschmidt constant n_0 is the number of particles (atoms or molecules) of an ideal gas in a given volume V (the number density), and is usually quoted at standard temperature and pressure:

$$n_0 = \frac{N}{V} = \frac{p}{kT} = \frac{p}{RT}N_A = \frac{N_A}{V_m}. \tag{2.24}$$

From equation (2.17), equation (2.20), equation (2.23) and equation (2.24), the following gas equations are derived:

$$pV = \frac{m}{m_m}kT \quad \text{and} \quad p = n_0 kT. \tag{2.25}$$

The model of an ideal gas can be applied with sufficient accuracy[9] to air with a mean pressure of 1 bar (variation about 0.65–1.35 bar) and a mean temperature near the earth's surface of 285 K (variation about 213–317 K). For a description of air as a gas mixture, Dalton's law (also called Dalton's law of partial pressures) is important. This law states that the total pressure p exerted by a gaseous mixture is equal to the sum of the partial pressures p_i of each individual component in a gas mixture: $p = \sum p_i$. It follows that $V = \sum V_i$ and $n = \sum n_i$. The gas density ρ (note the difference to the number concentration $c(N)$) can also be expressed by several terms:

$$\rho = \frac{m}{V} = c(N)m_m = \frac{N}{V}m_m = \frac{pM}{RT}. \tag{2.26}$$

In the gas mixture, we can now introduce all equations related to a specific compound (molecule) i or the sum of all gases; hence, it is also valid that $m = \sum m_i$ and $\rho = \sum \rho_i$. A mean molar mass $\overline{M}$ of air (or generally a gas mixture) can be defined based on mole-weighted fractions of individual molar masses:

$$\overline{M} = \frac{\sum n_i M_i}{\sum n_i} = \sum x_i M_i. \tag{2.27}$$

Equation (2.26) states that the molar mass of a gas is directly proportional to its densities in the case of the same pressure and temperature. Based on standard values (normally 0 °C and 1 bar), the density for different pressures and temperature can be calculated according to:

$$\rho_{T,p} = \rho_0 \frac{p}{p_0} \frac{T_0}{T}. \tag{2.28}$$

Furthermore, it is practicable to introduce a *relative* density ρ_r, which is the ratio of a gas density to the density of a gas selected as standard (ρ_s) because the ratio of densities does not depend on pressure and temperature. Oxygen and dry air are often selected as standards; they have the following densities, respectively (at 1 bar and 273.15 K):

$$\rho_{O_2} = 0.001429 \text{ g cm}^{-3},$$
$$\rho_{air} = 0.0012928 \text{ g cm}^{-3}.$$

The relative density is also equivalent to the ratio of the relevant mole masses:

$$\rho_r = \frac{\rho}{\rho_s} = \frac{M}{M_s}. \tag{2.29}$$

---

9 Because of the complexity of air, which is a multicomponent and multiphase system, we are only able to model roughly this system. It makes no sense to describe single processes with a complexity several orders of magnitudes more than the process with the lowest accuracy.

2.3.2.3 Mean free path and number of collisions

To understand heat conduction, diffusion, viscosity and chemical kinetics, the mechanistic view of molecule motion is of fundamental importance. The fundamental quantity is the *mean-free path*, i. e. the distance of a molecule between two collisions with any other molecule. The number of collisions between a molecule and a wall shown in Chapter 2.3.2.2 is $z = c(N)qvdt/6$. Similarly, we can calculate the number of collisions between molecules from a geometric view. We denote that all molecules have the mean speed $\bar{v}$ and their mean *relative speed* with respect to the colliding molecule is $\bar{g}$. When two molecules collide, the distance between their centres is d; in the case of identical molecules, d corresponds to the effective diameter of the molecule. Hence, this molecule will collide in the time dt with any molecule centre that lies in a cylinder of a diameter $2d$ with the area πd^2 and length $\bar{g}dt$ (it follows that the volume is $\pi d^2\bar{g}dt$). The area πd^2 where d is the molecule (particle) diameter is also called the *collisional cross section σ*. This is a measure of the area (centred on the centre of the mass of one of the particles) through which the particles cannot pass each other without colliding. Hence, the number of collisions is $z = c(N)\pi d^2\bar{g}dt$. A more correct derivation, taking into account the motion of all other molecules with a Maxwell distribution (see below), leads to the same expression for z but with a factor of $\sqrt{2}$. We have to consider the relative speed, which is the vector difference between the velocities of two objects A and B (here for A relative to B):

$$(\bar{g})^2 = \overline{(\vec{V_A} - \vec{V_B})^2} = \overline{\vec{V_A}^2} - \overline{2\vec{V_A}\vec{V_B}} + \overline{\vec{V_B}^2} = \overline{\vec{V_A}^2} + \overline{\vec{V_B}^2} \approx 2\bar{v}^2. \quad (2.30)$$

Since $\overline{\vec{V_A}\vec{V_B}}$ must average zero, the relative directions being random, the average square of the relative velocity is twice the average square of the velocity of A + B and, therefore, the average root mean square velocity is increased by a factor $\sqrt{2}$ (remember that $\sqrt{\bar{v}^2} = \bar{v}$), and the collision rate is increased by this factor ($\bar{g} = \sqrt{2}\bar{v}$). Consequently, the number of collisions increases by this factor of $\sqrt{2}$ when we take into account that all the molecules are moving:

$$z = \sqrt{2}c(N)\pi d^2\bar{v}dt. \quad (2.31)$$

To determine the distance travelled between collisions, the *mean-free path l*, we must divide the mean molecule velocity by the collision frequency or, in other words, the mean travelling distance $\bar{v}dt$ by the number of collisions. Furthermore, we must find an expression for the collision frequency ($v = z/dt$) and the *mean-free time* ($\tau = 1/v$):

$$l = \frac{\bar{v}dt}{z} = \frac{1}{\sqrt{2}c(N)\pi d^2} \quad \text{and} \quad \tau = \frac{1}{\sqrt{2}c(N)\pi d^2\bar{v}} \quad \text{and} \quad \bar{v} = \frac{l}{\tau}. \quad (2.32)$$

That fraction of gas molecules is of interest when velocity is within v and $v + dv$. This fraction $f(v)dv$ is time-independent from the exchange with other molecules. The

Boltzmann distribution states that at higher energies (=kT) the probability to meet molecules is exponentially less:

$$f(v)dv = \text{const} \cdot \exp\left(-\frac{m_m v^2}{2kT}\right)dv = \text{const} \cdot \exp\left(-\frac{Mv^2}{2RT}\right)dv. \tag{2.33}$$

The Maxwell distribution describes the distribution of the speeds of gas molecules at a given temperature:

$$f(v)dv = \sqrt{\frac{2}{\pi}}\left(\frac{m_m}{kT}\right)^{\frac{3}{2}} v^2 \cdot \exp\left(-\frac{m_m v^2}{2kT}\right)dv. \tag{2.34}$$

The fraction $f(v)dv$ denotes the probability dW and is also represented by dz/z (=d ln z):

$$f(v)dv = dW = \frac{dz}{z}. \tag{2.35}$$

To find the *most probable speed* v_p of molecules, which is the speed most likely to be possessed by any molecule (of the same mass m_m) in the system (in other words, it is the speed at a maximum of probability), we calculate $df(v)/dv$ from equation (2.34), set it to zero and solve for v, which yields:

$$v_p = \sqrt{\frac{2kt}{m_m}} = \sqrt{\frac{2RT}{M}}. \tag{2.36}$$

By contrast, the *mean speed* follows as the mathematical average of the speed distribution, equivalent to the weighted arithmetic mean of all velocities (N number of molecules):

$$\bar{v} = \frac{1}{N}\int_0^N v dN = \int_0^\infty vf(v)dv = \sqrt{\frac{2}{\pi}}\left(\frac{m_m}{kT}\right)^{\frac{3}{2}}\int_0^\infty \exp\left(-\frac{m_m v^2}{2kT}\right)v^3 dv = \sqrt{\frac{8kT}{\pi m_m}} = \sqrt{\frac{8RT}{\pi M}}. \tag{2.37a}$$

The *root mean square speed* v_{rms} is the square root of the average squared speed:

$$v_{rms} = \left(\int_0^\infty v^2 f(v)dv\right)^{\frac{1}{2}} = \sqrt{\frac{3kT}{m_m}} = \sqrt{\frac{3RT}{M}}. \tag{2.37b}$$

All three velocities are interlinked in the following ratio:

$$v_p : \bar{v} : v_{rms} = \sqrt{2} : \sqrt{\frac{8}{\pi}} : \sqrt{3}. \tag{2.38}$$

Now, we can set the expression equation (2.36) for the mean molecular velocity in equation (2.31) and ascertain the mean molecular number of collisions z and the mean-free path l, replacing n by p/kT (r molecule radius) in equation (2.32):

$$z = 2c(N)\pi d^2 dt \sqrt{\frac{kT}{m_m}} \quad \text{or collision frequency} \quad v = 2c(N)\pi d^2 \sqrt{\frac{kT}{m_m}}, \tag{2.39}$$

$$l = \frac{kT}{\sqrt{2}\pi d^2 p} = \frac{kT}{4\sqrt{2}\pi r^2 p}. \tag{2.40}$$

Note that these equations are valid only for like molecules. In the case of air, many different (unlike) molecules have to be considered with different radii or diameters, molar masses and molecular speeds. For the example of gas mixture A and B, the following expressions must be applied for mean molecule diameter and mass, respectively:

$$d_{AB} = (d_A + d_B)/2, \tag{2.41}$$

$$m_{AB} = m_A m_B/(m_A + m_B). \tag{2.42}$$

In the air, the mean-free path has an order of 10^{-7} m and does not depend on temperature but is inversely proportional to pressure. The expressions for collision number and mean-free path are useful for understanding chemical reactions (see collision theory in Chapter 3.3.3) but have only limited worth for applications because the molecular diameter (or radius) is not directly measurable. However, the molecular diameter is typically determined from viscosity measurements.

2.3.2.4 Viscosity

Air is a *viscous* medium; hence, we observe *friction* or *drag*. Friction converts kinetic energy into heat. The internal friction between two moving thin air layers in x direction results in a gradient of the speed in y direction (perpendicular to the flow); f_f is the *frictional force*, q the area between the moving layers:

$$f_f = q\eta\frac{dv}{dz} = q \cdot \tau, \tag{2.43}$$

where η is the *dynamic* or *absolute* viscosity (dimension kg/m·s) and the *shear stress* τ in Newtonian fluids is defined by $\tau = q(dv/dz) = f_f/q$. A *kinematic* viscosity v is defined as the ratio of the dynamic viscosity to the density of the fluid: $v = \eta/\rho$. The derivative (dv/dz) is the *shear velocity*, also called friction velocity. When compared with Newton's equation (2.10), the meaning of τ is clearly seen to be a flux density of momentum. The momentum flows between moving layers in the direction of decreasing shear velocity (from higher layers down to the earth's surface). More exactly, the frictional force must be regarded in all directions, i. e. being a vector. The viscosity of a gas is in direct relation to the mean-free path; mass density $\rho(m)$, see Table 2.4 (note

Table 2.4: Useful quantities in molecular kinetics (V – gas volume, d – molecule diameter, q – square).

quantity	meaning	description
m_m	molecule mass	$= M/N_A$
m	mass	$= N m_m$
M	molar mass	$= N_A m_m$
n	mole number	$= m/M$
N	number of molecules	
N_A	*Avogadro* constant	$= M/m_m$
n_0	*Loschmidt* constant[a]	$= N/V$
$\rho(N)$	(number) density	$\equiv c(N) = N/V$
$\rho(m)$	(mass) density	$\equiv c(m) = c(N)m_m = Nm_0/V = m/V$
$c(N)$	(number) concentration	$\equiv \rho(N) = N/V$
$c(m)$	(mass) concentration	$\equiv \rho(m) = m/V = n_0 m_0$
V_m	molar volume	$= V/n$
p	pressure	$= f/q = n_0 kT = nRT = \rho_m v^2/3$
k	*Boltzmann* constant	$= R(m_m/M) = R/N_A$
R	gas constant	$= k(M/m_m) = kN_A$
$\bar{v}$	mean molecule velocity[b]	$= \sqrt{v_{rms}^2}$
v_{rms}	root-mean-square speed	$= \sqrt{3kT/m_m}$
a	acceleration	$= dv/dt$
f	force	$= p \cdot q = m_m(dv/dt) = v(dm/dt) = ma$
F_d	diffusion flux	$= dn/dt$
z	collision number	$= \sqrt{2}\, n_0 \pi d^2 v dt$
l	mean free path	$= vdt/z = 1/(\sqrt{2}\, n_0 \pi d^2)$
η	(dynamic) viscosity	$= (1/\pi d^2)\sqrt{m_m kT/6}$
D_g	(gas) diffusion coefficient	$= (\bar{v} \cdot l)/3$

[a] it is a constant for pure gases (not mixtures)
[b] also transport velocity

the difference to the number density):

$$\eta = \frac{1}{3}\rho_m \bar{v} l = \frac{1}{3}c(N)m_m \bar{v} l = \frac{1}{\pi d^2}\sqrt{\frac{m_m kT}{6}}. \tag{2.44}$$

However, this equation is an approximation; with respect to intramolecular interactions, a more exact relation $\eta = 0.499\rho_m \bar{v} l$ is obtained. Viscosity is independent of pressure but increases with increasing temperature because of rising molecule speed. The temperature dependence of viscosity is described by an empirical expression; the *Sutherland* constant C and the constant B follow from equation (2.44):

$$\eta = \frac{B\sqrt{T}}{1 + C/T}. \tag{2.45}$$

2.3.2.5 Diffusion

In summary, we can state that no preferential direction exists in molecular motion. Hence, there is no transport of any quantity (the Brownian motion). Any flux (transport of material, heat or momentum) is caused by either turbulent diffusion (air parcel advection) or by laminar diffusion. Diffusion (when using this term in chemistry, only laminar transport is meant) is the flux (dn/dt) due to concentration gradients (dc/dz). Such concentration gradients occur near interfaces very close to the earth's surface (uptake by soils, waters and vegetation) and between air gases and dust particles or aqueous droplets (clouds, fog, rain) through phase transfer processes. Physically, diffusion means that the mean-free path of molecules (or particles[10]) increases in the direction of decreasing number concentration; therefore, diffusion is directed the Brownian motion. The diffusion flux $F_d = dn/qdt$ (mole per time and area) is proportional to the concentration gradient dc/dx; D is the *diffusion coefficient*, V volume $(q \cdot dz)$ and n mole number, known as Fick's first law:

$$F_d = \frac{1}{q}\frac{dn}{dt} = -D\frac{1}{V}\frac{dn}{dz} = -D\frac{dc}{dz}. \qquad (2.46)$$

Combining the basic Newton's equation (2.10) with equation (2.43) in terms of the general quantity ε, which can be a molecule mass m_m, velocity v, heat and so on, and where β means a proportionality coefficient, namely a *characteristic transfer coefficient*, it follows that:

$$\frac{d\varepsilon}{dt} = -\beta \cdot \varepsilon = -\frac{1}{\varepsilon}F_d = -q\frac{\eta}{m_m}\frac{d\varepsilon}{dz}. \qquad (2.47)$$

This is a *general flux equation* for material, energy and momentum. The negative sign indicates a decrease in the quantity ε in the direction of the gradient $d\varepsilon/dz$ (z is for vertical direction and x, y for any horizontal one). In terms of velocity from equation (2.47), an expression for the *fluidity* ϕ follows:

$$\frac{d\vec{v}}{dx} = -\frac{1}{\eta \cdot q}m_m\frac{d\vec{v}}{dt} = -j \cdot \tau, \qquad (2.48)$$

where $\varphi = 1/\eta$ (τ = shear stress). Hence, viscosity means the transport of momentum against a velocity gradient, and diffusion is material transport against a concentration gradient. Comparing equation (2.46) and equation (2.47) generates a simple expression for the gas diffusion coefficient D_g (which is different from the liquid-phase diffusion coefficient). Table 2.4 summarises the most important quantities for describ-

---

10 The diffusion of large particles (PM) compared with molecules is described in a different way. The term *particle* here involves molecules, atoms and molecule clusters.

ing molecular kinetics.

$$D_g = \frac{1}{3}\bar{v}l = \frac{1}{\pi d^2}\frac{1}{\rho(m)}\sqrt{\frac{m_m kT}{6}} = \frac{1}{\pi d^2}\frac{1}{n_0}\sqrt{\frac{N_A}{M}}\sqrt{\frac{kT}{6}}. \tag{2.49}$$

The diffusion coefficient for gases lies in the range of $0.1\,\mathrm{cm}^2\,\mathrm{s}^{-1}$ (CH_4 in the air) and $0.6\,\mathrm{cm}^2\,\mathrm{s}^{-1}$ (H_2 in the air). For particles (not molecules) larger than about $0.1\,\mathrm{nm}$ in diameter, the diffusion coefficient decreases with increasing size. Their diffusion is by several orders of magnitude smaller than for gas molecules, which is important for collision growth.

2.4 Water and waters

The heading of this Chapter would be 'liquids' in classical textbooks on physical chemistry. In the environment, however, the only liquid[11] is water in terms of aqueous solutions, called natural water or waters.

> Water is the most abundant compound in humanity's environment. It is the only substance in our environment that can exist in all three states of matter simultaneously: gaseous (vapour), liquid (natural waters) and solid (ice and snow).

The term 'water' is used in two senses. First, *pure* water is a chemical substance, and we call the discipline that studies it *water chemistry* (or chemistry of water), which deals with the chemical and physical properties of water as molecules, liquid and ice. The systematic name of water (H_2O) is dihydrogen monoxide. Second, *natural* water is a solution; the discipline that studies it is called hydrochemistry (sometimes hydrological chemistry), chemical hydrologyor aquatic chemistry. Water in nature is always a solution, mostly a diluted system, and in equilibrium or non-equilibrium but exchange with the surrounding medium, solids (soils, sediments, rocks, vegetation) and gases (atmospheric and soil air). The unique chemical and physical properties of water mean that it plays key roles in the environment as a:
- solvent (for compounds essential for life, but also pollutants, dissolved and non-dissolved),
- chemical agent (for photosynthesis in plants and photochemical oxidant formation in the air),
- reaction medium (aqueous-phase chemistry),
- transport medium (in the geosphere, for example, oceanic circulation, river run-off, cloud transportation; in the biosphere in plants, animals and humans),

---

11 We do not consider here extreme environmental conditions such as deep Earth (magma), deposits (crude oil) and the stratosphere (nitric acid droplets).

- energy carrier (latent heat: evaporation and condensation; potential energy in currents and falling waters),
- geological force (weathering, ice erosion, volcanic eruptions).

Natural water is an aqueous solution of gases, ions and molecules, but it also contains undissolved, suspended and/or colloidal inorganic particles of different sizes and chemical compositions and biogenic living and/or dead matter such as cells, plants, and so on, sometimes termed *hydrosol*. Water occurs in the environment in different forms:

- liquid bulk water (natural waters): in rivers, lakes, wetlands, oceans, and groundwater (held in aquifers),
- soil water (humidity): adsorbed onto soil particles,
- liquid droplet water: in clouds, fog, rain, but also as dew on surfaces,
- ice-particulate water in the atmosphere: snow, hail, grains,
- water vapour (humidity) in the atmosphere (one gaseous component among many other gases of air),
- hydrates: chemically bonding water molecules (as ligands) onto minerals,
- clathrate hydrates: water-based crystalline solids physically resembling ice, inside which small non-polar gas molecules are trapped (existing under high pressure in the deep ocean floor),
- bulk ice: snow cover, glaciers, icebergs.

2.4.1 Cycling and chemical composition of waters

> The water or *hydrological cycle* is the continuous circulation of water throughout the Earth and among its systems.

At various stages, water moves through the atmosphere, the biosphere, and the geosphere (Figure 2.2), in each case performing functions essential to the survival of the planet and its life forms. Thus, over time, water evaporates from the oceans; then falls as precipitation; is absorbed by the land; and, after some period, makes its way back to the oceans to begin the cycle again. The total amount of water on the Earth has not changed in many billions of years, though water distribution has changed. Only a very small percentage of all the water in the environment is actually present in the atmosphere. Of the atmospheric water, most is in the vapour phase; the liquid water content (LWC) of clouds is only in the order of $1\,\mathrm{g\,m^{-3}}$, the cloud ice water content (IWC) is even less, down to $0.0001\,\mathrm{g\,m^{-3}}$. However, clouds play a huge role in the climate system, whereas precipitation closes the cycle for water and substances dissolved in it (wet deposition).

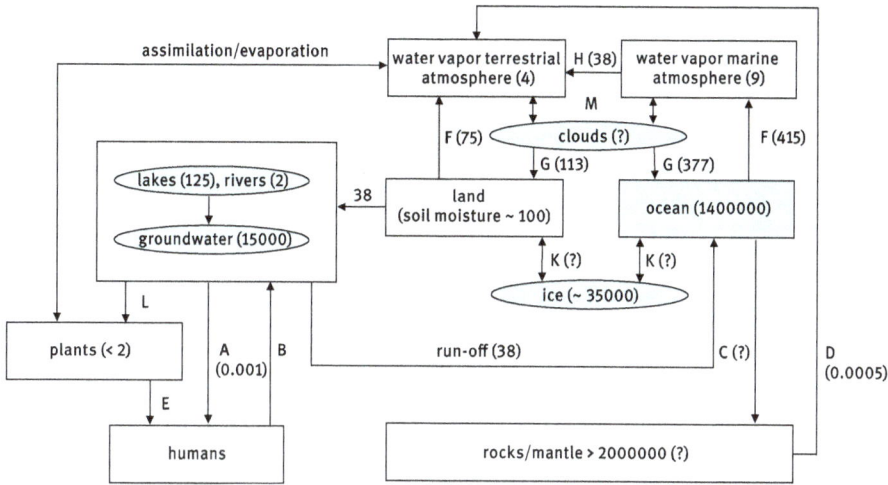

Figure 2.2: The global water cycle and water reservoirs; reservoir amounts in 10^{18} g and fluxes in 10^{18} g yr^{-1}. A: domestic and industrial water, B: wastewater, C: subduction, D: volcanisms, E: food, F: evaporation, G: precipitation, H: atmospheric transport (clouds and water vapour), K: freezing and melting, L: uptake including agriculture (0.003), M: cloud cycling (nucleation and evaporation). Data from different sources see Möller (2019).

The water cycle is driven by processes that force the movement of water from one reservoir to another. Evaporation from the oceans and land is the primary source of atmospheric water vapour (Figure 2.2). Water vapour is transported, often over long distances (which characterise the type of air masses), and eventually condenses into cloud droplets, which in turn develop into precipitation. Globally, there is as much water precipitated as is evaporated, but overland precipitation exceeds evaporation, and over oceans, evaporation exceeds precipitation. The excess precipitation over land equals the flow of surface and groundwater from continents to the oceans. Flowing water also erodes, transports and deposits sediments in rivers, lakes and oceans, affecting water quality.

The global water cycle includes sub-cycles, such as the cloud processing (condensation onto CCN and evaporation with forming new CCN), the plant water cycle (assimilation and transpiration), and the human water cycle (water processing for drinking and processing as well as wastewater treatment). As water cycles through the environment, it interacts strongly with other biogeochemical cycles, notably the cycles of carbon, nitrogen, and other nutrients. These linkages directly affect water quality and the availability of potable water and industrial water supplies. It is estimated that 70–80 % of worldwide water is used for irrigation in agriculture, while 15–20 % is for industry, and the rest of water is useed for household purposes (drinking, bathing, cooking, sanitation, and gardening).

Table 2.5: Mean composition of world's ocean (in mg L^{-1}) with 3.5 % salinity.

substance	concentration[a]	species	concentration[a]	dissolved gases	concentration
H_2O	965,000	SiO_2	2–6	CO_2	~80–90
Cl^-	19,353	F^-	1.3	N_2	~12.5
Na^+	10,781	Li	0.17	O_2	<7
SO_4^{2-}	2712	Rb	0.12	DMS[b]	13–660
Mg^{2+}	1284	Ba	0.021	Ar	~0.45
Ca^{2+}	412	P	0.09	NH_3	<0.06
K^+	399	I	0.064	H_2O_2	<0.003
HCO_3^-	126	Ba	0.021	Ne	0.00012
Br^-	67.3	Mo	0.01		
N^c	15.5	Ni	0.0066		
Sr	7.94	Fe	0.0034		
B^d	4.5	U	0.0033		

[a] all elements as ions, acids and hydroxo complexes
[b] dimethyl sulphide $(CH_3)_2S$ in surface water
[c] in the form of highly variable concentrations of NO_3^- (<0.7 ppm) NO_2^- (<0.02 ppb) and NH_4^+ (<10 ppb) as well as dissolved N_2 (12 ppm)
[d] as boric acid

Table 2.6: The composition of average river water and seawater (in mg L^{-1}); data from Langmuir (1997).

substance	river water	seawater	seawater to river water	ratios	river water	seawater
HCO_3^-	58.6	146	2.5	Ca/HCO_3	0.26	2.8
Ca^{2+}	15.0	412	27	Ca/SO_4	1.4	0.15
SO_4^{2-}	10.6	2707	255	Na/Cl	0.77	0.55
Cl^-	7.8	19,383	2485	Na/Mg	1.46	8.66
Na^+	6.3	10,764	1794	Na/K	2.22	102.5
Mg^{2+}	4.1	1243	303			
K^+	2.3	105	39			

Rivers and other forms of surface water actually account for a relatively small portion of the planet's water supply, but they loom large in the human imagination as the result of their impact on our lives. The chemical composition of river water is significantly different from that of seawater (Tables 2.5 and 2.6). At first approximation, seawater is mainly a solution of Na^+ and Cl^- while river water is a solution of Ca^{2+} and HCO_3^-. Interestingly the ratio Na/Cl is the only one that is relatively similar for rivers and oceans, suggesting that both components are within a global cycle from sea spray through cloud transportation to continents and precipitation. Carbonate is approximately in equilibrium with atmospheric CO_2, and the concentration difference between river water and seawater is determined by the pH. Ions transported by rivers are

Table 2.7: Chemical composition of different rivers (in mg L^{-1}); data from Livingstone (1963) and Holland (1984).

	HCO_3^-	SO_4^{2-}	Cl^-	SO_4^{2-}	Mg^{2+}	Na^+	K^+	Fe	SiO_2
North America	68	20	8	1	6	9	1.4	0.16	9
South America	31	4.8	4.9	0.7	1.5	4	2	1.4	11.9
Europe	95	24	6.9	3.7	5.6	5.4	1.7	0.8	7.5
Asia	79	8.4	8.7	0.7	5.6	9.3	–	0.01	11.7
Africa	43	13.5	12.1	0.8	3.8	11	–	1.3	23.2
Australia	32	2.6	10	0.05	2.7	2.9	1.4	0.3	3.9
World	58.1	11.2	7.8	1	4.1	6.3	2.3	0.67	13.1

the most important source of most elements in the ocean. Rivers collect dissolved matter from precipitation, weathering and pollution. Their chemical composition shows large seasonal and interannual variations as well as great differences between the continents, reflecting the processes at the sources (Table 2.5 to 2.7).

Most (99.99 % or more) of all atmospheric water remains in the gaseous phase of an air parcel. Hydrometeors may be solid (ice crystals in different shapes and forms) or liquid (droplets ranging from a few μm up to some tens of μm). We distinguish the phenomenon of hydrometeors into clouds, fog and precipitation (rain, drizzle, snow, hail, etc.). This atmospheric water is always a chemical aqueous solution where the concentration of dissolved trace matter (related to the bulk quantity of water) is up to several orders of magnitude higher than in the gaseous state of the air.

Clouds are the intermediate in the water cycle between vaporised water from the surface and precipitation back to the surface. Clouds also have several important functions in air chemistry: transportation, removal as well redistribution of atmospheric compounds, and the radiation budget. At any given time, clouds of one form or another mask about 60 % of the earth's surface. Despite their relatively low spatial occupancy of the troposphere, clouds provide an aqueous-phase medium for surface and bulk chemical reactions for unique transformations. The formation of clouds and fog droplets requires two conditions: first, the presence of aerosol particles acting as *cloud condensation nuclei* (CCN) for water vapour condensation, and second, a slight supersaturation. Table 2.8 shows the chemical composition of cloud water from different sites in the world. The chemical composition results from the soluble substances of the CCN and the scavenging of soluble gases from the air.

Not all clouds precipitate. Indeed, from only a very small portion of cloud, precipitations actually reaches the ground surface. The basic problem is that cloud water droplets or ice particles are frequently too small to fall from the cloud base or to survive on the way to the ground because they evaporate. Whereas a cloud droplet is on average 8 μm in diameter, a raindrop is between 500 and 5,000 μm (0.5–5 mm); this means that a small raindrop is as large in volume as 240,000 cloud drops. Assuming 240 cloud droplets per cm^{-3}, there is only one raindrop in one litre of air. Several mi-

Table 2.8: Chemical composition of cloud water at different regions of the world, in µeq L^{-1}. Data from different sources, see Möller (2003, 2019). A – altitude in m above sea level (a. s. l.)

site /region	A	Cl$^-$	NO$_3^-$	SO$_4^{2-}$	Na$^+$	NH$_4^+$	K$^+$	Ca^{2+}	Mg^{2+}	H$^+$
Seeboden (Switzerland)[a]	1030	77	440	380	20	920	15	88	21	5
Mt. Wilson (LA, USA)[b]	750	190	1486	859	241	580	21	139	90	1184
Whiteface Mt. (USA)[c]	1483	31	110	140	11	89	20	17	6	280
Rossia[d]	–	48	7	128	35	61	18	45	45	8
Kleiner Feldberg (Germany)[e]	826	242	779	409	107	1543	22	56	–	316
Mt. Brocken (1992–1995)[f]	1042	73	256	223	72	311	6	72	23	85
Mt. Brocken (1996–2002)[g]	1042	90	236	183	93	286	5	39	21	66
Mt. Brocken (2003–2009)[h]	1042	81	201	151	84	279	5	26	21	37
Åreskutan (Sweden)[i]	1250	5	9	32	6	10	<0.8	0.7	0.8	35
Åreskutan (Sweden)[j]	1250	0.8	2	6	<0.4	1	<0.8	0.3	<0.2	13
Mt. Norikura (Japan)[k]	3026	118	55	531	96	183	74	–	–	160

[a] winter 1990 until spring 1991
[b] 120 samples 1982–1983
[c] 28 samples 1976
[d] 655 aircraft samples 1960–1970
[e] 1983–1986
[f] 3512 samples
[g] 10,112 samples
[h] 9217 samples
[i] 80 summer samples 1983/1984 only NE + W
[j] 41 samples from NW (remote air), as "g"
[k] about 24,000 samples 1992–2009

crophysical processes occur in clouds dependent on temperature, vertical resolution, dynamic and other parameters that result in the growth of a particle and different precipitation forms; the chemical composition of precipitation results from the cloud water and from scavenging of gases and dust particles through falling drops (Table 2.9). Normally, precipitations water is less concentrated than cloud water due to the precipitation formation in clouds where clean drops and ice particles from upper layers 'dilute' the water. Only in highly polluted areas can precipitation be enriched due to sub-cloud scavenging.

As seen in Table 2.10, the rainwater chemical composition reflected the extreme city air pollution in the nineteenth century. The London values also show the high importance of NH_3 emissions in that time from the absence of wastewater management. As a result, the pH was not different from late twentieth century values. Interesting are the Beijing rainwater values, which were not significantly different from German background values but had a very high pH due to alkaline soil dust from the Gobi Desert. Later, however, pH became less than 5 due to the extreme increase of Chinese SO_2 emission. We see that the balance between acidic and alkaline constituents determines the rainwater acidity (see also Chapter 5.3.2).

Table 2.9: Chemical composition of precipitation in different world regions (in µeq L^{-1}, rounded). Data from different sources, see Möller (2019).

site	H$^+$	Na$^+$	K$^+$	NH$_4^+$	Ca^{2+}	Mg^{2+}	Cl$^-$	NO$_3^-$	SO$_4^{2-}$
China (1981–1984)[a]	0.2	44	21	160	460	–	39	34	162
India (1988–1996)[b]	0.1	18	8	40	56	–	32	18	36
Cape Grimm (1977–1985)[c]	1	1167	30	2	78	247	1372	3	152
Central Australia (1980–1984)[d]	17	4	1	3	2	1	8	4	4
Lancaster (NW England)[e]	34	57	4	55	15	15	79	28	89
Central Bohemia[e]	42	7	3	62	30	–	–	51	116
Hungary[f]	32	23	9	61	85	16	26	41	119
Seehausen (1983–1989), $n = 1185$	46	30	6	87	58	12	58	44	148
Seehausen (1991–1994), $n = 561$	35	26	4	45	23	7	37	38	52
Seehausen (1995–2000), $n = 768$	14	16	1	42	17	5	23	41	41
North Sweden (1983–1987)[g]	23	4	2	13	3	2	5	11	33
South Sweden (1983–1987)[g]	46	24	2	36	10	7	30	36	67
Seehausen, Germany (1996–2002)[h]	14	20	2	52	16	5	25	39	40
Melpitz, Germany (2000)[i]	13	8	1	44	13	3	8	31	35
Neuglobsow (2000)[j]	13	14	2	39	18	4	14	31	33
Staudinger, Germany (2007–2008)[k]	10	14	3	44	18	5	14	36	31
Peitz, Germany (2011–2012)[l]	6	30	17	78	30	9	28	58	47
Radewiese, Germany (2011–2012)[m]	4	31	7	52	483	12	20	41	469

[a] suburb of Beijing
[b] suburb in semiarid area
[c] Tasmania, south Pacific air
[d] Katherine, annual rainfall 75–136 mm
[e] Hradec, mean of 483 samples
[f] mean from 6 stations
[g] mean from 10 stations
[h] mean from 20 stations
[i] rural station near Leipzig, weekly wet-only samples
[j] background station 60 km north of Berlin, weekly wet-only samples
[k] Hanau (within power station Staudinger) 25 km east of Frankfurt, $n = 61$
[l] 15 km north of Cottbus, close to power station Jänschwalde, $n = 123$
[m] 15 km northeast from Cottbus, direct under the plume of Jäanschwalde power station, $n = 115$

Table 2.10: Historical comparison of precipitation chemical composition. Data from different sources, see Möller (2019).

site	pH	SO$_4^{2-}$	NO$_3^-$	NH$_4^+$	Cl$^-$	sum[a]	Δ[b]
		in mg L^{-1}				in µeq L^{-1}	
London 1870	4.8	57.7	8.9	10.6	10.5	1029	1014
Beijing 1981	6.5	16.6	5.0	4.0	2.1	254	254
Seehausen 1985	4.3	13.4	4.3	2.7	2.9	274	229

[a] $[SO_4^{2-}] + [NO_3^-] + [Cl^-] - [NH_4^+]$
[b] $[SO_4^{2-}] + [NO_3^-] + [Cl^-] - [NH_4^+] - [H^+]$ = missing neutralising cations such as Ca^{2+}

2.4.2 Physical and chemical properties of water

> **i** Water is unusual in *all* its physical and chemical properties.

Its boiling point (abnormally high), its density changes (maximum density at 4 °C, not at the freezing point), its heat capacity (the highest of any liquid except ammonia), and the high dielectric constant as well as the measurable ionic dissociation equilibrium, for example, not what one would expect when comparing water with other similar substances (hydrides). All the physical and chemical properties of water (Table 2.11) make our environment unique and have shaped the course of chemical evolution. Water is the medium in which the first cell arose and the solvent in which most biochemical transformations take place.

2.4.2.1 Water structure: hydrogen bond

Under certain conditions, an atom of hydrogen is attracted by rather strong forces to two atoms instead of only one so that it may be considered acting as a bond between them. This is called the *hydrogen bond* (see Chapter 3.3.1 for chemical bonding). This statement is from book by Linus Pauling *The Nature of the Chemical Bond* (1939). At that time, the hydrogen bond was recognised as mainly ionic in nature. The energy associated with the hydrogen bond is about $20 \, kJ \, mol^{-1}$. Due to hydrogen bonding, water molecules form dimers, trimers, polymers, and clusters. The hydrogen bonds are not necessarily linear (Figure 2.3). The ion mobility of H_3O^+ and OH^- is anomalously high: $350 \cdot 10^{-4}$ and $192 \cdot 10^{-4} \, cm^2 \, V^{-1} s^{-1}$ (25 °C), respectively, in comparison with $(50 - 75) \cdot 10^{-4} \, cm^2 \, V^{-1} s^{-1}$ for most other ions. Chapters 3.2.2.3 and 3.3.5.2 deal with the chemistry of the proton (H_3O^+) and aquated electron (H_2O^-) both are fundamental species in nature. The mobility results from the special structure of liquid water where the H_2O molecules are linked in chains, stabilised by hydrogen bonding, which also gives liquid water great internal cohesion (Figure 2.3). Figure 2.4 (in the scheme the tetrahedral structure shown in Figure 2.3 remained but is not clearly seen) illustrates that the proton (H^+) is not really transported but only the charge – similar to the OH^- transport – and thus explaining the high ionic mobility. The water grid (quasi-crystalline) is restructured. Therefore, the high evaporation heat and entropy, the high surface tension and relatively high viscosity are a result of the hydrogen bonding structure (Figure 2.3). Bernal and Fowler (1933) first proposed a water structure model in the sense of a 'quasi-crystalline structure'. Nowadays, many models of the structure of liquid water have been proposed (e.g. Rick 2004), but there is no consensus even on the number of H_2O molecules forming species ('polymers').

When water freezes, the crystalline structure is maintained (Figure 2.3) and determined by the prevailing condition; at least nine separable ice structures exist. Nor-

Table 2.11: Physical and chemical properties of water.

property	value	dimension
molar mass	18.015268	$g\,mol^{-1}$
freezing point at 1 bar	1.00	°C
boiling point at 1 bar	100.0	°C
vapour pressure at 25 °C	3.165	kPa
latent heat of melting at 1 bar	332.5	Jg^{-1}
latent heat of evaporation at 1 bar	2257	Jg^{-1}
specific heat capacity of water	4187	$Jg^{-1}K^{-1}$
specific heat capacity of ice	2108	JgK^{-1}
specific heat capacity of water vapour	1996	JgK^{-1}
critical temperature at 1 bar	647.096	K
critical pressure	220.64	bar
critical density	322	gL^{-1}
maximum density (at 3.98 °C)	1.0000	$g\,cm^{-3}$
density of water at 25 °C	0.99701	$g\,cm^{-3}$
density of ice at melting point (0 °C)	0.91672	$g\,cm^{-3}$
density of gas at boiling point (100 °C)	0.0005976	$g\,cm^{-3}$
viscosity, dynamic	0.8903	cP^a
viscosity, kinematic	0.008935	$stokes^b$
surface tension of water at 25 °C	72	$dyn\,cm^{-1}$
dielectric constant at 25 °C	78.39	–
Prandtl number at 25 °C	6.1	–
cryoscopic constant	1.8597	$K\,kg\,mol^{-1}$
O–H bond dissociation energy	492.2148	$kJ\,mol^{-1}$
bond energy, average at 0 K (H–O–H → O + 2 H)	458.9	$kJ\,mol^{-1}$
conductivity, electrolytic, at 25 °C	0.05501	$\mu S\,cm^{-1}$
conductivity, thermal, for water at 25 °C	0.610	$W\,m^{-1}K^{-1}$
conductivity, thermal, for ice at –20 °C	2.4	$W\,m^{-1}K^{-1}$
conductivity, thermal, for vapour at 100 °C	0.025	$W\,m^{-1}K^{-1}$
electron affinity at 25 °C	−16	$kJ\,mol^{-1}$
energy, internal (U) for water at 25 °C	1.8883	$kJ\,mol^{-1}$
enthalpy of formation, ΔH_f, at 25 °C	−285.85	$kJ\,mol^{-1}$
enthalpy (H = U + PV), at 25 °C	1.8909	$kJ\,mol^{-1}$
enthalpy of vaporis ation (liquid), at 0 °C	45.051	$kJ\,mol^{-1}$
enthalpy of sublimation (ice), at 0 °C	51.059	$kJ\,mol^{-1}$
Gibbs energy of formation^c, ΔG_f, at 25 °C	−237.18	$kJ\,mol^{-1}$
surface enthalpy (surface energy) at 25 °C	0.1179	$J\,m^{-2}$
surface entropy (= dγ/dT) at 25 °C	0.0001542	$J\,m^{-2}K^{-1}$
ionic dissociation constant, $[H^+][OH^-]/[H_2O]$, 25 °C	$1.821 \cdot 10^{-16}$	$mol\,L^{-1}$
O–H bond length (liquid, ab initio)	0.991	Å
H–O–H bond angle (liquid, ab initio)	105.5	°

Table 2.11: (continued).

property	value	dimension
redox potential E_0: water oxidation[d]	1.229 v	V
redox potential E_0: water reduction[e]	−0.8277	V

[a] acentipoise (=0.008903 $g\,cm^{-1}\,s^{-1}$)
[b] (=0.8935 · 10^{-6} · $m^2\,s^{-1}$)
[c] =chemical potential (μ)
[d] $2\,H_2O \rightarrow O_2\,(g) + 4\,H^+ + 4\,e^-$
[e] $2\,H_2O + 2\,e^- \rightarrow H_2\,(g) + 2\,OH^-$

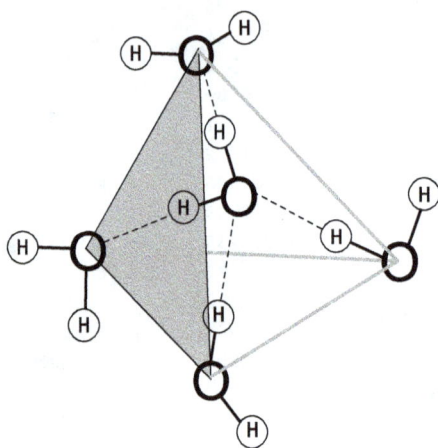

Figure 2.3: Water structure and hydrogen bonds (dotted lines) in a tetrahedral grid; note that the structure is not in a plane.

mally, ice has a hexagonal structure (En) when cooling down liquid water; each O atom is surrounded by a regular tetrahedron of a further four O atoms. The positioning of H is very complex. The four hydrogen bonds around an oxygen atom form a tetrahedron in a fashion found in the two types of diamonds. Thus, ice, diamond, and close packing of spheres are somewhat topologically related. Water ice is unusual because its density is less than that of the liquid water with which it is in equilibrium. This is an important property for the survival of life in water. When the ice melts, a few hydrogen bridges (probably every fourth one) begin to break, the H_2O molecules close ranks, and the density consequently increases.

Many salts crystallised from aqueous solutions are not water-free but take the form of well-defined hydrates. Other solid phases contain bound water in varying amounts. A classic case is a water coordinated onto oxoanions, for example, $CaSO_4 \cdot 2\,H_2O$. The most frequent are cation complexes with water, for example, alums such as $[Al(OH_2)_6]^{3+}$; this is an explanation for the large water content in rocks. Framework

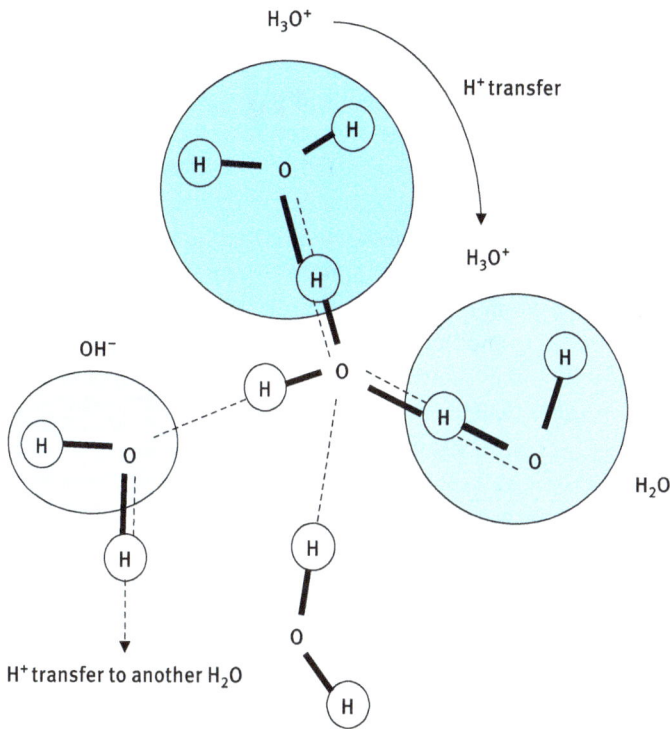

Figure 2.4: Formation and transport of hydronium (H_3O^+) and hydroxide (OH^-) ions in the water grid (as shown in Figure 2.3) by charge transfer, i. e. changing hydrogen bonding (dotted line) from strong covalent to weak interaction.

silicates (e. g. zeolites) hold huge amounts of water within their cavities; faujasite is the mineral with the highest water content: $Na_2Ca[Al_2Si_4O_{12}]^2 \cdot 16\,H_2O$. Similar to the zeolitic grids, water $(H_2O)_n$ can build up cage-like inclusion structures (clathrate hydrates) where, in a skeleton of $46\,H_2O$, there exist six cavities of the same size and two more, smaller ones. The guest molecules in high-pressure clathrates are Ar, Kr, CH_4 and H_2S.

Polywater (also called anomalous water), first described in the 1960s in the Soviet Union and controversially discussed in the 1970s, does not exist, however, and was probably a mixture of colloidal silicic acid.

2.4.2.2 Water as a solvent

The polarity of water gives it important properties that the biosphere needs to function: it is a universal solvent, and it adheres and is cohesive. Thus, water facilitates chemical reactions and serves as a transport medium. Even in a covalent bond, atoms may not share electrons equally. In H_2O, the unequal electron sharing creates two electric dipoles along each of the O–H bonds. The H–O–H bond angle is 104.5°, 5° less than the bond angle of a perfect tetrahedron, which is 109.5°. The structure shown in Figure 2.3 is idealised. Only in ice, it is fixed (crystalline), but in liquid water at any moment, depending on the temperature, each water molecule forms hydrogen bonds with an average of 3.4 other water molecules. They are in continuous motion in the liquid state; hence, hydrogen bonds are constantly and swiftly being broken and formed (Figure 2.4). The protolytic equilibrium is described in more detail in Chapter 3.2.2.2. Hydrogen bonding is unique for water. The bonds readily form between an electronegative atom (usually oxygen, nitrogen or sulphur) and a hydrogen atom covalently bonded to another electronegative atom in the same or another molecule: $H^{\oplus}-O^{\ominus}-H^{\oplus} \cdots O^{\ominus}-$ and $H^{\oplus}-O^{\ominus}-H^{\oplus} \cdots N-$. However, hydrogen atoms covalently bonded to carbon atoms (which are not electronegative) do not participate in hydrogen bonding; hence, hydrocarbons are insoluble in water. However, organic compounds with oxygen (and nitrogen) containing functional groups (like alcohols, aldehydes, acids, ketones, etc.) are water-soluble. The more oxygen groups and the fewer carbon atoms in a compound, the more soluble it is in water.

> ℹ️ It is the polarity *and* the hydrogen bond affinity that makes water a solvent for many chemically different substances: oxygenated and/or nitrogen-containing organic compounds (most biomolecules, which are generally charged or polar compounds), salts (electrostatic interacting solid grids), but also non-polar gases (biologically important CO_2, O_2 and N_2) and all polar gases (for example, SO_2, NH_3, HCl, HNO_3 which are important for the atmosphere).

It needs no further explanation that the solubility of non-polar molecules is much less than that of polar substances. The property to interact with water is also called hydrophilicity (affinity to water: attraction) and inversely hydrophobia (non-affinity to water: repulsion).

2.4.3 Properties of aqueous solutions

In this chapter, we only consider physical properties; chemical properties such as solubility and solid-aqueous equilibrium, acid-base reactions and aqueous chemistry will be discussed later in Chapters 3.2.1 and 4.3.3.

2.4.3.1 Surface tension and surface-active substances

An important property of the water surfaces is the *surface tension* that expresses the cohesion of water molecules (Chapter 2.4.2.1). On molecules existing close to the droplet surface, forces are directed to the inner of the droplet. Therefore, each liquid has the tendency to form spherical particles (if they are not counteracting forces such as gravitation and other outer forces). The reason is simple: a sphere of a given volume has the smallest surface of all bodies. Thus, a growing droplet needs to overcome the molecular cohesion. There are two equivalent definitions of surface tension:

$$\gamma = \frac{f}{l} \quad \text{and} \quad \gamma = \frac{dW_s}{dq}. \tag{2.50}$$

The force f concentrates a surface band of width l, and a dimension of surface tension is $N\,m^{-1}$ or $kg\,s^{-2}$). The other definition is the ratio of the surface energy to the surface (dimension: energy/surface but reduced on metric units to $kg\,s^{-2}$). Water has a surface tension (298 K) of $72.85 \cdot 10^{-3}\,N\,m^{-1}$.

Some organic substances dissolved in the droplets or transported from the gaseous surrounding to the surface can accumulate on the surface when they have hydrophilic and hydrophobic properties in one molecule (for example, aliphatic alcohols, aldehydes and acids). They form a liquid film and reduce the surface tension according to Gibbs's equation:

$$d\gamma = -RT \cdot \Gamma_q d\ln c, \tag{2.51}$$

where c is the concentration of the surface-active substance Γ_s, and surface excess is $n_s(q)/q$, i. e. the amount of matter per square unit. The importance of such films becomes clear when considering that all processes, linked with the free enthalpy G of the water body (evaporation, adsorption, desorption, surface reactions), result generally in a change of G through a change of T, p, S or n (see also Chapter 3.1.2.3 and equation (3.39)):

$$dG = -SdT + Vdp + \gamma dq + \sum \mu_i dn_i. \tag{2.52}$$

2.4.3.2 Vapour pressure lowering: Raoult's law

Compared with pure water, solutions have two different fundamental properties: lowering the vapour pressure[12] and freezing point depression.[13] In about 1886, François Marie Raoult (1830–1901) discovered that substances have lower vapour pressures in solution than in pure form and that the freezing point of an aqueous solution decreases in proportion to the amount of a non-electrolytic substance dissolved. The ratio of the partial vapour pressure of substance i in solution to the vapour pressure of the pure substance (subscript 0 denotes the pure substance) is equal to the mole fraction x of i:

$$p_i = x_i p_i^0. \tag{2.53}$$

---

12 In physical chemistry, the *boiling-point elevation* is connected with the vapour pressure lowering but as said, the only liquid is water in the natural environment and temperatures of about the water boiling point will never be reached.

13 In cloud microphysics, these changes are crucial for droplet growth and precipitation formation.

This law is strictly valid only under the assumption that the chemical interaction between the two liquids equals the bonding within the liquids: the conditions of an ideal solution. In the atmosphere, water is the solvent, and the dissolved matter is predominantly non-volatile. The vapour pressure of aqueous solution, p_w, is smaller than that of pure water, p_w^0, whereas the vapour pressure of the solution p_S is $p_S = p_w + p_i$. Because of $x_w + x_i = 1$, we rewrite Raoult's law as follows:

$$p_S = p_w^0 + (p_i^0 - p_w^0)x_i. \tag{2.54}$$

Assuming that $p_i^0 \to 0$ (the dissolved substance is non-volatile) we find for the *relative* lowering of the vapour pressure of the solution $(\Delta p/p)$, i. e. equal to the mole fraction of i:

$$\Delta p = p_w^0 - p_S = x_i p_w^0. \tag{2.55}$$

2.4.3.3 Freezing point depression

The freezing point depression follows from the lowering of vapour pressure. From the Clausius–Clapeyron equation (equation (3.50)) and Raoult's law (equation (2.55)), it follows ($\Delta_{sm}H$ enthalpy of smelting) that:

$$\Delta T = \left(\frac{RT^2}{\Delta_{sm}H} \right)x_i. \tag{2.56}$$

Taking the molality m (defined as the ratio of the amount of dissolved matter and the mass of water, expressed in mol kg^{-1} and in contrast to the molarity not depending on T (in dissolved solution, molarity is proportional to molality), we obtain:

$$\Delta T = K_f m_i, \tag{2.57}$$

where K_f is the cryoscopic constant, which is empirical and can be determined experimentally; for water $K_f = 1.86$ K kg mol^{-1}. The freezing (or solidification) temperature is the temperature at which the liquid has the same vapour pressure as the solid. Water, however, exists supercooled,[14] i. e. despite a certain freezing point depression, droplets remain liquid at lower temperatures because of the kinetic inhibition of crystallisation. The homogeneous process is that spontaneous freezing occurs only for $\leq$232 K or –41 °C (T_H) and saturation near that of liquid water (Koop et al. 2000).

---

14 Heterogeneous ice nucleation in clouds with supercooled water results in the subsequent efficient growth of the ice crystals because of the Bergeron–Findeisen process. It is assumed that this is the main initiation process of precipitation in the mid-latitudes.

However, the temperature at which water becomes ultimately a solid (T_S) may be deeper; computer studies based on classical nucleation theory using experimental data suggest that the crystallisation rate of water reaches a maximum around 225 K ($-48\,°C$), below which ice nuclei form faster than liquid water can equilibrate. The heterogeneous process requires the presence of ice nuclei (IN) with a hexagonal crystal structure similar to that of water ice, which allows freezing at temperatures as high as $-5\,°C$.

2.4.3.4 Diffusion in solutions

Similar to the diffusion of molecules in the air (or gases), see Chapter 2.3.2.4, the spontaneous transfer of a dissolved substance in solution from a site of higher concentration to the site of lower concentration is called diffusion and described by the first Fick's law, equation (2.46). As for gases, the driving force of diffusion is the concentration gradient. Assuming that the diffusing particle is spherical and much larger than the particles (molecules) of the solvent, the movement is described by the Stokes law, and it follows for the aqueous diffusion coefficient:

$$D_{aq} = \frac{RT}{6\pi N_A \eta r},\qquad(2.58)$$

where r is the radius of diffusing particle, and η is the viscosity of the solution; D lies in the order of $10^{-5}\ cm^2\ s^{-1}$ for many dissolved substances. Equation (2.58) also represents the diffusion of solid particles in gases; from measurement data in the air (particle range 10–10,000 nm), the empirical relationship $-\lg D = 1.48 + 2.33\ \lg r$ can be derived.

2.4.4 Water vapour

The water vapour in the air is a result of the vaporisation of water from the earth's surface. We can consider liquid water to be condensed gas. At any given time, a certain number of molecules can escape the liquid from the surface to the surrounding air (we call it *evaporation*). Because of air motion (turbulent mixing and advection), there is no equilibrium, i. e. transfer of water molecules from the air back to the surface (we call it *condensation*) in the same flux as evaporation. Such equilibrium can only be reached in a closed undisturbed chamber. If the equilibrium between condensed and vaporous water is reached, the pressure is called *saturation pressure* p^∞. Such conditions are important on a microscale for cloud formation but are also frequently observed in flue gases. The *relative humidity* RH is the ratio of the vapour pressure at temperature T to the saturation vapour pressure at the same temperature expressed as a percentage:

$$RH = \frac{100 \cdot p_{H_2O}}{p^\infty(T)}.\qquad(2.59)$$

The water vapour pressure p_{H_2O} is numerically identical to the mixing ratio x_{H_2O}. The *absolute humidity* (or water vapour concentration or density) is the mass of water vapour in a volume of air at a given temperature: m_{H_2O}/V. From equation (2.59), it follows:

$$\rho_{H_2O} = p_{H_2O}\frac{18}{RT} = p^{\infty}(T)\frac{RH}{RT}0.18. \tag{2.60}$$

Note that in all the equations, the saturation vapour pressure is measured above a plane that is an 'endless' water surface. Another important quantity is the *dew point*, the temperature when, at a given absolute humidity, the saturation vapour pressure is reached, i. e. the relative humidity becomes 100 %. Under these conditions, water vapour starts to condense – when surfaces are available.

The water vapour content in air varies in the range of 0–4 vol %, depending on temperature and saturation. Normally the chemical composition of air is based on dry air (see Table 2.3). Under normal conditions (20 °C and 60 % RH), air contains around 1 % water vapour (absolute humidity). The density of water vapour is less than that of other gaseous air constituents ($N_2 + O_2$); hence, wet air at the same temperature and pressure has a lower density than dry air. Consequently, at the same pressure, dry air has a slightly higher temperature (called *virtual* temperature) than wet air to obtain the same density. Without significant error, we calculate the mean mole mass of dry air only from the main constituents N_2, O_2, and Ar to be:

$$M_{dry\ air} = 0.78M_{N_2} + 0.21M_{O_2} + 0.01M_{Ar} = 28.96 \quad \text{(more exactly 28.9644)}.$$

The mean mole mass of wet air (x – mixing ratio of water vapour) is given by

$$M_{wet\ air} = (1 - x) \cdot 28.96 + x \cdot 18.0.$$

2.5 Solid matter

2.5.1 General remarks

Structural rigidity and resistance to changes of shape or volume characterise the solid state of matter. In contrast, gas and liquid fill the entire volume available such as the atmosphere and hydrosphere. A liquid we have introduced as condensed gas, hence water being condensed (water) vapour. A solid is a frozen liquid, i. e. each solid (if it is not evaporated or decomposed before) melts at a given temperature, the melting point (for example, ice or minerals in magma). The interatomic or intermolecular distance is much smaller in liquids than in gases, but not so in solids compared to liquids. The important difference is that the structure of solids is due to different chemical bonds (see Chapter 3.3.1).

It is remarkable that two elements from group 4, namely carbon (C) and silicon (Si), are the key elements forming solid matter in nature. It is their balanced harmonic affinities to electropositive and electronegative elements that additionally provide the largest quantity of different chemical compounds; the organic matter (hydrocarbons: $\equiv$C–H) and the inorganic matter (silicates: =Si=O), both forming chains, plates and cubic material with different physical properties maintaining life forms (biomass) and providing the foundation for life (the rocky world).

Solids in the environment are always composites containing more or less water and gases. They usually are inorganic (minerals) and organic materials (dead and decomposed biological matter as well as organic compounds from biogeochemical cycling and chemical use) and biological material (plants, myriads of microorganism, animals, etc.). In the environment (we again exclude biological matter[15]), the most common solid structure is a regular geometric lattice (*crystalline solids*, including metals and ordinary ice). However, pure metals are mostly very rare in nature. Non-crystalline (amorphous) solids are rare in nature (mineraloids, wax) or even unknown (glass, plastics). A *mineraloid* is a mineral-like substance (for example, obsidian, opal) but not a crystal; it is a natural amorphous glass. Crystalline solids may be found as single crystals, from microscopic to giant size. *Granular material* is formed by the erosion of rocks (stones, gravel and sand); other natural granular materials include snow and coal. *The powder material* is composed of very fine particles that are not cemented together. *Rock* is a naturally occurring solid aggregate of one or more minerals or mineraloids. Concerning its origin, we separate igneous, sedimentary and metamorphic rock. *The conglomerate* is a sedimentary rock formed from rounded gravel, and boulder-sized clasts cemented together in a matrix. However, in our treatise of environmental chemistry, we will not consider the *lithosphere* ('rocky world') but only the *pedosphere* filled by soil. The depth of bedrock (the interface between pedosphere and lithosphere) is between two and more than 20 metres. On average, soil thickness is only one metre; young soils have only a few centimetres thickness. Soils (the *pedosphere*) also interface with the hydrosphere (waters) and atmosphere (air).

Solids are also particulates in waters (insoluble or in saturated solution) and in the air (dust). This chapter can only describe in short some of the main features; the interested reader must study textbooks on soils and atmospheric aerosols.

2.5.2 Soils

Soil is the final product of the climate influence and biogeochemical cycling; it is the interface between the atmosphere and the lithosphere. Soil acts as an engineer-

---

15 The most important biological solid is wood, a fibrous structural tissue composite of cellulose fibers (which are strong in tension) embedded in a matrix of lignin, which resists compression.

ing medium, a habitat for soil organisms, a recycling system for nutrients and organic wastes, a water quality regulator, a modifier of atmospheric composition, and a medium for plant growth. The generalised content of soil components by volume is roughly 50 % solids (45 % mineral and 5 % organic matter) and 50 % voids, of which half (but very variable) is occupied by water and half by gas (mostly air). Trace gases, for example, CO_2, CH_4, NH_3, N_2O, H_2S and many others enrich soil air.

Without organisms, the soil would be *sand*, a naturally occurring granular material composed of finely divided rock and mineral particles under the influence of the weather. The composition of mineral sand is highly variable (Table 2.12), depending on the local rock sources and conditions (Table 2.13). Soil continually undergoes development through numerous physical, chemical and biological processes, including weathering (see Chapter 5.1.5) with associated erosion. Typical soil parent

Table 2.12: Variation of the chemical composition of soil-forming rocks (igneous, sandstone and limestone), in % from data from Clarke (1920).

element	concentration range
Al	0.8–16
Fe	0.5–3.1
Mg	1–8
Ca	3–43
Na	0.05–4

Table 2.13: The average composition of known terrestrial matter (in %); – no value given.

element	Wedepohl (1995)	Mason and Moore (1982)	Clarke (1920)
O	–	46.60	47.33
Si	28.8	27.72	27.74
Al	7.96	8.23	7.85
Fe	4.32	5.00	4.50
Ca	3.85	3.63	3.47
Na	2.36	2.83	2.46
Mg	2.20	2.09	2.24
K	2.14	2.59	2.46
Ti	0.40	0.44	0.46
H	–	0.14	0.22
P	0.076	0.105	0.12
Mn	0.072	0.095	0.08
F	–	0.0625	0.10
Ba	–	0.0425	0.08
Sr	–	0.0375	0.02
S	0.070	0.0260	0.12
C	–	0.0200	0.19

mineral materials are quartz (SiO_2), calcite ($CaCO_3$), feldspar ($KAlSi_3O_8$) and biotite ($K(Mg, Fe)_3AlSi_3O_{10}(OH)_2$).

The largest fraction of net primary production (NPP, see Chapter 5.2.3) is delivered to the soil as dead organic matter (litter), which is decomposed by microorganisms under the release of CO_2, H_2O, nutrients and a final resistant organic product, *humus*. NPP is the primary driver of the coupled carbon and nutrient cycles and is the primary controller of the size of carbon and organic nitrogen stores in landscapes. Hence, soils are large emitters of volatile organic compounds and various nitrogen and sulphur compounds.

Soil chemical reactions include adsorption/desorption, precipitation, polymerisation, dissolution, complexation and oxidation/reduction. Limiting factors for chemical and biochemical processes are soil humidity (rainfall) and porosity (gas exchange), providing reaction media and transportation of chemicals.

2.5.3 Dust

Dust comprises solid particles (containing variable amounts of water) suspended in air, also called *particulate matter* (PM) and (scientifically) *atmospheric aerosol* (Table 2.14). Note, however, that the term aerosol includes the dispersed matter (solid particles) *and* the dispersant (air); hence, not the aerosol is sampled but the particles, called atmospheric or aerosol particle (AP). APs are either directly emitted or produced from gaseous precursors via homogeneous nucleation (gas-to-particle conversion). However, atmospheric properties can only be understood as a colloidal system. The role of APs in the climate system is summarized here:
– Optical function: changing visibility (by reducing it).

Table 2.14: Origin and types of atmospheric aerosol particles (further classification possible concerning biogenic, geogenic and anthropogenic origin).

source characteristics			particle characteristics
direct	wind blow	inorganic	soil dust and sea salt
		organic	plant debris, degradation products
		biological	bacteria, viruses, pollen
	combustion	inorganic	ash
		organic	smoke, soot (BC)
	industrial	inorganic	dust
	volcanic	inorganic	ash
	extraterrestrial	inorganic	meteoric dust
indirect	gas emissions	inorganic	salts (sulphate, nitrate, ammonium, chloride)
		organic	SOA

- Radiation function: changing atmospheric heat budget (cooling tendency).
- Water cycle function: cloud and thereby precipitation formation (increasing and decreasing).
- Chemical function: providing a surface for heterogeneous chemistry.

Primary sources of atmospheric dust are soils (including vegetation, urban and agricultural areas), the ocean (sea salt), industrial and municipal plants (e.g. flue ash from power stations) and traffic (soot); secondary sources are chemical processes in air, namely sources of salt-forming gases such as NH_3, HCl, SO_3, HNO_3 (gas-to-particle conversion).

A large fraction of atmospheric dust (besides soil dust and sea salt) concerns secondary inorganic aerosol (ammonium, nitrate and sulphate) and secondary organic aerosol (SOA); the generalised content of dust components by volume is roughly 30–40 % insoluble minerals (silicates), 30–40 % soluble salts and 20–30 % organic matter (OC or OM) and soot (BC and EC), see Table 2.15. With respect to aerosol impacts on humans, different dust fractions have been defined (the sampling threshold, however, is not very sharp) according to the aerodynamic diameter (in µm): PM_1, $PM_{2.5}$ and PM_{10}. The fraction $PM_{2.5}$ contributes about 70 % to PM_{10}.

Table 2.15: Principal chemical composition of particulate matter of different origin.

origin	≥95 %	1–5 %	<1 %
soil dust	O, Si, Al	Fe, Ca, K	all other elements
sea salt	Na, Cl	SO_4^{2-}, Mg	all other elements
industrial dust	Ca, O, C	Fe, elements	all other elements
secondary inorganic	SO_4^{2-}, NH_4^+, NO_3^-	Cl	–
secondary organic	C, O	H	S, N

Whereas secondarily produced particles are < 1 µm, soil dust is > 1 µm. Sea salt particles can range in the size of the CCN (~0.2 µm) but are dominantly in the lower m range. Industrial dust can range from nm particles (soot from combustion) to coarse particles. Biological particles also range from sub-µm (bacteria) to 10–30 µm.

2.5.3.1 Soil and sea salt particles

Soil particles are entrained into the air by wind erosion caused by strong winds over bare ground. While large sand particles quickly fall to the ground, smaller particles (less than about 10 µm) remain suspended in the air as mineral (or soil) dust aerosol. Billions of tons of mineral dust aerosols are released each year from arid and semi-arid regions into the atmosphere. Mineral dust particles are estimated to be the most common aerosol by mass; estimates of its global source strength range from 1,000 to

5,000 Mt per year^{-1}. Locally, especially in urban areas where traffic is on the streets, re-suspension of soil dust by moving vehicles contributes to about one-third of the PM10 levels; this fraction is mainly between 2.5–10 μm. Various physical processes generate sea salt aerosols, especially the bursting of entrained air bubbles during whitecap formation, resulting in a strong dependence on wind speed (see also Chapter 4.7.2). Sea salt particles cover a wide size range (about 0.05–10 mm diameter) and have a correspondingly wide range of atmospheric lifetimes. Thus, it is necessary to analyse their emissions and atmospheric distribution in a size-resolved model as for soil dust. The chemical composition of sea salt corresponds roughly to the seawater composition (Table 2.5). Several studies in the last few years have shown that sea salt aerosol actually contains a substantial amount of organic matter, consisting of both insoluble material (biological debris, microbes, etc.) and water-soluble constituents.

2.5.3.2 Organic matter and soot

Carbonaceous material includes organic compounds ranging from very soluble to insoluble, plus elemental carbon and biological species. Organic compounds cover a very wide range of molecular forms, solubilities, chemical reactivities and physical properties, which makes complete characterisation extremely difficult, if not impossible.

Organic aerosol (OA) originates as either primary organic aerosol (POA) or secondary organic aerosol (SOA), see Table 2.14. POAs are emitted into the atmosphere from biomass burning, fossil fuel and biofuel use, and sea spray. SOA is defined as products of gas-phase oxidation from volatile organic compounds (VOC), emitted from biogenic or anthropogenic sources, such as vegetation and combustion emissions (e. g. aldehydes such as nonanal[16] and polycyclic aromatic hydrocarbons such as pyrene and naphthalene), which have partitioned from the gas to the aerosol phase. Organic aerosols have a large impact on air quality, biogeochemistry, and the climate through interactions with reactive trace gases, water vapour, clouds, precipitation and radiation. They can affect biogeochemistry through either their deposition of nutrients on land or ocean or by changing climate. They can influence climate by changing earth's energy budget by scattering and absorbing the radiation or acting as cloud condensation nuclei. However, they represent one of the largest uncertainties in climate science, being, for the most part, a climate cooling species.

The formation of SOA, leading to the formation of nanoparticles of blue haze over forested areas, is highly complex and not fully understood. Organic compounds (gases and particles) comprise a mixture of an extremely large number of different molecules, with each compound further undergoing atmospheric chemical reactions to produce a range of oxidized products, hence myriads of secondary products. These products are likely involved in the earliest stages of particle nucleation and growth.

---

16 Also called nonanaldehyde, pelargonaldehyde or aldehyde C-9 ($C_9H_{18}O$).

At present, the term bioaerosol or biological aerosol is used, which is scientifically unsound. Clearly, the role of biogenically derived APs (to call them *biological particles* is correct) is reconsidered after the insights first given by Louis Pasteur (1822–1895) 150 years ago in a new light, i. e., not as a transmitter of diseases but playing a likely role as CCN and IN.

A large fraction of PM is soot, the historical *symbol* of air pollution. There has been a long dispute in the literature on the definition of soot, also called elemental carbon (EC), black carbon (BC) and graphitic carbon. Surely, soot is the best generic term for impure carbon particles resulting from the incomplete combustion of a hydrocarbon (EM – elemental matter is also found in literature and might 'integrate' EC and BC). The formation of soot depends strongly on the fuel composition. It spans carbon from graphitic (EC) through BC to organic carbon fragments (OC). The surface coating of primary hydrophilic soot particles play a great role, likely via SO_2 absorption and sulphuric acid formation and via NMVOC absorption, with its oxygenation possibly providing a matrix for the stabilization of very fine SOA particles.

Each of the available methods (optical, thermal, and thermo-optical) refers to a different figure; it remains a simple question of definition. Hence, the comparison of different soot methods is senseless. In Europe, carbonaceous matter ranges from $0.17\ \mu g\,m^{-3}$ (Birkenes, Norway) to $1.83\ \mu g\,m^{-3}$ (Ispra, Italy) for EM and for OC from $1.20\ \mu g\,m^{-3}$ (Mace Head, Ireland) to $7.79\ \mu g\,m^{-3}$ (Ispra, Italy). The percentage of TC to PM10 in rural backgrounds amounts to 30 %: 27 % OM and 3.4 % EM, respectively. Within this range are values measured in Berlin and surrounding areas: $1.3\text{-}2.2\ \mu g\,m^{-3}$ EM and $2.8\text{-}3.4\ \mu g\,m^{-3}$ OC, whereas TC contributes 20 % to PM10 (Table 2.16).

Table 2.16: Composition of particulate matter PM10 in Berlin and environment (daily samples in one year: 2001/2002); in $\mu\,gm^{-3}$; unpublished data.

species	kerbside	urban	rural
total PM$_{10}$	34.5	24.4	20.1
residual[a]	15.3	9.4	6.8
OC	4.3	3.4	2.8
BC	4.3	2.2	1.3
sulphate	4.1	3.6	3.6
nitrate	3.4	3.0	3.0
ammonium	2.0	1.8	1.8
chloride	0.50	0.19	0.21
sodium	0.34	0.29	0.34
calcium	0.42	0.19	0.16
potassium	0.20	0.14	0.09
magnesium	0.05	0.04	0.05
iron	0.7	0.2	0.1

[a]insoluble minerals as difference PM$_{10}$ and measured species

2.5.3.3 Condensation nuclei and cloud chemistry

In Chapter 3.2.1.2 will be stated that the formation of droplets only from water vapour is impossible in air.

However, some products of gas-phase reactions such as SO_3/H_2SO_4 (sulphuric acid), CH_3HSO_3 (methane sulphonic acid) and many oxygenated organic compounds have small vapour pressures but partly a high affinity to H_2O. These molecules can accommodate each other (single component nucleation) and among different species (multicomponent nucleation) and form a cluster of molecules as a metastable phase. After reaching a critical radius, they become stable. It is believed that in the atmosphere, *gas-to-particle conversion* occurs not by condensation of single species but rather by involving at least two, probably more, different species. Students know well from laboratory praxis that condensed fine matter forms when opened near bottles of aqueous ammonia solution and sulphuric acid, nitric acid and/or hydrochloric acid. Clearly, single and combined reactions occur with different numbers of species, depending on its gaseous concentration and the stability of the embryo formed (g – gaseous, p – particulate):

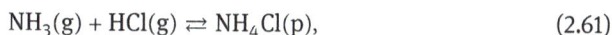

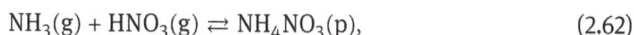

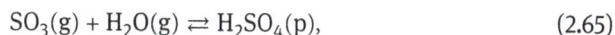

$$NH_3(g) + HCl(g) \rightleftharpoons NH_4Cl(p), \tag{2.61}$$

$$NH_3(g) + HNO_3(g) \rightleftharpoons NH_4NO_3(p), \tag{2.62}$$

$$NH_3(g) + H_2SO_4(g) \rightleftharpoons NH_4HSO_4(p), \tag{2.63}$$

$$2\,NH_3(g) + SO_3(g) + H_2O \rightleftharpoons (NH_4)_2SO_4(p), \tag{2.64}$$

$$SO_3(g) + H_2O(g) \rightleftharpoons H_2SO_4(p), \tag{2.65}$$

$$NH_3 + HNO_3 + SO_3 + H_2O \rightarrow (NH_4)_2NO_3HSO_4. \tag{2.66}$$

Recently was found that amines are far more efficient than ammonia in forming particles with sulphuric acid. Furthermore, organic species likely play a key role in nucleation and growth. Thus, it was found that the formation of SOA is accelerated in the presence of SO_2 and sulphuric. The first step in atmospheric SO_2 oxidation is OH addition (Chapter 4.5.3), and this radical can react with alkoxyl radicals (RO) to form sulphonic acid and further with organic peroxy radicals to form dialkyl sulphates:

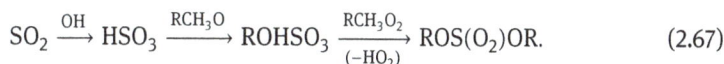

$$SO_2 \xrightarrow{OH} HSO_3 \xrightarrow{RCH_3O} ROHSO_3 \xrightarrow[(-HO_2)]{RCH_3O_2} ROS(O_2)OR. \tag{2.67}$$

These results strongly suggest the importance of particle-phase reactions that lower the volatility of organic species via accretion (oligomerization) processes. Organosulphates, including nitrated derivatives (e. g. nitroxy organosulphates) were detected in ambient aerosols. It was shown that the interaction between biogenic organic acids and sulphuric acid enhances nucleation and the initial growth of those nanoparticles.

The fate of nuclei is particle coagulation, a process in which small particles (assumed spherical) collide and coalesce completely to form larger spherical particles. Small particles are indeed spheroidal, and the assumption of spherical particles seems reasonable. After a certain size, however, the particles cannot coalesce completely and start to form long chains, which eventually grow into three-dimensional fractal-like structures.

The ability of a particle to act as a CCN depends strongly on its size and chemical composition, which implies that the knowledge of both parameters would suffice to provide an accurate prediction on ambient CCN concentrations. Particle hygroscopicity plays a key role in understanding the mechanisms of haze formation and particle optical properties. The chemical composition of CCN determines the water chemistry (see Chapter 2.5.2) of individual droplets and the ability to absorb gases. In the short period (on average about one hour) of the existence of drops of different sizes and chemical composition, the chemical composition of the interstitial gas phase and that of the droplet phase is changing. Depending on the cloud microphysics and cloud dynamics, the droplets can have one of two fates, they evaporate or precipitate. Only in one out of ten cases, on average, does the cloud precipitate. Subsequent evaporation and condensation again until final dissipation back to water vapour is the more frequent process, called *cloud processing*. During that process, the aerosol particles, acting as nuclei and coming to residues, are generally growing and becoming more water-soluble. The cloud amplifies the production of CCN. Recent research suggests that clouds can take up water-soluble organic molecules, which then are oxidized and form SOA after evaporation of cloud droplets. The term *cloud chemistry* is considered here to comprise

- cloud water chemical composition),
- reactions that take place in cloud water droplets (aqueous chemistry) and
- reactions that take place in the gas (interstitial) phase between the droplets (gas-phase chemistry).

Therefore, cloud chemistry (such as aerosol chemistry) is also termed multiphase chemistry. Even though the volume fraction of liquid water in clouds (LWC) rarely exceeds 10^{-6} (1 mL in 1 cubic meter of air), the fundamental role of cloud droplets as a medium for the chemical reaction has long been recognized. Clouds influence the photochemistry of the atmosphere the radiation budget and redistribute emitted trace compounds to other regions and from the boundary layer to the free troposphere. In the cloud, chemical transformations of droplets occur, which otherwise would not take place or would proceed much more slowly. Cloud processes are responsible for more than 70 % of sulphate formation from gas-phase SO_2 (Chapter 4.5.3). More and more, it is accepted that clouds also influence the ozone budget regionally and even global between 10 % and 30 % (Chapter 5.3.2.2). Despite the short lifetime of single cloud droplets, cloud systems may exist many hours and even a few days and may transport pollutants over distances of several hundreds of kilometres. Finally, clouds produce precipitation that acts as a very efficient removal mechanism of chemical substances (Chapter 5.3.4).

Interactions among aerosols, clouds and precipitation are critical in shaping the climate system. Aerosols, cloud and precipitation are intrinsically linked. Aerosol particles (APs) are the condensation and freezing nuclei on which cloud particles form (heterogeneous nucleation process).

3 Fundamentals of physical chemistry

Transport and transformation of *chemical species* constantly occur in the environment, within soils, waters and air, as well as crossing the reservoir interfaces. As already stated, the atmosphere is the global reservoir, characterised by the highest rates of turnover. The earth's surface (soils, waters and vegetation) is the source of atmospheric constituents, but it is also the disposal site of air pollutants (after deposition) and direct pollution from wastewater, landfills, agrochemicals, and other human activities. The ocean is not only the final depot of all waste via river run-off and atmospheric deposition but also a potential humans' mineral deposit within a global sustainable cycling economy (see Chapter 6). As often already mentioned in this book, we cannot separate chemical and physical processes, but this book will not refer to transport processes except for molecular diffusion to interfaces.

> Physical chemistry describes particulate phenomena in chemical systems in terms of laws and concepts of physics. ℹ️

Normally, the complex term physical chemistry also covers the key properties of 'environmental' materials we have discussed in Chapter 2: states of matter (Chapter 2.2), ideal gases (Chapter 2.3.2) and aqueous solutions (Chapter 2.4.3). In Chapter 3, we will discuss the gist of environmental chemistry, i. e. equilibriums (Chapter 3.2) and chemical reactions (Chapter 3.3). To understand their nature, origin and mechanisms, the fundamentals of thermodynamics, thermochemistry and reaction kinetics will be presented.

3.1 Chemical thermodynamics

Thermodynamics was originally the study of the energy conversion between heat and mechanical work but now tends to include macroscopic variables such as temperature, volume, pressure, internal energy and entropy.

> In physics and chemistry, and thereby in the environment, thermodynamics includes all equilibrium processes between liquids, gases and solids. These processes occurring in energetic changes are the key factor for understanding environmental states and thereby environmental changes. ℹ️

Changes in heat and kinetic energy can be measured by work done. Specifically, chemical thermodynamics[17] is the study of the interrelation of heat and work with chemi-

---

[17] Unfortunately, professors like to introduce many disciplinary terms such as technical thermodynamics, not to confuse students but to make their own field more important. In my understanding, there is only thermodynamics, which you can apply to many (physical) systems (including chemical).

https://doi.org/10.1515/9783110735178-003

cal reactions or physical state changes. The structure of chemical thermodynamics is based on the first two laws of thermodynamics. Four equations called the 'fundamental equations of Gibbs' can be derived from the first and second laws of thermodynamics. Then using these four equations and relatively simple mathematics, a multitude of equations relating to the thermodynamic properties of the thermodynamic system can be obtained. Thermodynamics deals with four types of systems:

1. *Isolated system*: neither energy (work and heat) nor matter can be exchanged with the surrounding.
2. *Adiabatic isolated system*: neither heat nor matter can be exchanged with the surrounding; other kinds of energy (work) may be exchanged.
3. *Closed system*: energy exchange with the surrounding is possible but not matter exchange.
4. *Open system*: Contrary to the closed system, all kinds of exchange with the surrounding are allowed.

In previous Chapters, we have already introduced the following *state functions*: temperature T, pressure p, volume V and amount n. In the following Chapters, we introduce further state functions to describe the energetic state of the system, internal energy U, entropy S, enthalpy H, free energy F (Helmholtz energy) and free enthalpy G (Gibbs energy). State functions are values that depend on the state of the substance and not on how that state was reached. A state function is also called *state variable* and *state quantity*. Remember that we have already introduced *extensive* (proportional to the amount) and *intensive* (independent from the amount) state quantities in Chapter 2.2.2.

3.1.1 First law of thermodynamics and its applications

The first law of thermodynamics is a special case of the law of energy conservation.

> **i** This law says: within an isolated system, the sum of all kinds of energy remains constant. Energy can be transformed from one form to another but can be neither created nor destroyed.

3.1.1.1 Internal energy

Rudolf Julius Emanuel Clausius (1822–1888) introduced the term *internal energy*. Note that thermodynamic functions in capital letters (U) refers to molar quantities (denoted to one mol of substance) and small letters (u) to a given amount: $u = nU$.

The *internal energy* of a system or body (for example, a unit of air or water volume) with well-defined boundaries, denoted by U, is the total kinetic energy due to the motion of particles (translational, rotational and vibrational) and the potential energy associated with the vibrational and electric energy of atoms within molecules or any matter state. This includes the energy in all chemical bonds and that of free electrons (for example, hydrated electrons in water and photons in the air).

The change $\Delta U = U_2 - U_1$ as a result of a state change means, according to the law of energy conservation, that energy is either taken up from the environment ($\Delta U > 0$) or released into the surroundings ($\Delta U < 0$). The first case is called the *endothermic process* (for example, the evaporation of water), and the second case is the *exothermic process* (for example, oxidation). *Heat Q* takes a special place among different kinds of energy[18] (which are summarised behind the term *work W*). Hence, the change of internal energy (we only can measure the change but not the absolute value of internal energy) is defined by:

$$\Delta U = W + Q \quad \text{or in differential form} \quad dU = dW + dQ. \tag{3.1}$$

Equation (3.1) is the mathematical expression of the first law of thermodynamics: The change of internal energy of a system is equal to the sum of gathered or released energy in the form of work and heat. In a closed system, the internal energy remains constant.

In the gas phase, work is carried out primarily[19] as *pressure-volume work* (in the condensed phase, such as droplets and solid particles, surface work, electrical work and expansion work also occur). The change of volume occurs at constant pressure

---

18 Work, energy and heat are well defined quantities in physics and should not be confused with the adequate terms use in daily life. In physics, the term *energy* is defined as the amount of work, carried out by the physical system. Performing *work* at a body increases its energy content. Hence, energy is 'stored work'. Work is a process or body quantity whereas energy is a state quantity. To carry out work, we need *power* (remember: energy = work · power). We separate between mechanical and electrical work. Mechanical work again is subdivided into acceleration work, displacement work (belonging are lifting work, volume work, surface work and elastic work) and frictional work. Different forms of energy are distinguished: kinetic energy (stored acceleration work) and potential energy (stored displacement work). Belonging potential energy (in sense of a positioning energy) are also electric energy (motion of electric charged particles; analogous to lifting energy), magnetic energy and gravitation energy. Potential energy can be regarded as the capability to carry out work. Energy, producing through frictional work, is random microscopic motion energy, called *heat*. Heat cannot fully be transformed into other forms of energy in contrast to mechanical energy. The transferred heat ΔQ is a process quantity such as work. Heating means that work is dissipated. When frictional work (heat) appears, then is the real work carried out at the system larger or in other terms, the usable work is lower than the reversible work. Radiation energy (electromagnetic wave) is a mixture of electric and magnetic energy. The only incoming energy to the Earth is solar radiation energy, which will be transformed into all other forms of energy. Large amounts of heat are stored in the earth's interior as a result of planetary formation and continuous ongoing nuclear reaction (radioactive decay).

19 Further transformed into accelerational and frictional work.

(*isobaric* change of state), and so it is valid that

$$-W = p\Delta V. \tag{3.2}$$

If no other work is carried out, the differential change of internal energy is described by

$$dU = dQ - pdV. \tag{3.3}$$

The internal energy of an ideal gas[20] depends only on temperature (the Gay-Lussac law). During an isothermal expansion, when air performs positive work through the overbearing external pressure, it must uptake an equivalent amount of heat to meet a constant temperature ($-W = Q > 0$). There is no heat exchange with the surroundings ($Q = 0$); this is defined as an *adiabatic* change of state. Consequently, the gas (air) cools, and the internal energy decreases by the amount equivalent to the work performed ($-W = -\Delta U$).

> In air parcels, pressure and volume change with each shift in height. With an ascent, the volume increases (expansion) and pressure decreases. As long as there is no heat exchange with the surrounding air, the internal energy remains constant, and the altitude change is adiabatic. Therefore, adiabatic air mass changes are an important condition for the condensation of water vapour onto the cloud condensation nuclei. While adiabatic, rising air cools by 0.98 °C per 100 m; this is called the dry adiabatic lapse rate (DALR), or dT/dz. As soon as the air parcel is saturated by water vapour, it partly condenses and is then heated by the released heat. Then, the wet adiabatic temperature gradient (lapse rate) is observed, which lies between 0.4 °C at large temperatures and 1 °C at low temperatures. During adiabatic changes, the potential temperature remains constant, i. e. an air parcel with 10 °C in 1000 m altitude contains about the same heat as a near-surface air parcel at 20 °C. The temperature gradient determines the atmospheric layering. It is called a stable atmospheric boundary layer[21] (SBL) if the air temperature decreases less with altitude than in the case of adiabatic layering. The lifting air becomes cold faster than its environment and sinks down again so that only small vertical displacements occur. By contrast, if the air temperature decreases faster than the adiabatic lapse rate (the rising air is warmer than the environment), another buoyant force evolves – it becomes a labile layering.

3.1.1.2 Molar heat capacity

In an isochoric process (no volume change) the heat needed to heat one mol of gas, liquid or solid body by 1 ° is called *molar heat capacity* C_V, defined to be $C_V = (dQ/dT)V$. At constant volume, work can not be carried out ($p\Delta V = 0$), hence $dU = dQ = C_V dT$

---

20 For real gases the internal energy also depends on volume.

21 The boundary layer is the lowest part of the troposphere in contact with the earth's surface and therefore determined by extensive exchange processes and the friction.

or:

$$C_V = \left(\frac{\Delta Q}{\Delta T}\right)_V = \left(\frac{\Delta U}{\Delta T}\right)_V. \qquad (3.4)$$

However, most processes proceed not at constant volume but at constant pressure (isobaric process), and we define the molar heat capacity C_p, defined as $C_p = (dQ/dT)p$. Using equation (3.3) and equation (3.4) and taking into account that $pdV = RdT$ (from equation (2.23) – note that we regard molar quantities, i. e. $V - V_m$), it follows:

$$C_p dT = dU + RdT = C_V dT + RdT \quad \text{or} \quad C_p = C_V + R. \qquad (3.5)$$

The gas constant R is a constant numerically equal to the work of expansion of one mole of an ideal gas in an isobaric process with an increase in temperature by 1.

Because isobaric processes are frequent in chemistry, to calculate the change of internal energy at constant pressure and the performed work, a new state function H, denoted *enthalpy* has been introduced:

$$H = U + pV = U + nRT. \qquad (3.6a)$$

In differential form:

$$dH = dU + pdV + Vdp. \qquad (3.6b)$$

The last equation can be integrated for isobaric changes ($dp = 0$):

$$\Delta H = \Delta U + p\Delta V = Q \quad \text{or in differential form} \quad dH = dU + pdV = dQ. \qquad (3.6c)$$

The heat capacity at constant pressure is thus defined:

$$C_p = \left(\frac{\Delta H}{\Delta T}\right)_p. \qquad (3.7)$$

The change of enthalpy is equal to the heat change in isobaric processes.

From equation (3.7), we can derive the relationships between the change of enthalpy and the (infinitesimal) change of temperature at a constant pressure in the case of constant heat capacity within a certain range of temperature (note however that molar heat depends on T; see special books on physical chemistry):

$$\Delta H = \int_{T_1}^{T_2} C_p dT. \qquad (3.8a)$$

However, if within ΔT a change of the state occurs (melting, evaporation), the associated heat of melting or evaporation must be considered (T_u temperature of change):

$$\Delta H = \int_{T_1}^{T_u} C_p dT + \Delta H_u + \int_{T_u}^{T_2} C_p dT. \tag{3.8b}$$

3.1.1.3 Thermochemistry: heat of chemical reaction

Thermochemistry is the study of the heat associated with chemical reactions. During chemical reaction energy can be either released (*exothermic reaction*) or absorbed (*endothermic reaction*). A general form of a chemical reaction is given by equation (3.9):

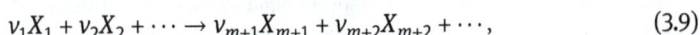

$$v_1 X_1 + v_2 X_2 + \cdots \rightarrow v_{m+1} X_{m+1} + v_{m+2} X_{m+2} + \cdots, \tag{3.9}$$

where X_i with $i = 1 \ldots m$ are the parent substances and the products with $i = m+1 \ldots n$, and v_i are stoichiometric coefficients. In terms of thermodynamics, the reaction may also be written as:

$$0 = \sum_i v_i X_i. \tag{3.10}$$

This is called the *first thermochemical law*, i. e. the molar amount of energy evolved or absorbed during a chemical change always remains the same for the same quantities of reacting substances. The change in internal energy is then given by:

$$\Delta U = \sum_i v_i U_i. \tag{3.11}$$

The reaction heat is given by the difference in internal energy before and after the chemical reaction. More frequently, reactions proceed at constant pressure, and we can describe the energy change as the enthalpy (see equation (3.6a)–(3.6c)):

$$\Delta H = \Delta U + \Delta nRT. \tag{3.12}$$

According to the first law of thermodynamics, the change of internal energy does not depend on the pathway and transforming that on chemical reactions. We state that the reaction enthalpy is always the same, independent from the pathway of parent compounds to the products (initial and terminal status). This is called Hess's law (second thermochemical law): The reaction heat is equal to the sum of the reaction heats of all subsequent partial reactions, outgoing from the same parent compounds to the same products. Based on this law, we can calculate reactions heats, whose measurement is impossible or at least extremely complicated. It makes sense to introduce the *standard state function* related to the standard state (1 bar and 298.15 K). The standard enthalpy

$H^{\ominus}$ of reaction equation (3.9) is now given by:

$$\Delta H^{\ominus} = v_{m+1} H^{\ominus}_{X_{m+1}} + v_{m+2} H^{\ominus}_{X_{m+2}} - v_1 H^{\ominus}_{X_1} - v_2 H^{\ominus}_{X_2} = \sum_i v_i H^{\ominus}_i, \tag{3.13}$$

where $H^{\ominus}_i$ is the standard enthalpy of formation for compound i, listed in special tables. A similar procedure is possible for all changes of state, such as evaporation, melting, condensation, crystallisation, dissolution, ionisation, dissociation, neutralisation, and so on. From equation (3.7), it follows that the temperature dependency of the enthalpy (without phase transfer):

$$H(T_2) = H(T_1) + \int_{T_1}^{T_2} C_p dT. \tag{3.14}$$

This equation can be applied to each compound participating in the chemical reaction; written in standard enthalpies of formation, called Kirchhoff's law:[22]

$$\Delta_R H^{\ominus}(T_2) = \Delta_R H^{\ominus}(T_1) + \int_{T_1}^{T_2} \Delta_R C_p dT, \tag{3.15}$$

where $\Delta_R C_p = \sum_i v_i C_{p,i}$. For the gas phase reaction (in parenthesis standard enthalpy of formation in kJ/mol)

$$CO(-110.5) + H_2O(-241.8) \rightleftharpoons CO_2(393.5) + H_2(0),$$

it follows for the reaction enthalpy: $\Delta_{rxn} H_{298} = (-393.5 + 110.5 + 241.8)\,kJ/mol = -41.2\,kJ/mol$.

3.1.2 Second law of thermodynamics and its applications

From the first law of thermodynamics, we have learned that heat and work are mutually convertible. Nevertheless, there was no statement whether this conversion is possible without limits. Moreover, no information about the direction of chemical processes was provided. For describing chemical conversions, this question is of crucial importance. The second law of thermodynamics is an empirical finding that has been accepted as an axiom of thermodynamic theory. With this law, we can calculate the state of equilibrium and hence the chemical yield of a reaction.

---

22 Not to be confused with the same-named radiation law of Kirchhoff.

> The basic empirical finding is that all spontaneously natural processes go in a specified direction, the simplest of these is that heat will naturally flow from a hotter to a colder one (and never vice versa).

All processes (changes of state and chemical reactions) can be subdivided into reversible and irreversible ones. All voluntary processes in a given direction are irreversible. For example, the volcanic eruption of SO_2 into air leads through diffusion to mixing – a separation of SO_2 from the air is per se impossible but only when processing outer work.

3.1.2.1 Entropy and reversibility

The quantities ΔU and ΔH might also be expressed as heat because all kinds of energy, which the system exchanges with its surroundings, can be completely transferred into heat, in agreement with the law of energy conservation. The general driving force can be quantified as *entropy*. With such quantification, it can be studied whether a process runs voluntarily. Voluntary processes in the environment are of crucial interest because it is nearly[23] impossible to trigger the intended changes of pressure and temperature. Only *voluntary* chemical processes can be observed in nature. The finding that all processes can be grouped into voluntary and non-voluntary processes leads to the second law of thermodynamics.

> It is impossible to carry out a process with uptake of heat from a reservoir and its complete transfer into work (there is no *perpetuum mobile*).

In a slightly different phrasing, heat is low-grade energy, i. e., whereas heat always degrades, the heat of a lower grade will always remain (for example, in the form of infrared radiation). This 'loss' can be called *dissipated work*. The 'value' of heat is determined by the temperature of the system; the more elevated the temperature, the larger the part of the heat that is transferable into work (useful energy). This property of heat is characterised by the entropy, which is a state function, hence independent from the way of the process:

$$dS = \frac{dQ_{rev}}{T}.$$

(3.16)

In the case of isothermal processes, we can rewrite it as follows:

$$\Delta S = \frac{Q_{rev}}{T},$$

(3.17)

---

23 It is not impossible, for example in weather modification (rainmaking, hail prevention and fog dissipation), but mostly through 'catalytic' triggering.

where Q_{rev} is the reversible heat taken up by the system at a given temperature. In a closed system,[24] the total energy remains constant, and thereby the direction of a process is associated with the redistribution of energy.

> Experience has shown that voluntary processes always result in a greater disorder of the system ($\Delta S > 0$). This is a condition for an *irreversible process* in which neither the system nor the surroundings cannot return to their original conditions.

Consequently, a reversible process is an isolated adiabatic system, as characterised by $\Delta S = 0$. This is an ideal abstraction because all processes in nature are spontaneous and thereby voluntary and irreversible. This does not exclude cycling processes, for example, biogeochemical material cycles where all single processes are irreversible but directed in a cycle, not returning the system to its original conditions but keeping a stationary state (see Chapter 3.2.3). This also does not exclude small intermediate steps that are reversible. The Nernst heat theorem (also called the third law of thermodynamics)[25] says that the entropy of each pure solid body goes to zero at the absolute zero (0 K):

$$\lim_{T \to 0} S = 0.$$

Hence at each T ($T > 0$) must $S > 0$. The absolute value of S is calculated according to:

$$S = \int_0^T C_p \frac{dT}{T}. \tag{3.18}$$

Considering spontaneous processes from the view of probability f (with a range of values $0 \ldots 1$), irreversible processes operate as transfers from a less probable to a more probable state. Ludwig Boltzmann (1844–1906) derived the equation:

$$S = k \cdot \ln f + \text{constant}; \tag{3.19}$$

k is the Boltzmann constant (not to be confused with the reaction rate constant k, which is variable in contrast to the fundamental Boltzmann constant). With the assumption by Max Planck (1858–1942) that the constant is zero, equation (3.19) can also

---

24 In nature a closed system is a fiction or a model approximation. The atmosphere is open to space and the earth's surface. The earth system is open regarding energy flux and only apparently closed regarding mass, when not considering cosmic epochs.

25 Discovered in 1905 by Walther Herman Nernst (1864–1941) while lecturing in the Walther-Nernst-Hörsaal (lecture hall) of the physicochemical institute at Bunsenstr. 1, Berlin, where the Author (DM) attend lectures on physical chemistry in the years 1968–1969.

be written as:

$$\Delta S = k \cdot \ln \frac{f_2}{f_1}, \tag{3.20}$$

where f_1 and f_2 denote the probabilities of the initial and final state, respectively.

3.1.2.2 Thermodynamic potential: Gibbs–Helmholtz equation

The equilibrium condition for an isolated system is $dS = 0$, i. e. change of entropy is zero in equilibrium.

During the reaction, the entropy increases to a maximum when approaching the equilibrium. Isolated systems (no exchange of energy and matter with the surrounding), however, do not occur in nature. For non-closed systems, we need another condition of equilibrium. For isothermal and isobaric processes, the total entropy, that is the entropy of the system dS and that of the surrounding dS_S, remains constant: $dS_S = -dS$. As for the reversible heat change Q_{rev}, we use the enthalpy H, we get – $dS = dH/T$ or $dH + TdS = 0$. The temperature is constant (isothermal process). Hence, we can write the difference in the form:

$$d(H - TS) = 0. \tag{3.21}$$

This is the condition for equilibrium under isotherm-isobaric conditions; the function $(H - TS)$ reaches a minimum at equilibrium (but its change is zero) and has been introduced as thermodynamic potential or *free enthalpy* G (also called the Gibbs energy):

$$G = H - TS. \tag{3.22}$$

Equilibrium condition: $dG = 0$ (T = constant, p = constant, G minimum). Rearranged equation (3.22) reads as $H = G + TS$, i. e. the enthalpy is the sum of free enthalpy G and a bonded energy TS. Free enthalpy is that part of enthalpy, which can be fully transformed during a reversible process into any kind of other energy, i. e. the maximum work carried out. The bonded energy TS cannot be extracted from the system at constant temperature.

To characterise the equilibrium condition for an isotherm-isochoric process (constant volume), the *free energy* (also called Helmholtz energy) F has been introduced in a similar way:

$$F = U - TS. \tag{3.23}$$

Changes in the state at a constant temperature can be written as $dF = dU - TdS$ and $dG = dH - TdS$. With the condition of 'voluntariness' of the process, that is $dS \geq 0$, another important thermodynamic condition follows: $dF_{T,V} \leq 0$ and $\Delta G_{T,p} \leq 0$, respectively.

In a spontaneous operating process, the change of free energy is negative, whereas in equilibrium $dW_{T,V} = 0$ is valid. The change of free energy corresponds to the maximum possible work that can be carried out ($dW = -p\Delta V$). A more general criterion for the 'voluntariness' of processes, however, is the attempt to garner a maximum from the sum of entropy changes of the system (dS) and the surroundings ($-dU/T$) or, in other words, to gain a small total entropy. The criterion $\Delta G_{T,p} \leq 0$ is in chemistry in this sense interpreted as a chemical reaction at a constant temperature and constant pressure if it is connected with a decrease in free enthalpy. Hence, it makes sense to introduce *free standard enthalpies* $\Delta_R G^\ominus$ to calculate reactions and equilibriums:

$$\Delta_R G^\ominus = \Delta_R H^\ominus - T\Delta_R S^\ominus. \tag{3.24}$$

It is logical to treat the change of free enthalpy as a function of p and T:

$$dG = dH - TdS - SdT. \tag{3.25}$$

Because $H = U + pV$, it is $dH = dU + pdV + Vdp$ and using the fundamental equation $dU = TdS - pdV$, it follows finally that:

$$dG = Vdp - SdT. \tag{3.26}$$

The free enthalpy is a function of pressure and temperature $G(p, T)$ or $dG = (\partial G/\partial p)_T dp + (\partial G/\partial T)_p dp$, and it follows:

$$\left(\frac{\partial G}{\partial T}\right)_p = -S \quad \text{and} \quad \left(\frac{\partial G}{\partial p}\right)_T = V. \tag{3.27}$$

As S takes positive values, G must decline if T is increases in a system at constant pressure and composition. In gases, G responds more sensibly to pressure variation than in condensed phases (because gases have a large molar volume). From equation (3.27), the temperature dependency of free enthalpy can be derived. Due to $S = (H - G)/T$, after a few steps, we get the well-known Gibbs–Helmholtz equation:

$$\left(\frac{\partial}{\partial T}\left(\frac{G}{T}\right)\right)_p = -\frac{H}{T^2}. \tag{3.28}$$

Relating this equation to the initial and final states of a chemical reaction or physical change of state ($\Delta G = G_2 - G_1$), it follows:

$$\left(\frac{\partial}{\partial T}\left(\frac{\Delta G}{T}\right)\right)_p = -\frac{\Delta H}{T^2}. \tag{3.29}$$

3.1.2.3 Chemical potential

The thermodynamic potential of a pure substance only depends on T and p. In mixtures, another state function, the amount n must be considered: $G(T, p, n_1, n_2 \ldots)$. In a binary system, the differential change is given by:

$$dG = \left(\frac{\partial G}{\partial p}\right)_{T,n_1,n_2} dp + \left(\frac{\partial G}{\partial T}\right)_{p,n_1,n_2} dT + \left(\frac{\partial G}{\partial p}\right)_{T,n_1,n_2} dp + \left(\frac{\partial G}{\partial n_1}\right)_{T,p,n_2} dn_1 + \left(\frac{\partial G}{\partial n_2}\right)_{T,p,n_1} dn_2.$$

The first and second differential quotient (pressure and temperature gradient, respectively) we have already known; the last two differential quotients ($\partial G/\partial n$) of any chemical species are called partial molar thermodynamic potential (partial free enthalpy) or, according to Gibbs, *chemical potential μ*.

> **i** The chemical potential denotes how the free enthalpy of a system changes with changing chemical composition.

Therefore, the equilibrium of all kinds can be clearly described, especially in mixed phases or multiphase. A substance between two phases (e. g. aqueous/dissolved and gaseous) has a certain chemical potential in each phase – equilibrium is reached when $\mu_1 = \mu_2$ or $\Delta\mu = 0$.

> **i** The driving force of all voluntary processes is the compensation of gradients, such as temperature, pressure, concentration, electric charge, and so on.

This phrase is essential for understanding objects carrying out work in our environment, e. g. volcanisms due to geothermal energy flow, solar irradiation photosynthetic organisms, wind and water.[26]
For pure substances (because of $dn_i = 0$) it is valid that $\mu = G = H - TS$. The molar-free enthalpy for solids and liquids depends little on pressure, but for gases, this dependency is large. The total derivative dG follows from equation (3.25) with consideration of the gas equation for molar quantities ($V = RT/p$; we disclaim here the exact marking as a molar quantity V_m):

$$dG = V dp = RT d \ln p, \tag{3.30}$$

---

[26] Nernst was strongly opposed to the "heat death" scenario of a dying universe that seemed to follow from the incessant rise of entropy on a cosmic scale; he was not the only scientist who worried about it. Max Planck wrote that the phrase "entropy of the universe" has no meaning because it admits of no accurate definition. Today's view is that universe and its major constituents that have never been in equilibrium in their entire existence, and that universe is infinite in time and space.

where $dp/p = d \ln p$ ($= d \ln(p/p_0)$) more exactly; p_0 is set to 1 bar but might earn any reference value. After integration and $\mu = G$, we get the important equation:

$$\mu = \mu^{\ominus} + RT \ln p, \tag{3.31}$$

where $\mu^{\ominus}$ denotes the *chemical standard potential*. The difference $\mu - \mu^{\ominus}$ is equal to the molar work when transferring the ideal gas reversible and isothermal from standard pressure on p.

In an open system, where its chemical composition does not need to be constant, the change of G must be described with the variation of p and T as well with n, the composition. Besides the terms $(\partial G/\partial p)_{T,n} = V$ and $(\partial G/\partial T)_{p,n} = -S$, it further follows equation (3.32) as a general definition of the chemical potential:

$$\left(\frac{\partial G}{\partial n_i} \right)_{p,T,n_j} = \mu_i. \tag{3.32}$$

With n_j (besides p and T) the constancy of the chemical composition is expressed. From the fundamental thermodynamic equation, the following now follows for H and U:

$$\mu_i = \left(\frac{\partial U}{\partial n_i} \right)_{S,V,n_j} \quad \text{and} \quad \mu_i = \left(\frac{\partial H}{\partial n_i} \right)_{S,p,n_j}, \quad \text{respectively.} \tag{3.33}$$

In an ideal mixing, i. e. the components do not interact through intermolecular forces, the chemical potential of each component is equal to that of the pure component if its pressure is identical to the partial pressure in the composition (valid in the air):

$$\mu_i = \mu_i^{\ominus} + RT \ln p_i. \tag{3.34}$$

Because of $p_i = p \cdot x_i$, it follows that:

$$\mu_i = \mu_i^{\ominus} + RT \ln p + RT \ln x_i = [\mu_i^{\ominus}] + RT \ln x_i, \tag{3.35}$$

where $[\mu_i^{\ominus}]$ is another standard potential, based on the total pressure. Similar to gases for diluted solutions (otherwise, activities a_j must be used) we write $\mu_i^{\ominus}$ but note that $\mu_i^{\ominus}$ is identical to $[\mu_i^{\ominus}]$, but we neglect now the brackets. This denotes a new standard potential of the dissolved compound i at a concentration $1 \, \text{mol L}^{-1}$:

$$\mu_i = \mu_i^{\ominus} + RT \ln c_i. \tag{3.36}$$

With the presence of a multiphase system (e. g. droplets or particles in air, particles in water), several new types of energy enlarge the inner energy U of the system (equation (3.3)). The work of mixing μdn, surface work γdq (q surface) and electrical work

φdQ (Q charge) are the most important forms:

$$dU = TdS - pdV + \mu dn + yds + \varphi dQ + \cdots. \tag{3.37}$$

Considering only the phase transfer and change of temperature, pressure and amount of substance, equation (3.22) transforms into:

$$G = G(T, p, n_i) = H(T, p, n_i) - TS = U + pV - TS \tag{3.38}$$

and

$$dG(T, p, n) = \frac{\partial G(T, p, n)}{\partial T} dT + \frac{\partial G(T, p, n)}{\partial p} dp + \sum_i \frac{\partial G(T, p, n_i)}{\partial n} dn_i. \tag{3.39}$$

It follows that at constant pressure and temperature

$$dG_{p,T} = \sum \mu_i dn_i. \tag{3.40}$$

Again, the system is in equilibrium when $dG = 0$. Without further discussion, it is clear that such a condition is hardly achievable in environments.

3.1.2.4 Chemical potential in real mixtures: activity

We stated that under environmental conditions, air can always be regarded as an ideal gas (otherwise, the pressure p must be exchanged by the *fugacity* ψ); this is also valid for exhaust gas treatment such as flue gas scrubbing carried out under atmospheric pressure. However, in real aqueous solutions at very high concentrations, for example, the scrubbing solution in flue gas treatment, nucleation of CCN (droplet formation) and saturated deep or salty waters, the chemical potential can be expressed by the same equations as for an ideal solution, if you exchange concentration c (equation (3.36)) by *activity a*:

$$\mu_i = \mu_i^\ominus + RT \ln a_i. \tag{3.41}$$

In real solutions, the intermolecular (or *interionic*, depending on the nature of solute) distance is smaller and hence attractive forces (the Coulomb interaction) occur, making the apparent concentration (which determines all thermodynamic relationships) smaller than in the true solutions, expressed by the *activity coefficient* γ_i:

$$a_i = \gamma_i c_i. \tag{3.42}$$

The standard condition $\mu_i = \mu_i^\ominus$ is given when $RT \ln a_i = 0$, i.e. $a = 1$ (fraction), $a = 1\,\text{mol/L}$ or $a = 1\,\text{mol/kg}$ (concentration). The real solution approaches the ideal case when $\gamma_i \rightarrow 1$:

$$\mu_{\text{real}} = \mu_{\text{ideal}} + RT \ln \gamma_i. \tag{3.43}$$

Activity coefficients depend on temperature, chemical composition (i. e. from all electrolytes in solution) and ionic strength. In mixtures, the charge (valence) z of different ions is important for the value of a. The *electrolytic dissociation* (see Chapter 3.2.2.2) of a binary salt is described by the general expression:

$$A_{v_+}^{z+} B_{v_-}^{z-} \rightleftharpoons v_+ A^{z+} + v_- B^{z-}. \tag{3.44}$$

For example:

$$\text{NaCl} \rightleftharpoons \text{Na}^+ + \text{Cl}^-,$$

$$(\text{NH}_4)_2\text{SO}_4 \rightleftharpoons 2\,\text{NH}_4^+ + \text{SO}_4^{2-}.$$

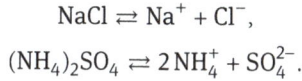

It is convenient to define a mean ionic activity $a_\pm$:

$$\mu_\pm \equiv \frac{v_+ \mu_{A_{v+}} + v_- \mu_{B_{v-}}}{v} = \mu_\pm^\ominus + RT \ln a_\pm, \tag{3.45}$$

with $v = v_+ + v_-$. It is not possible to estimate the activity coefficient of single ions, only the *mean activity coefficient*[27] $\gamma_\pm$ (Debye–Hückel's equation):

$$\gamma_\pm = \sqrt[\Sigma z]{\prod \gamma_i^z}. \tag{3.46}$$

This general equation simplifies for one-one-valence ions (NaCl, for example):

$$\gamma_\pm = \sqrt{\gamma_+ \gamma_-}. \tag{3.47}$$

The theory of activity coefficients in solution is very complex (the interested reader should refer to special literature, e. g. Wright 2007); mean binary activity coefficients can be measured (Table 3.1) and calculated according to the Pitzer theory.

3.2 Equilibrium

The term *equilibrium* generally means the condition of a system under which all competing influences are balanced. It is used differently in biology, physics, chemistry and economics (and also in other disciplines).

---

27 The term 'mean' is here not used in its common sense of an average.

Table 3.1: Mean activity coefficients of electrolytes in aqueous solution at 298.15 K.

molality (in mol/kg)	0.001	0.005	0.01	0.05	0.1	0.5	1.0
KCl	0.966	0.927	0.902	0.818	0.771	0.655	0.611
NaCl	0.966	0.930	0.906	0.779	0.736	0.689	0.664
H_2SO_4	0.737	0.646	0.543	–	0.379	0.221	0.186
HCl	0.966	0.929	0.904	0.730	0.796	0.757	0.809
Na_2SO_4	0.887	0.778	0.714	0.536	0.453	–	–
Na_2CO_3	0.891	0.791	0.729	0.565	0.488	0.288	–
NH_4Cl	0.961	0.911	0.880	0.790	0.792	0.620	0.579
NH_4NO_3	0.959	0.912	0.882	0.783	0.726	0.558	0.471

Here we consider the *chemical equilibrium* (the state in which the concentrations of the reactants and products have stopped changing in time), the *gas-aqueous equilibrium* (where the rates of condensation and vaporisation of material are equal), and the *solubility equilibrium* (any chemical equilibrium between solid and dissolved states of a compound at saturation). In environmental chemistry, the *dynamic equilibrium* (the states in which two reverse[28] processes occur at the same rate) is often regarded. However, in nature as an open system, equilibrium exists only on microscale or approximated – the general tendency of all chemical processes is irreversibility. In the previous chapters, we dealt with the conditions for equilibrium, and we summarise here:

Isolated system: $dS = 0$
Open system (isotherm-isochoric process): $dF = 0$
Open system (isotherm-isobaric process): $dG = 0$
Mixtures: $d\mu = 0$

3.2.1 Phase equilibrium

In the natural environment, the following phase equilibriums occur:
- gas-liquid (pure substance): Almost water – water vapour concerns plane surfaces (rivers, lakes, ocean) and droplets (sea spray, cloud, fog, rain) but under special conditions, other liquids such as crude oil/vapour (evaporation and condensation),
- solid-liquid (pure substance): Ice – water (melting and freezing),
- solid-gas (pure substance): Ice – water vapour (sublimation and deposition),
- solid-gas (binary system): Adsorption of gases on solids and desorption,

---

28 Note that *reversible* is not meant (often misused in literature), but instead opposed processes such as influx and outflow.

- solid-liquid (binary system): Adsorption of liquids or dissolved substances on solids and desorption,
- gas-liquid (binary system): Gas dissolution in water (dissolution and evaporation),
- solid-liquid (binary system): Solid dissolution in water (dissolution and precipitation).

The last two-phase transfer processes are also called dissolution equilibrium, and they play a crucial role in the environment: dissolution of gases from the air in waters and dissolution of solids in natural waters. The first three-phase transfer processes determine the climate on Earth, the distribution of water among its liquid, vaporous and icy states. Adsorption (and desorption) is an important process in soils and the atmosphere on aerosol particles.

3.2.1.1 Gas-liquid equilibrium: evaporation and condensation

We can consider each liquid as a condensed gas. Above every liquid, a vapour forms until the equilibrium between both phases is reached. At each temperature, a some of the molecules in the liquid transfer to the surrounding air, consuming energy (*enthalpy of evaporation*). The vapour contains more energy than the liquid. The liquid is in equilibrium with gas when the flux of condensation is equal to the flux of evaporation. The equivalent vapour pressure p (in a closed volume or close to the liquid surface) is the vapour pressure equilibrium. In a closed system, it corresponds to the saturation vapour pressure. The vapour pressure equilibrium depends neither on the amount of liquid nor vapour but only on temperature (and droplet size if not bulk water, see next chapter. However, in nature, gas-liquid equilibrium occurs only under very limited conditions: close to the water surface and within clouds. Evaporating water from the surface of a river, lake or sea is advected by wind, hence shifting the equilibrium towards more evaporation. Theoretically, the ocean would slowly evaporate until atmospheric water vapour saturation is reached. Fortunately, water vapour condenses in the air and precipitates back to the sea (and land) by rain and snow, resulting in a global dynamic equilibrium (see Chapter 2.4.1).

Condensation and evaporation occur at any vapour pressure. When the vapour pressure becomes smaller than the equilibrium value, water evaporates. In equilibrium in both phases exists the same chemical potential:

$$- S_{aq}dT + V_{aq}dp = -S_g dT + V_g dT.\tag{3.48}$$

From this, we derive the molar evaporation enthalpy at temperature T ($\Delta_V H / T = \Delta_V S$):

$$\frac{dp}{dT} = \frac{\Delta_V S}{\Delta T} = \frac{\Delta_V H}{T \Delta_V V},\tag{3.49}$$

which is called Clapeyron's equation. Since the molar volume of air is much larger than that of water, we can approximate $\Delta_V V \approx V_m RT/p$ and get the Clausius–Clapeyron equation, describing the change of vapour pressure with temperature:

$$\frac{d\ln p}{dT} = \frac{\Delta_V H}{RT^2} \quad \text{or} \quad p_2 = p_1 \exp\left\{\frac{\Delta_V H}{R}\left(\frac{1}{T_1} - \frac{1}{T_2}\right)\right\}. \tag{3.50}$$

In laboratory praxis, from the Clausius–Clapeyron plot of $\ln p$ against $1/T$, the enthalpy can be derived. The phase equilibrium in mixtures (remember that water in nature is not pure) is more precisely described by Raoult's law (see Chapter 2.4.3.2).

3.2.1.2 Gas-liquid equilibrium: special case for droplets

In the air,[29] we have to consider droplets and not a bulk solution as in rivers, lakes or oceans. From experience, we know that dispersed small droplets combine into larger drops. That is because the enthalpy also depends on the surface: the aqueous amount in the form of droplet possesses a higher chemical potential than the same amount of liquid after coalescence (bulk solution). Another consequence consists in the higher partial pressure droplets have $(\bar{p} + \Delta p = p)$ compared with a bulk volume or with a flat surface $(\bar{p})$. Assuming spherical particles, the change in vapour pressure Δp can be simply derived. The change in free enthalpy $dG_{T,n} = dW_V + dW_s$ is expressed $(T, n =$ constant) by the changing pressure-volume work $dW_V = dpdV = \Delta p \cdot 4\pi r^3/3$ and the surface energy change dW_s (r particle radius, γ surface tension, s surface) $dW_s = \gamma ds = \gamma 4\pi r^2$. Equilibrium gains when $\partial(dW_V/\partial r) = \partial(dW_s/\partial r) = 0$. It follows that $\Delta p 4\pi r^2 = \gamma 8\pi r$ and finally $\Delta p = 2\gamma/r$. Now, we calculate the change in the chemical potential with changing droplet size, where $d\mu = RT \ln p$ (equation (3.31)) and molar volume, defined by $V_m = RT/p$ (equation (2.22)), through the dispersion of a bulk liquid on droplets:

$$\int_{p^\infty}^{p} d\mu = \int_{p^\infty}^{p} RT \ln p = V_m \int_{p^\infty}^{p} pd\ln p = V_m \int_{p^\infty}^{p} dp. \tag{3.51}$$

After integrating and using the general definition of the chemical potential (equation (3.34)), we obtain:

$$\Delta\mu = RT \ln \frac{p}{p^\infty} = V_m \Delta p. \tag{3.52}$$

---

[29] In many technical processes (for example exhaust gas cleaning via wet scrubbing) droplets occur.

Now replacing the expression for Δp, we get the Kelvin equation; the equation is named in honour of William Thomson (1824–1907), commonly known as Lord Kelvin:

$$\ln \frac{\overline{p + \Delta p}}{p^\infty} = \ln \frac{p}{p^\infty} = \frac{2\gamma V_m}{rRT} = \ln \mathfrak{S}, \tag{3.53}$$

where $\mathfrak{S}$ is the saturation ratio. This equation[30] is valid only for pure water, but in air, we always meet diluted aqueous solutions. In combination with Raoult's law, we can consider the influence of dissolved matter on lowering the vapour pressure (not shown here). Equation (3.53) says that the formation of droplets is possible only for immense supersaturation; a droplet with $r = 10\,nm$ is stable only if supersaturation is 120 % ($p/p^\infty = 1.12$). The small water droplets are thermodynamically instable because of their large vapour pressure. This agrees with the observation that droplets in the air are formed only through condensation onto nuclei. By contrast, when droplets exist in air, the Kelvin equation says that larger droplets grow via vapour condensation at the expense of smaller droplets, which evaporate.

3.2.1.3 Absorption of gases in water: Henry's law

Each gas in contact with a liquid develops equilibrium with the dissolved component as found first by William Henry (1772–1836) in 1803:

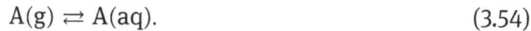

$$A(g) \rightleftharpoons A(aq). \tag{3.54}$$

The equilibrium constant is called Henry's law constant H (also reciprocal ratios are in use); dimensions of [A] in $mol\,L^{-1}$ and of p in Pa (see Table 3.2):

$$H_c = \frac{[A(aq)]}{[A(g)]} \quad \text{or expressed by the partial pressure of } A: \quad H_p = \frac{[A(aq)]}{p_A}. \tag{3.55}$$

A common problem with Henry's law constants is caused by the wide variety of possible units of measure. Because Henry's law is used in many different disciplines, many different usages and conventions have developed. Particular attention is paid to the description of the gas-phase concentrations by partial pressure, mass concentration, molar concentration, etc., while units for liquid-phase composition can include molality, molarity and weight fractions. When mole or weight fractions is used for the liquid phase, it is not always obvious which is meant from the units of Henry's law constant since fractions are dimensionless by convention.[31] The temperature dependency

---

30 This equation is also called Gibbs–Thompson equation and the effect (surface curvature, vapour pressure and chemical potential) is also called the Gibbs–Kelvin effect or Kelvin effect.

31 If the concentration units applied for both phases are the same, then the Henry constant itself becomes dimensionless. This is unfortunate, because depending on which units were used, different dimensionless Henry's law constants with different values, exist for the same solute.

of equilibrium constants is described by the van't Hoff equation (see equation (3.85)) where the reaction enthalpy must be replaced by the enthalpy of dissolution $\Delta_{diss}H$:[32]

$$\Delta_{diss}G^{\ominus} = -RT \ln H = \Delta_{diss}H^{\ominus} - T\Delta_{diss}S^{\ominus},$$

$$\frac{d\ln H}{dT} = \frac{\Delta_{diss}H^{\ominus}}{RT^2} \quad \text{or in another form} \quad \frac{d\ln H}{d(1/T)} = -\frac{\Delta_{diss}H^{\ominus}}{R}. \tag{3.56}$$

This equation is valid only within a limited temperature range; for larger T changes, the standard dissolution enthalpy must be expressed by Kirchhoff's law (see text-books of physical chemistry), taking into account the T dependency of molar heats, given by empirical interpolations formulas (A and I are constants):

$$\ln H = -\frac{\Delta_{diss}H^{\ominus}}{RT} - \frac{1}{R}\left(A_1 \ln T + \frac{A_2}{2}T + \frac{A_3}{6}T^2 + \cdots\right) + I. \tag{3.57}$$

In close relation to Henry's law constant are:
-   Bunsen's absorption coefficient α (the volume of gas absorbed by one volume of water at a pressure of 1 atm);
-   Ostwald's solubility β (the quantity of solvent needed to dissolve a quantity of gas at a given temperature and pressure):

$$\alpha \cdot RT = \beta = H_p RT. \tag{3.58}$$

However, these solubility coefficients have only been used in older literature. Hence, the Bunsen coefficient α is identical to the partial pressure-related Henry constant H_p (dimension in $mol\,L^{-1}Pa^{-1}$). Another quantity used is the gas solubility S (the mass of a gas dissolved in 100 g of pure water under standard conditions, that is, the partial pressure of the gas and the water saturation pressure is equal to 1 atm or 101.325 Pa). Approximately, (without considering the density of the solution) it follows (M molar mass of the gas, H_p in $mol\,L^{-1}\,Pa^{-1}$) that:

$$S = M \cdot H_p \cdot 10.1325 \quad \text{(exactly in g of solute per 100 g of water).} \tag{3.59}$$

It is important to note that few simplifications have been considered in the application of Henry's law (sometimes also called Henry–Dalton's law): the validity of the ideal gas equation, ideal diluted solution and the partial molar volume of the dissolved gas is negligible compared with that in the gas phase. However, the range of its validity under environmental conditions is appreciable. Only under very specific conditions (saturation level) of heterogeneous nucleation, evaporating droplets and dew, salty

---

32 Also termed enthalpy of solution (see also next Chapter 3.2.1.4). More exactly expressed, it is the change of enthalpy (respectively heat) associated with the dissolution of a substance in a solvent at constant pressure resulting in infinite dilution.

Table 3.2: Henry's law constant (in $mol\,L^{-1}\,atm^{-1}$) and temperature dependency (equation (3.55)) for selected gases (a large data source is found in Sander 2015).

species	H	$-dlnH/d(1/T)$ in K	species	H	$-dlnH/d(1/T)$ in K
H_2O_2	$1.0 \cdot 10^5$	6300	NO_2	$1.2 \cdot 10^{-2}$	2500
HO_2	$4.0 \cdot 10^3$	5900	NO	$1.9 \cdot 10^{-3}$	1400
OH	$3.0 \cdot 10^1$	4500	N_2	$6.1 \cdot 10^{-4}$	1300
O_2	$1.3 \cdot 10^{-3}$	1500	NH_3	$6.1 \cdot 10^1$	4200
O_3	$0.94 \cdot 10^{-3}$	2400	SO_2	1.2	2900
H_2	$7.8 \cdot 10^{-4}$	500	HOCl	$6.6 \cdot 10^2$	5900
HNO_3	$2.1 \cdot 10^5$	8700	Cl_2O	$1.7 \cdot 10^1$	1700
N_2O_5	∞		ClO_2	1.0	3300
HNO_2	$5.0 \cdot 10^1$	4900	Cl	0.2	
NO_3	1.8		Cl_2	$9.5 \cdot 10^{-2}$	2100
N_2O_4	1.4		$ClNO_3$	∞	
N_2O_3	0.6		NH_2Cl	$9.4 \cdot 10^1$	4800
N_2O	$2.5 \cdot 10^{-2}$	2600	H_2S	$8.7 \cdot 10^{-2}$	2100
			Hg	$9.3 \cdot 10^{-2}$	

waters, Henry's law is not valid. Absolute values of solubility cannot be found from thermodynamic considerations. Nevertheless, general rules are valid for all gases:
- decreasing solubility with increasing temperature;
- decreasing solubility with increasing salinity of waters (same ratio for all gases);
- increasing volume of aqueous solution with gas dissolution.

The term "apparent Henry's law constant" is also often found in literature, describing the ratio between the amount of to "total dissolved" substance and its equilibrium amount in the gas phase. However, it is often unclear what "total dissolved" meant beyond the physical dissolved substance (the *true* Henry's law constant), which can include dissociation products (which is best termed *effective* Henry's law constant, see below), adsorption onto surfaces and/or suspended solids (namely for immiscible organic substances), and even for cases where the physical dissolved substance undergoes chemical transformations. Moreover, Henry's law is defined as an infinite-dilution limit (fugacity ψ and activity a for substance i):

$$H_i = \lim_{a_i \to 0} \frac{\psi_i}{a_i}.$$

Thus, the term 'apparent' is also considered because the exact value of fugacity and activity coefficients of vapour and liquid phases are unknown. Nevertheless, it is recommended neither to use the term 'apparent' nor 'true', knowing that the equilibrium equation (3.54) only describes the physical dissolved gas species for gases without subsequent chemical hydrolysis, such as O_2, O_3, N_2, NO, NO_2. However, many environmentally important gases (CO_2, SO_2, HCl, NH_3, HNO_2, HNO_3, organic acids) undergo hydrolysis with subsequent electrolytic dissociation immediately after aqueous dissolution, shown here for CO_2 (it is an anhydride):

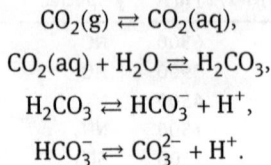

$$CO_2(g) \rightleftharpoons CO_2(aq),$$
$$CO_2(aq) + H_2O \rightleftharpoons H_2CO_3,$$
$$H_2CO_3 \rightleftharpoons HCO_3^- + H^+,$$
$$HCO_3^- \rightleftharpoons CO_3^{2-} + H^+.$$

For acids, it is simpler:

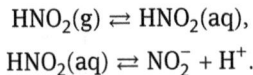

$$HNO_2(g) \rightleftharpoons HNO_2(aq),$$
$$HNO_2(aq) \rightleftharpoons NO_2^- + H^+.$$

Of interest, however, is the total solubility of the gas, including the hydrolysis products. Hence an effective Henry's law constant (basically, it is not a constant because it depends on the pH of solution) has been introduced, where the dissolved matter comprises the anhydride (for example, CO_2 or SO_2) and the acid (H_2CO_3, H_2SO_3). The acid can dissociate and thereby increases the total solubility of gas A, as described by the effective Henry's law constant H_{eff}:

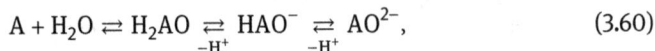

$$A + H_2O \rightleftharpoons H_2AO \underset{-H^+}{\rightleftharpoons} HAO^- \underset{-H^+}{\rightleftharpoons} AO^{2-}, \tag{3.60}$$

$$H_{eff} = \frac{[A(aq)]_{total}}{[A(g)]} = \frac{[A(aq)] + [H_2AO] + [HAO^-] + [AO^{2-}]}{[A(g)]}. \tag{3.61}$$

Excluded from the total solubility or effective Henry's law are subsequent reactions, for example, the oxidation of dissolved SO_2 into sulphuric acid (sulphate). The oxidation increases the flux into the water and thereby the phase partitioning but cannot be described by equilibrium conditions (Chapter 3.2.2).

For very soluble gases such as HCl and HNO_3, the 'physical' dissolved molecule does not exist (or is in immeasurably small concentrations, i. e. $c(aq) \rightarrow 0$) because of full dissociation:

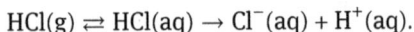

$$HCl(g) \rightleftharpoons HCl(aq) \rightarrow Cl^-(aq) + H^+(aq).$$

Therefore, Henry's law constant for such gases represents the equilibrium between gas and the first hydrolysis states:

$$H = H_{eff} = \frac{[HAO^-]}{[A(g)]}. \tag{3.62}$$

More limitations must be considered in the application of Henry's law under atmospheric conditions. In droplet dispersion (clouds, fog, rain, wet scrubber), because of the different chemical composition of the CCN and thereby initial droplets after its formation through heterogeneous nucleation, different equilibriums occur on a micro-scale. It is important to note that then Henry's law is valid only for each droplet with its gaseous surrounding, taking into account the gas concentration close to the droplet's surface and the aqueous phase concentration (neglecting here further limitations through mass transport). This is likely the main reason (besides others such as surface-active components influencing the gas-liquid equilibrium) why in bulk experimental approaches (integral collecting the droplets and analysing the cloud water), deviations from Henry's law has always been found. In modelling, the spectral resolution of particles and droplets (concerning size and chemical composition) is the only way to closer to reality.

3.2.1.4 Solubility equilibrium: solid-aqueous equilibrium

If a solid substance (the *solute*) is in contact with a liquid (the *solvent*, however, we only regard water in the environment) without chemical reaction with the water molecule, the dissolution proceeds only to a certain limit, called a *saturated solution*. This is in connection with the solid deposit of a heterogeneous system where the equilibrium is described by the concentration (or mixing ratio) of the solid substance in an aqueous solution:

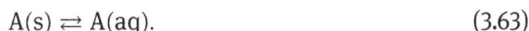

$$A(s) \rightleftharpoons A(aq). \tag{3.63}$$

According to IUPAC termination (IUPAC 2006), different terms such as dissolution, solution and solvation must be distinguished. *Dissolution* is the kinetic process, the transfer of the (soluble) solid substance into the liquid solvent (e. g. water). Dissolution occurs because the solvent (water) is dipolar, attracts, and associates with molecules or ions of a solute. Dissolution is always accompanied by *solvation*, i. e. the dissolved molecule becomes surrounded by solvent molecules:

$$A(aq) \rightleftharpoons A(H_2O)_n. \tag{3.64}$$

The index (aq) also often includes that the species is *hydrated* (solvated by H_2O molecules). Additionally, the solubility can be increased by *complexation* (see Chapter 3.2.2.7). *Solubility* quantifies the dynamic equilibrium state (equation (3.64)) achieved when the rate of dissolution equals the rate of *precipitation*. The solubility depends on temperature. It can be derived in a similar equation as for evaporation (equation (3.50)) where $\overline{\Delta_{diss} H^\infty}$ dissolution enthalpy for indefinite dilution:

$$\frac{d \ln c}{dT} = \frac{\overline{\Delta_{diss} H^\infty}}{RT^2}. \tag{3.65}$$

If the saturation concentration is exceeded (for example, by changing T and pH), the solute precipitates and forms a solid *precipitate* (or *sediment*). Salts dissociate (the

Table 3.3: Solubility product constants in $(mol/L)^2$ for 25 °C.

electrolyte	$CaSO_4$	$Ca(OH)_2$	$MgCO_3$	$CaCO_3$[a]	$FeCO_3$	FeS	$AlPO_4$	$Fe(OH)_2$
K_{sp}	$4.9 \cdot 10^{-5}$	$5.0 \cdot 10^{-6}$	$6.8 \cdot 10^{-6}$	$3.4 \cdot 10^{-9}$	$3.1 \cdot 10^{-11}$	$8 \cdot 10^{-19}$	$9 \cdot 10^{-21}$	$2.8 \cdot 10^{-39}$

[a]calcite

charge is here not important for describing the solubility and therefore not regarded, see equation (3.44)):

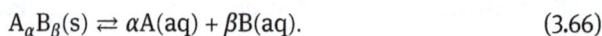

$$A_\alpha B_\beta(s) \rightleftharpoons \alpha A(aq) + \beta B(aq). \tag{3.66}$$

The equilibrium is described by (see Chapter 3.2.2.1):

$$K_{diss} = \frac{[A]^\alpha [B]^\beta}{[AB]_s}. \tag{3.67}$$

The term $[AB]_s$ denotes the concentration of AB in the pure substance AB and is therefore constant (e. g. dimension in mol per litre or kilogram). We define a *solubility product constant* K_{sp} and can write (see Table 3.3 for some examples):

$$K_{sp} = [A]^\alpha [B]^\beta = K_{diss}[A_\alpha B_\beta]. \tag{3.68}$$

The T dependency in the case of electrolytes is now given by (see equation (3.56)):

$$\frac{d \ln K_{sp}}{dT} = \frac{\overline{\Delta_{diss} H^\infty}}{RT^2}. \tag{3.69}$$

Note that in equation (3.67) and equation (3.68), instead of concentration c, activity a (see Chapter 3.1.2.4) should be used more accurately. However, in the case of diluted solutions and difficultly soluble substances (e. g. SiO_2, $CaCO_3$, $AgCl$), hence very small aqueous-phase concentrations, the activity coefficients are practically equal to 1, and we can use concentrations instead of activities.

3.2.1.5 Adsorption and desorption

A measure for the adsorption of a substance on the surface is the coverage degree θ:

$$\theta = \frac{\text{number of occupied adsorption sites}}{\text{number of available adsorption sites}}. \tag{3.70}$$

In the case of multilayer formation, it is useful to express the coverage degree by $\theta = V/V_{ads}$ where V_{ads} denotes the volume of the adsorbed substance in a monolayer. Adsorption and desorption can be described as well as the kinetic and equilibrium process. Equilibrium is described by the general condition $f(n, p, T) = 0$ in the gas phase and $f(n, c, T)$ in the aqueous phase, respectively. The process is described at constant pressure (*adsorption isobar*) or constant temperature (*adsorption isotherm*). There is no difference in the kinetic description for adsorption from the gas (g) or aqueous (aq) phase. The most simple adsorption isotherm, according to Freundlich and Langmuir, is based on the equilibrium:

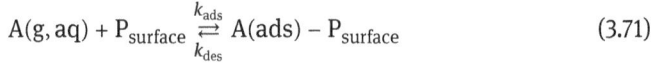

$$A(g, aq) + P_{surface} \underset{k_{des}}{\overset{k_{ads}}{\rightleftharpoons}} A(ads) - P_{surface} \tag{3.71}$$

or simplified $A(g, aq) \rightleftharpoons A(ads)$. It is assumed that the coverage degree changes over time and is proportional to the partial pressure of A and the number N of free adsorption sites; similarly, we set equation (3.73) for the desorption kinetic:

$$\left(\frac{d\theta}{dt}\right)_{ads} = k_{ads}p_A N(1 - \theta), \tag{3.72}$$

$$\left(\frac{d\theta}{dt}\right)_{des} = k_{des}N\theta. \tag{3.73}$$

In equilibrium, $(K = k_{ads}/k_{des})$ is $d\theta/dt = 0$, and we get for gases (change p into c for dissolved substances in aqueous phase):

$$\theta = \frac{Kp_A}{1 + Kp_A}. \tag{3.74}$$

When $Kp_A \gg 1$, $\theta \to 0$. Normally, however, it is the condition $Kp_A = 1$ (low gas concentration) that leads to $\theta = Kp_A$ and $\theta \ll 1$. As a consequence, equation (3.72) simplifies to $R_{ads} = k'_{ads}p_A$; the adsorption site number N we include in the adsorption coefficient. Because $\theta = n/n_{max}$ (n is the amount of the adsorbed substance) and setting $1/K = \beta$ (in the sense of an adsorption coefficient, see equation (3.58)), we finally obtain:

$$\frac{p_A}{n} = \frac{\beta}{n_{max}} + \frac{p_A}{n_{max}}. \tag{3.75}$$

Another adsorption isotherm, according to Brunauer, Emmet and Teller (the so-called BET isotherm), is useful for the formation of multilayer adsorption, i. e. there is no principal saturation or coverage degree limitation:

$$\frac{p_A}{(p_A^\infty - p_A)V} = \frac{1}{EV_{mon}} + \frac{(E - 1)p_A}{EV_{mon}p_A^\infty}, \tag{3.76}$$

where p^∞ denotes the saturation pressure of the pure liquid phase of the adsorbed gas A; V is the total volume of the adsorbed substance A (V_{mon} that of the monolayer); E is an empirical constant.

A special phenomenon is the adsorption of gases in porous media, such as water adsorption in soils, called *adsorption and capillary condensation*. Conventional models of liquid distribution, flow and solute transport in partially saturated porous media are limited by the representation of media pore space as a bundle of cylindrical capillaries (BCC); they ignore the dominant contribution of adsorptive surface forces and liquid films at low potentials. For small capillaries (radius 2–30 nm), the adsorbed water layers can be unified, i. e. the cylinder is filled with water, a meniscus immediately forms at the liquid-vapour interface that allows for equilibrium below the saturation vapour pressure p^∞. Meniscus formation is dependent on the surface tension of the liquid (see Chapter 2.4.3.1) and the shape of the capillary. The Kelvin equation (equation (3.53)) describes the vapour pressure lowering:

$$\ln \frac{p}{p^\infty} = \frac{2\gamma V_m}{rRT}.$$

3.2.2 Chemical equilibrium

We stated above that a reversible process between two states characterises equilibrium A $\rightleftarrows$ B, whereas no limiting conditions are expressed for the states A and B. Thus, it can be a chemical (reversible) reaction or any phase transfer (gas-liquid, solid-liquid, solid-gas). It is stated that each chemical reaction tends to reach an equilibrium. However, the kinetics of the transfer process A $\rightarrow$ B can be so slow that under environmental conditions, the equilibrium will never be approached because other (faster) processes such as mixing and transport permanently interrupt the process A $\rightarrow$ B. Hence, only a few types of (fast) chemical reactions represent measurable equilibrium: the acid-base reaction, adduct formation, complexation and addition–dissociation. Phase equilibrium is often observed in nature: dissolution–precipitation, evaporation–condensation and absorption–desorption. All non-chemical processes in multiphase systems can be subdivided into partial steps considering each chemical species separately transferring to the interface. In equilibrium, each substance has the same chemical potential in all phases. When the state variable changes (pressure, temperature, mole fraction) but the equilibrium remains, the chemical potentials change. However, it is valid that the changes are the same in all phases: $d\mu_i = d\mu_i'$. Any chemical reaction or phase transfer, including the chemical species A_i and B_j, is described by (ν stoichiometric coefficient):

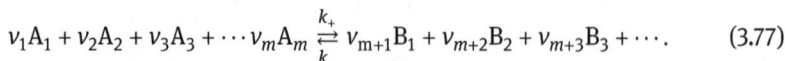

$$\nu_1 A_1 + \nu_2 A_2 + \nu_3 A_3 + \cdots \nu_m A_m \underset{k_-}{\overset{k_+}{\rightleftarrows}} \nu_{m+1} B_1 + \nu_{m+2} B_2 + \nu_{m+3} B_3 + \cdots. \qquad (3.77)$$

3.2.2.1 Mass action law

The *law of mass action* establishes the relationship between states A and B via the *equilibrium constant K*; the brackets denote the concentration (or for non-ideal systems, the activity and fugacity, respectively). By convention, the products form the numerator:

$$K = \frac{[B_1]^{\nu_{m+1}}[B_2]^{\nu_{m+2}}[B_3]^{\nu_{m+3}}\cdots}{[A_1]^{\nu_1}[A_2]^{\nu_2}[A_3]^{\nu_3}\cdots}. \tag{3.78}$$

In equation (3.77), k_+ and k_- represent the rate constants of the partial processes, i. e. for the forward (k_+) and backward (k_-) reaction or transfer. Equilibrium also means that the fluxes of the forward and backward processes are equal: $F_+ = F_-$ because of $d\mu_i = d\mu_i'$, and therefore, $dW_{T,V} = 0$ and $\Delta G_{T,p} = 0$. Hence, it is valid that:

$$F_+ = \left(\frac{dn}{dt}\right)_+ = k_+[A_1]^{\nu_1}[B_2]^{\nu_2}\cdots = F_- = \left(\frac{dn}{dt}\right)_- = k_-[B_1]^{\nu_{m+1}}[B_2]^{\nu_{m+2}}\cdots. \tag{3.79}$$

It follows that:

$$K = \frac{k_+}{k_-}. \tag{3.80}$$

In 1886, Jacob van't Hoff (1852–1911) derived thermodynamically the law of mass action based on the work from the initial substances A_i and final substances (products) B_j as well as the following relationship between free energy and enthalpy and the equilibrium constant either for constant pressure or constant volume but always at constant temperature:

$$\left(\Delta_R F^\ominus\right)_{V,T} = \left(\Delta_R G^\ominus\right)_{p,T} = -RT \ln K. \tag{3.81}$$

From the general definition of the chemical equilibrium ($d\mu_i = d\mu_i'$), it follows that when arriving at the equilibrium (depending on the rate constants, time is needed to achieve the equilibrium),[33] the conditions (lower cases denote mass-related and not molar quantities as for capital letters) must be the following:

$$(dg)_{p,T} = 0 \quad \text{(isobar)} \quad \text{and} \quad (df)_{V,T} = 0 \quad \text{(isochoric–isothermal)}. \tag{3.82}$$

Another general condition for equilibrium follows:

$$(g)_{p,T} = (f)_{V,T} = \sum \mu_i dn_i = 0. \tag{3.83}$$

---

33 Modellers treating fast equilibrium according to the computer integrating time sometimes forget this. However, when the time to achieve the equilibrium is larger than the numeric time step, nonsense is calculated.

Because of $\sum dn_i = \nu_i$, it also follows that $\Delta_R G = \sum \nu_i \mu_i = 0$. Now, the thermodynamic derivation of K becomes coherent:

$$0 = \Delta_R G = \sum \nu_i \mu_i = \sum (\nu_i \mu_i^{\ominus} + RT \ln c_i^{\nu_i}) = \Delta_R G^{\ominus} + RT \ln K. \tag{3.84}$$

We now transfer from the infinitesimal change d to the difference Δ, and it follows from $\Delta_R G^{\ominus} = \Delta_R H^{\ominus} - T\Delta_R S^{\ominus} = -RT \ln K$ through a transformation with respect to $\ln K$ and the derivative with respect to T. Using the Gibbs–Helmholtz equation (3.28), we derive the universal van't Hoff's reaction isobar:

$$\frac{d \ln K}{dT} = \frac{\Delta_R H^{\ominus}}{RT^2}. \tag{3.85}$$

Transforming this equation, a practicable expression for the temperature-dependent (of every type) equilibrium constant then follows:

$$K(T) = K_{298} \exp \left[\frac{\Delta_R H^{\ominus}}{R} \left(\frac{1}{298} - \frac{1}{T} \right) \right]. \tag{3.86}$$

3.2.2.2 Electrolytic dissociation

Wilhelm Ostwald and Svante August Arrhenius (1859–1927) propounded the electrolytic dissociation theory in the 1880s. The principal feature of this theory is that certain compounds, called *electrolytes*, dissociate in solution to give ions as shown in equation (3.44). The positively charged ions are called *cations*, and the negatively charged ions are *anions*. This results in the *conductivity* of the solution. The theory of electrolytes is not limited to aqueous solutions, but because of the aspect of environmental chemistry, we only consider electrolytes in water. In contrast to the thermal dissociation $AB \rightleftharpoons A + B$, the reaction time (kinetics) plays no role: electrolytic dissociation occurs immediately. The law of mass action for the reaction $AB \rightleftharpoons A^+ + B^-$ is given by

$$K = \frac{[A]^+ [B^-]}{[AB]}. \tag{3.87}$$

Note that K in equation (3.87) is an *apparent dissociation constant*, dependent on the electrolyte concentration and only valid for weak electrolytes; the equilibrium $AB \rightleftharpoons A^+ + B^-$ expressed by the activities represents the true dissociation constant. With the degree of dissociation $\alpha = [A^+]/[AB] \equiv [B^-]/[AB]$, it follows (compare with the degree of dissociation in equation (3.95)):

$$K = \frac{\alpha^2 [AB]}{(1 - \alpha)}. \tag{3.88}$$

A *weak electrolyte* is only partially dissociated, i. e. beside the ions A^+ and B^-, a substantial amount of the undissociated compound AB is in solution; almost all organic acids and bases are weak electrolytes and a few salts such as $HgCl_2$, $Hg(CN)_2$ and $[Fe(SCN)_3]_2$. *Strong electrolytes* are practically in an aqueous solution fully dissociated, i. e. $[AB] \rightarrow 0$; these are almost all salts and strong acids and bases.

We will not consider here the influence of electric current on electrolytes (called *electrolysis*); this is a task of *electrochemistry*. This is a very important field of industrial chemistry and (water electrolysis based on solar energy) future sustainability chemistry (Chapter 6). In biochemistry, also, many electric phenomena occur; we will only shortly consider photosynthesis (Chapter 5.2.2). However, in aqueous systems in the environment, many electron transfer processes occur, discussed later concerning redox processes (Chapter 3.3.5.1) and photocatalysis (Chapter 3.3.6.4). Ions in aqueous solution are always *hydrated*, i. e. surrounded by the water molecules $A^+(H_2O)_n$ according to the polar structure; the charge is not fixed in that water complex.

3.2.2.3 Acids, bases and the ionic product of water

Acids (HA) are species that produce during their aquatic dissolution hydrogen ions (H^+), whereas bases (BOH) produce hydroxide ions (OH^-):

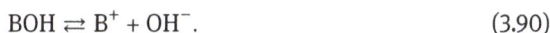

$$HA \rightleftharpoons H^+ + A^-, \tag{3.89}$$

$$BOH \rightleftharpoons B^+ + OH^-. \tag{3.90}$$

This definition (Arrhenius–Ostwald's theory) excludes ions to be acids. Johannes Nicolaus Brønsted (1879–1947) modified the Arrhenius definition:

Acids are chemical species that donate H^+, whereas bases accept H^+:

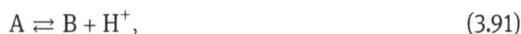

$$A \rightleftharpoons B + H^+, \tag{3.91}$$

with $K_a = [B][H^+]/[A]$ the acidity constant. According to this definition, H^+ is not an acid, whereas OH^- is a base. Bivalent (or multivalent) acids (H_2CO_3, H_2SO_4, H_3PO_4) and bases dissociate stepwise:

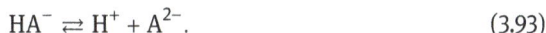

$$H_2A \rightleftharpoons H^+ + HA^-, \tag{3.92}$$

$$HA^- \rightleftharpoons H^+ + A^{2-}. \tag{3.93}$$

The Brønsted theory includes the Arrhenius–Ostwald theory and is the most useful for the environmental applications of diluted aqueous systems with the gas–liquid interaction. The proton theory of acids and bases was independently from Brønsted introduced by the English chemist Thomas Martin Lowry (1874–1936) in 1923, thus also called Brønsted–Lowry theory (namely in the English speaking community). In the same year, 1923, Gilbert Newton Lewis (1875–1946)[34] introduced his acid-base theory, which uses electrons instead of proton transfer and specifically states that an acid is a species that accepts an electron pair while a base donates an electron pair: A^+ (Lewis acid: electron acceptor) + B^- (Lewis base: electron donator) → A–B (coordinate covalent bond). Lewis's theory includes Brønsted's theory, for example, H^+ (Lewis acid) + H_2O (Lewis base) $\rightleftharpoons H_3O^+$.

> The Lewis theory of acids and bases is regarded by the scientific community as the most versatile and rigorous concept for explaining not only acid/base behaviour but polar covalent reactions in general. The Lewis concept is simple yet powerful in its scope and can be used to help beginning students understand reaction mechanisms more fully. While proton transfer is a structural result of the reaction, it does not address the why or how of reaction. The Lewis concept provides answers to these fundamental questions, serving as the overarching concept of polar reaction chemistry. For students of chemistry, the greater priority of Lewis's theory can bring a greater understanding of why reactions occur. However, for environmental scientists, it seems appropriate to focus on proton transfer described by the Brønsted-Lowry concept.

According to the Brønsted definition, Table 3.4 lists the most abundant acids and bases. Note that the listed species in PM (aerosols) do not occur in ionic form or free acids (e. g. H_2SO_4) but only as salts (e. g. NH_4NO_3 and NH_4HSO_4); O^{2-} denotes oxides (this 'ion' does not exist in aqueous solutions) and RCOOH organic acids. Hydroper-

Table 3.4: Acids and bases in the environment.

	strong acids	week acids	strong bases	weak bases
gases (in the air)	HI HBr HCl H_2SO_4 HNO_3 HF HNO_2 H_3PO_4	RCOOH ROOH HO_2 H_2S	none	NH_3 RNH_2
particulate matter, soils and aqueous solutions (waters)	HSO_4^-	HCO_3^- NH_4^+ HSO_3^- $H_2PO_4^-$ HPO_4^{2-}	CO_3^{2-} HCO_3^- OH^- $[O^{2-}]^a$ $RCOO^-$ PO_4^{3-}	SO_4^{2-} HSO_4^- SO_3^{2-} HSO_3^- NO_3^- NO_2^- $H_2PO_4^-$ HPO_4^{2-}

^{a}metal oxides

---

34 In 1926 Lewis coined the term "photon" for the smallest unit of radiant energy.

oxides are weak acids because of the following equilibriums forming radical ions: $H_2O_2 \leftrightarrow H^+ + HO_2^-$, $HO_2 \leftrightarrow H^+ + O_2^-$ and $ROOH \leftrightarrow H^+ + ROO^-$ (see Chapter 4.3.3).

The advantage of the *Brønsted* theory is that the formation of acids and bases occurs via protolysis reactions, including the corresponding acids and bases.

The *degree of dissociation* α describes the position of the equilibrium equation (3.91) and therefore the *acid strength*:

$$\alpha = \frac{[B]}{[A] + [B]}. \tag{3.94}$$

For $\alpha = 0.5$, it follows $[A] = [B]$. For binary solutions (acid + water), the degree of dissociation was originally defined to be:

$$\alpha' = \frac{[H^+]}{[A]} = \frac{[B]}{[A]}. \tag{3.95}$$

Therefore, it follows from equation (3.94) and equation (3.95) that:

$$\alpha = \frac{\alpha'}{1 - \alpha'}. \tag{3.96}$$

The solvent H_2O itself is *amphoteric*, that is, can act both as an acid and as a base.

We can specify equation (3.92) as follows:

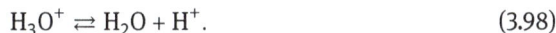

$$H_2O \rightleftharpoons OH^- + H^+, \tag{3.97}$$
$$H_3O^+ \rightleftharpoons H_2O + H^+. \tag{3.98}$$

Liquid water dissociates according to equation (3.97) only to a small percentage (0.00000556 % – one litre of water contains 55.6 mol H_2O and produces 10^{-7} mol $L^{-1} H^+$). The *acid constant*, often called the *acidity*[35] constant K_a, is defined as $([B][H^+])/[A]$. Including the corresponding base reaction:

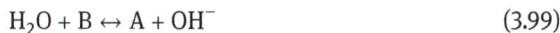

$$H_2O + B \leftrightarrow A + OH^- \tag{3.99}$$

defined as $K_b = [A][OH^-]/[B]$. An example for reaction equation (3.99) is the dissociation of ammonia:

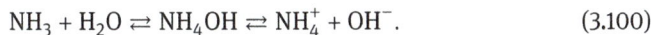

$$NH_3 + H_2O \rightleftharpoons NH_4OH \rightleftharpoons NH_4^+ + OH^-. \tag{3.100}$$

---

35 It is recommended to only use the term *acid* constant to avoid mistakes with the term *acidity* (see Chapter 5.3.2), which is different from the equilibrium constant K_a based on the mass effect law.

What we see is that ammonium (NH_4^+) is an acid. Whereas gaseous ammonia NH_3 is (besides organic amines RNH_2) the only important gaseous base in air, it will be removed by wet deposition to soils after dissolution as an acid (NH_4^+) making the soil more acidic.

Equation (3.97) does not represent a chemical reaction since the hydrogen ion (or proton) H^+ does not exist in *free* form in aqueous solutions. Instead, it 'reacts' with H_2O according to equation (3.98) to the hydronium ion in a first step and then hydration occurs, for example, to $H_9O_3^+$. However, according to recommendations given by IUPAC, only the symbol H^+ should be used. Hence, we have more exactly to write any acid dissociation in the form:

$$HA + H_2O \rightleftharpoons H_3O^+ + A^-, \tag{3.101}$$

$$H_2O + H_2O \rightleftharpoons H_3O^+ + OH^-. \tag{3.102}$$

For simplification, we disregard that $H^+ + H_2O \rightarrow H_3O^+$ (it is a convention to use the symbol H^+). The high mobility of the proton in aqueous solutions (contribution to the conductivity) is provided by *tunnel transfer* along hydrogen bridges within the H_2O clusters. The dissociation of neutral water is very low ($\alpha' = 1.8 \cdot 10^{-9}$) and that is why the water *activity* $\lg[H_2O] = 1.745$ is constant and is included in all equilibrium constants. Using the equilibrium expressions of equation (3.91) and equation (3.97), the relationships:

$$K_a K_b = K_w \quad \text{or} \quad pK_a + pK_b = pK_w \tag{3.103}$$

are valid, where the water *ionic product* $K_w = [H^+][OH^-] \approx 10^{-14}$ or, written in logarithmic form $pK_w = -\lg K_w \approx 14$. In contrast to the hydrogen ion (H^+), the oxonium ion H_3O^+ ($H^+ \cdot H_2O$ – also called protonised water) is an acid (hydronium is an old but still frequently used term). With $pK_a(H_3O^+) = -1.74$ the ion H_3O^+ is the strongest acid that exists in an aqueous solution. That means acids with $K_a > K_a(H_3O^+) \approx 55$ are totally protolysed ($\alpha \rightarrow 1$) and do not exist as acids in aqueous solutions. K_a values of these 'very strong' acids are only inexactly detectable because $K_a \rightarrow \infty$ (Table 3.5). With $pK_a(H_2O) = ([H^+][OH^-])/[H_2O] \approx 15.74$ is the hydroxide[36] ion OH^- the strongest base existing in aqueous solutions and a one-electron donor, according to $OH^- + X \rightarrow OH + X^-$; note that $[H_2O] \approx 55 \, mol \, L^{-1}$ has been used in the equations above.

We now formulate the general reaction equation for corresponding acids and bases:

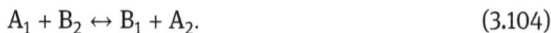

$$A_1 + B_2 \leftrightarrow B_1 + A_2. \tag{3.104}$$

---

36 It should never be named hydroxyl because that is the name of the OH radical.

Table 3.5: pK_a values (in mol-L units) of different acids (298 K).

acid		H^+ + base	pK_a
HI	$\rightleftharpoons$	$H^+ + I^-$	−9.3
HBr	$\rightleftharpoons$	$H^+ + Br^-$	−9.0
HCl	$\rightleftharpoons$	$H^+ + Cl^-$	−6.23
HO_2NO_2	$\rightleftharpoons$	$H^+ + {}^-O_2NO_2$	−5
H_2SO_4	$\rightleftharpoons$	$H^+ + HSO_4^-$	−2.0
HNO_3	$\rightleftharpoons$	$H^+ + NO_3^-$	−1.34
$H_2O_2^+$	$\rightleftharpoons$	$H^+ + HO_2$	−1
$HOCH_2SO_3H$	$\rightleftharpoons$	$H^+ + HOCH_2SO_3^-$	<0
HSO_4^-	$\rightleftharpoons$	$H^+ + SO_4^{2-}$	1.92
$SO_2 (+H_2O)$	$\rightleftharpoons$	$H^+ + HSO_3^-$	1.76
H_3PO_4	$\rightleftharpoons$	$H^+ + H_2PO_4^-$	2.14
$Fe(H_2O)_6^{3+}$	$\rightleftharpoons$	$H^+ + Fe(H_2O)_5OH^{2+}$	2.2
HF	$\rightleftharpoons$	$H^+ + F^-$	3.3
HNO_2	$\rightleftharpoons$	$H^+ + NO_2^-$	3.3
$Fe(H_2O)_5OH^{2+}$	$\rightleftharpoons$	$H^+ + Fe(H_2O)_4(OH)^+$	3.5
$CO_2 (+H_2O)$	$\rightleftharpoons$	$H^+ + HCO_3^-$	3.55
HCOOH	$\rightleftharpoons$	$H^+ + HCOO^-$	3.74
CH_3COOH	$\rightleftharpoons$	$H^+ + CH_3COO^-$	4.75
HO_2	$\rightleftharpoons$	$H^+ + O_2^-$	4.7
HSO_3^-	$\rightleftharpoons$	$H^+ + SO_3^{2-}$	7.2
H_2S	$\rightleftharpoons$	$H^+ + HS^-$	7.2
$H_2PO_4^-$	$\rightleftharpoons$	$H^+ + HPO_4^{2-}$	7.2
NH_4^+	$\rightleftharpoons$	$H^+ + NH_3$	9.23
HCO_3^-	$\rightleftharpoons$	$H^+ + CO_3^{2-}$	10.3
H_2O_2	$\rightleftharpoons$	$H^+ + HO_2^-$	11.7
$HOCH_2SO_3^-$	$\rightleftharpoons$	$H^+ + {}^-OCH_2SO_3^-$	11.7
HPO_4^{2-}	$\rightleftharpoons$	$H^+ + PO_4^{3-}$	12.4
HS^-	$\rightleftharpoons$	$H^+ + S^{2-}$	12.9
H_2SiO_4	$\rightleftharpoons$	$H^+ + HSiO_4^-$	13.3

The definition of acids can be extended according to the solvent theory as follows (this definition is symmetric concerning the acid-base relationship in contrast to the Brønsted theory):

Acids/bases increase/decrease H^+ or increase/decrease OH^-:

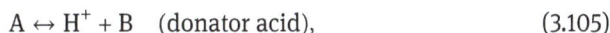

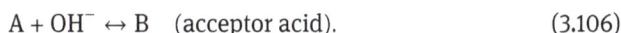

$$A \leftrightarrow H^+ + B \quad \text{(donator acid)}, \tag{3.105}$$
$$A + OH^- \leftrightarrow B \quad \text{(acceptor acid)}. \tag{3.106}$$

According to the pK_a value rank, acids could be subdivided (this is not an objective ranking) into strong and weak acids (Table 3.5); remember $pK = -\lg K$:

very strong		pK_a	$\leq pK_a(H_3O^+) = -1.74$
strong	$-1.74 <$	pK_a	≤ 4.5
weak	$4.5 \leq$	pK_a	≤ 9.0
very weak		pK_a	≥ 9.0

Figure 3.1 shows for carbonic acid the two-step dissociation with dependence on pH; between pH 4.8 and 9.3, only bicarbonate (HCO_3^-) exists in solution, below pH 2.8 only undissociated H_2CO_3 and above pH 11.6 only carbonate (CO_3^{2-}).

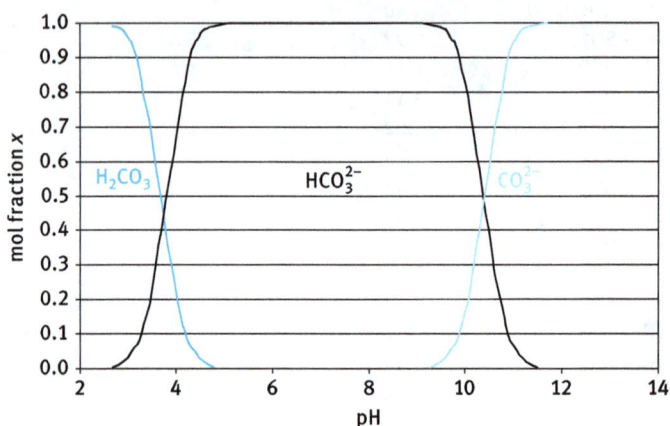

Figure 3.1: Dissociation diagram for carbonic acid; $pK_1 = 3.76$ and $pK_2 = 10.38$.

3.2.2.4 pH value

As we have seen, $[H^+]$ is an important chemical quantity in the diluted aqueous phase; Søren Peder Lauritz Sørensen (1868–1939) defined in 1909 the pH (derived from Latin *pondus hydrogenii*) to be:

$$pH = -\lg[H^+]. \tag{3.107}$$

> ℹ Today, the pH as an *acidity measure* is defined based on the activity as pH ($\equiv$ paH) $= -\lg a_{H^+}$.

As stated in Chapter 3.1.2.4, single activities, however, are not measurable. Therefore, using mean activities $a_\pm$, a conventional pH scale is defined based on fixed buffer solutions with known pH values. Another acidity measure is the Hammett acidity function H_0:

$$H_0 = pK_a + \lg[B] - \lg[A], \tag{3.108}$$

where $H_0 = -\lg[H^+] \equiv pH$. In the following, no difference is made between various pH definitions based on the assumption that in most cases in environmental chemistry $a_{H+} = a_{\pm} = [H^+]$ and $y = 1$ (activity coefficient). A generalised acid-base equation follows from equation (3.89), equation (3.97) and equation (3.98):

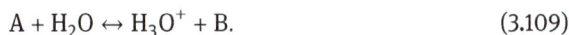

$$A + H_2O \leftrightarrow H_3O^+ + B. \qquad (3.109)$$

Therefore, it follows that:

$$pK_a = pH - \lg([B]/[A]) \quad \text{or} \quad pH = pK_a - \lg([A]/[B]), \qquad (3.110)$$

which is also called the Henderson–Hasselbalch equation and describes the derivation of pH as a measure of acidity (using pK_a) in biological and chemical systems. The (mean) hydrogen ion activity in terms of pH can be estimated directly using the hydrogen electrode (against a reference electrode) and measuring the electric potential. Based on the Nernst equation (see equation (3.159)), $E_H = RT \ln a_{\pm}/F$ it follows that (F is the Faraday constant):

$$pH = \frac{F(E - E_B)}{2.303RT}, \qquad (3.111)$$

where E_B is the reference electrode potential ($E = E_H - E_B$). Equation (3.111) does not reflect the diffusion potentials that make pH measurements more complicated. The accuracy of pH estimations in natural water samples (pH range 3–8) is no better than ±0.1 pH units despite the *quality assurance* of the pH measurement against standards to be 0.02 in this range of pH.

3.2.2.5 Hydrolysis of salts and oxides

It is known that aqueous solutions of certain salts react not neutrally but as acid or alkaline, depending on how the salt contains the cation of a weak base or the anion of a weak acid. This effect occur because the ions of the weak base or acid, which are in an aqueous solution besides hydrogen and hydroxide ions, respectively, must be in equilibrium with the corresponding undissociated molecule. We consider a salt, formed by the neutralisation of a weak acid with a strong base, for example, calcium carbonate $CaCO_3$. Despite the fact that it is hardly soluble, the dissolved fraction is fully dissociated into calcium and carbonate ions, but carbonate is in equilibrium with bicarbonate (HCO_3^-):

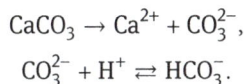

$$CaCO_3 \rightarrow Ca^{2+} + CO_3^{2-},$$
$$CO_3^{2-} + H^+ \rightleftharpoons HCO_3^-.$$

As a result, new water molecules dissociate according to $H_2O \rightleftharpoons H^+ + OH^-$ and the concentration of produced bicarbonate is practically equal to the concentration of hydroxide ions; hence, we can write the overall equilibrium in terms of:

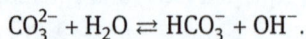

$$CO_3^{2-} + H_2O \rightleftharpoons HCO_3^- + OH^-.$$

This is the acid neutralisation property of natural limestone (soil-derived dust) and flue ashes from coal-fired power stations. Other salts with similar properties (such as alkali salts of organic acids) do not occur under natural environmental conditions. An example of a salt formed from a weak base and a strong acid and naturally found in the environment is ammonium nitrate NH_4NO_3, which is fully dissociated into ammonium and nitrate ions, but ammonium is in equilibrium with ammonia making the aqueous solution slightly acidic:

$$NH_4^+ \rightleftharpoons NH_3 + H^+.$$

Soluble oxides do not occur under natural environmental conditions but are constituents of industrial dust (flue ash, cement), namely calcium oxide CaO, which will be fully transformed into the hydroxide:

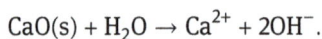

$$CaO(s) + H_2O \rightarrow Ca^{2+} + 2OH^-.$$

The elementary steps behind the gross reaction are (just after dissociation of CaO into $Ca^{2+} + O^{2-}$):

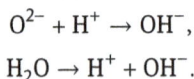

$$O^{2-} + H^+ \rightarrow OH^-,$$
$$H_2O \rightarrow H^+ + OH^-.$$

3.2.2.6 Buffer solutions

When adding an acid or base to an aqueous solution, normally, pH changes drastically. Many biological processes (in soils and organisms), however, proceed only within certain and limited pH ranges. To obtain such insensitivity, the system needs buffer solutions.

> A *buffer solution* is an aqueous solution consisting of a mixture of a weak acid and its conjugate base, or vice versa. The pH value of such solutions remains relatively constant against added acids or bases and dilution.

The effectiveness of soil buffering systems depends on numerous physical, chemical and biological properties of soils. Equation (3.110) is also useful for estimating the pH of a buffer solution and finding the equilibrium pH in acid-base reactions. The ratio [A]/[B] or [acid]/[conjugate base] is termed in the *buffer ratio*. The hydrogen activity (assuming $f = 1$) is calculated in a buffer for acid and bases, respectively:

$$a_{H^+} = K\frac{[A]}{[B]} \quad \text{and} \quad a_{OH^-} = K\frac{[B]}{[A]}.$$

The *buffer capacity* is a measure to express the amount of acid or base to effect a pH change; it is defined by $\beta = dc/dpH$. For the case of a weak acid and its salt, the buffer capacity is obtained by differentiation of the Henderson–Hasselbalch equation (equation (3.110)) in its form:

$$pH = pK + 0.4343\ln\frac{c}{c_A - c},$$

where c_A is the concentration of the acid, and c is the concentration of its salt. It also follows:

$$\beta = \frac{dc}{dpH} = 2.303c\left(1 - \frac{c}{c_A}\right). \tag{3.112}$$

Weak acids such as almost all organic acids can play the role of buffering strong acids in nature. Solutions containing acid and its salt (for example, acetic acid and acetate) convert H^+ from strong acids (nitric and sulphuric acid) and from CH_3COO^- into undissociated CH_3COOH without changing the pH (in certain limits). A buffer solution for bases contains ammonia (NH_3) and ammonium (NH_4^+). In nature, phosphates such as KH_2PO_4 and Na_2HPO_4 are buffer solution for pH range 5.4–8.0 and borax, a natural mineral (sodium tetraborate $Na_2B_4O_7$) together with hydrogen chloride (HCl) for pH range 7.6–9.2. Organic acids (e. g. citric acid) with Na_2HPO_4 provide a buffer solution in the pH range 2.2–8. Blood is protected against pH changes by a buffer from $NaHCO_3$ and Na_2HPO_4. Buffer solutions of defined composition also have an exactly known pH (and are therefore references for pH electrodes).

3.2.2.7 Complex ions
We have already stated that electrolytic dissociation in water is always combined with hydration, i. e. the formation of water complexes $X(H_2O)_n$, where H_2O is called the *ligand*.

A *complex ion* has a metal ion at its centre with a number of other molecules or ions surrounding it.

These can be considered to be attached to the central ion by coordinate (dative covalent) bonds.[37] Cations can add atoms, ions and molecules, which already have eight

---

37 In some cases, the bonding is actually more complicated than that.

electrons in the *valence* shell (octet). The ligand donates electron pairs to the cation, which are then jointly used, for example:

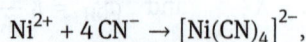

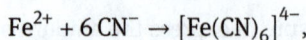

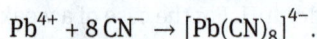

$$Ni^{2+} + 4\,CN^- \rightarrow \left[Ni(CN)_4\right]^{2-},$$

$$Fe^{2+} + 6\,CN^- \rightarrow \left[Fe(CN)_6\right]^{4-},$$

$$Pb^{4+} + 8\,CN^- \rightarrow \left[Pb(CN)_8\right]^{4-}.$$

The *coordination number* is given by the number of monovalent ligands, above 4, 6 or 8. Complex ions can be mixed, i. e. contain different ligands, see Table 3.6. The name of the cation (metal) is derived from Latin, for example, cuprate (Cu), ferrate (Fe), plumbate (Pb). In the case that the complex contains different ligands, first the anionic and then the neutral species are written in alphabetical order. The valence of the metal is named in parenthesis, for example: $[CuCl_4]^{2-}$ = tetrachlorocuprate(II) ion, $Na_2[CuCl_4]$ = disodium tetrachlorocuprate.

Table 3.6: Common ligands in complex ions.

anion ligands						neutral ligands	
F^-	fluoro	SO_3^{2-}	sulphito	NH_2^-	amino	O_2	dioxygen
Cl^-	chloro	SO_4^{2-}	sulphato	$-NO_2^-$	nitro	N_2	dinitrogen
Br^-	bromo	$S_2O_3^{2-}$	thiosulphato	$-ONO^-$	nitrito	NO	nitrosyl
O^{2-}	oxo	S^{2-}	thio	NO_3^-	nitrato	CO	carbonyl
O_2^{2-}	peroxo	$C_2O_4^{2-}$	oxalato	CN^-	cyano	NH_3	ammine
OH^-	hydroxo	CO_3^{2-}	carbonato	SCN^-	thiocyano	H_2O	aqua

An excess of ligands can dissolve many metals; AgCN is practically insoluble in water and precipitates when adding cyanide to a solution of Ag^{2+}. However, when adding more cyanide, AgCN is again dissolved:

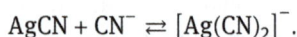

$$Ag^{2+} + CN^- \rightleftharpoons AgCN,$$

$$AgCN + CN^- \rightleftharpoons \left[Ag(CN)_2\right]^-.$$

Transition metals such as Fe, Mn and Cu are soil components and play an important role in redox reactions. Its solubility depends on pH (hydroxo complexes) and the concentration of available ligands. The equilibrium constants for complex ions must be known to describe the chemical status because different ferrato complexes also have different specific reaction rates in redox processes.

3.2.3 Dynamic equilibrium and steady state

In environmental literature, the term *equilibrium* is often used to describe a situation of no changing amount of a substance in a reservoir in terms of $dn/dt = 0$ or $F_+ = F_-$, or in other words that the
forward and backward fluxes (or inflow and outflow) of a process are equal. However, this is not a
chemical but a dynamic equilibrium and is better to be named a *steady state*.

As noted, one of the general conditions for chemical equilibrium is $\mu_+ = \mu_-$. Although
a chemical equilibrium occurs when two or more reversible processes occur at the
same rate, and such a system can be said to be in a steady state, a system that is in
the steady state might not necessarily be in a state of equilibrium, because some of
the processes involved are not reversible. A system in a steady state has numerous
properties that do not change over time. The concept of steady state has relevance
in many fields, in particular thermodynamics. Hence, the steady state is a more general situation than dynamic equilibrium. If a system is in the steady state, then the
recently observed behaviour of the system will continue in the future. In stochastic
systems, the probabilities that various states will be repeated will remain constant.
We will generalise it as follows:

$$\left(\frac{dn}{dt}\right) = \left(\frac{dn}{dt}\right)_+ - \left(\frac{dn}{dt}\right)_- = 0 \quad \text{or} \quad F = F_+ - F_- = 0. \tag{3.113}$$

A global biogeochemical cycle may be derived from the budget of the composition of
the individual reservoirs, with a (quasi) steady state being considered to exist. The
substances undergoing the biogeochemical cycle pass through several reservoirs (atmosphere, hydrosphere, pedosphere, lithosphere and biosphere) where certain concentrations accumulate due to flux rates, determined by transport and reaction.

In many systems, the steady state is not achieved until some time has elapsed after the system is
started or initiated. This initial situation is often identified as a transient state, start-up or warm-up
period.

The term steady state is also used to describe a situation where some, but not all of the
state variables of a system are constant. For such a steady state to develop, the system
does not have to be a flow system. Therefore, such a steady state can develop in a
closed system where a series of chemical reactions take place. Literature on chemical
kinetics usually refers to this case, calling it *steady-state approximation*. Steady-state
approximation, occasionally called stationary-state approximation, involves setting
the rate of change of a reaction intermediate in a reaction mechanism equal to zero.
Steady-state approximation does not assume the reaction intermediate concentration
is constant (and therefore, its time derivative is zero). Instead, it assumes that the variation in the concentration of the intermediate is almost zero. The concentration of the
intermediate is very low, so even a considerable relative variation in its concentration

is small if considered quantitatively. These approximations are frequently used because of the substantial mathematical simplifications this concept offers. Whether or not this concept can be used depends on the error the underlying assumptions introduce. Therefore, even though a steady state, from a theoretical point of view, requires constant drivers (e. g. constant inflow rate and constant concentrations in the inflow), the error introduced by assuming steady state for a system with non-constant drivers might be negligible if the steady state is approached fast enough (relatively speaking).

This is often the case for very reactive chemical species, especially radicals such as hydroxyl (OH). Steady state OH concentration is then expressed by the condition $d[OH]/dt = 0$ from which follows:

$$\sum_i k_i[A_i][B_i] = [OH] \sum_j k_j[C_j], \tag{3.114}$$

where the terms on the left side represent all reactions producing OH and those on the right side all reactions consuming OH; k is the reaction rate constant. However, this is actually true only for a short time, depending on the error being taken into account (a few minutes, but this time is large compared with the atmospheric OH lifetime in the order of seconds). The approximation of OH steady state is also useful for adopting a mean 'constant' OH concentration (independent of diurnal cycles) to reduce secondary-order reactions to pseudo-first-order reactions (see Chapter 3.3.3).

The principle of stationary (or instationary) is also applied to the atmospheric budget of trace species, regarding F_+ the source term (emission Q) and F_- the total removal term R (deposition and chemical conversion). With the definition of the residence time (see Chapter 3.3.3 for more details), it follows from equation (3.114) that:

$$\ln \frac{n_0}{n(\Delta t)} = \frac{\Delta t}{\tau}, \tag{3.115}$$

where n_0 is the (initial) amount of a substance in the reservoir at $t = 0$. To fit the steady-state condition $Q = R$ over a given unit of time (for example, a year), there is no condition concerning the quantity of the residence time. Normally, it is believed that a short residence time (let's say $\tau \ll$ one year) corresponds to a large removal capacity; if $\tau = 1$ day (for example, formic acid), after seven days, 99.9 % of the initial amount n_0 is removed from the atmosphere independent of the absolute amount. Consequently, the yearly removal capacity is some 50 times higher. As another consequence, the mean atmospheric concentration of this substance remains very low (but depends on the influx and the emission).

Regarding the example of a large residence time $\tau = 10$ years (for example, methane), after one year, only about 10 % of n_0 is removed from air as it follows from equation (3.116). Yet we have to consider that n_0 represents the total amount of the substance at an arbitrary time (here set to $t = 0$); this consists of the actual emission flux and past cumulative emission versus the removal rate. Let's say an experiment with an emission process starts (year 1) with continuous emission over time (100 units per year); after a year, 90 % remains in the air due to the slow removal capacity (τ = 10 yr). In year 2, the fresh emission (100 units) will be added to the remaining emission from year 1, but the absolute removal (10 % relatively) is larger: $(100 + 90) \cdot 0.1 = 19$ units. Hence, the atmospheric amount (and concentration) increases from year to year, but the absolute removal amount also rises until reaching the equilibrium, when the yearly removal becomes equal to the emission (100 units) – that is, for the example of CH_4, in 10 years. In all following years, the stationary $Q = R$ remains and a stationary concentration is achieved. Finally, choosing a substance having a very large residence time (between hundreds and thousand years, such as CO_2 and some halogenated organic compounds, CFC), the steady state is never reached on human timescales. For CFC, due to its banning, the emission has been practically zero for a few years, and its concentration no longer increases but – depending on the removal time – slightly decreases. A full stop of CO_2 emission would result in an atmospheric concentration no longer increasing, but also no measurable decrease for several hundreds of years. Air pollution control, in this case, would not result in recovery but would stop further catastrophic development.

The higher the residence time (and emission), the larger the atmospheric amount of the substance. Because of $dn/dt = F_- = R = n/\tau$ (equation (3.113)) and the condition $Q = R$ it follows that:

$$n_{atm} = \tau \cdot Q. \tag{3.116}$$

However, this experiment simply implies that a steady state is achieved sometime after the system is started and that Q is constant, which is the normal case for natural processes on a climatological timescale. With yearly rising emissions, as it is typical for anthropogenic processes such as fossil fuel combustion, the system remains out of stationary, and the time to achieve the steady state is endless while the emissions continue to increase. Hence, the atmospheric concentration increases (as seen for the greenhouse gases). This increase is simply calculated from:

$$\text{concentration increase (in \%)} = 100 \cdot \frac{Q_i - R_i}{n_i + Q_i}, \tag{3.117}$$

where the subscript i denotes the year regarded, n total amount in air and Q and R the yearly amount of emission and removal, respectively.

3.3 Theory of chemical reactions

One of the key concepts in classical chemistry is that all chemical compounds can be described as groups of atoms bonded together, and chemical reactions can be described as the making and breaking those bonds.

Hence, we introduce this chapter with a short excursion into chemical bonding. To understand chemical bonds deeper, it is obligatory to take adequate chemical textbooks in your hand. Chapter 3.3.1 represents knowledge reduced to a minimum only to illustrate the principles. On the other hand, it is my experience (even as a chemist) that it is sufficient (for non-chemists).

3.3.1 Chemical bonding

Chemical reactions are connected (with the exception of nuclear reactions, see Chapter 5.3.5) only with changes in the electronic structure of atoms. Chemists are interested only in three elementary particles, electrons, protons and neutrons. The last two particles, we have introduced as parts of the atomic nucleus, characterise the elements (Chapter 2.2.1). The structure of the atomic *electronic shell* (remember, the number of electrons is equal to the number of protons in an uncharged atom) is characterised by the so-called *orbitals*, synonymous with the term *wave function*, but is has no illustrative meaning (as shown by pictures in many textbooks as also here in Figure 3.3). The electron cannot be particulate described in space and time – it is a probability function of energy density, characterised by so-called quantum numbers. For the purpose of this book, to apply chemistry to environmental processes, it is not necessary to go into details of the atomic structure.

> ℹ️ Chemical bonding occurs when one or more electrons are simultaneously attracted to *two* nuclei. This is the most important fact about chemical bonding that you should know.

In its most fundamental sense, the structure of a molecule is specified by the identity of its constituent atoms and the sequence in which they are joined together, that is, by the *bonding connectivity*. This, in turn, defines the *bonding geometry* – the spatial relationship between the bonded atoms (Figure 3.2). The energy of a system of two atoms depends on the distance between them. At large distances, the energy is zero, meaning 'no interaction'. At distances of several atomic diameters, attractive forces dominate, whereas at very close approaches, the force is repulsive, causing the energy to rise. The attractive and repulsive effects are balanced at the minimum point in the curve (Figure 3.2). The internuclear distance at which the potential energy minimum occurs defines the *bond length*. This is more correctly known as the *equilibrium bond length* because thermal motion causes the two atoms to vibrate about this distance. In general, the stronger the bond, the smaller the bond length will be. The *bond energy* is the amount of work that must be done to pull two atoms completely apart; in other words, it is the same as the depth of the 'well' in the potential energy curve shown in Figure 3.2. This is almost, but not quite the same as the *bond dissociation energy* actually required to break the chemical bond (we will meet it later with the photodissociation); the difference is the very small *zero-point energy*.

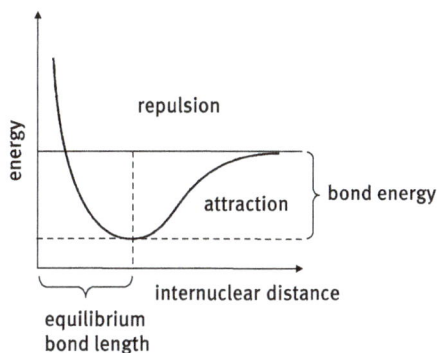

Figure 3.2: Schema of potential energy between two atoms (chemical bond).

In general, the bonds may be *ionic* or *covalent*. The transfer of electrons from one atom to the other forms an ionic bond. If electrons are lost, the atom becomes a positively charged ion (Na^+ for example), and if electrons are gained, the atom becomes a negatively charged ion (Cl^- for example). The ion pair that results is held together loosely by electrostatic attraction. Both ions form a crystal, sodium chloride or common salt $(NaCl)_n$. Metals release electrons forming a microcrystalline structure (*metallic bond*), where the electrons move freely in the conduction band, explaining electric and thermal conductivity. In other cases, electrons are not transferred but shared between atoms (*covalent bond*). In elementary molecules with identical atoms, such as N_2, O_2 and Cl_2, the electrons are shared equally. On the other hand, in heteronuclear molecules that consist of unlike atoms (such as H_2O), the electrons forming the bond are shared unequally; this case of bonding is termed *polar covalent*.

The number of (so-called valence) electrons in the outer shell determines the group in the periodic table of the elements based on the octet rule. Only noble gases (in group 8 or 0, hence most right or left in the periodic table of elements) hold complete shells with eight electrons (He in s-orbital two electrons) and are extremely stable and exist as atoms only (gaseous), see Table 3.7. All other elements, having between one and seven electrons in the outer shell, must form molecules (covalent or metallic bond). Elements with few outer electrons (1–3) tend to be electropositive, i. e. to release the outer electron to become stable with the lower octet shell. Elements with more electrons (5–7) tend to be electronegative, i. e. to take up electrons to become stable with the full octet shell. Only elements from group 4 (C and Si) hold balanced electron affinity, i. e. they provide 4 electrons but also take up 4 electrons. This property makes the chemistry of carbon (organic world) and silicon (mineral world) so unique concerning the diversity of molecules and structures.

Table 3.7: Electronic configuration of the first 20 atoms of the periodic table of the elements at the ground level; z ordinal number.

z	element		config	1s	2s	2p			3s	3p			4s
1	H	hydrogen	$1s^1$	↑									
2	He	helium	$1s^2$	↑↓									
3	Li	lithium	$2s^1$	↑↓	↑								
4	Be	beryllium	$2s^2$	↑↓	↑↓								
5	B	boron	$2s^2\,2p^1$	↑↓	↑↓	↑							
6	C	carbon	$2s^2\,2p^2$	↑↓	↑↓	↑	↑						
7	N	nitrogen	$2s^2\,2p^3$	↑↓	↑↓	↑	↑	↑					
8	O	oxygen	$2s^2\,2p^4$	↑↓	↑↓	↑↓	↑	↑					
9	F	fluorine	$2s^2\,2p^5$	↑↓	↑↓	↑↓	↑↓	↑					
10	Ne	neon	$2s^2\,2p^6$	↑↓	↑↓	↑↓	↑↓	↑↓					
11	Na	sodium	$3s^1$	↑↓	↑↓	↑↓	↑↓	↑↓	↑				
12	Mg	magnesium	$3s^2$	↑↓	↑↓	↑↓	↑↓	↑↓	↑↓				
13	Al	aluminium	$3s^2\,3p^1$	↑↓	↑↓	↑↓	↑↓	↑↓	↑↓	↑			
14	Si	silicon	$3s^2\,3p^2$	↑↓	↑↓	↑↓	↑↓	↑↓	↑↓	↑	↑		
15	P	phosphorus	$3s^2\,3p^3$	↑↓	↑↓	↑↓	↑↓	↑↓	↑↓	↑	↑	↑	
16	S	sulphur	$3s^2\,3p^4$	↑↓	↑↓	↑↓	↑↓	↑↓	↑↓	↑↓	↑	↑	
17	Cl	chlorine	$3s^2\,3p^5$	↑↓	↑↓	↑↓	↑↓	↑↓	↑↓	↑↓	↑↓	↑	
18	Ar	argon	$3s^2\,3p^6$	↑↓	↑↓	↑↓	↑↓	↑↓	↑↓	↑↓	↑↓	↑↓	
19	K	potassium	$4s^1$	↑↓	↑↓	↑↓	↑↓	↑↓	↑↓	↑↓	↑↓	↑↓	↑
20	Ca	calcium	$4s^2$	↑↓	↑↓	↑↓	↑↓	↑↓	↑↓	↑↓	↑↓	↑↓	↑↓

> **i** The number of electrons that an element can accept, lose or share with other atoms (Table 3.8) determines the valency or oxidation number of an atom.

The chemical bond is described by two different theories, which, however, show for simple structures very similar results, the *valence bond theory* (VB) and the *molecular orbital theory* (MO). It is again useful to emphasise that there exist no different electrons (in some books named σ or π electron); they are all equal but occupy different electron configurations according to the energetic level where the position of the electron is unknown at a given time (Heisenberg's uncertainty principle). Figure 3.3 shows different orbitals.

- σ *orbitals* result from the overlapping of s–s, s–p_z and p_z–p_z atomic states; they have a cylindrical, i. e. rotational symmetry along the bond axis z (vertical arranged in a diagram).
- π *orbitals* result from lateral overlapping of p_x–p_x and p_z–p_z atomic states; they form two lobes above and below the two nuclei. Along the bond axis, z, the electron residence probability is zero.

As seen from Table 3.7, carbon with its valence structure $2s^2 2p^2$ should operate only two bonds, but we know that it almost provides four bonds (such as in CH_4). This

Table 3.8: Oxidation numbers (valences) of some important elements (Note: the number 0 – zero – is the neutral elemental state).

element	most common oxidation number
O	−2, −1
H	+1
Ca, Mg	+2
K, Na	+1
C	−4, −3, −2, −1, +1, +2, +3, +4
N	−3, +1, +3, +4, +5
S	−2, +3, +4, +6
Cl	−1, +1, +3, +4, +5, +7
Mn	+2, +3, +4, +6, +7
Fe	+2, +3
Cu	+1, +2
Cr	+3, +6

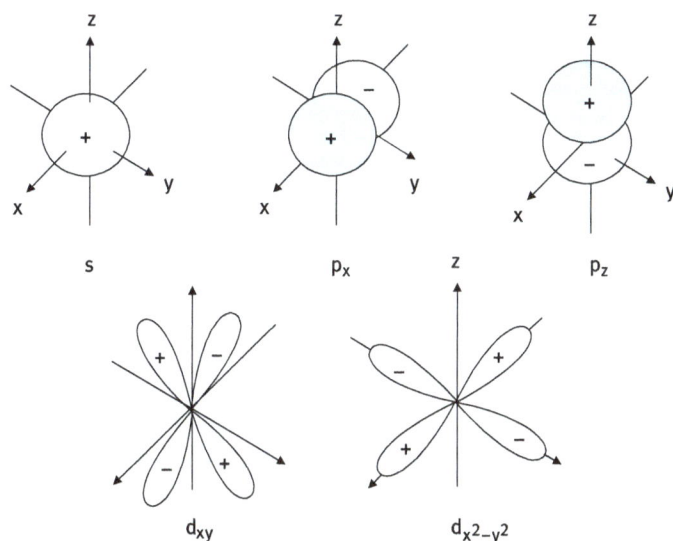

Figure 3.3: Symmetry of electron configurations for s, p and d levels.

is explained by the concept of valence hybridisation: mixing of atomic orbitals into new *hybrid orbitals* (with different energies, shapes, etc., than the component atomic orbitals). In methane, one s orbital is mixed with three p orbitals, resulting in four equivalent sp^3 hybrids, giving the characteristic tetrahedral coordinated carbon (Figure 3.4).

For another example, ethene (C_2H_4) has a double bond between the carbons. For this molecule, carbon will sp^2 hybridise because one π bond is required for the double

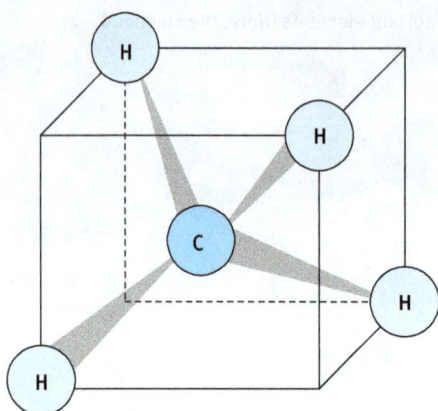

Figure 3.4: Tetrahedral coordinated carbon in methane, explained by sp^3 hybridisation (angle between two H atoms 109 °).

bond between the carbons, and only three σ bonds are formed per carbon atom. In ethene, the two carbon atoms form an σ bond by overlapping two sp^2 orbitals and each carbon atom forms two covalent bonds with hydrogen by s–sp^2 overlap all with 120 ° angles (trigonal-planar). The π bond between the carbon atoms perpendicular to the molecular plane is formed by 2p–2p overlap. The hydrogen–carbon bonds are all of equal strength and length, which agrees with experimental data. The free rotation as it is found in alkanes (single bond) is lost; hence isomers were formed.

Friedrich August Kekulé (1829–1896) proposed in 1865 the formula of benzene (C_6H_6), shown in Figure 3.5 with alternating single and double bonds, often simplified by the following structural formula:

The isomers proposed by Kekulé are conjugated double bonds; Figure 3.5 shows two different formula schemes. Conjugated double bonds in a molecule mean that the single and double bonds alternate. This enables the electrons to be delocalised over the whole system and so be shared by many atoms. In other terms, the delocalised elec-

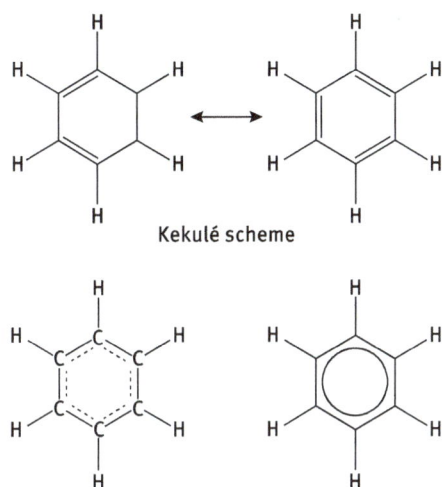

Kekulé scheme

Figure 3.5: Formula of benzene C_6H_6.

trons may move around the whole system. Benzene, a natural constituent of crude oil, is the best-known example.

Finally, in ethyne (C_2H_2) a triple bond is realised by sp hybridisation giving a linear structure: H–C≡C–H; Table 3.9 shows different hybrid types with important geometric assembly.

Table 3.9: Valence hybrids.

coordination number	hybrid	geometric assemble	examples
2	sp	linear, <180°	ethine C_2H_2
3	sp^2	trigonal planar, <120°	ethene C_2H_4
4	sp^3	tetrahedral, <109°28'	methane CH_4
5	sp^3d	trigonal bipyramidal, <120° inside the mean plane and <90° to the head	phosphorous pentachloride PCl_5
6	sp^2d^2	octahedral, <90°	sulphur hexafluoride SF_6

In any stable structure, the potential energy of its atoms is lower than that of the individual isolated atoms. Thus, the formation of methane from its gaseous atoms (a reaction that cannot be observed under ordinary conditions but for which the energetic is known from indirect evidence)

$$4\,H + C \rightarrow CH_4$$

is accompanied by the release of heat and is thus an exothermic process. The quantity of heat released is related to the stability of the molecule. The smaller the amount of energy released, the more easily can the molecule absorb thermal energy from the environment, driving the above reaction in reverse and leading to the molecule's decomposition. A highly stable molecule such as methane must be subjected to temperatures of more than 1,000 °C for significant decomposition to occur.

Many molecules are energetically stable enough to meet the above criterion but are so *reactive* that their lifetimes are too brief to make their observation possible. The molecule CH_3, *methyl*, is a good example (it is a radical): it can be formed in the air by the reaction $CH_4 + OH$, but it is so reactive that it combines with almost any molecule it strikes within a few collisions.

3.3.2 Types of chemical reactions

> ℹ️ A *chemical reaction* is a process that results in the interconversion of chemical species.

Table 3.10 summarises the base reaction types. Chemical reactions might be elementary reactions or stepwise reactions. A stepwise reaction consists of at least one reaction intermediate and involves at least two consecutive elementary reactions. Parallel reactions are several simultaneous reactions that form different respective products from a single set of reactants.

Table 3.10: Types of chemical reactions; A, B and C = any atomic or molecular entity, H = hydrogen, e^- = electron.

term	reaction scheme
addition	$A + B \rightarrow AB$
insertion	$A + BC \rightarrow AB + C$
abstraction	$AB + C \rightarrow A + BC$
dissociation[a]	$AB \rightarrow A + B$
radioactive decay[b]	$A \rightarrow B + C$
proton transfer	$A + H^+ \rightarrow AH^+$ or $A + BH \rightarrow AH^+ + B^-$
electron transfer	$A + e^- \rightarrow A^-$ or $A + B^- \rightarrow A^- + B$

[a] thermal or irradiative
[b] A, B and C are atoms only

Thermodynamics is important for describing chemical reactions. As seen before, it explains whether a reaction is 'voluntary' and in which direction and in what state equilibrium it goes. Hence, thermodynamics also describes the mechanisms without providing information on the explicit pathway. Many reactions can be parallel and thereby competitive. In quantifying the overall chemical processes (production and/or decay of substances), because of the substantial mathematical simplifications, it is important to delete reactions of minor importance (i. e. those with reaction rates about two orders of magnitude less than the fastest reaction).

The task of chemical kinetics is to describe the speed at which chemical reactions occur.

The reaction rate is the change in the number of chemical particles per unit of time through a chemical reaction. This is the term:

$$R_N = \frac{dN}{dt} \text{ [number/time].}$$

As $n = N/N_A$ (N_A is the Avogadro constant) the rate (or speed) also can be expressed in mole per time:

$$R_n = \frac{dn}{dt} \text{ [mole/time].}$$

Finally, because $c = N \cdot M/V$ (M is the molar mass) is the change of concentration per time:

$$R_c = \frac{dc}{dt} \text{ [mass/volume} \cdot \text{time].}$$

In chemistry, the last concentration-related term is usually used, and equation (3.118) shows the recalculation:

$$R = \frac{dN}{dt} = N_A \frac{dn}{dt} = \frac{V}{M} \frac{dc}{dt}. \tag{3.118}$$

In the natural environment, the investigation of chemical reactions is very limited. All early attempts to estimate reaction rates were unsuccessful because too many variables existed. However, concentration measurements of different trace species – together with physical parameters that describe transport, mixing and phase state conditions (in so-called complex field experiments) – are extremely helpful in several directions. These directions include: a) establishing empirical relationships between substances and another reservoir (air, soil and water) parameters to gain insights into mechanisms; b) recognising still unidentified or not yet considered substances under the specific measurement conditions; c) providing data sets for model evaluations.

Kinetics, specifically studying rate laws and measurement of rate constants, can only be done under laboratory conditions, whereas reaction conditions could be simulated in special reactors closely resembling the environment. Once established, the k-value of an elementary reaction is universally applicable, or in other words, 'pure chemistry' is independent of the reservoir and geographical specifics, but the conditions for reactions (pressure, temperature, radiation, humidity) and the concentration field depends on location.

The kinetics can also help estimate the reaction mechanisms. As mentioned, thermodynamics only describes the difference between the initial and final rate but not the pathway. Kinetics, according to the above equations, can also describe the gross reaction but show whether there is a complex reaction when the reaction order (see below) is not an integer. As an example, the (photochemical) ozone formation in tropospheric air from CH_4 can be written with the following gross equation:

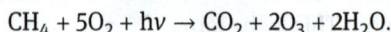

$$CH_4 + 5O_2 + h\nu \rightarrow CO_2 + 2O_3 + 2H_2O.$$

It is self-evident that this cannot be an elementary reaction. Studying the intermediates, one can detect CH_3, CH_3O_2, CH_3O, HCHO, HCO, CO, OH and HO_2, relatively stable intermediates such as formaldehyde (HCHO) and carbon monoxide (CO), whereas all other species are very short-lived radicals. Moreover, in the process of methane oxidation, some other products can be formed. The larger the molecule (e. g., ethane C_2H_6) the more reaction pathways exist and hence products. It is a great challenge to study short-lived intermediates and the kinetics of elementary steps. This is necessary for modelling and finally for environmental management such as pollution abatement.

The process behind ozone formation (valid for each oxidative degradation of organic molecules) is *complex*, i. e. it consists of a sequence of events, called elementary acts or unit steps, constituting the reaction mechanism (*subsequent reactions*). The *rate-determining step* (RDS), sometimes also called the limiting step, is a chemistry term for the slowest step in a chemical reaction sequence. In a multistep reaction, the steps nearly always follow each other so that the product(s) of one step is/are the starting material(s) for the next. Therefore, the rate of the slowest step governs the rate of the whole process. In a chemical process, any step that occurs after the RDS will not affect the rate and, therefore, does not appear in the rate law. Intermediate states between the steps usually involve some unstable intermediate species with higher energy content than those of the reactants or the reaction products.

3.3.3 Chemical kinetics: reaction rate constant

Even complex chemical reaction mechanisms can be separated into several definite *elementary reactions*, i. e. the direct electronic interaction process between molecules and/or atoms when colliding. To understand the total process A → B – for example, the oxidation of sulphur dioxide to sulphate – it is often adequate to model and budget calculations in the environment to describe the *overall reaction*, sometimes called the *gross reaction*, independent of whether the process A → B is going via a reaction chain A → C → D → E → ⋯ → Z → B. The complexity of mechanisms (and thereby the rate law) is significantly increased when parallel reactions occur: A → X beside A → C, E → Y beside E → F. Many chemical processes in our environment are complex. If only one reactant (sometimes called educt) is involved in the reaction, we call it a unimolecular reaction, where the reaction rate is proportional to the concentration of only one substance (*first-order reaction*). Examples are all radioactive decays, rare thermal decays (almost autocatalytic) such as PAN decomposition and all photolysis reactions, which are very important in air and surface water.

The most frequent type of chemical reactions are bimolecular reactions (*second-order reaction s*: A + B); for example, simple molecular reactions ($NO+O_3 \rightarrow NO_2+O_2$) and frequent radical reactions ($NO+OH \rightarrow HNO_2$). However, trimolecular (or termolecular) reactions are also common in the atmosphere ($O + O_2 + N_2 \rightarrow O_3 + N_2$) because the primary collision complex A...B is energetically unstable and must transfer excess energy onto a third collision partner (*third-order reaction*). Collision partners are the main constituents of air (N_2 and O_2) because of a collision's probability, but also H_2O (for example, $HO_2 + HO_2 + H_2O \rightarrow H_2O_2 + O_2 + H_2O$), probably because of its specific role in the transfer complex (charge transfer, hydrogen bonding or steric reasons).

The number of reactants on which the concentration of the reaction rate depends determines the *reaction order*. The third-order reactions might be considered elementary (despite the fact that the collision partner certainly does not collide at the same time with A and B). All reactions with higher or non-integer numbers principally reflect the kinetics of complex mechanisms. The bimolecular reaction can be written in three different schemes:

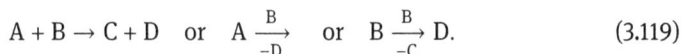

$$A + B \rightarrow C + D \quad \text{or} \quad A \xrightarrow[-D]{B} \quad \text{or} \quad B \xrightarrow[-C]{B} D. \tag{3.119}$$

The following rate law is valid (remember that the brackets denote the concentration of the substances A etc. and k reaction rate constant:

$$-\frac{d[A]}{dt} = -\frac{d[B]}{dt} = \frac{d[C]}{dt} = \frac{d[D]}{dt} = k[A][B] = R. \tag{3.120}$$

All given reaction rate constants for bimolecular reactions in this book are in cm^3 molecule^{-1} s^{-1} at 298 K if not otherwise mentioned. An important quantity to describe the state of the chemical reaction equation (3.119) is the *reaction quotient Q*. In contrast to equation (3.78), the concentrations are not expressed as equilibrium (when the reaction reaches equilibrium, the condition $Q = K$ is fulfilled):

$$Q = \frac{[C][D]}{[A][B]}. \tag{3.121}$$

To solve the differential equation (3.120) analytically, we substitute time-dependent concentrations by ($[A]_0 - x$) and ($[B]_0 - x$) where subscript 0 denotes the initial concentration for $t = 0$ and x expresses the concentration of any of the products (assuming that neither C nor D is included in other reactions):

$$\int_0^x \frac{dx}{([A]_0 - x)([B]_0 - x)} = k \int_0^t dt. \tag{3.122}$$

By partial fraction decomposition, we get:

$$k = \frac{1}{t([A]_0 - [B]_0)} \ln \frac{[A]_0([B]_0 - x)}{[B]_0([A]_0 - x)}. \tag{3.123}$$

Alternatively, for the simple case that A = B (2 A → products), we get:

$$k = \frac{1}{t}\left(\frac{1}{[A]_0 - x} - \frac{1}{[A]_0}\right) = \frac{1}{t}\frac{1}{[A]_0([A]_0 - x)}. \tag{3.124}$$

Often in nature, there are situations where the concentration of the second reactant B can be considered constant in the given period or its relatively timely change $\Delta[B]/[B]$ can be neglected. This we most often assume when the concentration of B is greater than that of A, and thereby, the numerical error remains small; this is the case in many reactions when O_2 is the partner. Another case is the steady state of B ($[B]/dt = 0$). Then, the concentration of B can be included in the rate constant, and the reaction type is reduced to a *pseudo-first-order* reaction with mathematical simplifications in further treatment:

$$-\frac{d[A]}{dt} = k[A][B] = k'[A]. \tag{3.125}$$

Similar trimolecular reactions (remember that the third collision partner is often in high excess) can be simplified to pseudo-second-order and eventually even to pseudo-first-order. The solution of a first-order rate law is simple:

$$\int_{[A]_0}^{[A]} \frac{d[A]}{[A]} = \int_{[A]_0}^{[A]} d\ln[A] = k \int_0^t dt. \tag{3.126}$$

It follows the exponential time dependency of concentration of A:

$$[A] = [A]_0 \exp(-kt). \tag{3.127}$$

From equation (3.127), useful characteristics of times such as the half-life $\tau_{1/2}$ and residence or lifetime τ can be derived. The half-life is defined as the time during which the initial concentration is reduced by half: $[A] = 0.5[A]_0$:

$$\tau_{1/2} = \frac{1}{k} \ln 2. \tag{3.128}$$

More important for environmental reservoirs is the *residence time*, which is simply the reciprocal of k, but only in the case of a first-order process: $\tau = 1/k$. The residence time τ corresponds to the condition $[A]/[A]_0 = e$ (this follows from equation (3.127)), where e is a mathematical constant, sometimes called Euler's number (e ≈ 2.7183). In other words, the initial concentration of A is reduced by about 37 %.

Arrhenius empirically found that between the logarithm of the reaction rate k and the reciprocal temperature a linear relation exists, which is similar to van't Hoff's reaction isobar (equation (3.85)):

$$\frac{d \ln k}{dT} = \frac{E_A}{RT^2} \tag{3.129}$$

and termed now the Arrhenius equation, E_A is an activation energy:

$$k = k_m \exp\left(-\frac{E_A}{RT}\right). \tag{3.130}$$

The term $\exp(-E_A/RT)$ simply expresses the fraction of chemical species with a per mol higher energy than E_A. The Boltzmann distribution follows from equation (2.33), which shows the probability of molecules having (or exceeding) a certain speed or energy when taking into account that $3RT/M$ is equivalent (equation (2.18)) to the mean molar kinetic energy of the molecules. Thus, E_A terms the (molar) kinetic energy required for the transfer from A + B to the transition state $AB^{\ddagger}$. The factor k_m expresses the maximum reaction rate constant ($E_A \to 0$ and/or $T \to 0$), also called the Arrhenius constant. The interpretation of k_m is possible from both the collision theory and the theory of the transition state; in reality k_m depends on T (often $\beta = 1/2$), B constant:

$$k_m = BT^{\beta}. \tag{3.131}$$

Exceptions of the validity of equation (3.131) are trimolecular reactions where k decreases with increasing T. This becomes clear when we consider that the transition complex $AB^{\ddagger}$ can decompose thermally back to A + B instead of transforming to C + D:

$$A + B \underset{k_b}{\overset{k_a}{\rightleftarrows}} [AB^{\ddagger}] \xrightarrow{+M(k_c)} C + D + M. \tag{3.132}$$

The reaction scheme equation (3.132) is principally valid for all bimolecular reactions:

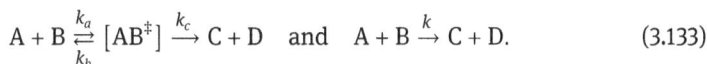

$$A + B \underset{k_b}{\overset{k_a}{\rightleftarrows}} [AB^{\ddagger}] \xrightarrow{k_c} C + D \quad \text{and} \quad A + B \xrightarrow{k} C + D. \tag{3.133}$$

According to the kinetic theory of gases, the reaction rate (here expressed as the number N of molecules changing per time) is equal to the number of collisions z between A and B (equation (2.39)) where the factor 2 means that at each collision two molecules disappear:

$$R_N = \frac{dN}{dt} = k_m N_A N_B = \frac{2z}{dt}. \tag{3.134}$$

Now, expressing z in terms of two different molecules (equation (2.42)) with a radius r_A and r_B as well as a molar mass M_A and M_B it follows that:

$$z = 4(r_A + r_B)^2 n_A n_B dt \sqrt{\pi RT \left(\frac{1}{M_A} + \frac{1}{M_B} \right)} \tag{3.135}$$

and

$$k_m = 8(r_A + r_B)^2 \sqrt{\pi RT \left(\frac{1}{M_A} + \frac{1}{M_B} \right)}. \tag{3.136}$$

Thus, in $k_m = f(T^{1/2})$, k_m clearly represents the maximum reaction rate constant. Not all collisions result in a successful reaction, i. e. turnover according to equation (3.119) because the transition state AB is formally in equilibrium with the reactants with a pseudo equilibrium constant $K^{\ddagger} = [AB^{\ddagger}]/[A][B]$. This circumstance is described by the introduction of a probability factor y (also called a steric factor, but not to be confused with the accommodation coefficient despite some similarities) in the Arrhenius equation:

$$k = y \cdot k_m \exp\left(-\frac{E_A}{RT} \right) = k_0 \exp\left(-\frac{E_A}{RT} \right). \tag{3.137}$$

The rate law concerning the transition state is

$$\frac{d[AB^{\ddagger}]}{dt} = k_a[A][B] - k_b[AB^{\ddagger}] - k_c[AB^{\ddagger}]. \tag{3.138}$$

Assuming a steady state for the short-lived intermediate $d[AB^{\ddagger}]/dt = 0$ (Bodenstein principle) and further assuming that $AB^{\ddagger}$ once formed will preferably react to C + D because of the much higher free reaction enthalpy $\Delta_- H^{\ddagger}$ compared with $\Delta_+ H^{\ddagger}$ (Figure 3.6), and therefore, $k_c \gg k_b$, it follows that:

$$\frac{[AB^{\ddagger}]}{[A][B]} = \frac{k_a}{k_b + k_c} \approx \frac{k_a}{k_c}. \tag{3.139}$$

This reliable approach, however, is in contradiction to the Eyring approach, which assumes that $AB^{\ddagger}$ is within equilibrium with the initial reactants. This is derived by using the general relationships equation (3.78) and equation (3.81) and treating the transition state in pseudo equilibrium (equation (3.132)) as follows:

$$\frac{k_a}{k_b} = K^{\ddagger} = \frac{[AB^{\ddagger}]}{[A][B]} = \exp\left(-\frac{\Delta_+ G^{\ddagger}}{RT} \right) = \exp\left(-\frac{\Delta_+ H^{\ddagger}}{RT} \right) \exp\left(\frac{\Delta_+ S^{\ddagger}}{R} \right), \tag{3.140}$$

where the state variables are related to the transition state $AB^{\ddagger}$. Note the difference between equation (3.139) and equation (3.140). It would follow from both equations

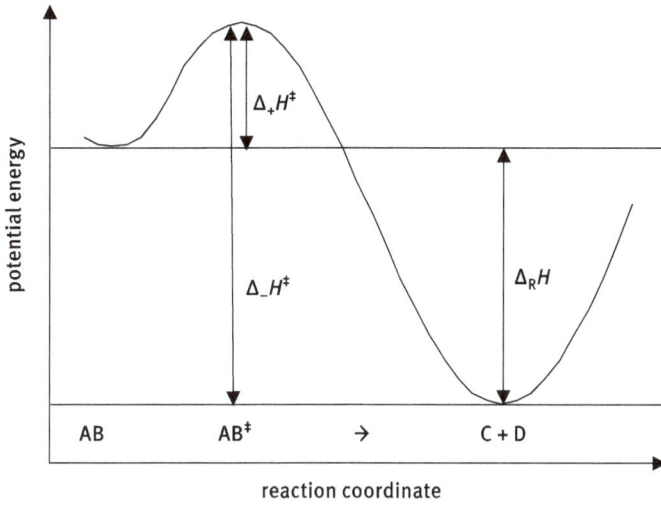

Figure 3.6: Energetic scheme of a bimolecular reaction; instead of enthalpies (ΔH), free enthalpies can also be used (ΔG) in this scheme.

that $k_b = k_c$, which means that each 50 % of $AB^{\ddagger}$ will turn back to the reactants and turn forward to the products – an unlikely condition. The rate of product formation ('successful' reaction) is equal to the rate of $AB^{\ddagger}$ right-hand transformation in equation (3.133) but also to the disappearance (turnover) of A (or B); τ represents the residence time of the transition state:

$$R = \frac{d[\text{products}]}{dt} = k[A][B] = k_c[AB^{\ddagger}] = k_c K^{\ddagger}[A][B] = [AB^{\ddagger}]\tau^{-1}. \tag{3.141}$$

From equation (3.141), it follows that:

$$k = k_c \frac{k_a}{k_b} = k_c K^{\ddagger} = k_c \exp\left(-\frac{\Delta_+ H^{\ddagger}}{RT}\right)\exp\left(\frac{\Delta_+ S^{\ddagger}}{R}\right). \tag{3.142}$$

According to the kinetics theory, the residence time $\tau = 1/k_c$ represents the ratio of the mean translation velocity of $AB^{\ddagger}$ and the displacement vdt the transition state passes; with $\tau = h/kT$, where h is Planck's constant. It follows:

$$k = \frac{kT}{h} \exp\left(-\frac{\Delta_+ H^{\ddagger}}{RT}\right)\exp\left(\frac{\Delta_+ S^{\ddagger}}{R}\right). \tag{3.143}$$

Figure 3.6 shows the chemical reaction of type equation (3.129) schematically. It follows that for the reaction enthalpy of the total process (A + B = C + D), $\Delta_R H = \Delta_- H^{\ddagger} - \Delta_+ H^{\ddagger}$ is the difference between the transition state enthalpies of the backward and forward reactions (Figure 3.6). $\Delta_+ H^{\ddagger}$ is the difference between the enthalpy of the transition state and the sum of the enthalpies of the reactants in the ground state. This is called *activation enthalpy* (Figure 3.6). The transition state $AB^{\ddagger}$ is formed at maximum energy. $\Delta_R G$ is the free reaction enthalpy of the overall process or, in other words, the formation of $(C + D)$. When comparing the right-hand term of equation (3.137) with the middle term of equation (3.140) and applying equation (3.81) to the total process, it follows that:

$$k = k_c \exp\left(-\frac{\Delta_+ G^{\ddagger}}{RT}\right) = k_0 \exp\left(-\frac{E_A}{RT}\right) = \exp\left(-\frac{\Delta_R G}{RT}\right) \tag{3.144}$$

and

$$E_A = \Delta_+ G^{\ddagger} - RT \ln\frac{k_0}{k_c}. \tag{3.145}$$

Hence, the activation energy is not (fully) identical to the activation enthalpy $\Delta_+ H^{\ddagger}$ as is often reported in the literature. The so-called Eyring plot $\ln(k/T)$ versus $1/T$ gives a straight line with the slope $-\Delta_+ H^{\ddagger}/R$ from which the enthalpy of activation can be derived and with intercept $\ln(k/h) + \Delta S^{\ddagger}/R$ from which the entropy of activation is derived.

The reaction scheme equation (3.133) is valid but with the most likely belief that $k_c \gg k_b$ there is no flux from the transition state $AB^{\ddagger}$ back and thereby no equilibrium. The Arrhenius factor simply explains that the formation of $AB^{\ddagger}$ is *inhibited*, i. e. only a fraction of collisions $A + B$ turn to $AB^{\ddagger}$ despite the reacting molecules providing the needed activation energy E_A. Let us compare the following reactions between the OH radical and a simple organic compound (k in cm^3 molecule^{-1} s^{-1}):[38]

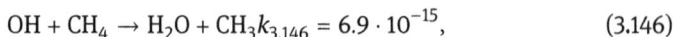

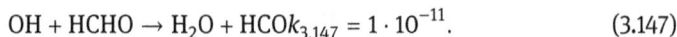

$$OH + CH_4 \rightarrow H_2O + CH_3 \qquad k_{3.146} = 6.9 \cdot 10^{-15}, \tag{3.146}$$
$$OH + HCHO \rightarrow H_2O + HCO \qquad k_{3.147} = 1 \cdot 10^{-11}. \tag{3.147}$$

Both reaction rate constants are considerably different (ratio more than 10^4). Both transition states represent a hydrogen bonding and H abstraction $H_3CH\cdots OH$ and $HO\cdots HCHOH$, and it is not likely that in the case of methane, the transition state turns with a probability of more than a factor of 10^4 back to the initial substances

---

38 The SI unit for a reaction rate constant is m^3 mol^{-1} s^{-1}, but for gas-phase reactions in literature almost the unit cm^3 molecule^{-1} s^{-1} is given. Note that 'molecule' is not a unit, but often included for clarity (it is meant number of molecules per cm^3, i. e. the unit cm^{-3} would be sufficient). For aqueous-phase reactions, often the unit L mol^{-1} s^{-1} or M^{-1} s^{-1} is used (1 M = 1 mol/L). For recalculation, remember that 1 mol = $6.023 \cdot 10^{23}$ molecules/atoms (Avogadro's number). It is 1 L/mol = $1.8 \cdot 10^{19}$ cm^3/molecule and 1 cm^3/molecule = $6 \cdot 10^{-20}$ L/mol, respectively. For illustration, $k = 10^{10}$ L mol^{-1} s^{-1} is a very fast aqueous-phase reactions (corresponds to $0.18 \cdot 10^{-10}$ cm^3 molecule^{-1} s^{-1}) whereas $k = 10^{-10}$ cm^3 molecule^{-1} s^{-1} is a very fast gas-phase reactions. In this book, k-values are cited as given in literature.

compared with formaldehyde. Hence, k_m is not significantly different in these reactions from y. However, the activation energy of the methane attack is much larger than that for HCHO, explaining the difference in k values.

The Arrhenius plot $\ln k$ versus $1/T$ gives a straight line with a slope $-E_A/R$ from which the activation energy can be derived and with intercept $\ln k_0$. Normally, in chemical standard reference books or tables, rate constants are given for 25 °C (about 298 K), and the following equation can be used for recalculations:

$$\ln \frac{k(T)}{k_{298}} = \frac{E_A}{R}\left(\frac{1}{298} - \frac{1}{T}\right) \quad \text{or} \quad k(T) = k_{298} \exp\left(\frac{E_A}{R}\left(\frac{1}{298} - \frac{1}{T}\right)\right). \tag{3.148}$$

The activation energy of chemical reactions lies in the range 20–150 kJ mol^{-1}. The time of processing $A + B \rightarrow C + D$ is in the order of only 10^{-12} s. This corresponds to $k = k_m$, and thereby $y = 1$ (sometimes in the literature, the steric or probability coefficient is denoted by A); this means that each collision will turn the state $A + B$ into $C + D$. Controversially, when $y = 0$, no collision results in a chemical conversion. The reality is somewhere in between.

In the case of (pseudo-) first-order reactions, the dimension of the reaction rate constant k is a reciprocal time (for example, $1/s$). To obtain a better understanding of the rate of disappearance of a pollutant, it is often useful to take a rate r in $\% \, h^{-1}$; the recalculation is then made with the following expression:

$$r = 100\left[1 - \exp(-k \cdot 3600)\right] \quad \text{and} \quad k = \frac{-\ln(1 - \frac{r}{100})}{3600}. \tag{3.149}$$

For illustration, a reaction with $k = 10^{-5} \, s^{-1}$ has a specific conversion of $r = 3.5 \% \cdot h^{-1}$ and a residence time of about one day (exactly 27.7 h). For air pollutants, therefore, after one day of air mass travelling, which is equivalent to (typical wind speed 5 m s^{-1}) a distance of about 500 km, the initial concentration has decreased to about 37 %. For aqueous pollutants in rivers, the chemical residence time must be compared with the flow rate.

It is important not to confuse the reaction rate constant (sometimes referred to as the specific reaction rate) with the absolute reaction rate R (= $dc/dt = kc$), which is often called the simple turnover flux (mass per volume and time). In environmental chemistry, this reaction rate is often given in ppb $\cdot h^{-1}$. Recalculation between concentration c and mixing ratio x is then given by equation (2.6).

Finally, there exist zero-order reactions where the reaction rate does not depend on the concentration of the reactant and remains constant until the substance has disappeared. Zero-order reactions are always heterogeneous.

3.3.4 Catalysis

Almost all reactions in the environment, living organisms and chemical reactors are catalytic, i. e. a substance called a *catalyst* increases the rate of a reaction without modifying the overall standard enthalpy change in the reaction.

This process is called *catalysis*, according to the IUPAC (1981). An example is the oxyhydrogen reaction that needs at 1 bar and 280 K about 10^{11} years to produce 1 mol H_2O. However, in the presence of finely divided platinum, the reaction becomes explosive as was discovered by Johann Wolfgang Döbereiner (1780–1849) in 1823. Berzelius introduced the term catalysis in 1835.

The catalyst is both a reactant and a product of the reaction. Catalysis can be classified as homogeneous catalysis, in which only one phase is involved (for example, stratospheric ozone decay), and heterogeneous catalysis (for example, exhaust gas cleaning), in which the reaction occurs at or near an interface between phases. Both types of catalysis proceed in gas and aqueous phases. In biological (aquatic) systems, the catalyst is called an *enzyme* (biocatalyst), controlling the metabolisms. Sometimes a process where the rate of the reactions is lowered is called *negative catalysis*. However, it should strictly be named *inhibition*, and substance is an *inhibitor*.[39]

The term catalysis is also often used when the substance is consumed in the reaction (for example, base-catalysed hydrolysis of esters). Strictly, such a substance should be called an *activator*. A bimolecular reaction of the type A + B → C + D proceeds catalytically via (K catalyst):

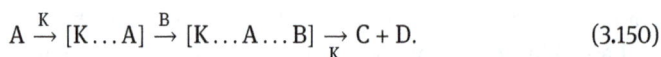

$$A \xrightarrow{K} [K \ldots A] \xrightarrow{B} [K \ldots A \ldots B] \xrightarrow[K]{} C + D. \tag{3.150}$$

The reactant(s) and catalyst produce the reactive intermediates that react faster than the reactants alone. The position of the thermodynamic equilibrium is not influenced; only the reaction rate is increased by choosing other reaction mechanisms.

3.3.5 Electrochemistry

We extensively discussed that chemical reactions are always connected with energy turnover; in most cases, the energy appears as heat. However, in electrolytes (see

---

[39] To stabilise some products (foods, cosmetics), radical scavengers are added. There also was the idea in *geoengineering*, to add chemicals to the atmosphere to avoid stratospheric ozone depletion. However, in my mind any attempt to 'control' atmospheric chemistry is unsuccessful due to the stochastic character of transport and mixing processes (control is impossible) and due to the large volume (and subsequent time) for chemical processing.

Chapter 3.2.2.2), the chemical species (ions), having a charge, can spontaneously transport electrical energy and produce an electrical current, called a galvanic (or voltaic) cell, such as in a *battery*. This is the conversion of chemical energy into electrical one. Conversely, when a chemical reaction is caused by an externally supplied current, as in *electrolysis*, electrical energy is converted into chemical energy. Both processes are called *electrochemical* reactions. *Electrical energy* is the energy carried by moving electrons in an electric conductor, such as an electrolyte. It is out of the scope of this book to deal with electrochemistry in galvanic cells because they only exist in technical systems and biological cells. Here we only consider chemical reactions where electrons are transferred directly between molecules and/or atoms, called oxidation-reduction reactions and the hydrated electron, an extra electron solvated in liquid water.

In general, electrochemistry describes the overall reactions when individual redox reactions are separate but connected by an external electric circuit and an intervening electrolyte.

3.3.5.1 Oxidation–reduction reaction (redox process)

Oxidation and reduction are key processes in chemical evolution and, thereby, in the environment.

Oxidation is always coupled with reduction (therefore, *redox* as shorthand for the reduction–oxidation reaction):

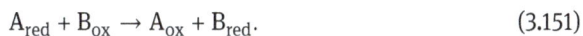

$$A_{red} + B_{ox} \rightarrow A_{ox} + B_{red}. \tag{3.151}$$

The reaction equation (3.151) describes all chemical reactions in which atoms have their oxidation number (oxidation state) changed. Oxidation state is defined as the charge an atom might be imagined to have when electrons are counted according to the following agreed set of rules (Table 3.11):
1. the oxidation state of a free element (uncombined element) is zero;
2. for a simple (monatomic) ion, the oxidation state is equal to the net charge on the ion;
3. hydrogen has an oxidation state of +1, and oxygen has an oxidation state of –2 when they are present in most compounds;[40] and
4. the algebraic sum of the oxidation states of all atoms in a neutral molecule must be zero, whereas, in ions, the algebraic sum of the oxidation states of the constituent atoms must be equal to the charge on the ion.

---

40 Exceptions to this are that hydrogen has an oxidation state of –1 in hydrides of active metals, e. g. LiH, and oxygen has an oxidation state of –1 in peroxides.

Table 3.11: Oxidation states of elements in stable compounds in the environment (in radicals, further states occur, see Table 3.8).

-4	-3	-2	-1	0	+1	+2	+3	+4	+5	+6	+7
			Hᵃ	H	H						
C				C		C		C			
	N			N	N	N	N	N	N		
	P			P			P	P	P		
		O	Oᵇ	O							
		S		S				S		S	
			Cl	Cl	Cl		Cl	Cl	Cl		
			Br	Br	Br			Br	Br		
			I	I	I			I	I		I
				Fe		Fe	Fe				
				Mn	Mn	Mn					
				Cu	Cu	Cu					

ᵃin metal hydrides, which are unstable under normal conditions
ᵇonly in peroxide

Hence, oxidation and reduction can be defined by the following three criteria, where all oxidations meet criteria 1 and 2 and many meet criterion 3, but this is not always easy to demonstrate:
1. the complete, net removal of one or more electrons from a molecular entity (also called 'de-electronation');
2. an increase in the oxidation number of any atom within any substrate; and
3. the gain of oxygen and/or loss of hydrogen of an organic substrate.

The redox process equation (3.151) can be formally divided into oxidation and reduction:

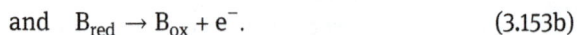

$$A_{red} \rightarrow A_{ox} + e^- \tag{3.152a}$$

$$\text{and} \quad A_{ox} + e^- \rightarrow A_{red}, \tag{3.152b}$$

$$B_{ox} + e^- \rightarrow B_{red} \tag{3.153a}$$

$$\text{and} \quad B_{red} \rightarrow B_{ox} + e^-. \tag{3.153b}$$

The reducing agent (also called a reductant or reducer) is an *electron donator*, and the oxidising agent (also called an oxidant or oxidiser) is an *electron acceptor*. With the extension of criterion 3, oxygen is an oxidising agent and hydrogen a reducing agent according to:

$$O_2 + e^- \rightarrow O_2^-, \tag{3.154}$$

$$H \rightarrow H^+ + e^-. \tag{3.155}$$

Because of the general principle of electroneutrality and mass conservation, oxidation and reduction must be balanced within the completely natural environment. The gross 'reaction' of sequences equation (3.154) and equation (3.155) is $H_2O \rightleftharpoons H_2O$. Water is the source substance of oxidising agents (oxygen and related substances) and reducing agents (hydrogen and related substances), whereas in the biosphere, hydrogen is enriched and oxygen is depleted, and vice versa in the atmosphere. This simple difference (gradients in 'reservoir' chemical potential) creates biogeochemical driving forces. As biology is relatively ubiquitous in the hydro- and pedosphere, the most ubiquitous elements that readily undergo redox transformations – iron in the ferrous/ferric (Fe^{2+}/Fe^{3+}) system and oxygen – are commonly encountered. Many microbial species, when starved of oxygen (anoxic conditions), turn to metals to shuttle their electron transport chains along and store energy via organophosphates for later use in living processes, such as reproduction or building protective coatings.

In atmospheric chemistry, the term *oxidant* is frequently used. This is a qualitative term that includes all trace gases with a greater oxidation potential than oxygen (for example, ozone, peroxyacetyl nitrate, hydrogen peroxide, organic peroxides, NO_3, etc.). Hence, oxygen is excluded (therefore, the term *reactive oxygen species* (ROS) has been introduced). This 'definition' of an oxidant is incorrect and should not be used in this narrow sense. As reaction equation (3.151) suggests, the determination of whether a chemical species is a reducing or oxidising species is relative and depends on the specific redox potential of the pairs of half-reactions equation (3.152a)–(3.152b) and equation (3.153a)–(3.153b). If a half-reaction is written as a reduction equation (3.152b) and equation (3.153a), the driving force is the *reduction potential*. If a half-reaction is written as an oxidation equation (3.152a) and equation (3.153b), the driving force is the *oxidation potential* related to the reduction potential by a sign change. So the redox potential is the reduction-oxidation potential of a compound measured under standard conditions against a standard reference half-cell. In biological systems, the standard redox potential is defined at pH = 7.0 versus the hydrogen electrode and partial pressure of $p_{H2} = 1$ bar. However, the concept of current flow is only applicable to aqueous systems.[41] In the gaseous phase of air, electron exchange occurs within the transition state of two molecular entities, in a broad sense of charge-transfer complexes.

Any substance in a specified phase has an electrochemical potential $(\mu_i)_{el}$ consisting of the chemical potential μ_i (partial molar free enthalpy) and a specified electric potential:

$$(\mu_i)_{el} = \mu_i + zFE, \tag{3.156}$$

---

[41] In the natural environment, only aqueous systems occur, but under laboratory conditions any other solution can be regarded (for example, organic solvents), thereby providing carriers for electrons.

where z is the number of elementary charges (e) exchanged in the oxidation-reduction process and F is the Faraday constant (molar electric charge $F = e \cdot N_A$). The *electromotive force E* (the equivalent of the term redox potential) is the energy supplied by a source divided by the electric charge transported through the source: $-zFE = \Delta_R G$. By contrast, the *electric potential* is the work required to bring a charge from infinity to that point in the electric field divided by the charge (for a galvanic cell, it is equal to the electric potential difference for zero current through the cell). The molar charge $zF \cdot n$ (n mole number) is defined as the *electric charge Q* (base unit ampere) and the current I is defined as the rate of charge flow:

$$I = \frac{dQ}{dt} = zF\frac{dn}{dt}.$$
(3.157)

The current has the rate dimension of moles per unit of time. This equality is important for understanding the equivalence between mass conversion (dn/dt) and current flow (dQ/dt). By using equation (3.79) and equation (3.81), we can describe the free enthalpy of the redox process equation (3.151) by:

$$\Delta_R G = -RT \ln K + RT \ln \frac{[A_{ox}][B_{red}]}{[A_{red}][B_{ox}]} = -zFE,$$
(3.158)

whereas $-RT \ln K = \Delta_R G^{\ominus}$ is the standard free enthalpy (in equilibrium). The important Nernst equation then follows:

$$E = E^{\ominus} + \frac{RT}{zF} \ln \frac{[A_{ox}][B_{red}]}{[A_{red}][B_{ox}]} \quad \text{or} \quad E = E^{\ominus} + \frac{0.05916}{z} \log \frac{[A_{ox}][B_{red}]}{[A_{red}][B_{ox}]}.$$
(3.159)

Standard redox potentials are listed in Table 3.12. From equation (3.159), when only considering the half-reaction oxidation or reduction, an adequate reduction potential E_{red} and oxidation potential E_{ox} can be derived:

$$E_{ox} = E^{\ominus}_{ox} + \frac{RT}{zF} \ln \frac{[A_{ox}]}{[A_{red}]},$$
(3.160)

$$E_{red} = E^{\ominus}_{red} + \frac{RT}{zF} \ln \frac{[B_{red}]}{[B_{ox}]}.$$
(3.161)

It follows for the coupled process that:

$$E = E_{red} + E_{ox}.$$
(3.162)

Equation (3.160) describes the (oxidation) reaction equation (3.152a); therefore, and the reverse (reduction) process equation (3.152b) is expressed by equation (3.161) when substance B is replaced by A. Because $E_{ox}(A) = E_{red}(A)$, it also follows then that $E^{\ominus}_{red} = E^{\ominus}_{ox}$. This explains why 'only' reduction standard potentials are listed in the literature, mostly termed the electrode potentials of half-reactions (Table 3.12).

Table 3.12: Standard reduction potentials (in V) of selected half-reactions in the aqueous H_xO_y system at 25 °C. e^- electron transferred from the electrode, e_{aq}^- hydrated electron, (g) gaseous, (aq) dissolved.

reactants	products	E	pH
e^-	$\rightarrow$ e_{aq}^-	−2.89	−
$H^+ + e^-$	$\rightarrow$ $H(aq)$	−2.32	−
$1/2\,H_2 + e^-$	$\rightarrow$ H^-	−2.25	−
$2\,H_2O + 2\,e^-$	$\rightarrow$ $H_2 + 2\,OH^-$	−0.828	pH = 14
$2\,H^+ + 2\,e^-$	$\rightarrow$ H_2	−0.414	pH = 7
$O_2 + 4\,H^+ + 4\,e^-$	$\rightarrow$ $2\,H_2O$	+1.229	pH = 0
$O_2 + 4\,H^+ + 4\,e^-$	$\rightarrow$ $2\,H_2O$	+0.816	pH = 7
$O_2(g) + 2\,H^+ + 2\,e^-$	$\rightarrow$ H_2O	+2.430	acid
$O_2 + 2\,H^+ + 2\,e^-$	$\rightarrow$ H_2O_2	+0.695	acid
$O_2 + 2\,H_2O + 4\,e^-$	$\rightarrow$ $4\,OH^-$	+0.401	pH = 14
$O_2(g) + H_2O + 2\,e^-$	$\rightarrow$ $2\,OH^-$	+1.602	pH = 14
$O_2(g) + e^-$	$\rightarrow$ O_2^-	−0.35	pH = 7
$O_2(g) + H^+ + e^-$	$\rightarrow$ HO_2	−0.07	acid
$O_2(aq) + H^+ + e^-$	$\rightarrow$ HO_2	+0.10	acid
$O_2(aq) + e^-$	$\rightarrow$ O_2^-	−0.18	pH = 7
$O_2 + e^-$	$\rightarrow$ O_2^-	−0.563	pH = 14
$O_2 + H_2O + 2\,e^-$	$\rightarrow$ $OH^- + HO_2^-$	−0.0649	pH = 14
$O + e^-$	$\rightarrow$ O^-	+1.61	pH = 7
$O^- + H_2O + e^-$	$\rightarrow$ $2\,OH^-$	+1.59	pH = 14
$O_2^- + H_2O + e^-$	$\rightarrow$ $HO_2 + OH^-$	−0.105	pH = 14
$O_2^- + 2\,H_2O + e^-$	$\rightarrow$ $H_2O_2 + 2\,OH^-$	+0.695	pH = 14
$O_3 + 2\,H^+ + 2\,e^-$	$\rightarrow$ $O_2 + H_2O$	+2.075	pH = 0
$O_3 + H^+ + e^-$	$\rightarrow$ $O_2 + OH$	+1.77	acid
$O_3 + H_2O + 2\,e^-$	$\rightarrow$ $O_2 + 2\,OH^-$	+1.246	pH = 14
$O_3 + H_2O + e^-$	$\rightarrow$ $O_2 + OH + OH^-$	−0.943	neutral
$OH + H^+ + e^-$	$\rightarrow$ H_2O	+2.85	pH = 0
$OH + e^-$	$\rightarrow$ OH^-	+1.985	pH = 0
$OH + H^+ + e^-$	$\rightarrow$ H_2O	+2.30	pH = 0
$HO_2 + H^+ + e^-$	$\rightarrow$ H_2O_2	+1.495	pH = 0
$HO_2^- + H_2O + 2\,e^-$	$\rightarrow$ $3\,OH^-$	+0.867	pH = 14
$H_2O_2 + H^+ + e^-$	$\rightarrow$ $OH + H_2O$	−1.14	pH = 0
$H_2O_2 + 2\,H^+ + 2\,e^-$	$\rightarrow$ $2\,H_2O$	+1.763	pH = 0

For solutions in protic solvents (water is the most important one), the universal reference electrode for which the standard electrode potential is zero *by definition* under standard conditions (25 °C, 1 bar H_2) and at all temperatures, is the hydrogen electrode (H^+/H_2). In Table 3.12, the absolute electrode potentials for hydrogen are given, which can be interpreted in the following way: A redox couple more negative than 0.414 V should liberate hydrogen from water and a couple more negative than 0.828 V should liberate H_2 from $1\,mol\,L^{-1}\,OH^-$ solution.

3.3.5.2 Hydrated electron: a fundamental species

> **i** Besides the hydrogen ion (H^+) and hydroxide ion (OH^-), another fundamental species exist in aqueous solutions, the hydrated electron e_{aq}^-.

The hydrated electron e_{aq}^-, also written as H_2O^-, was first postulated by radiation chemists in 1952 and characterised in 1962 by recording its absorption spectrum (Hart and Anbar 1970). However, within the natural environment, there is no liquid water where radiation smaller than 195 nm exists to provide the reaction

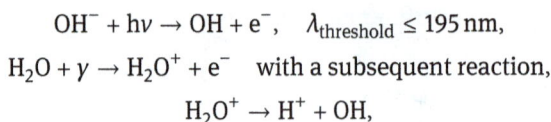

$$OH^- + h\nu \rightarrow OH + e^-, \quad \lambda_{threshold} \leq 195\,nm,$$
$$H_2O + \gamma \rightarrow H_2O^+ + e^- \quad \text{with a subsequent reaction,}$$
$$H_2O^+ \rightarrow H^+ + OH,$$

explaining the biological impact of such kinds of radiation (note that the OH radicals react with all biomolecules; as seen from Table 3.12, OH is the strongest oxidising agent).

The interaction of highly energetic photons or charged particles (generally speaking, the interaction with ionising radiation) with water results, in general, in the ejection of a quasi-free electron from the valence shell, leaving behind a positively charged water radical cation, $H_2O^{(\bullet)+}$, which then becomes stabilized as a cluster, $(H_2O)_n^{(\bullet)+}$. In pure water, the hot electrons generated after the ionization of the water relax into solvent molecules and become trapped as hydrated electron species, while H_2O^+ rapidly forms oxidizing OH radicals via proton transfer, which is fully accomplished in less than 1 ps.

The solvated electron,[42] especially the hydrated electron, have since been found to be an extremely important reactive species. When the solvent medium is water, the hydrated electron becomes essential to myriad physical, chemical and biological processes. In a simple picture of an electron in a cavity, the description of the hydrated electron state structure is analogous to that of a hydrogen atom, with a ground state of s-type and an excited state of p-type character. They absorb between 600 nm and

---

[42] Many papers refer to the first observation of a solvated electron in 1864 by Weyl (*Pogg. Ann.* 123, 350) who studied the dissolution of some alkali and alkaline earth metals in liquid ammonia and who observed the characteristic deep blue color of the solution (owing to the light absorption of solvated electrons) and formation of hydrogen, called 'hydrogen ammonia'. However, Kraus (*J. Amer. Chem. Soc.* 30, 1323–1344) first stated in 1908 the existence of "an electron, surrounded by an envelope of solvent molecules". Liquid ammonia dissociates similar to H_2O: $2\,NH_3 \rightleftharpoons NH_2^- + NH_4^+$. The reaction of potassium with liquid ammonia is thereby very similar to that of sodium with water (see below in text): $K + NH_3(l) \rightarrow KNH_3 \rightarrow KNH_2 + H$. Amides ($-NH_2$) are well-known compounds; today, we know that K reacts with $NH+_4$ via electron transfer, forming the ammonium radical NH_4, which decomposes into $NH_3 + H$.

800 nm and appear blue. However, the hydrated electron is far more complex because of the ultrafast dynamics of structural change, solvation and recombination. It is the simplest electron donor, and all of its reactions are, in essence, electron transfer reactions. What is known is that their presence enhances the reactivity of water molecules with other molecules in a number of important chemical, physical and biological processes. Hydrated electrons form when an excess of electrons is injected into liquid water. In fact, hydrated electrons and hydrogen atoms constitute a conjugated acid-base pair (see equation (3.165)), the former being the basic and more strongly reducing species.

Each student of chemistry knows the dangerous experiment when one places elemental sodium (or potassium) into liquid water:

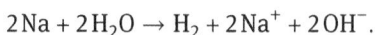

$$2\,Na + 2\,H_2O \rightarrow H_2 + 2\,Na^+ + 2\,OH^-.$$

In detail, however, this reaction becomes heterogeneous when Na encounters H^+ and H_2O. The elementary step of the above reaction is $Na(s) + H^+ \rightarrow Na^+ + H(g)$, an electron transfer from Na onto H^+; however, more exactly, first an electron transfer on H_2O occurs:[43]

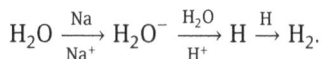

$$H_2O \xrightarrow[Na^+]{Na} H_2O^- \xrightarrow[H^+]{H_2O} H \xrightarrow{H} H_2.$$

Elemental sodium acts as a strong reducing agent (like a cathode) – it corresponds to the electrochemical H_2O reduction:

$$H_2O + 2\,e^- \rightarrow H_2 + 2\,OH^-.$$

In the following, we will terminate the hydrated electron with the symbol e_{aq}^- or H_2O^- unless it exists only in H_2O clusters where $n = 6$–50:

$$e^- + n\,H_2O \rightarrow (H_2O)_n^-. \tag{3.163}$$

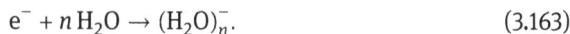

In the natural environment, there are only two possibilities to produce e_{aq}^-, either chemically or under solar light conditions via so-called photosensitisers (see Chapter 3.3.6.4). The only known direct chemical production (Hughes and Lobb 1976) is:

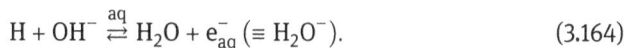

$$H + OH^- \underset{aq}{\rightleftharpoons} H_2O + e_{aq}^- \;(\equiv H_2O^-). \tag{3.164}$$

Therefore, in an anoxic medium and (for example, biogenic) *in situ* hydrogen production, this is an important pathway to initiate reduction chains by electron transfer

---

[43] Note the style of writing chemical reactions A + B: this is in line of the fate of B (= H_2O) where above the arrow the other reactant is written (Na, H^+ and H, respectively) and below the arrow the corresponding products (Na$^+$, H_2O). Compare with equation (3.119).

processes. The hydrogen atom and hydrated electron are interconvertible. Reaction equation (3.164) represents a conjugate acid-base pair; in other terms, H is a (very weak) acid in an equilibrium where $K = 2.3 \cdot 10^{-10}$ mol L^{-1} (p$K_a = 9.6$); $k_{3.164} = 2 \cdot 10^7$ L mol^{-1} s^{-1} and $k_{-3.164} = 16$ L mol^{-1} s^{-1}. The back reaction equation (3.164) is so slow that it can be neglected compared with other reactions of the hydrated electron. Reaction equation (3.165) shows that the electron is the ultimate Lewis base:

$$H \rightleftharpoons H^+ + e_{aq}^-. \tag{3.165}$$

The most studied and likely formation of e_{aq}^- under normal environmental conditions goes via photosensitisation by reversible electron transfer (see Chapter 3.3.6.4). Photoinduced electron transfer has been demonstrated in many molecules where the donor and acceptor are linked together intramolecularly. The efficiency of intramolecular electron transfer is strongly influenced by the separation distance between the donor and acceptor and the structure of the molecular link. The donor-acceptor groups can form collision complexes or exchange an electron at a long distance. In 'electron hopping' mechanisms, an electron proceeds from the donor to acceptor via a series of consecutive 'hops' to various acceptor groups. This mechanism plays a crucial role in the primary stages of photosynthesis, where consecutive electron transfer takes place.

The hydrated electron reacts rapidly with many species with a reduction potential more positive than −2.9 V. Tunnelling[44] between solvent traps can also explain the mobility of e_{aq}^- since it is much higher than expected for a singly charged ion with a radius of 0.3 nm. Once e_{aq}^- is produced, then the timescale for the formation of subsequent reactive species (H, HO$_2$, OH, H$_2$O$_2$) is in the order of 10^{-7} s. Table 3.13 lists some reactions of e_{aq}^-; many reactions are so fast that they are at the diffusion-controlled limit ($\sim 10^{10}$ L mol^{-1} s^{-1}). The lifetime of e_{aq}^- is in the order of 1 ms, but e_{aq}^- does not react with OH$^-$ and H$_2$. The back reaction of equation (3.165) leads to atomic hydrogen: $k_{3.166} = 2.26 \cdot 10^{10}$ L mol^{-1} s^{-1}:

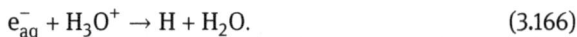

$$e_{aq}^- + H_3O^+ \rightarrow H + H_2O. \tag{3.166}$$

This reaction only proceeds in acid solution, whereas in neutral and alkaline solutions e_{aq}^- reacts with O$_2$ (for the fate of O$_2^-$ see equation (4.35)):

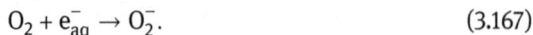

$$O_2 + e_{aq}^- \rightarrow O_2^-. \tag{3.167}$$

---

44 Tunneling or quantum tunnelling refers to the quantum mechanical phenomenon where a particle tunnels through a barrier (see Figure 3.6) that it classically could not surmount.

Table 3.13: Aqueous chemistry of the hydrated electron; X – halogen; R – organic rest.

reaction			k (in L mol^{-1} s^{-1})
$e_{aq}^- + O_2$	$\rightarrow$	O_2^-	$1.9 \cdot 10^{10}$
$e_{aq}^- + H_3O^+$	$\rightarrow$	$H + H_2O$	$2.3 \cdot 10^{10}$
$e_{aq}^- + e_{aq}^-$	$\rightarrow$	$H_2 + 2\,OH^-$	$0.54 \cdot 10^{10}$
$e_{aq}^- + OH$	$\rightarrow$	OH^-	$3.0 \cdot 10^{10}$
$e_{aq}^- + H + H_2O$	$\rightarrow$	$H_2 + OH^-$	$2.5 \cdot 10^{10}$
$e_{aq}^- + CO_2$	$\rightarrow$	CO_2^-	$7.7 \cdot 10^{9}$
$e_{aq}^- + NO_3^-$	$\rightarrow$	$NO_3^{2-} \xrightarrow{H_2O} (NO_2)_{aq} + 2\,OH^-$	$1.0 \cdot 10^{10}$
$e_{aq}^- + N_2O$	$\rightarrow$	$N_2 + O^-$	$0.87 \cdot 10^{10}$
$e_{aq}^- + RX$	$\rightarrow$	$RX^- \rightarrow R + X^-$	
$e_{aq}^- + Mn^{2+}$	$\rightarrow$	Mn^+	$3.8 \cdot 10^{7}$
$e_{aq}^- + Fe^{3+}$	$\rightarrow$	Fe^{2+}	$3.5 \cdot 10^{8}$
$e_{aq}^- + Cu^{2+}$	$\rightarrow$	Cu^+	$3 \cdot 10^{10}$
$e_{aq}^- + HCOOH^a$	$\rightarrow$	products	$1.4 \cdot 10^{8}$
$e_{aq}^- + CH_3COOH$	$\rightarrow$	products	$1.8 \cdot 10^{8}$

^{a} alcohols are unreactive

Interestingly, the above-discussed reaction of elemental sodium with water is now considered a source of hydrated electrons:

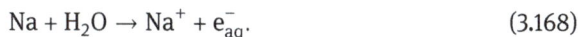

$$Na + H_2O \rightarrow Na^+ + e_{aq}^-. \tag{3.168}$$

Again, considering our principle of main pathways according to the overall rate, the fate of e_{aq}^- in natural waters and hydrometeors is the formation of superoxide/hydroperoxyl radicals (Figure 3.7), but whether the electron transfer goes directly onto O_2 or via H^+ plays no role because all reactions proceed very fast. The gross water splitting process is given by:

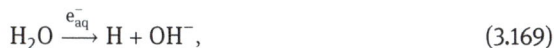

$$H_2O \xrightarrow{e_{aq}^-} H + OH^-, \tag{3.169}$$

where the hydrated electron is produced by reaction equation (3.183a). We see that this primary process of photosynthesis occurs in the plant cell as well as in abiotic environments. Only the oxygen content determines the further fate of hydrogen, either excess hydrogen for a reducing environment or the formation of O_xH_y in an oxic environment. In natural waters, electron scavengers reduce the lifetime of e_{aq}^- (see equations (4.68)–(4.71) for the fate of O^-):

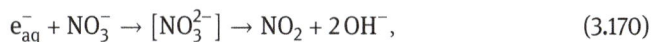

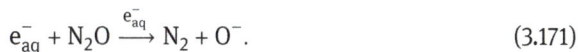

$$e_{aq}^- + NO_3^- \rightarrow [NO_3^{2-}] \rightarrow NO_2 + 2\,OH^-, \tag{3.170}$$

$$e_{aq}^- + N_2O \xrightarrow{e_{aq}^-} N_2 + O^-. \tag{3.171}$$

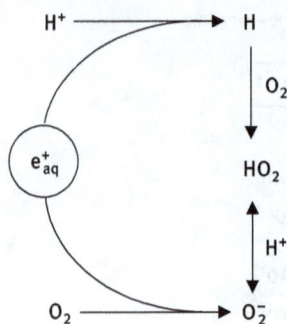

Figure 3.7: Simplified scheme of the hydrated electron fate in natural water (possible scavenging by ozone not shown, see Figure 5.22).

3.3.6 Photochemistry

> Photochemistry is the branch of chemistry concerned with chemical reactions caused by the absorption of light (from far UV to IR).

There are many excited states of a molecule that do not result in chemical conversions, such as photodissociation (see Figure 3.8). Photochemical paths offer the advantage over thermal methods of forming thermodynamically disfavoured products, overcoming large activation barriers in a short time and allowing reactivity otherwise inaccessible by the thermal method. In atmosphere and surface water, however, photodissociation (often also termed *photolysis*), which is the cleavage of one or more covalent bonds in a molecular entity resulting from absorption of light, is of importance, especially for producing radicals. Solar driven photochemistry is the only way in the environment to create reactive species such as radicals, which then initiate thermal chemical conversions.

> Without photochemistry (i. e., no radiation), our planet would be almost chemically inactive.

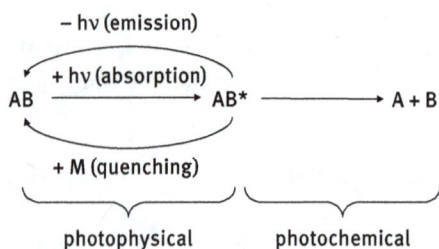

Figure 3.8: Scheme of the photophysical and photochemical process.

3.3.6.1 Solar radiation transfer to the earth's surface

Radiation describes any process in which energy emitted by a body travels through a medium or through space, ultimately to be absorbed by another body. Radiant energy is the energy of electromagnetic waves. Sunlight (solar radiation), in the broad sense, is the total spectrum of the electromagnetic waves given off by the Sun.

The irradiative surface of the Sun, or photosphere, has an average temperature of about 5,800 K. Most of the electromagnetic radiation emitted from the Sun's surface lies in the visible band centred at 500 nm, although the Sun also emits significant energy in the ultraviolet and infrared bands, and small amounts of energy in the radio, microwave, X-ray and gamma-ray bands. The total quantity of energy emitted from the Sun's surface is approximately $6.3 \cdot 10^7 \, \mathrm{W \, m^{-2}}$. The energy emitted by the Sun passes through space until planets, other celestial objects, or interstellar gas and dust intercept it. A physical law known as the inverse-square law determines the intensity of solar radiation striking these objects. This law merely states that the intensity of the radiation emitted from the Sun varies with the squared distance from the source. For example, the intensity of radiation from the Sun is $9140 \, \mathrm{W \, m^{-2}}$ at the distance of Mercury; but only $1370 \pm 5 \, \mathrm{W \, m^{-2}}$ at the distance of Earth – a threefold increase in distance results in a nine-fold decrease in intensity of radiation. This quantity is called the *solar constant* I_K. It is important to note that this quantity is related to a plane perpendicular to the radiation beam. Therefore, the Earth receives solar radiation only hemispherical ($I_K \pi r_{\mathrm{Earth}}^2$), and on a global average of $I_K/4$ ($343 \, \mathrm{W \, m^{-2}}$). The *actinic radiation* (and hence related quantities such as actinic flux) is the solar radiation that can initiate photochemical reactions. The term 'spectral' simply means that a quantity is measured per wavelength per interval (Figure 3.9).

Only about 30 % of the solar energy intercepted at the top of the earth's atmosphere passes directly through the surface. On the way through the atmosphere, direct solar radiation undergoes *scattering*, *absorption* and *reflection* on molecules and

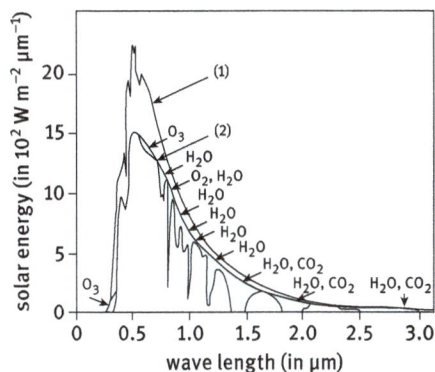

Figure 3.9: Solar spectrum at the top of the earth's atmosphere (1) and the earth's surface (2); absorbing gases are denoted.

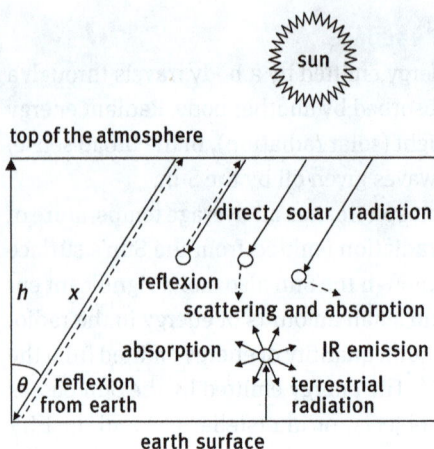

Figure 3.10: Scheme of solar radiation transfer through the Earth's atmosphere; θ is the solar zenith angle; h is the height of the atmosphere; x is the length of the radiation beam through the atmosphere.

suspended particles, such as dust particles and cloud droplets (Figure 3.10). The atmosphere reflects and scatters some of the received visible radiation. Gamma rays, X-rays, and ultraviolet radiation less than 200 nm in wavelength are selectively absorbed in the upper atmosphere by oxygen and nitrogen and turned into heat energy. Most of the solar ultraviolet radiation with a range of wavelengths from 200 to 300 nm is absorbed by ozone (O_3) and oxygen found in the stratosphere.[45] Infrared solar radiation with wavelengths greater than 700 nm is partially absorbed by carbon dioxide, ozone, and water present in the atmosphere in liquid and vapour forms (Figure 3.9). Thus, the troposphere penetrates only radiation with wavelength > 290–300 nm and initiates photochemical processes (see Chapter 3.3.6.3). The process of scattering is elastic and occurs when small particles and gas molecules diffuse part of the incoming solar radiation in random directions (Figure 3.10) without any alteration to the wavelength of the electromagnetic energy, i. e. no energy transformation results. Hence, scattering reduces the amount of incoming radiation reaching the Earth's surface. The sky is bright also from directions where there is no direct sunlight because of scattering; that light is called diffuse solar radiation (skylight). Scattering occurs on particles when their diameter is similar to or larger than a wavelength (Mie's scattering) and on molecules generally much smaller than the wavelength of the light (Rayleigh's scattering). The sky has a blue appearance in the daytime because the Rayleigh scattering is inversely

---

45 The stratosphere is the atmospheric layer above the troposphere. It is characterised by the ozone layer (maximum between 17 and 25 km) and an increase of T with altitude (due to radiation absorption); water vapour is extremely low. The stratosphere is limited in about 50 km by the stratopause (T maximum about 0 °C).

proportional to the fourth power of wavelength, which means that the shorter wavelength of blue light will scatter more than the longer wavelengths of green and red light. When the Sun is near the horizon, the light passes a longer distance through the atmosphere, and red light remains after the scattering out of blue light.

Absorption is defined as a process in which solar radiation is retained by a substance and converted into heat energy, or in other words, into *inner energy* (rotation, vibration, and translation) but also into dissociation (photolysis, see Chapter 3.3.6.3). The absorbed light will finally be emitted as long-wave radiation (dissipated heat). Reflection is the third process of altering direct solar radiation through the atmosphere, where sunlight is redirected by 180 after it strikes very large particles (cloud droplets, liquid and frozen) and the earth's surface.

Sunlight reaching the earth's surface unmodified by any of the above atmospheric processes is termed direct solar radiation. Roughly 30 % of the Sun's visible radiation (wavelengths from 400 nm to 700 nm) is reflected back to space by the atmosphere or the Earth's surface. The reflectivity of the Earth or any body is referred to as its *albedo*, defined as the ratio of light reflected the light received from a source, expressed as a number between zero (total absorption) and one (total reflectance).

Due to the interaction of solar radiation with molecules and particles of the atmosphere, the radiant flux decreases with the path x through the atmosphere (Figure 3.10). Johann Heinrich Lambert (1727–1777) showed in 1760 that the reduction of light intensity is proportional to the length of path x (or layer thickness) and the light (radiant flux) itself, $\Delta E = x \cdot E$, from which we derive the equation:

$$\frac{dE}{dx} = -m' \cdot E, \tag{3.172}$$

where m' is the *extinction module*. The minus sign denotes that the radiation decreases. Now, expressing the path x by the solar zenith angle θ, we obtain $x = h \cdot \sec\theta$ and finally

$$E = E_0 \exp(-m \cdot \sec\theta), \tag{3.173}$$

where $m = m' \cdot h$ and E_0 the irradiation. For $\theta > 60$, equation (3.173) must be corrected due to the curvature of the Earth. August Beer (1825–1863) found in 1848 that the extinction of light further depends on the concentration of substances within the irradiated medium, i. e., $m = \kappa \cdot c$, where κ is the extinction coefficient (fraction of light lost due to scattering and absorption per unit distance in a participating medium). The extinction coefficient, depending on wavelength, is further separated into coefficients for absorption and scattering for molecules as well as particles: $\kappa = \kappa_{abs}^{gas} + \kappa_{abs}^{particle} + \kappa_{scat}^{gas} + \kappa_{scat}^{particle}$. By combination of Lambert's and Beer's relationships, we obtain the Lambert–Beer law:

$$E = E_0 \exp(-\kappa \cdot c \cdot \sec\theta). \tag{3.174}$$

The amount of light absorbed by a substance, depending on the wavelength, can be calculated according to:

$$E_{abs}(\lambda) = E(\lambda)\sigma(\lambda) \cdot c. \tag{3.175}$$

The *absorption cross section* σ, depending on wavelength and temperature and specific for each substance, characterises the effective area of a molecule for scavenging of a photon. Equation (3.175) is used to calculate the photolysis rate (see Chapter 3.3.6.3).

Between 0.3 and 0.7 μm (visible range) and 8–12 μm, with the exception of ozone bands, there is virtually no absorption in the atmosphere – therefore, these ranges are called *atmospheric windows*; solar and terrestrial radiation can penetrate the atmosphere unopposed. Between 1 and 8 μm H_2O (2.5–3.5 μm and 4.5–7.5 μm) and CO_2 (2.2–3.5 μm and 4–4.5 μm as well as 15-20 μm) absorb terrestrial radiation partially, and at >15 μm nearly completely.

3.3.6.2 Photoexcitation: electronic states

In Chapter 3.3.1, we introduced the orbitals, synonymous with the term wave function, as a spatial residence of the electrons in the shell of atoms and molecules, which form the chemical bond. Between the *ground state*, which is the level of the lowest potential energy of a species, and the photodissociation, which is the broken chemical bond, i. e. from one molecule two chemical species forms, are many well-defined *excited states* of the molecule (but also the atom) according to quantum theory. Whereas in atoms, solely changes of the energy by different electronic transfers are possible, in molecules occur also different rotational and vibrational states (see Figure 3.11), visible in spectra as bands. Excitation of rotation and vibration takes place in infrared (absorption and emission). The description of the equivalent orbitals is made by quantum numbers. The *principal quantum number n* (=1, 2, 3, ± . . .) determines the energy of the electron (shell). The principal quantum number determines the shell, often called K, L, M, N, . . . (from inner to outer shells), equivalent to n = 1, 2, 3, 4, ... The *magnetic quantum number* m_l (=0, ±1, ±2, ±3, . . .) follows from the magnetic moment of the electron, rotating around the nucleus. Together with the *orbital quantum number*[46] l (=0, 1, 2, . . . , n – 1) the orbital angular momentum of an electron is determined. The orbital quantum number characterises the different types of orbitals: s, p, d, f, . . . for l = 0, 1, 2, 3, ... Hence, it follows 1s, 2p, . . . electrons (see Table 3.7). Finally, the *spin quantum number* s, which only has one figure (= ± 1/2, i. e. spin down or up), is associated with the self-rotation of electrons (spin). It follows as a fours (main) quantum number the *spin projection quantum number* m_s (=s, s – 1, s – 2, . . . , –s or in other terms, 1/2, –1/2, –2/3, . . . , –1/2), Table 3.14.

---

46 Also called azimuthal and secondary quantum number.

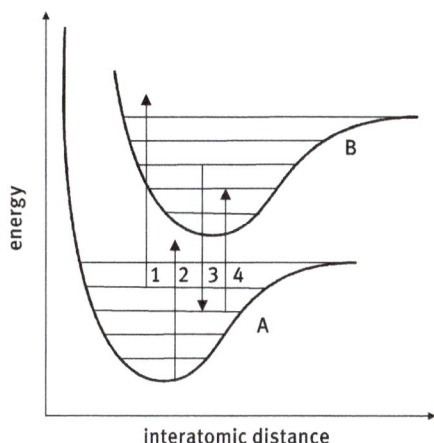

energy

interatomic distance

Figure 3.11: Scheme of electronic excitation of a diatomic molecule also called potential energy curve (anharmonic oscillator). A is a ground state, B is the first excited state, 1 direct dissociation from the exited ground state, 2 direct dissociation from the ground state, 3 radiation transfer from excited state, 4 excitation from ground state.

Table 3.14: Relationship between quantum numbers.

orbital	$l =$	values $m_l =$	number of values of m_l
s	0	0	1
p	1	$-1, 0, +1$	3
d	2	$-2, -1, 0, +1, +2$	5
f	3	$-3, -2, -1, 0, +1, +2, +3$	7
g	4	$-4, -3, -2, -1, 0, +1, +2, +3, +4$	9

The important Pauli principle says that each orbital never contains more than two electrons, and if there are two electrons in the same orbital, they have paired spin (symbol ↑↓). Before electrons occupy an orbital twice, the first different orbitals in subshells are occupied. The ground state is the configuration with the maximum unpaired electrons or spins (Hund's rule). Paired spins (↑↓) compensate the single spins and result in zero total spin, a configuration called a *singlet*. The angular momentums of two parallel spins (↑↑) add together to a total spin different from zero, a configuration called a *triplet*. Generally, the energetic level of triplet configuration is lower than that of a singlet. It is caused by the Coulomb rejection, which decreases between electrons with spin correlation.

In atoms and molecules, total quantum numbers result from individual electron quantum numbers. Here we only consider the total orbital quantum number L, which results from the single orbital quantum number l according to $L = l_1 + l_2, l_1 + l_2 - 1, \ldots, |l_1 - l_2|$ as integer numbers $(0, 1, 2, 3, \ldots)$ and is denoted by capital letters S, P, D,

F, …. The total spin S is for two electrons $S = 1, 0$. For three electrons (because each electron has the spin $s = 1/2$), it follows for all states (also called *terms*) 3/2, 1/2, 1/2. The *multiplicity* M of a term is given by the value $2S + 1$, i. e. the number of possible energetic levels regarding total spin. For a closed shell, $S = 0$ is valid, and it follows $M = 1$ (*singlet*). For a single electron, $S = s = 1/2$ is valid, and it follows $M = 2$ (*doublet*). For two unpaired electrons, $S = 1$ is valid, and it follows $M = 3$ (*triplet*). We will meet such terms for two different oxygen atoms (singlet and triplet in Chapter 4.3.2.1). In the symbols ^3P and ^1D, the left upper index denotes the multiplicity (here triplet or singlet), and the capital letter denotes the total orbital quantum number (P for $L = 1$ and D for $L = 2$).

Similar molecules are described where instead of Latin letters (S, P, D, F), Greek letters (Σ, Π, Δ, Φ) are used. The general term symbol of a molecule is

$$^{2S+1}\Lambda_{g/u}^{+/-},$$

where the quantum number $\Lambda = 1, 2, 3, 4\ldots$ or Σ, Π, Δ, Φ, respectively. The left upper index denotes the total spin quantum number (from $S = 1$ singlet and $S = 1$ triplet follows), the right upper index has the symbol + or − and denotes the reflection symmetry along an arbitrary plane along the internuclear axis (− means change and + means retention) and the right lower index (symbol g or u)[47] denotes the parity. The symbol for dioxygen O_2 in its ground state is $^3\Sigma_g^-$. However, normally, diatomic molecules are situated as a singlet.

Diatomic homoatomic molecules (such as O_2) show no vibration terms (i. e. they do not absorb and emit in IR); electromagnetic radiation only can interact with an oscillating dipole, but homoatomic molecules do not change their dipoles. Depending on the radiation wavelength, molecules show different excited states (vibrational-rotational) from which a lower energetic and the ground state is gained either by light emission (*phosphorescence* and *fluorescence*) or collision with another molecule (*quenching*), where energy is transferred to the collision partner. Because oxygen is (besides nitrogen) the most abundant collision partner in air, O_2 can transfer from the (unusual) triplet ground state into the energetic higher singlet state $O_2(^1\Sigma_g^+)$ and $O_2(^1\Delta_g)$, see Chapter 4.3.2.1.

3.3.6.3 Photodissociation: photolysis rate coefficient

Figure 3.7 shows a photochemical process schematically. It is beyond the scope of this book to go into detail about photoexcitation. According to the quantum structure of the electronic molecular system, such photons are absorbed corresponding to existing bands of rotational and vibrational states (Figure 3.11). The excited molecule AB^* can

---

47 From German g = *gerade* (even) and u = *ungerade* (uneven).

turn back to the ground state through light emission (fluorescence and/or phosphorescence) and also via collision with any molecule, called quenching. Only when the absorbed light energy (corresponding to a photolysis threshold wavelength) is large enough to overcome the intermolecular distance, the excited state can turn into breakdown products A + B. The photolysis is represented by

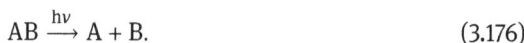

$$AB \xrightarrow{h\nu} A + B. \tag{3.176}$$

The term $h\nu$ symbolises the energy of a photon (h is Planck's quantum, and ν is frequency; remember that $c = \nu \cdot \lambda$, where λ is the wavelength, and c is the speed of light). A first-order law describes the rate of process equation (3.176):

$$\frac{d[AB]}{dt} = -j_{AB}[AB], \tag{3.177}$$

where j is the *photolysis rate coefficient*, often also called the *photolysis rate constant* (sometimes termed *photolysis frequency* – which physically is the most correct term), but in contrast to the reaction rate constant k (which only depends on T), j depends on many parameters and is not constant but changes permanently over time. The dimension of j is reciprocal time, called photolytic residence time $\tau = j^{-1}$. For a daily mean estimate of the photochemical conversion rate according to equation (3.177), it is useful to determine an average rate coefficient as:

$$\bar{j} = \frac{1}{t_{sunset} - t_{sunrise}} \int_{t_{sunrise}}^{t_{sunset}} j(\Delta t)dt, \tag{3.178}$$

where Δt is the time interval of the integration step. The photolysis rate coefficient j is calculated by integrating the product of the *spectral actinic flux* $S(\lambda)$, *spectral absorption cross section* $\sigma(\lambda)$ and the photodissociation *quantum yield* $\Phi(\lambda)$ over all relevant wavelengths (Madronich 1987):

$$j(\lambda, T) = \int_{\lambda_{min}}^{\lambda_{max}} S_\lambda(\lambda)\sigma(\lambda, T)\Phi(\lambda, T)\lambda. \tag{3.179}$$

Because of the quantum characteristics of light absorption in consideration of photochemistry, we use here *spectral* quantities, i. e. per unit of wavelength intervals (in nm normally given despite the SI recommendation of m).

> The actinic flux S_λ, in terms of photons per time, called in traditional terms light intensity, photon flux, irradiance or radiant flux and simply radiation, which has caused some confusion, is the quantity of light available to molecules at a particular point in the atmosphere and which, on absorption, drives photochemical processes in the atmosphere. Actinic flux describes the number of photons (or radiation) incident on a spherical surface, such as the molecule of the atmospheric

species, and is the suitable radiation quantity for photolysis rate coefficient determination. The actinic flux and the irradiance are split into diffuse and direct part. Actinic flux measurements are not trivial. They require spectroradiometers with specially configured optics to enable measurements of radiation equally weighted from all directions. The actinic flux does not refer to any specific orientation because molecules are oriented randomly in the atmosphere. This distinction is of practical relevance: the actinic flux (and thereby a j-value) nears a brightly reflecting surface (e. g., over snow or above a thick cloud) can be a factor of three times higher than that near a nonreflecting surface. Hence, the presence of clouds can drastically change the actinic flux throughout the atmosphere.

The *absorption cross section* $\sigma(\lambda)$ is a measure of the area (dimension: cm^2 $molecule^{-1}$) of the molecule (given by the electron density function), through which the photon cannot pass without absorbing when its energy is equivalent to a molecular quantum term (otherwise it will be reflected) in contrast to the definition of a collisional cross section. Furthermore, the dimensionless *quantum yield* $\Phi(\lambda)$ is the ratio between the number of excited (or dissociated) molecules and the number of absorbed photons (dimensionless with values 0...1). This quantity depends on the wavelength and approaches one at the so-called *threshold wavelength*. The quantum yield is numerically dimensionless but formally denotes molecules per photon.

The expression equation (3.179) for the j-value can be transformed (through integration) into a more applicable form by using the mean values of σ and Φ for given wavelength intervals $\Delta\lambda$ (as tabulated in standard books and sources, for example, Finlayson-Pitts and Pitts 2000):

$$j = \sum_{\lambda=290\,nm}^{\lambda_i} \overline{S_\lambda}(\lambda)\overline{\sigma}(\lambda)\overline{\Phi}(\lambda). \tag{3.180}$$

The values for $\overline{\sigma}(\lambda)$ and $\overline{\Phi}(\lambda)$ have been determined from laboratory investigations. For the actinic flux with respect to the solar zenith angle θ, height z and wavelength λ either measurements and/or calculations are needed. All radiative transfer modelling is ultimately based on the fundamental equation of radiative transfer, or the transfer of energy as photons. Analytic solutions to the radiative transfer equation (RTE) exist for simple cases; however, for more realistic media with complex multiple scattering effects, numerical methods are required. Hough (1988) gives another useful expression for j in parameterised form:

$$j = a_i \exp(b_i \cdot \sec\theta), \tag{3.181}$$

where a_i and b_i are substance-specific constants (Table 3.15). It is clear that equation (3.181) reflects the Lambert–Beer law (equation (3.174)) and is analogue to the Arrhenius equation. Figure 3.12 shows two examples of photolysis frequency for two days with different cloudiness. As mentioned, clouds scatter solar radiation and can reduce as well as enhance photodissociation.

3.3 Theory of chemical reactions

Table 3.15: Some important photolysis reactions in the air (after Hough 1988); residence time $\tau = 1/j$.

no.	reactant	products	photolysis rate (in s^{-1})	τ (in h)[a]
(1)	O_3	$O(^1D)$	$2 \cdot 10^{-4} \exp(1.4 \sec\theta)$	0.3
(2)	NO_2	$O(^3P) + NO$	$1.45 \cdot 10^{-2} \exp(0.4 \sec\theta)$	0.01
(3)	HNO_2	$OH + NO$	$0.205 \cdot j_2 3 \cdot 10^{-3} \exp(0.4 \sec\theta)$	0.06
(4)	HNO_3	$OH + NO_2$	$3 \cdot 10^{-6} \exp(1.25 \sec\theta)$	38
(5)	NO_3	different	$3.29 \cdot j_2$	0.003
(6)	HCHO	$H + HCO$	$6.65 \cdot 10^{-5} \exp(0.6 \sec\theta)$	2.1
(7)	HCHO	$CO + H_2$	$1.35 \cdot 10^{-5} \exp(0.94 \sec\theta)$	6.9
(8)	CH_3CHO	$H + CH_3CO$	$= j_6$	2.1
(9)	$ClONO_2$	$ClO + NO_2$	$2.9 \cdot 10^{-5}$	9.4
(10)	Cl_2	$2Cl$	$= j_1$	0.3

[a] calculated for $\theta = 30°$ (about maximum value in the Central Europe)

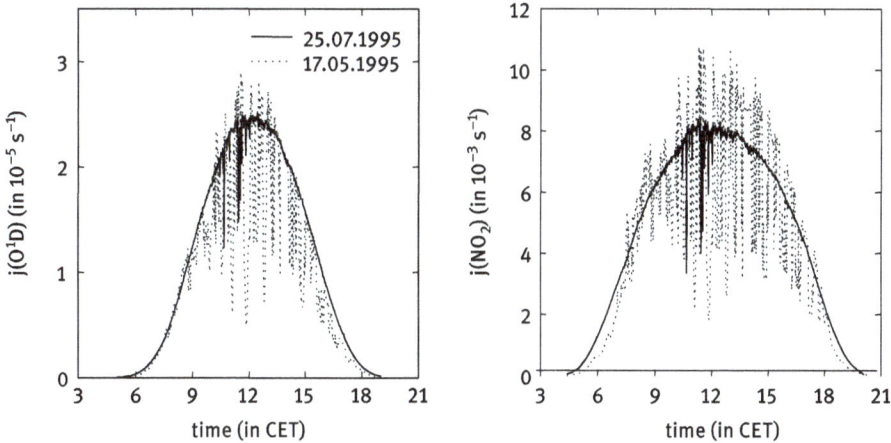

Figure 3.12: Diurnal variation of j-O (^1D) and j-NO_2 at two different days in summer 1995, measured at the airport Munich (Germany). July 25 represents a cloud-free day with few cumuli at noon, whereas May 17 was cloudy between 4/8 and 7/8; after Reuder (1999).

3.3.6.4 Photocatalysis: photosensitising and autoxidation

Photocatalysis is the acceleration of a photoreaction in the presence of a catalyst, which absorbs light, producing reactive species going in subsequent reactions with reactants. It is incorrect to call the direct photolysis (for example of ozone) with subsequent formation of identical reactive species as a photocatalytic process.

The generic reaction scheme is given here, where W represents a chromophoric substance (the *photocatalyst*) and A an electron acceptor:

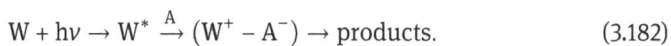

$$W + h\nu \rightarrow W^* \xrightarrow{A} (W^+ - A^-) \rightarrow products. \tag{3.182}$$

There are several pathways in product formation:

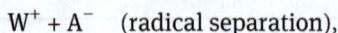

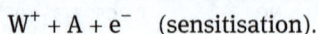

$$W^+ + A^- \quad \text{(radical separation)},$$
$$W + A \quad \text{(ground state reversible reaction)},$$
$$WA \quad \text{(coupling reaction)},$$
$$W^+ + A + e^- \quad \text{(sensitisation)}.$$

It has been shown that in all natural waters, chromophoric substances exist (e. g. chlorophyll) that can produce hydrated electrons after illumination and, consequently, reduce dissolved O_2:

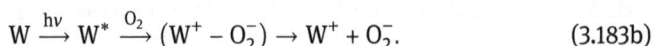

$$W \xrightarrow{h\nu} W^+ + e^- \xrightarrow{aq} e^-_{aq}, \tag{3.183a}$$

$$W \xrightarrow{h\nu} W^* \xrightarrow{O_2} (W^+ - O_2^-) \rightarrow W^+ + O_2^-. \tag{3.183b}$$

Many oxides (and partly sulphides) of transition state metals (Wo, No, Ir, Ti, Mn, V, Ni, Co, Fe, Zn and others) have been characterised as inorganic semiconductors to be able to support photosensitising, see Table 3.16. The generic reaction $M + h\nu \rightarrow M^+ + e^-$ has been known for more than 100 years. Their oxides (for example, TiO_2, ZnO, Fe_2O_3) produce hole-electron pairs when absorbing photons with energy equal to or greater than the band gap energy E_b of the semiconductor (Figure 3.13)

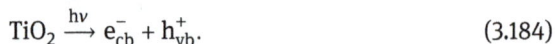

$$TiO_2 \xrightarrow{h\nu} e^-_{cb} + h^+_{vb}. \tag{3.184}$$

In the absence of suitable electron and hole scavengers adsorbed to the surface of a semiconductor particle, recombination occurs within 1 ns:

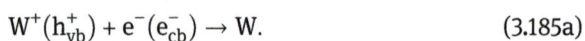

$$W^+(h^+_{vb}) + e^-(e^-_{cb}) \rightarrow W. \tag{3.185a}$$

Table 3.16: Electrochemical series of metals (standard electrode potential in V).

metal	potential	reduced species
Mg	−2.372	Mg^{2+}
Al	−1.662	Al^{3+}
Ti	−1.63	Ti^{2+}
Mn	−1.185	Mn^{2+}
Cr	−0.913	Cr^{2+}
Zn	−0.762	Zn^{2+}
Fe	−0.447	Fe^{2+}
Cd	−0.403	Cd^{2+}
Co	−0.280	Co^{2+}
Ni	−0.257	Ni^{2+}
Pb	−0.126	Pb^{2+}

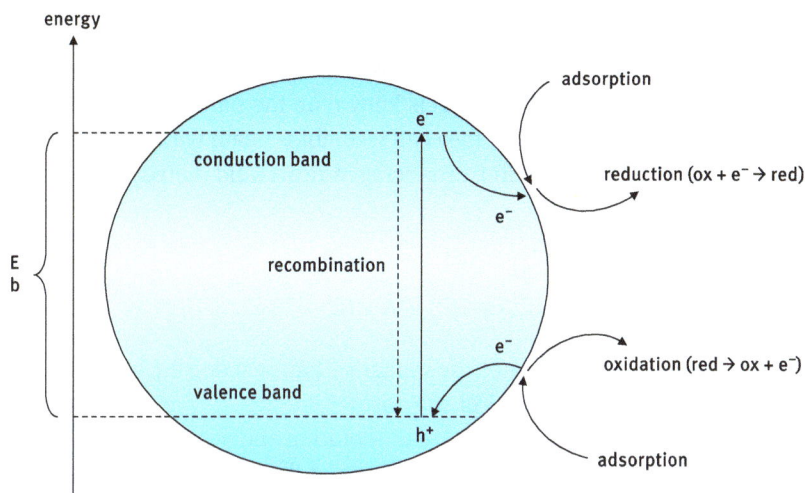

Figure 3.13: Schema of photosensitising of a semiconductor; E_b energy of the band gap.

Many organic substances have been found to act as photosensitisers such as phenones (mixed aliphatic-aromatic ketones), porphyrins (organic dyes), naphthalenes and pyrenes (polycyclic aromatic hydrocarbons), and cyano compounds as well as inorganic species such as transition metal complexes and many others. In the presence of sunlight, they produce hydrated electrons e_{aq}^- or directly O_2^- according to equation (3.183a) – the key role of natural water, mainly at the interface with air, is the formation of ROS (reactive oxygen species), thereby providing oxidation processes, including corrosion and autoxidation. This looks at first confusing because e_{aq}^- works as a reducing species but the key oxidising species in solution is the OH radical (a strong electron acceptor similar to the atmospheric gas phase), which is produced in a chain of electron transfer processes (for more details, see Chapter 4.3.3):

$$O_2 \xrightarrow{e^-} O_2^- \xrightarrow{H^+} HO_2 \xrightarrow{e^-} HO_2^- \xrightarrow{H^+} H_2O_2 \xrightarrow{e^-} OH + OH^-.$$

It is well known that the formation of singlet dioxygen in biochemistry (which can afterwards add on olefins) generates:

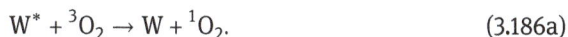

$$W^* + {}^3O_2 \rightarrow W + {}^1O_2. \tag{3.186a}$$

As a transient step, this reaction can first proceed with the following reaction sequence:

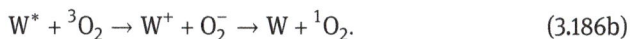

$$W^* + {}^3O_2 \rightarrow W^+ + O_2^- \rightarrow W + {}^1O_2. \tag{3.186b}$$

However, when appropriate scavengers are present, the valence band holes h_{vb}^+ function as powerful oxidants, whereas the conduction band electrons e_{cb}^- function as moderately powerful reductants. In an aqueous solution, e_{cb}^- can transfer to H_2O, gaining the hydrated electron H_2O^- or e_{aq}^-, which moves along the water structure until scavenging by electron receptors. Much less is known on the nature and fate of W^+ and h_{vb}^+, respectively. We speculate that they can recombine again with free electrons or react with main anions in the solution, producing radicals (and thereby amplifying the oxidation potential:

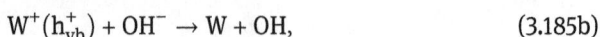

$$W^+(h_{vb}^+) + OH^- \rightarrow W + OH, \tag{3.185b}$$

$$W^+(h_{vb}^+) + Cl^- \rightarrow W + Cl. \tag{3.185c}$$

Nonetheless, it is clear that the rate of the oxidative half-reaction involving h_{vb}^+ is closely related to the effective removal of the partner species, i. e., e_{cb}^-, by suitable electron scavengers, which must be present in the solution and available at the semiconductor interface. By contrast, in processes aiming at the oxidative destruction of a target organic substrate, any hole scavengers present in the same environment are potential competitors for the consumption of h_{vb}^+. As far as electron scavenging is concerned, the most common electron scavenger, in aerated aqueous solution, is O_2 and its role has already been emphasised (equation (3.183a)).

More than 100 years ago, the rusting of iron was studied, and the 'hydrogen peroxide theory' was published (Dunstan et al. 1905):

$$Fe + H_2O \rightarrow FeO + H_2 \quad \text{and} \quad 2FeO + H_2O_2 \rightarrow Fe_2O_2(OH)_2 \quad \text{(rust).}$$

It was shown, however, that iron exposed to water and oxygen, with the exclusion of carbon dioxide, underwent rusting. It was concluded that pure oxygen and liquid water alone are essential to the corrosion of iron, the presence of acid is unnecessary. The detection of hydrogen peroxide during the corrosion of many metals gave rise to the idea that it acted as an intermediary in corrosion processes. Traube (1882) proposed the H_2O_2 formation in autoxidation processes despite nothing being known about formation mechanisms. First, Schönbein (1861) showed that H_2O_2 is formed during the slow oxidation of metals in atmospheric air. In modern terms, we write this sequence as:

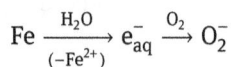

$$Fe \xrightarrow[(-Fe^{2+})]{H_2O} e_{aq}^- \xrightarrow{O_2} O_2^-$$

or (Me – metal):

$$Me \xrightarrow{O_2} [Me^+-O-O^-] \xrightarrow[(-OH^-)]{H_2O} [Me^+-O-O-H]$$

$$\rightarrow [Me^{2+} \cdots {}^-O-O-H] \xrightarrow[(-OH^-)]{H_2O} Me^{2+} + H_2O_2.$$

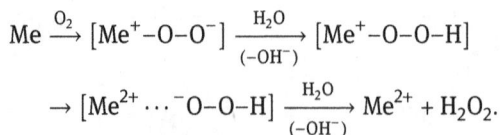

Once oxides of iron are formed (they are semiconductors), the process accelerates, similar to equation (3.184) due to photosensitising.

Autoxidation, defined as the very slow oxidation process where dioxygen adds on some bodies, can be written as (B – chemical body or substrate, A – other compounds, reactive to oxygen, e. g. NO)

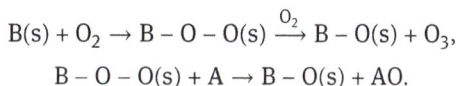

$$B(s) + O_2 \rightarrow B - O - O(s) \xrightarrow{O_2} B - O(s) + O_3,$$
$$B - O - O(s) + A \rightarrow B - O(s) + AO.$$

In nature, it is known that freshly mined coal, being exposed to the air can be slowly oxidised where the reaction heat can lead to self-ignition. Corrosion of metals is termed autoxidation, but this can also apply to any material (wood, biomass, textiles, plastics etc.) aged in the air where solar light enhances the process. However, the role of superoxide anions in autoxidation processes only became clear after the 1950s, interestingly because of studying bleaching processes. The bleaching and germicidal properties of H_2O_2 had been found very early and opened the door to many industrial, wastewater treatment and medical applications.

3.3.7 Heterogeneous chemistry

Heterogeneous chemistry denotes processes ongoing in a multiphase system, where the reactions proceed at the surface of the more condensed phase or interface, respectively.

Hence, it is also called *surface chemistry* and (more fashionable) *interfacial chemistry*. It occurs in soils, waters and air where the following interfaces are available: solid–gas, solid–aqueous, and aqueous–gas.

Surface reactions are always enhanced, namely due to the increased concentration of reactants after adsorption at the surface compared to the bulk phase (remember that the reaction rate increases with concentration according to equation (3.120). Moreover, the surface often provides catalytical effects. All gas-phase reactions proceed faster at low temperatures on surfaces; only the increase of T (that is very limited under natural environmental conditions) finally increases the total rate. On water surfaces, specific arrangements of molecules are possible to favour conversions. To carry out a heterogeneous reaction, several partial steps occur:

- turbulent transport of molecules close to the surface (to the diffusion layer interface),
- molecular diffusion to the surface (interface),
- adsorption onto the surface,
- chemical surface reaction,
- desorption into the gas (or aqueous phase, respectively), or
- diffusion and mixing within the other phase (solid or aqueous, respectively), and
- chemical reaction in the other phase (if possible).

As mentioned for subsequent reactions, the slowest partial process determines the overall kinetics. Often the process is mass transport limited, i. e. adsorption and surface chemistry are faster than the transport to the interface (*mass accommodation*).

Of crucial importance are reactions in air at surfaces of droplets (cloud, fog), surface water and particulate matter. Atmospheric heterogeneous chemical transformation depends on the available particle surface. But also, with respect to the very low surface to volume ratio in air, the enhancement of reactions and even specific reactions not occurring in the gas phase makes heterogeneous chemistry an important conversion pathway (for details, see Möller 2019).

3.3.8 Radicals, groups and nomenclature

In nature, where the modification of reaction conditions is limited and not under the control of the chemical system itself, reactive *radicals* play a crucial role (often referred to as free radicals, but the word 'free' in front of 'radical' is obsolete and should not be used).

Justus von Liebig and Friedrich Wöhler founded the theory of radicals in 1832. At that time, radicals have been regarded as the 'elements' of organic chemistry, which in today's understanding rather correspond to characteristic groups. Today we define a radical as an atom or group of atoms with one or more unpaired electrons. A radical may have a positive, negative or zero charge. The unpaired electrons cause them to be highly chemically reactive. Although radicals are generally short-lived because of their reactivity, long-lived radicals exist. The prime example of a stable radical is molecular dioxygen O_2 in the triplet state. Oxygen is also the most common molecule in a diradical state. Multiple radical centres can exist in a molecule. Other common atmospheric substances of low-reactive radicals are nitrogen monoxide NO and nitrogen dioxide NO_2 (Table 3.17).

Metals and their ions or complexes often possess unpaired electrons, but, by convention, they are not considered to be radicals. According to IUPAC recommendations (Koppenol 2000), a radical is indicated in a formula by a superscript dot, which precedes any charge, e. g., $O_2^{\cdot-}$ (not $O_2^{-\cdot}$). The dot is placed as an upper right superscript to the chemical symbol so as not to interfere with indications of mass number, atomic number, or composition. In the case of diradicals, the superscript dot is preceded by the appropriate superscript number, e. g., $O_2^{2\cdot}$. To avoid confusion, the number and the radical dot can be placed within parentheses, as in $NO_2^{(2\cdot)-}$. In these formulas, the dot symbolises the unpaired electron of the atom; however, we will use it only in this chapter so as not to make symbols more complicated. In polyatomic radicals, the central atom is the atom that binds other atoms or groups (called *ligands*) to itself, thereby occupying a central position in the radical. The most important central atoms in environmental chemistry (hydrogen is always regarded as a ligand) are: C, N, S, I,

Table 3.17: Some important radicals in environmental chemical processes.

formula	systematic name	trivial name
$H^{\bullet}$	hydrogen, monohydrogen(•)	(atomic) hydrogen
$O^{\bullet}$	oxygen, monooxygen(2•)(triplett)	(atomic) oxygen[a]
$O^{\bullet-}$	oxide(• –)	–
$O_2^{\bullet-}$	dioxide(•1–)	superoxide anion
$O_3^{\bullet-}$	trioxide(•1–)	ozonide anion
$HO^{\bullet}$	hydridooxygen(•)	hydroxyl radical
$HO_2^{\bullet}$	hydridodioxygen(•), hydrogen dioxide	hydroperoxy radical
$HO_3^{\bullet}$	hydridotrioxygen(•), hydrogen trioxide	trioxidane
$RO^{\bullet}$	alkoxyl, e. g. CH_3O = methoxyl radical[b]	(alkoxy is obsolete)
$RO_2^{\bullet}$ ($ROO^{\bullet}$)	alkyldioxyl	alkyl peroxyl radical
$HCO^{\bullet}$	hydridooxidocarbon(•)	oxomethyl, formyl radical
$^{\bullet}CH_3$	methyl	methyl radical
$CN^{\bullet}$	nitridocarbon(•)	cyanyl radical
$SCN^{\bullet}$	nitridosulphidocarbon(•)	thiocyanate radical
$CS^{\bullet}$	thiocarbonyl	–
$HS^{\bullet}$	hydridosulphur(•)	sylfanyl
$SO_3^{\bullet-}$	trioxidosulphate (•1–)	sulphite radical
$SO_4^{\bullet-}$	tetraoxidosulphate (•1–)	sulphate radical
$CO_2^{\bullet-}$	dioxidocarbonate(•1–)	carbon dioxide anion radical
$CO_3^{\bullet-}$	trioxidocarbonate(•1–)	carbonate radical
HNO	oxidanimine, azanone	nitroxyl
$HN^{2\bullet}$	hydridonitrogen(2•)(triplett)	azanediyl
$NH_2^{\bullet}$	dihydridonitrogen(•)	azanyl
$PO^{\bullet}$	oxidophosphorus(•)	phosphorus monoxide
$ClO^{\bullet}$	oxidochlorine(•)	chlorine monoxide
$NO^{\bullet}$	oxidonitrogen(•)	nitrogen monoxide[c]
$NO_2^{\bullet}$	dioxidonitrogen(•)	nitrogen dioxide
$NO_3^{\bullet}$	trioxidonitrogen(•)	nitrogen trioxide, nitrate radical
$Cl_2^{\bullet-}$	dichloride(•1–)	–
$ClO^{\bullet}$	oxidochlorine(•)	chlorine monoxide
$ClO_2^{\bullet}$	dioxidochlorine(•)	chlorine dioxide

[a] the systematic name for molecular oxygen (O_2) is dioxygen and for ozone (O_3) trioxygen
[b] the letter 'R' is a form of abbreviation for the Remainder or Rest of the organic molecule, e. g. CH_3 (methyl)
[c] nitric oxide is obsolete
[d] almost hydroperoxyl radical named (although according to the IUPAC nomenclature for inorganic peroxides, it should be named hydroperoxo radical)
[e] this very important radical is also written as $^{\bullet}SH$ and further named mercapto radical and hydrosulphide radical

Br, Cl, O, F. Examples of ligands: –OH, –CH_3, –CN, –CO, –CS, –NH, –NH_2, –SH, –O_2, –OOH (see Table 3.17 and 3.18). If the unpaired electron occupies an orbital with considerable s or more or less pure p character, the respective radicals are termed σ- or π-radicals.

Table 3.18: Names of some groups as ligands, as the prefix for substituents in organic compounds and as ions.

formula	as ligand	as prefix	as ion formula	name
$-H$	hydrido	–	H^+	hydrogen
			H^-	hydride
$-OH$	hydroxo	hydroxy	OH^-	hydroxide
$-OOH$	hydridodioxido	hydridodioxido	–	–
$-O-$	–	oxy	O^{2-}	oxide
$-O^-$	–	oxido		
$=O$	oxo	oxo		
$-O-O^-$	dioxygen	–	O_2^-	hyperoxide
$-O-O-$	peroxy	dioxy	O_2^{2-}	peroxide
$-O-O-O-$	–	trioxy	O_3^-	ozonide
$-S-$	thio, sulphido	thio	S^{2-}	sulphide
$-S^-$	–	sulphido		
$=S$	–	thioxo		
$-S-H$	mercapto[f]	mercapto	HS^-	hydrogen sulphide
$-S-S-$	disulphido	dithio, sulphinyl	S_2^{2-}	disulphide
$=NH$	imido[d]	imido	NH^{2-}	imide
$-NH_2$	amido[e]	amino	NH_2^-	amide
$-N=O$	nitrosyl	nitroso	–	
$-NO_2$	nitro	nitro	NO_2^-	nitrite
$>C=O$	carbonyl[a]	carbonyl	–	
$>C=S$	thiocarbonyl[b]	thiocarbonyl	–	
$-COOH$	carboxyl	carboxo	–	
H_3C-S-	methylthio	methylthio	CH_3S^-	methanethiolate
$-C\equiv N$	cyano[c]	cyano	CN^-	cyanide
$-O-C\equiv N$	cyanato	cyanate	–	
$-C\equiv N-O$	fulminato	–	CNO^-	fulminate
$-N=C=O$	–	isocyanate	–	
$-S-C\equiv N$	thiocyanato	thiocyanato	SCN^-	thiocyanate[g]
$-CH_3$	methanido	methyl	–	

[a] new systematic name: oxidocarbonato
[b] new systematic name: sulphidocarbonato
[c] new systematic name: nitridocarbonato
[d] new systematic name: hydridonitrodo
[e] new systematic name: dihydridonitrido
[f] new systematic name: hydridosulphido
[g] old: rhodanide

Radicals exist in the gas and liquid (aqueous) phase. They play an important role in many chemical transformations. In natural waters, radical ions, a radical that carries an electric charge, exist. Those positively charged are called radical cations and those negatively charged radical anions, but the most important is the superoxide anion O_2^-.

Radicals are produced by a) electron transfer, b) thermolysis and c) photolysis or radiolysis (A and B represent atoms and molecular entities as well):

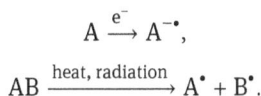

$$A \xrightarrow{e^-} A^{-\bullet},$$

$$AB \xrightarrow{\text{heat, radiation}} A^\bullet + B^\bullet.$$

There are presently three fields where a more detailed knowledge of the thermo-dynamic and kinetic properties of radicals would be extremely useful. The first is biomedicine. The discovery of superoxide dismutase and nitrogen monoxide as messengers has led to an explosive growth in articles in which one-electron oxidations and reductions have been explored. Organic radicals play an important role in the treatment of cancers. The other is the environmental (atmospheric and aquatic, including wastewater treatment) chemistry, where the modelling of reactions requires accurate reduction potentials.

Radicals play a key role in chain reactions, in which one or more reactive reaction intermediates (frequently radicals) are continuously regenerated, usually through a repetitive cycle of elementary steps (the 'propagation step'). The propagating reaction is an elementary step in a chain reaction in which one chain carrier is converted into another. The chain carriers are mostly radicals. Termination occurs when the radical carrier reacts otherwise. An example of one of the possible ozone destructions is shown below (R–Cl – chloro-organic compound):

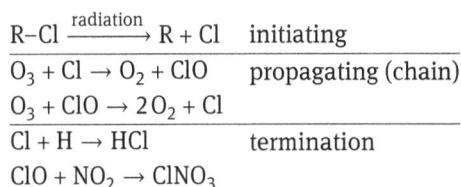

$R–Cl \xrightarrow{\text{radiation}} R + Cl$	initiating
$O_3 + Cl \rightarrow O_2 + ClO$	propagating (chain)
$O_3 + ClO \rightarrow 2O_2 + Cl$	
$Cl + H \rightarrow HCl$	termination
$ClO + NO_2 \rightarrow ClNO_3$	

Without termination, the gross propagating step results in $2O_3 \rightarrow 3O_2$ and can very often be cycled depending on parallel reactions. In the example above, the products of termination (HCl, $ClNO_3$) can act as source molecules and provide Cl radicals through photodissociation.

Many molecular entities (groups) exist as radicals, groups in organic compounds and ions, for example, HS• (sylfanyl), –SH (mercapto) and HS⁻ (hydrogensulphide). Besides, in practice often used trivial names, the IUPAC developed systematic names to derive the formula. However, trivial names are widespread in use. Atoms or groups of atoms other than the central atom are named anionic ligands (Table 3.18).[48]

---

[48] The IUPAC nomenclature is rather complicated but systematic; the most important feature of nomenclature is that given a complete name, a single unique structure can be drawn. However, most chemists use trivial names and do not know always the systematic IUPAC name.

4 Chemistry of elements and their compounds in the environment

As mentioned at the beginning of this book, we will treat the chemistry *for* application to the *environment* according to the elements and its compounds depending on the conditions of reactions and whether the substance exists in the gaseous or condensed (aqueous and solid) phase. Normally, in standard chemical textbooks, this Chapter would outline elements and their compounds according to the groups of the periodic table of the elements. Here we do it according to its abundance *and* importance in the environment (Tables 4.1 and 4.2). There is no need (as in standard textbooks) to characterise the properties and use of the pure substance and its formation under laboratory/industrial conditions. Short information is given on the occurrence in the environment, and the main part describes the chemical behaviour under environmental conditions. As of 2022, the periodic table of the elements has 118 confirmed elements; a total of 98 elements occur naturally.

Table 4.1: The chemical abundance of elements in mass %; according to conventions, the oceans are included in the earth's crust (for the chemical composition of air, see Table 2.3).

space		meteorites		Earth		earth's core		earth's mantle		earth's crust	
H	[a]	O	52.80	Fe	32.1	Fe	88.8	O	44.8	O	46.1
He	[a]	Si	15.37	O	30.1	Ni	5.8	Mg	22.8	Si	28.2
O	49.3	Mg	13.23	Si	15.1	S	4.5	Si	21.5	Al	.23
C	24.6	Fe	11.92	Mg	13.9			Fe	5.8	Fe	5.63
Ne	6.4	S	2.10	S	2.9			Ca	2.3	Ca	4.15
Fe	5.4	Al	1.15	Ni	1.8			Al	1.2	Na	2.36
N	5.9	Ca	0.90	Ca	1.5			Na	0.3	Mg	2.33
Si	3.4	Ni	0.65	Al	1.4					K	2.09
Mg	3.0	Na	0.62								
S	2.5	Cr	0.19								
		K	0.13								
sum 100		sum 97.91		sum 98.8		sum 99.1		sum 99.7		sum 99.65	

[a] H and He represent 98 % of total element mass (H 76.5 % and He 23.5 %); the other elements are set to 100 % as a sum

4.1 General remarks

As seen from Table 2.1, under 'normal' environmental conditions, chemistry takes place only under small changes of temperature, whereas the change of (atmospheric) pressure plays no role. Both reaction parameters are essential in industrial chemistry. Therefore, photochemical radical formation is crucial to initiate conversions.

https://doi.org/10.1515/9783110735178-004

Table 4.2: Reservoir distribution (in 10^{19} g element, except for the water molecule); after Schlesinger (1997).

reservoir	C	O	H_2O	S^g
atmosphere	0.075	119	1.7	negl.
ocean	3.8/0.07[a]	12,500[b]	14,000	128
land plants	0.06	negl.	negl.	negl.
soils, organic matter	0.15	? (negl.)	? (negl.)	negl.
fossil fuels	0.7	negl.	negl.	0.001
sediments	~5000/1500[d]	4745[c]	1500	247[h]
clathrates[e]	1.1	–	–	–
rocks	3200–9300	1200[f]	~2,000,000	?

[a] carbonate/dissolved organic carbon (DOC)
[b] in water molecules
[c] held in Fe_2O_3 and evaporitic $CaSO_4$
[d] held in $CaCO_3$
[e] methane hydrates
[f] held in silicates
[g] after Möller (1983)
[h] held in $CaSO_4$

Two criteria always help make the chemistry as simple as possible. First, we only consider the main reactions of species A with competitive chemical pathways. Let us regard as an example the two reactions (A = $O(^3P)$), where the k value is not very different between (a) and (b):

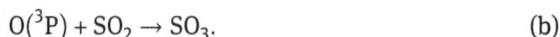

$$O(^3P) + O_2 \rightarrow O_3, \tag{a}$$
$$O(^3P) + SO_2 \rightarrow SO_3. \tag{b}$$

There are hundreds more reactions of $O(^3P)$ described in literature but under normal conditions in air, the pathway (a) is the only relevant reaction as the overall rate is many orders of magnitude larger than all other competitive reactions because of the size of the oxygen concentration compared with all other trace gas concentrations (remember: $R = k \cdot c$). Second, we only consider a limited number of chemical species that play a significant role in the environment. We do not regard the huge number of organic compounds of different origin and different fate in the environment (this is a special task for toxicological environmental chemistry). We focus on chemical species contributing to environmental problems in air, water and soil, such as:
– acidity/alkalinity (acidifying potential);
– oxidising/reducing agents (oxidation capacity);
– particulate matter formation (global cooling);
– greenhouse effect (global warming); and
– ozone depletion (UV radiation effects).

Biogeochemical cycling, where the emissions from plants and microorganisms (in soils and waters) in terms of specific fluxes and substances specify the chemical regime, drives environmental chemistry. Most of the natural chemical compounds also occur as pollutants from anthropogenic activities. Only a few pollutants are unknown in nature. The key question is the exceedance of the natural concentration level, called pollution, and the resulting adverse effects. The task of environmental chemistry is to detect the substances and their fates.

Life determines the global biogeochemical cycles of the elements of biochemistry, especially C, N, P and S, but the key process of plant life is the water-splitting process into H and O (Chapter 5.2.2) and thereby creating oxic (oxidising) and anoxic (reducing) environments. Figure 4.1 shows the principal three groups of compounds and their biogeochemical cycling schematically. Hydrides (methane, ammonia and hydrogen sulphide) represent the lowest oxidation state of the elements C, N and S that are bound in biomass (living organisms) as biomolecules. Nevertheless, they are released from living and dead biomass into the abiogenic surroundings (soil, water and air). In the air, these substances (and many more such as hydrocarbons, amines and or-

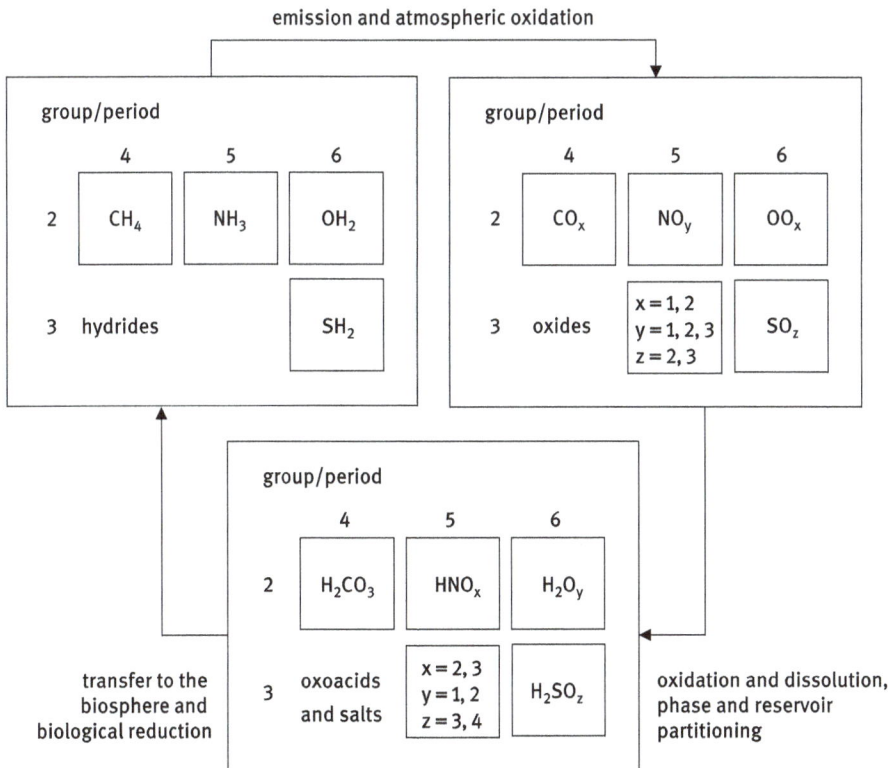

Figure 4.1: Characteristic groups for C, N, O and S and their roles in biogeochemical cycling.

ganic sulphides) are oxidised, finally to oxides of C, N and S. Many oxides represent anhydrides and form oxoacids and subsequent salts (such as carbonate, nitrate and sulphate) when reacting with water. These oxidised chemicals are again the stock for plants and microorganisms, which they reduce back to the lowest oxidation state. The important role of acidity in weathering we discuss in Chapter 5.1.5. A special role the water molecule plays, here written as hydride[49] OH_2 and oxoacid H_2O respectively, being a source of oxygen O_2 and hydrogen (via photosynthesis) as well *reactive oxygen species* (ROS) such as the hydroxyl radical OH and peroxides (HO_2, H_2O_2). Oxygen itself splits into atomic oxygen O and trioxygen (ozone) O_3.

There are other important elements. First, hydrogen, which in its molecular form (H_2) is relatively non-reactive but exists atomically as a short-lived atom in air and solution and as a long-lived proton in solution. Its role (Figure 4.1) alters between water, hydrides and acids. Second, phosphorus is also a key bioelement; P has only a few unstable, volatile compounds and plays only a minor role in atmospheric chemistry. Moreover, there are halogens (F, Cl, Br, I) that we could also include in Figure 4.1 because they form hydrides (acids such as HCl), oxides (e. g. ClO) and oxoacids (e. g. HOCl). Finally, many trace elements are biogeochemically important, either because of their roles as redox elements or as bioelements with very specific properties. However, some trace elements (for example, Cd and Hg) have no biological functions but are extremely poisonous. Table 4.3 shows the distribution of main compounds in the environment among different states of matter. It is remarkable that only one substance (H_2O) exists in all three phases simultaneously. Not all gases exist in dissolved form in the aqueous phase, but the main inorganics (essential for life such as sulphurous, nitrous, carbonaceous, chloride, ammoniacal) exist gaseous, dissolved and particulate. All dissolved species also exist in solid/particulate matter. All insoluble species exist either only as gases or as particulates. However, some organic substances can be transformed into solids (homogeneous nucleation process) but must be unsoluble.

Excluding noble gases, the number of gaseous elements is very limited: N, O, H, and Cl (Table 4.4). All these elements form many gaseous compounds. It is remarkable that the number of further elements (which are not gaseous) forming gaseous compounds is also limited: C, S, Br and I (Table 4.4). Table 4.4 lists more elements (Si, P, As, Se) that also form gaseous compounds but play no role[50] in the environment. In addition, many compounds and a few other elements (such as Br, I and Hg) can

---

49 The name hydride (see also Chapter 4.2) is manifold used; generic for a simple compound of an element and hydrogen (also named binary hydrides), separated in salty compounds (oxidation state −1), metalloid (half-metal) compounds (oxidation state 0), and covalent compounds (oxidation state +1). In a stronger sense, hydrides are only salty compounds between H and alkali and earth-alkali elements.

50 P may be an exception (see Chapter 4.8).

Table 4.3: Principle main compounds in the environment. PM – particulate matter, SOA – secondary organic aerosol, OA – (primary) organic solid matter (biogenic), EC – elemental carbon, red. S – H_2S, COS, CS_2, DMS and others, NMVOC – non-methane volatile organic compounds; H_2 and noble gases not listed.

gas	liquid	dissolved[a]	solid
H_2O	$(H_2O)_n$	$H^+ + OH^-$	$(H_2O)_n$ (ice, snow)
CO_2	–	$H^+ + HCO_3^-$	
HCl	–	$H^+ + Cl^-$	
NH_3	–	$NH_4^+ + OH^-$	PM (chloride, ammonium, carbonate, nitrate, sulphate)
NO_y	–	$H^+ + NO_2^- / NO_3^-$	
SO_2	–		
red. S	–	$H^+ + HSO_3^-/SO_4^{2-}$	
NMVOC	–	NMVOC[b]	SOA
–	–	OA[b]	OA
–	–	–	EC + BC (soot)
–	–	$Na^+ + Cl^{-c}$	seasalt $(NaCl)^c$
–	–	$K^+ + Ca^{2+} + Mg^{2+d}$	dust[e]
CH_4	–	–	clathrate

[a] partly after oxidation
[b] if soluble, organic acids protolysed
[c] about 90 % NaCl but including other compounds (sulphate, magnesium, calcium, potassium and many other)
[d] and many others, including anions (carbonate, sulphate)
[e] insoluble main components: Si, Al

Table 4.4: Main-group elements in gaseous compounds in the environment. In the air, only noble gases exist elementally, nitrogen (N_2), oxygen (O_2), hydrogen (H_2), and in traces halogen atoms (Cl, Br and I), atomic oxygen (O) and the transition metal mercury (Hg), in blue. Fluor (F) is gaseous but exists in nature only bonded due to its high reactivity. Bromine and mercury are the only elements that are liquid under normal temperature.

	4	5	6	7	8
1				H	He
2	C	N	O	F	Ne
3	Si	P	S	Cl	Ar
4		As	Se	Br	Kr
5				I	Xe
6					Ra

be detected as gaseous in the atmosphere because they have volatile properties. The number of primarily emitted important substances, intermediates and final products is relatively limited (excluding organic compounds; Tables 4.4 and 4.5).

Table 4.5: Principal main educts and products (simplified).

element/group	educts and intermediates	final products
oxygen	O_3, OH, HO_2, H_2O_2	O_2, H_2O
inorganic carbon	CO_2, H_2O	CO_2, H_2CO_3, HCO_3^-
sulphur	H_2S, DMS, COS, CS_2, SO_2	H_2SO_4, $SO_{2/4}^-$
nitrogen	NO, NO_2, NO_3, N_2O_5, HNO_2	HNO_3, NO_3^-
chlorine[a]	Cl_2, Cl, HCl	HCl, Cl^-
organic carbon	RCH_3, RCHO, RCOOH	$H_2O + CO_2$

[a]representative for other halogens (F, Br, I)

4.2 Hydrogen

Hydrogen, the lightest element, is characterised by its existence one-fold positive (like the alkali metals) and one-fold negative (like the halogens) hence it moves between the main groups 1 and 7.

i Hydrogen forms with all elements the largest number of compounds. There is another particularity: If a hydrogen atom is oxidised, an elementary particle, the proton H^+ results; its diameter is in the order of 10^5 smaller than 'normal' cations. However, the ionisation energy of H is so high that protons do not exist freely.

4.2.1 Natural occurrence

Hydrogen is the most abundant element in space and represents 76.5 % of total element mass. It is the only element that can escape the upper atmosphere into space by diffusion due to its small mass. Because of the excess of hydrogen in space, the hydrides (OH_2, CH_4, NH_3, and SH_2) should have the highest molecular abundance of compounds derived from such elements. On Earth, hydrogen occurs elementally (H_2) in the atmosphere. Most of the hydrogen is bound in water (H_2O). Most minerals of the earth's upper mantle contain small amounts of hydrogen, structurally bound as hydroxyl (OH). Another – and possibly most dominant – reservoir of hydrogen are hydrocarbons (C_xH_y), fossil fuels and organic compounds delivered to Earth after its formation and stored deep in the lithosphere. Under oxygen-free conditions, the product of thermal dissociation of hydrocarbons (note that T and p are extremely high in depths of 20 km and more) is $C + CO_2 + H_2$. Hydrogen can transform deep carbon into CH_4 and H_2O.

In 1900, Armand Gautier (1837–1920) first detected hydrogen as permanently present in the air. As the second most abundant reactive gas in the troposphere after CH_4, hydrogen is present at about 500 ppb. The troposphere has an estimated 155 Tg of hydrogen gas (H_2), with approximately a two-year lifetime. The dissociation of water (into hydrogen and oxygen), however, is only possible under natural conditions in the upper stratosphere (see Chapter 5.3.2.1). Ultimately, hydrogen comes from the water via thermal dissociation deep in the Earth (thus as natural hydrogen emanations on the earth's surface) and via water splitting by photosynthesis in plants (see Chapter 5.2.2).

The photochemical production from photolysis (Table 4.8) accounts for about 45 % of the total source of H_2. It returns by combining with the OH radical in the air back to the water. Soil uptake (55 Tg yr^{-1}) represents a major loss process for H_2 and contributes 80 % of the total destruction. H_2 oxidation by OH in the troposphere contributes the remainder. Obviously, the atmospheric trend of H_2 was continuous since the 1990s; these increases originate from anthropogenic sources: industry, transportation and other fossil fuel combustion processes, biomass burning, nitrogen fixation in soils. Natural sources include volcanism and soil emanation.

Large amounts of hydrogen are yearly industrially produced almost via reforming from fossil fuels; although presenting great potential for several applications, 52 % of total H_2 worldwide is used for refining, 42 % is used for ammonia production, and only 6 % is used for other applications, including its use as a clean and renewable fuel. Undoubles the challenges consists of future hydrogen production via water electrolysis based on solar energy, prevalently by photovoltaic and photothermal power stations.

As hydrogen reacts with tropospheric hydroxyl radicals, emissions of hydrogen to the atmosphere perturb the distributions of methane and ozone, the second and third most important greenhouse gases after carbon dioxide. Hydrogen is, therefore, an indirect greenhouse gas with a global warming potential GWP of 5.8 over a 100-year time horizon. A future hydrogen economy would therefore have greenhouse consequences and would not be free from climate perturbations. If a global hydrogen economy replaced the current fossil fuel-based energy system and exhibited a leakage rate of 1 %, then it would produce a climate impact of 0.6 % of the current fossil fuel-based system (Derwent et al. 2006).

Careful attention must be given to reduce to a minimum the leakage of hydrogen from the synthesis, storage and use of hydrogen in a future global hydrogen economy if the full climate benefits are to be realised. Hence, today's discussed use of hydrogen (by industrial water electrolysis using 'green' electricity) as fuel[51] seems not to be

---

[51] A strategic application of the H_2 is to consider it as a fuel, being able to be applicable for direct combustion, by itself or in some blends with natural gas, and also in fuel cells, where it can provide a reliable and efficient energy power, that can be used in stationary power stations and also as a good candidate for transportation vehicles. However, because of its very low volumetric energy density in a gaseous form under atmospheric conditions, hydrogen needs to be stored and transported effectively in any form, with high gravimetric and volumetric hydrogen densities.

sustainable, but its use to gain 'green' fuels via hydrogenation of carbon dioxide (see Chapter 6.2.4).[52]

> **i** Hydrogen is an energy carrier, not an energy source.

4.2.2 Compounds of hydrogen

Table 4.6 lists compounds of hydrogen found in the natural environment. Hydrides with metals do not exist free due to their reactivity (it could be possible that they were during the formation of the earth's formation). The hydrides of carbon (CH_4), nitrogen (NH_3), sulphur (H_2S), and phosphorus (PH_3) are most important within global cycling of the life-essential elements; they present the most reduced compound (produced in the biosphere) compared to oxidised compounds (produced in the atmosphere):

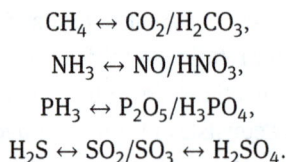

$$CH_4 \leftrightarrow CO_2/H_2CO_3,$$
$$NH_3 \leftrightarrow NO/HNO_3,$$
$$PH_3 \leftrightarrow P_2O_5/H_3PO_4,$$
$$H_2S \leftrightarrow SO_2/SO_3 \leftrightarrow H_2SO_4.$$

Table 4.6: Principal compounds of hydrogen; ROS – reactive oxygen species.

binary acids	CH_4	NH_3	PH_3	H_2O	H_2S	HF	HCl	HBr	HI
oxoacids	$O_mX(OH)_n$		X = C, N, P, S, Si						
carboxylic acids	$R-(COOH)_n$								
aquacomplexes	$[M(H_2O)_n]^{m+}$								
hydroxo complexes	$[M(OH)_n]^{(m-n)+}$								
hydrocarbons	RH								
ROS	OH	HO_2	H_2O_2						

Nearly all environmentally important chemical compounds contain hydrogen (Table 4.6); these compounds and their chemistry will be discussed in later Chapters concerning the other elements (O, N, S, C, P, halogens). The high importance of the H^+ (proton and hydronium ion, respectively) in terms of acidity has already been shown in Chapter 3.2.2.3 (acids and bases), and the relationships between H_2O, H, H^+ and e^-_{aq} in Chapter 3.3.5.2. Hydrogen is the ultimate reducing species (and therefore the

---

[52] The word 'green' includes manifold implications but generally being environment friendly and in line with nature. Specifically, it means *not* fossil-fuel-based technologies, energy etc. The term green chemistry, however, is slightly different used, see Chapter 6.

key in photosynthesis) and the species with the highest energy density[53] (liberated when oxidising to H_2O). The most important compound of hydrogen is water (H_2O); its properties were presented in Chapters 2.4.1 and 2.4.2.

4.2.3 Chemistry

The gross reaction $2\,H_2 + O_2 = 2\,H_2O$ is highly exothermic but does not proceed at a measurable rate at ambient temperatures because of the high activation energy needed to break the $O-O$ ($493\,kJ\,mol^{-1}$) and $H-H$ ($436\,kJ\,mol^{-1}$) bonding (Table 4.7).

Table 4.7: The oxyhydrogen reaction; M – collision partner.

starting phase	O_2 + **energy** $\rightarrow$ O + O
chain propagation	$O + H_2 \rightarrow OH + H$
	$H + O_2 \rightarrow HO_2$
	$H + HO_2 + M \rightarrow H_2O + O + M$
	$H + HO_2 + M \rightarrow 2\,OH + M$
	$OH + OH + M \rightarrow H_2O_2 + M$
	$OH + H_2 \rightarrow H_2O + H$
	$HO_2 + HO_2 \rightarrow H_2O_2 + O_2$
	$HO_2 + H_2 \rightarrow H_2O_2 + H$
	$H_2O_2 + OH \rightarrow H_2O + HO_2$
	$H_2O_2 + h\nu \rightarrow 2\,OH$
	$H_2O_2 + h\nu \rightarrow H_2O + O$
	$H_2O_2 + h\nu \rightarrow H + HO_2$
termination	$H + HO_2 + M \rightarrow H_2 + O_2 + M$
	$H + H + M \rightarrow H_2 + M$
	$H + OH \rightarrow H_2O$
	$OH + HO_2 \rightarrow H_2O + O_2$
	$O + O \rightarrow O_2$

There are several reactions producing free H atoms (Table 4.8); however, under aerobic conditions (always in air and surface waters), H combines quickly with oxygen into HO_2 ($k_{4.1} \sim 2 \cdot 10^{10}\,L\,mol^{-1}\,s^{-1}$), Figure 4.2:

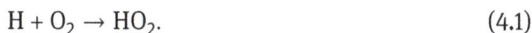

$$H + O_2 \rightarrow HO_2. \tag{4.1}$$

Once e^-_{aq} is produced, then the timescale for the formation of subsequent reactive species (H, HO_2, OH, H_2O_2) is in the order of $10^{-7}\,s$ (Figure 4.2). Only under anaerobic

---

[53] Followed by CH_4 and liberated when oxidising to CO_2.

Table 4.8: Reactions producing H under environmental conditions.

educts		products	comment
$H_2O + hv$	$\rightarrow$	$H + OH$	$\lambda < 242\,nm$
$CH_4 + hv$	$\rightarrow$	$H + CH_3$	$\lambda < 230\,nm$
$HCHO + hv$	$\rightarrow$	$H + HCO$	$\lambda < 370\,nm$ very important
$RCHO + hv$	$\rightarrow$	$H + RCO$	$\lambda < 310\,nm$ unimportant
$HNO_2 + hv$	$\rightarrow$	$H + NO_2$	$\lambda < 361\,nm$ unimportant
$CO + OH$	$\rightarrow$	$H + CO_2$	very important
$NH_2 + NO_2$	$\rightarrow$	$H + OH + N_2$	after $NH_3 + OH \rightarrow NH_2 + H_2O$ (unimportant)
$H^+ + e_{aq}^-$	$\rightarrow$	H	after photosensitising (important)

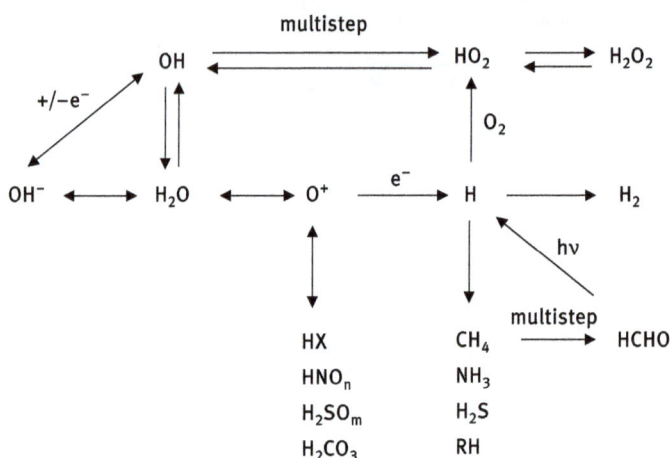

The first three rows (H_2O, CH_4, $HCHO$) are braced together with the comment: only in upper atmosphere.

Figure 4.2: Main chemical pathways of hydrogen and its compounds; X = Cl, Br, F, I; n = 2, 3; m = 3, 4. Note: Formation of CH_4, NH_3, H_2S and RH is only possible under anaerobic conditions by microorganisms and plants (assimilation).

conditions (soils, swamps, water bottom), H can be enriched; however, such environmental conditions are not favoured for the photocatalytic formation of e_{aq}^-. In plants, photosynthesis splits H_2O into H to reduce CO_2 and finally to build up hydrocarbons (see Chapter 5.2.2).

The main pathway for the production of hydrogen atoms in the air is methane (CH_4) oxidation by the OH radical and subsequent photolysis of formaldehyde (HCHO); see reactions equation (4.289) to equation (4.296). This process accounts for about 26 Tg H yr^{-1}:

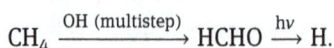

$$CH_4 \xrightarrow{\text{OH (multistep)}} HCHO \xrightarrow{hv} H.$$

The main sink (besides soil uptake) of molecular hydrogen is the relatively slow reaction with OH; $k_{4.2} = 6.7 \cdot 10^{-15}$ cm^3 molecule^{-1} s^{-1} at 298 K):

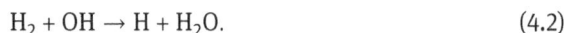

$$H_2 + OH \rightarrow H + H_2O. \tag{4.2}$$

Hence, if there is more H$_2$ in the stratosphere, it will react with hydroxyl radicals gaining more H$_2$O. Modelling studies show that this increase in H$_2$O cools the lower stratosphere. There are several excited species of molecular hydrogen, but there is no direct photodissociation of molecular hydrogen in the interstellar medium because atomic hydrogen depletes the spectrum above 13.6 eV, which corresponds to a wavelength smaller than 90 nm. At temperatures of 2,000 K, only 0.081 % of H$_2$ dissociates and this fraction increases to 7.85 % at 3,000 K (95.5 % at 5,000 K).

In Chapter 3.2.2.3, we met the species H$^+$ (proton) in the aqueous phase; however, this does not freely exist and is only present in combination with H$_2$O as the hydronium ion H$_3$O$^+$. In the gas phase, H$^+$ (but not under atmospheric conditions because the ionisation energy is very high: 1,311 kJ mol^{-1}) can be produced from atomic hydrogen. Atomic hydrogen has a high affinity to electrons (H$^-$) but this species exists only in hydrides. Most elements produce hydrides from very stable – the best known is water OH$_2$ – to very unstable compounds. It is likely that at an early stage of chemical evolution, many unstable metal hydrides (such as NaH) are formed. In the accretion phase of the Earth, they then decompose to hydrogen (H$_2$) in contact with water:

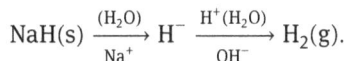

$$NaH(s) \xrightarrow[\text{Na}^+]{(H_2O)} H^- \xrightarrow[\text{OH}^-]{H^+(H_2O)} H_2(g).$$

Hydrogen undergoes several reactions in the aqueous phase. Atomic hydrogen is the major reducing radical in some reactions.[54] It effectively reacts as an oxidant forming hydride intermediates such as:

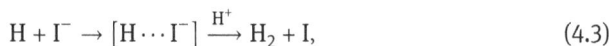

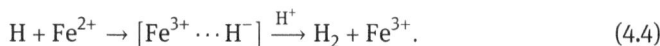

$$H + I^- \rightarrow [H \cdots I^-] \xrightarrow{H^+} H_2 + I, \tag{4.3}$$

$$H + Fe^{2+} \rightarrow [Fe^{3+} \cdots H^-] \xrightarrow{H^+} H_2 + Fe^{3+}. \tag{4.4}$$

In its reactions with organic compounds, the hydrogen atom generally abstracts H from saturated molecules and adds to the centres of unsaturated; the fate of the organic radicals is well known (Chapter 4.6.3.1):

$$H + CH_3OH \rightarrow H_2 + CH_2OH, \tag{4.5}$$

$$H + H_2C = CH_2 \rightarrow CH_3CH_2. \tag{4.6}$$

---

54 Fast radical–radical reactions (in parenthesis k in 10^{10} L mol^{-1} s^{-1}) such as H + H → H$_2$ (1.3), OH + OH → H$_2$O$_2$ (0.53) and OH + H → H$_2$O (3.2) play no role in natural waters because of other available reactants in concentrations orders of magnitude higher.

4.3 Oxygen

i It is agreed that free oxygen in an atmosphere is the result (and hence the indication) of biological life. Oxygen is the most abundant element on Earth and in space, besides hydrogen and helium (Table 4.1). Therefore, water (H_2O) is the most abundant compound in space.

The evolution of atmospheric oxygen is described in Chapter 5.1.4. Without photosynthesis by plants and other organisms, free oxygen would only exist in traces in the air due to the photodissociation of H_2O. However, without respiration, oxygen would rise above the present level and cause oxidative stress. The cycling of oxygen is inextricably linked with that of carbon (CH_4 also stands for any hydrocarbon) (Figure 4.3):

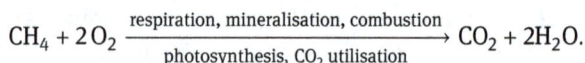

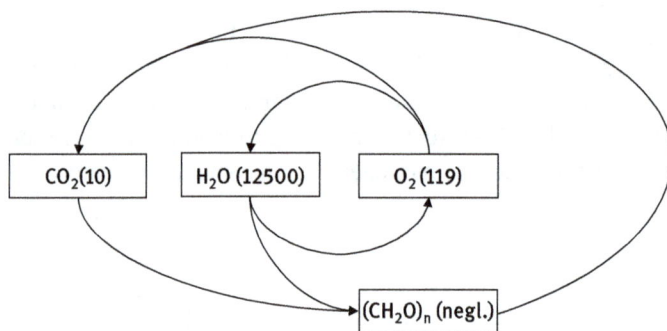

$$CH_4 + 2O_2 \xrightarrow[\text{photosynthesis, } CO_2 \text{ utilisation}]{\text{respiration, mineralisation, combustion}} CO_2 + 2H_2O.$$

Figure 4.3: Oxygen in chemical reservoirs and chemical cycling; numbers represent the total mass of oxygen in 10^{19} g (CO_2 is mostly dissolved as carbonate in the ocean and O_2 being molecular in the air), $(CH_2O)_n$ represents organic compounds, including biomass.

In chemical bonds, oxygen is derived mostly from monooxygen (O), less from dioxygen (O_2) and only as an unstable intermediate from trioxygen (O_3):

=O (oxo group) and –O– (oxy group), specifically: –OH (hydroxy group)
–O–O– (dioxy group, peroxo and peroxy, resp.)
–O–O–O– (trioxy group, ozonide)

There is some confusion in literature when using 'oxo' or 'oxy' as suffix (some authors use oxo in inorganic and oxy in organic chemistry, and others use oxo and oxy as a synonym). However, according to the IUPAC nomenclature, oxo compounds contain an oxygen atom, =O, doubly bonded to carbon or another element. The term thus embraces aldehydes, carboxylic acids, ketones, sulphonic acids, amides and esters. Oxy compounds contain the bivalent functional group –O–O–, found in ethers. Oxoacids (and their variants oxyacids, oxo acids, oxy-acids, oxiacids, oxacids) is a traditional name for any acid having oxygen in the acidic group. The term stands in contradiction to 'hydracids' (e. g. HCl) lacking oxygen. In the IUPAC nomenclature, the prefix peroxo is used for inorganic compounds and peroxy for organic compounds.

Atmospheric oxygen in its different forms (O, O_2 and O_3) provides oxidising agents for the decomposition of organic compounds (symbolised in Figure 4.3 as $(CH_2O)_n$) where it turns back into H_2O via several reactive (and biologically and atmospherically important) H_xO_y species (Chapter 4.3.2.2). Oxygen is only known in electronegative bonding, except for peroxides (–1), mostly in the oxidation state –2. Only fluorine is more electronegative than oxygen, but elemental F does not exist in nature. Therefore, oxygen forms oxides with all other elements.[55] In the past few decades, the role of molecular oxygen in forming metal complexes (superoxo O_2^- and peroxo –OO–), which play the role of oxygen carriers in biochemistry, has been studied. Molecular oxygen in its $3\Sigma_g^-$ ground state has triplet multiplicity but not singlet multiplicity, unlike most natural compounds. The unpaired electrons in two different molecule orbitals account for the paramagnetism of molecular oxygen. The high O_2 dissociation energy ($493.4\,\text{kJ mol}^{-1}$) does not allow, under tropospheric conditions, photolytic decay into oxygen atoms. Triplet multiplicity is the reason why most reactions of oxygen with organic substances, although exergonic, do not proceed at normal temperature but upon heating or in the presence of catalysts. Thus, reactions of organic compounds with dioxygen are kinetically inhibited. The other main and very stable oxygen compound is water (H_2O) because the O–H bonding cannot be broken by photodissociation within the environment.

Therefore, oxygen comprises four chemical groups from O_1 to O_4 (Figure 4.4). Transfers between these groups provide very few special reactions. The variety of

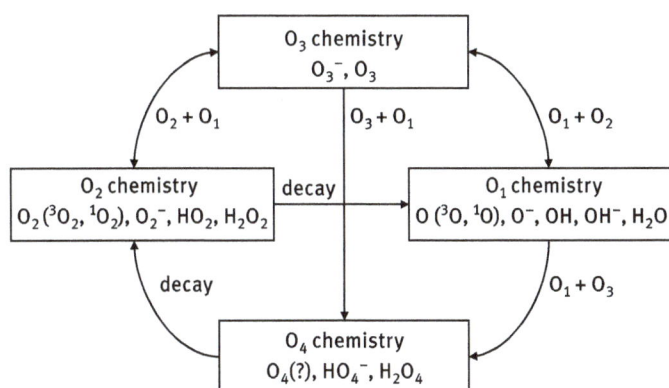

Figure 4.4: Chemical species and schematic relationships between O_1–O_4 chemistry.

55 Xenon (Xe) also forms oxides (and fluorides); from He, Ne and Ar stable compounds are unknown.

species within groups O_1 and O_2 is large, whereas group O_3 plays the role of connector between groups O_1 and O_2. Group O_4 only plays a (hypothetical) transient state from O_3 to O_2. Hydrogen radicals (H) and ions (H$^+$) are closely connected with oxygen chemistry; in Chapter 3.3.5.1, we stated that H and O are the 'symbols' for reduction and oxidation, respectively. The chemistry of oxygen species is very complex; all oxygen species are interlinked with all other chemical species found in the environment.

4.3.1 Natural occurrence

Our present atmosphere contains $119 \cdot 10^{19}$ g molecular oxygen (20.94 % in dry air), which represents only 0.006 % of total oxygen on the Earth almost completely fixed in oxides (see Table 4.2), very likely of primordial origin. As just stated, oxygen forms oxides with all elements (with the exception of He, Ne and Ar), but naturally, only quartz (SiO_2), ferric oxide (Fe_2O_3) and alumina (Al_2O_3) are dominant as different minerals in the earth's crust. Soluble oxides (from alkali and alkaline earth metals) do not exist naturally. In the air, gaseous oxides of sulphur, nitrogen and carbon play an important role, whereas oxides of halogens are in traces only but important in atmospheric ozone chemistry. As listed in Table 4.6, oxygen is bound in oxoacids from which the stable and environmentally important anions sulphate (SO_4^{2-}), nitrate (NO_3^-), carbonate (CO_3^{2-}) and phosphate (PO_4^{3-}) are derived. Nitrates and phosphates are a result of microbiological activities. Carbonates and sulphates (as calcium) are sediments; however, sulphate is mostly found dissolved in seawater. Oxygen forms the environmentally important carboxylic acids (Table 4.6) and is within oxygenated hydrocarbons (alcohols, aldehydes, and ketones).

In the air, besides molecular oxygen O_2, there are ozone (O_3) molecules and atomic oxygen (O) in small traces (only intermediary). Under natural conditions, O_3 is produced only in the stratosphere and transported down to the earth's surface. It is estimated by modelling that about 2,000 Mt O_3 is transported from the stratosphere into the troposphere; most of it is destructed in the free troposphere. Hence, the O_3 concentration increases with altitude and preindustrial surface O_3 mixing ratios were around 10 ppb and less. At present, human activities have resulted in an additional about 2,000 Mt O_3 in the troposphere and the near-surface O_3 concentration rose to 30–40 ppb.

4.3.2 Gas-phase chemistry

4.3.2.1 Atomic and molecular oxygen: O, O_2 and O_3

As mentioned, O_2 cannot be photochemically destructed in the troposphere because of the missing short wavelength (see below and Chapter 5.3.2.2). The O_2 photodissociation and subsequent O_3 formation in the lower stratosphere is the most important source of O_3 in the troposphere via the stratospheric-tropospheric exchange. However, the O_2 photolysis also plays an important role in environmental engineering (water treatment); there are three main pathways dependent on radiation energy:

$$O_2 + h\nu \rightarrow O(^3P) + O(^3P), \quad \lambda_{threshold} = 242\,nm, \tag{4.7}$$

$$O_2 + h\nu \rightarrow O(^3P) + O(^1D), \quad \lambda_{threshold} = 175\,nm, \tag{4.8}$$

$$O_2 + h\nu \rightarrow O(^1D) + O(^1D), \quad \lambda_{threshold} = 137\,nm. \tag{4.9}$$

The main fate of O_2 in the troposphere is the addition onto several radicals X to produce peroxo radicals (reaction equation (4.1) and below), an important step in the oxidation of trace gases:

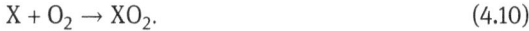

$$X + O_2 \rightarrow XO_2. \tag{4.10}$$

A specific reaction of that type and the only formation pathway of ozone, as well as the ultimate fate of $O(^3P)$, is the combination with oxygen to produce O_3; $k_{4.11} = 5.6 \cdot 10^{-34}[N_2]\,cm^3\,molecule^{-1}\,s^{-1}$ at 298 K; M – collision partner:

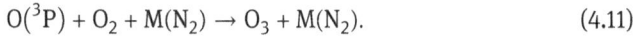

$$O(^3P) + O_2 + M(N_2) \rightarrow O_3 + M(N_2). \tag{4.11}$$

In all such reactions with molecular oxygen, we can assume a constant O_2 concentration of $5.6 \cdot 10^{18}$ molecules cm^{-3} under standard conditions in the air. Furthermore, taking into account the constant N_2 concentration ($2.08 \cdot 10^{19}$ molecules cm^{-3}), a pseudo-first-order rate constant $k_{4.11} = 6.4 \cdot 10^4\,s^{-1}$ follows and a lifetime of only $1.5 \cdot 10^{-5}\,s$ for $O(^3P)$. About 90 % of $O(^1D)$ turns by quenching into the triplet oxygen; $k_{4.12} = 4.4 \cdot 10^{-11}\,cm^3\,molecule^{-1}\,s^{-1}$ (M = N_2) at 298 K:

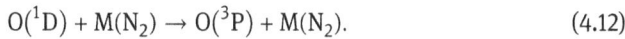

$$O(^1D) + M(N_2) \rightarrow O(^3P) + M(N_2). \tag{4.12}$$

The only other (and this is one of the very few initial reactions producing ROS) important reaction of $O(^1D)$ is with H_2O – producing OH radicals (see equation (4.19) and is, therefore, the key reaction in atmospheric chemistry.

In the gas phase, the photoexcitation formation of singlet from triplet molecules are quantum-chemically forbidden. There are several ways to form singlet dioxygen. The role of singlet dioxygen ($^1\Delta_g$ and $^1\Sigma_g^+$), where two electrons in the outer shell (2p4) are in the degenerated π-symmetric orbitals with the antipodal spin (total zero spin), is well known in organic chemistry. Singlet dioxygen is a metastable species, and because of its excitation energy of 94 kJ mol^{-1} it is chemically extraordinarily reactive. It is known for oxidative stress on living organisms and the ageing of a material's surface (autoxidation). However, in air chemistry, singlet dioxygen plays no role (as believed in the 1970s in smog chemistry). The photodissociation of ozone produces several excited oxygen species, including the triplet O (reaction equation (4.13)) and the singlet

O (reaction equation (4.14)); reaction equation (4.15), however, only occurs in the upper stratosphere:

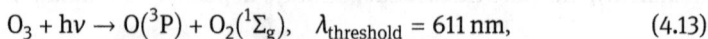

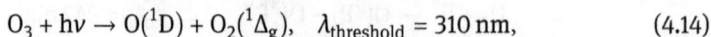

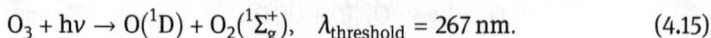

$$O_3 + h\nu \rightarrow O(^3P) + O_2(^1\Sigma_g), \quad \lambda_{threshold} = 611 \, nm, \tag{4.13}$$

$$O_3 + h\nu \rightarrow O(^1D) + O_2(^1\Delta_g), \quad \lambda_{threshold} = 310 \, nm, \tag{4.14}$$

$$O_3 + h\nu \rightarrow O(^1D) + O_2(^1\Sigma_g^+), \quad \lambda_{threshold} = 267 \, nm. \tag{4.15}$$

The formation of $O(^3P)$ and O_2 (and subsequently again O_3) in the ground state is already possible at very large wavelengths from O_3 photodissociation:

$$O_3 + h\nu \rightarrow O(^3P) + O_2(^3\Sigma_g), \quad \lambda_{threshold} = 1180 \, nm. \tag{4.16}$$

The different absorption bands are called Chappuis bands (440–850 nm), Huggins bands (300–360 nm) and Hartley bands (200–310 nm); the strongest absorption occurs at 250 nm. These different absorption bands through to ozone play a key role in the atmosphere: a) to prevent radiation in the lower atmosphere with wavelengths lower than about 300 nm, which destroys life; and b) to provide $O(^1D)$ from O_3 photolysis (reaction 4.14), which subsequently forms other ROS (Chapter 4.3.3).

The tropospheric net O_3 formation is another important source, especially because of anthropogenic enhancement (Chapter 5.3.2.2). Besides the O_3 photolysis (equation (4.14) and equation (4.16)), there is only one other important $O(^3P)$ source that provides the only net source for O_3 formation in the troposphere:

$$NO_2 + h\nu \rightarrow O(^3P) + NO, \quad \lambda_{threshold} = 398 \, nm. \tag{4.17}$$

In a pure oxygen gas system, we only consider four reactions (equation (4.13), equation (4.14), equation (4.16) and equation (4.11)), forming a steady state:

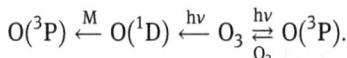

$$O(^3P) \xleftarrow{M} O(^1D) \xleftarrow{h\nu} O_3 \underset{O_2}{\overset{h\nu}{\rightleftarrows}} O(^3P).$$

It follows for the steady-state ozone concentration that:

$$[O_3] = \frac{k_{4.11}}{j_{4.13} + j_{4.14}} [O_2][O(^3P)]. \tag{4.18}$$

Without discussing here in detail, O_3 has several chemical sinks, where reactions with NO and NO_2 are the most important (Chapter 4.4.5), but alkenes also react with O_3 (Chapter 4.6.3.4) and heterogeneous loss is important (Chapter 5.3.2.1). Figure 4.5 shows all principal oxygen reactions; note that O_2 photolysis can be excluded in the environment in the narrower sense close to the earth's surface. The exit pathways (sinks) to products (Figure 4.3) are only possible if there are additional elements and compounds are available (Chapters 4.4 and 4.5).

other sources

O_xH_y chemistry

H

hν O_2

$O(^1D)$ O_3 HO_2

H_2O → OH ⇄ HO_2 → H_2O_2
         O_3

H_2O

other sources and sinks   other sources and sinks

$O(^1D)$

O_x chemistry

hν   M   hν

hν

O_2 ⤏ $O(^3P)$ ⇄ O_3 → other sinks
         hν
         O_2

other sources

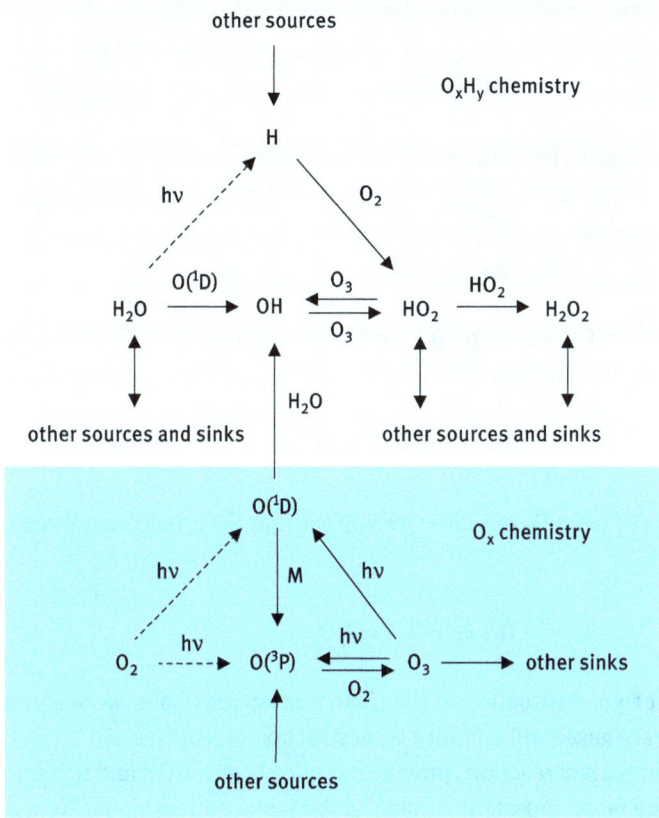

Figure 4.5: Scheme of H_xO_y gas-phase chemistry; dotted lines denote photolysis only in the stratosphere; other sources and sinks occur in the presence of other species than H_xO_y.

4.3.2.2 Reactive hydrogen-oxygen compounds: OH, HO_2 and H_2O_2

Besides reaction equation (4.12), $O(^1D)$ reacts in another channel (to only about 10 %) with H_2O to hydroxyl radicals (OH); $k_{4.19} = 2.2 \cdot 10^{-10}$ cm^3 molecule^{-1} s^{-1} at 298 K:

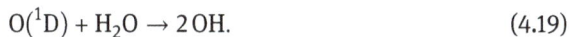

$$O(^1D) + H_2O \rightarrow 2\,OH. \tag{4.19}$$

The OH radical is the most important species in the atmosphere. This is often called the 'detergent' of the troposphere because it reacts with many pollutants, often acting at the first step of their removal.

Since OH (this is the neutral form of the hydroxide ion OH⁻, i. e. OH is a weak acid, see equation (4.67)) is from H_2O, it returns to water by the abstraction of H from nearly all hydrocarbons RH; this pathway is the definitive sink for OH. However, the organic radical R combines with O_2 (equation (4.10)) and in many subsequent reactions produces organic oxygen radicals and recycles HO_2 (see equations (4.277)–(4.285)):

$$OH \xrightarrow[-R]{RH} H_2O.$$

A second important pathway is the addition of OH on different species (SO_2, NO, NO_2) producing oxo acids:

$$OH \xrightarrow{X} XOH.$$

A third pathway is the direct OH transformation into the hydroperoxyl radical HO_2 by ozone:

$$OH \xrightarrow[-O_2]{O_3} HO_2.$$

HO_2 can turn back into OH (with O_3 but later, we will see that NO is most important), resulting in the scheme:

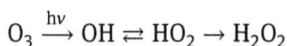

$$O_3 \xrightarrow{hv} OH \rightleftarrows HO_2 \rightarrow H_2O_2$$

and thereby in overall ozone destruction in the clean atmosphere. Later we see that the presence of other trace gases will enhance O_3 destruction in NO_x free air.

The following two important reactions provide recycling between OH and HO_2 and ozone destruction as the most important O_3 sink in the remote atmosphere. Hence, OH production is proportional to O_3 singlet oxygen photolysis. From measurements, the very robust relationship $[OH] \approx 0.1[HO_2]$ has been derived, where a maximum $[OH] \approx 5 \cdot 10^6$ molecules cm^{-3} has been found. In the presence of other trace gases, other dominant pathways exist, and in NO-containing air, a net formation of O_3 rather than destruction occurs (see Chapter 5.3.2.2); $k_{4.20} = 7.3 \cdot 10^{-14}$ cm^3 molecule^{-1} s^{-1} and $k_{4.21} = 2.0 \cdot 10^{-15}$ cm^3 molecule^{-1} s^{-1} at 298 K:

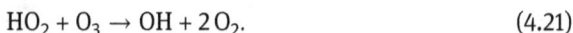

$$OH + O_3 \rightarrow HO_2 + O_2, \tag{4.20}$$

$$HO_2 + O_3 \rightarrow OH + 2O_2. \tag{4.21}$$

The net result of this reaction sequence is O_3 destruction according to $2O_3 \rightarrow 3O_2$, whereas the radicals recycle OH $\rightleftarrows$ HO_2. HO_2 produces via dimerisation hydrogen peroxide H_2O_2: $k_{4.22} = 1.6 \cdot 10^{-11}$ cm^3 molecule^{-1} s^{-1} and $k_{4.23} = 5.2 \cdot 10^{-32}$ [N_2] cm^3 molecule^{-1} s^{-1} (M = N_2) at 298 K:

$$HO_2 + HO_2 \rightarrow H_2O_2, \tag{4.22}$$

$$HO_2 + HO_2 + M \rightarrow H_2O_2 + M. \tag{4.23}$$

The dimerisation of HO_2 proceeds around 298 K by two channels: one bimolecular and the other trimolecular. A cyclic hydrogen bond intermediate $\cdots H-O-O \cdots H-O-O \cdots$ is very likely, which explains the 'abnormal' negative dependence of T and the pressure dependence. Enhancement in the presence of H_2O (M = H_2O) was observed; $k_{4.24}$ = $k_{4.22}[1 + 1.4 \cdot 10^{-21}[H_2O] \exp(2200/T)]$ cm^3 molecule^{-1} s^{-1}, for 1 vol-% H_2O. It follows that $k_{4.22} = 1.6 k_{4.20}$ at 298 K (exceeded H_2O_2 concentration were observed above cloud layers):

$$HO_2 + HO_2 + H_2O \rightarrow H_2O_2 + H_2O. \tag{4.24}$$

Note that $[M]/[H_2O] \geq 100$. H_2O_2 is rather stable in the gas phase, i. e. it does not undergo fast photochemical and gas-phase reactions (Figure 4.5). The only important sinks in the boundary layer are dry deposition and scavenging by clouds (with subsequent aqueous phase chemistry) and precipitation (wet deposition). In the free troposphere, however, H_2O_2 photolysis is regarded as important radical feedback.

To complete the O_xH_y chemistry, further reactions should be noted, which are unimportant near the earth's surface but become of interest in the free troposphere and upper atmosphere. The photolysis of H_2O_2 is very slow and heterogeneous sinks (scavenging by clouds, precipitation and dry deposition) can be neglected in the upper atmosphere:

$$H_2O_2 + h\nu \rightarrow 2\,OH, \quad \lambda_{threshold} = 557\,nm, \tag{4.25a}$$

$$H_2O_2 + h\nu \rightarrow H_2O + O(^1D), \quad \lambda_{threshold} = 359\,nm, \tag{4.25b}$$

$$H_2O_2 + h\nu \rightarrow H + HO_2, \quad \lambda_{threshold} = 324\,nm. \tag{4.25c}$$

The quantum yield is very low for $\lambda > 300$ nm, resulting in a slow photodissociation with j in the order of 10^{-6}–10^{-5} s^{-1} (a residence time of around one week). More important in the upper atmosphere is the reaction with OH; $k_{4.26}$ = 1.7 · 10^{-12} cm^3 molecule^{-1} s^{-1} at 298 K. Reactions equation (4.26) and equation (4.22) together represent a net radical sink: $OH + HO_2 \rightarrow H_2O + O_2$:

$$H_2O_2 + OH \rightarrow H_2O + HO_2. \tag{4.26}$$

4.3.3 Aqueous-phase chemistry

From the gas phase, the oxygen species (listed by decreasing solubility) H_2O_2, HO_2, OH, O_2 and O_3 can be transferred into the aqueous phase (droplets in air, surface water) by transport to the interface and scavenging. The residence time of atomic oxygen is too short to be transported. When not produced close to the interface, OH radicals also remain in the gas phase. Ozone (O_3) is very weekly soluble, however, due to interfacial and aqueous chemical reactions, the net flux can be significantly increased.

Despite the low solubility of O_2, the large concentration, when saturated in natural waters compared with trace species (pollutants), favours the recombination of many radicals with O_2 as in the air.

Figure 4.6 shows that in water, under the light influence (photosensitising) and contact with air, all the oxygen species listed in Table 4.9 can be produced. It can also clearly be seen that the presence of dioxygen (O_2) and ozone (O_3) opens competing pathways for gaining peroxides. The formation of aquated electrons (see Chapter 3.3.5.2) via photosensitising (see Chapter 3.3.6.4) is obviously a natural process, giving reactive oxygen species such as hydrogen dioxide (HO_2), hydrogen peroxide (H_2O_2) and the hydroxyl radical (OH) via oxygen and ozone (see the key reaction equation (3.183a)):

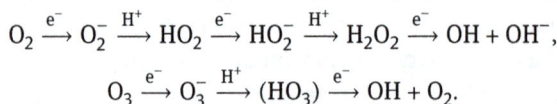

$$O_2 \xrightarrow{e^-} O_2^- \xrightarrow{H^+} HO_2 \xrightarrow{e^-} HO_2^- \xrightarrow{H^+} H_2O_2 \xrightarrow{e^-} OH + OH^-,$$

$$O_3 \xrightarrow{e^-} O_3^- \xrightarrow{H^+} (HO_3) \xrightarrow{e^-} OH + O_2.$$

The formation of water radical ions (H_2O^+) is impossible under natural conditions because of the lack of ionising radiation but opens (this radical is regarded as the strongest oxidiser – but the subsequent produced OH radical is *the* oxidiser) many ways in water treatment. In pure water (what does not exist in nature), OH would

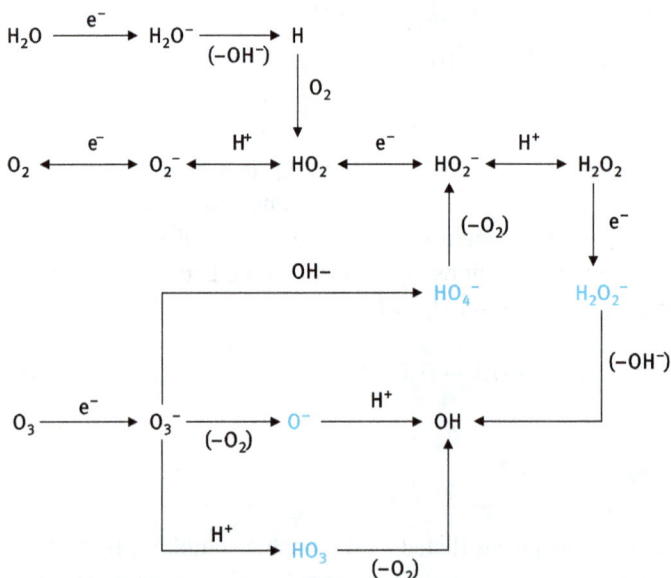

Figure 4.6: Water, oxygen and ozone redox chemistry ($\xrightarrow{e^-}$), including acid-base equilibriums ($\xrightarrow{H^+}$). Blue: short-lived (or speculative) species. The fate of OH is normally oxidation of other species (e. g. hydrocarbons), and the fate of H_2O_2 is normally S(IV) oxidation. Note that some electron transfers are non-reversible.

Table 4.9: All important O_xH_y components (O_x special case when $y = 0$). Note that ions only exist freely in the aqueous phase and ionic bonds in the solid phase. Some species are formed only under extreme conditions.

oxidation state	species	formation/destruction	name
−3	H_2O^-	$H_2O + e^-$	hydrated electron
−2	H_2O	$OH + H$ / $OH^- + H^+$	water
	H^+ (H_3O^-)	$H_2O \rightleftharpoons H^+ + OH^-$	proton (hydrogenium)
	OH^-	$H_2O \rightleftharpoons OH^- + H^+$	hydroxide anion
	O^{2-}	$O^{2-} + H_2O \rightarrow 2\,OH^-$	oxide ion
−1	H_2O^+	$H_2O - e^-$ / $OH + H^+$	water radical cation
	O^-	$O - e^-$	oxide (1-)
	OH	$H_2O - H$	hydroxyl radical
	O_2^{2-} (−O−O−)	$O_2 - 2e^-$	peroxide
	HO_2^-	$HO_2 + e^-$	hydrogen peroxide anion
	H_2O_2	$OH + OH$	hydrogen peroxide
−1/ ± 0	O_2^-	$O_2 + e^-$	superoxide anion
	HO_2	$O_2 + H$ / $O + OH$	hydroperoxyl radical
	O_3^-	$O_3 + e^-$	ozonide anion
	HO_3	$OH + O_2$	hydrogen ozonide[b]
	HO_4^-	$O_3 + OH^-$	tetraoxidane anion[b]
	H_2O_4	$HO_2 + HO_2$ / $2\,OH + O_2$	tetraoxidane[a,b]
±0	H	−	hydrogen
	O	−	oxygen
	O_2	$O + O$	dioxygen
	O_3	$O + O_2$	trioxygen (ozone)

[a] a higher hydrogen peroxide
[b] speculative or only very short-lived intermediates

dimerise into H_2O_2 or react with O_3 to HO_2 (see equation (4.52)). Thus, all in all, irradiation of water results in the formation of hydrogen peroxide.

Principally, almost all gas-phase reactions also take place in the aqueous phase. Moreover, many additional reactions, not occurring in the gas phase, have to be considered in the solution. Due to the fact that in solution, the concentration of many chemical species is enlarged compared to the gas phase, the total rate of chemical conversion increases, making aqueous-phase chemistry very important in the environment. Table 4.10 shows the main oxygen chemistry in biological systems (enzymatic processes) but also occurring in abiotic environments (catalytic processes).

Table 4.10: Basic O_xH_y reactions in the aqueous phase.

catalase	$2\,H_2O_2$	$\rightarrow$	$2\,H_2O + O_2$
Fenton reaction	$H_2O_2\ (+Fe^{II})$	$\rightarrow$	$OH + OH^-\ (+Fe^{III})$
dismutase	$HO_2 + O_2^-\ (+H^+)$	$\rightarrow$	$H_2O_2 + O_2$
enzymatic reduction	$O_2\ (+e^-)$	$\rightarrow$	O_2^-
deactivation	$O_2^-\ (+Fe^{III})$	$\rightarrow$	$O_2\ (+Fe^{II})$
ozone decay	$O_3 + O_2^-\ (+H^+)$	$\rightarrow$	$OH + 2\,O_2$

4.3.3.1 Aqueous-phase oxygen chemistry

In the aqueous phase, singlet dioxygen is directly produced from triplet oxygen by collision with so-called photosensitisers A^* (electronically excited molecules), which transfer energy onto O_2 in the ground state; the third collisional partner M stabilises the transfer complex:

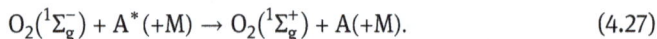

$$O_2(^1\Sigma_g^-) + A^*(+M) \rightarrow O_2(^1\Sigma_g^+) + A(+M). \tag{4.27}$$

This reaction also occurs in air (for example, with excited SO_2) but is of high importance in condensed phases, especially cells. Once singlet dioxygen forms, it undergoes interconversion reactions:

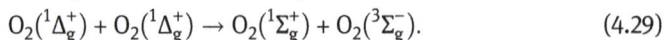

$$O_2(^1\Sigma_g^+) + A^*(+M) \rightarrow O_2(^1\Delta_g^+) + A(+M), \tag{4.28}$$

$$O_2(^1\Delta_g^+) + O_2(^1\Delta_g^+) \rightarrow O_2(^1\Sigma_g^+) + O_2(^3\Sigma_g^-). \tag{4.29}$$

Finally, singlet dioxygen can be produced chemically in the aqueous phase:

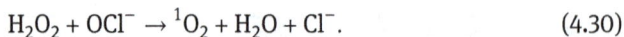

$$H_2O_2 + OCl^- \rightarrow {}^1O_2 + H_2O + Cl^-. \tag{4.30}$$

Hypochlorite plays a minor role in the free environment (Chapter 4.7.2), but hydrogen peroxide is ubiquitous in all natural systems and hypochlorite is enzymatically produced in cells.

The most important initial reaction is equation (3.167), which we can also write in the generic form of equation (4.31), where e^- means any kind of electron donor. The reduction potential of the donor must be smaller than $-0.33\,V$ in neutral solution or $-0.125\,V$ in acid solution:

$$O_2 \xrightarrow{e^-} O_2^-. \tag{4.31}$$

The limiting step is the electron transfer to O_2 from the photosensitiser (see reaction equation (3.183b)). This means that an electron source adequate for the reduction of O_2 will produce all the other reduced forms of dioxygen (O_2^-, HO_2, $HOOH$, HO_2^-) and finally OH via reduction, hydrolysis (or proteolysis) and disproportionation steps (Figure 4.6).

Thus, the most direct means to activate O_2 is the addition of an electron (or hydrogen atom), which results in significant fluxes of several ROS. The superoxide anion O_2^- is the protolytic dissociation product of the weak acid HO_2 ($pK_a = 4.8$). O_2^- is dominant in cloud and rainwater because of pH values normally around 5.5:

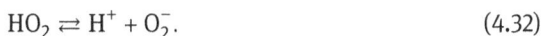

$$HO_2 \rightleftharpoons H^+ + O_2^-. \tag{4.32}$$

The uptake of HO_2 from the air is an important source of O_2^-, besides its aqueous-phase photocatalytic formation (equation (4.31) and equation (4.32)). In strong alkaline solution, spontaneous radical formation in alkaline solution was suggested; this pathway might be of interest in wastewater treatment:

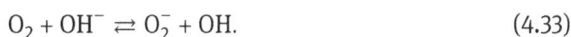

$$O_2 + OH^- \rightleftharpoons O_2^- + OH. \tag{4.33}$$

Finally, the following interesting pathway of O_2^- formation from photolysis of iron oxalate complexes is known:

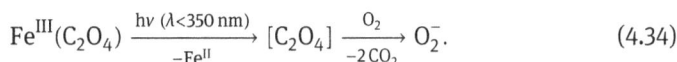

$$Fe^{III}(C_2O_4) \xrightarrow[-Fe^{II}]{h\nu \ (\lambda<350\,nm)} [C_2O_4] \xrightarrow[-2\,CO_2]{O_2} O_2^-. \tag{4.34}$$

Although superoxide ion is a powerful nucleophile in aprotic solvents, in water, it has less reactivity, presumably because of its strong hydration. Hence, within the water, superoxide anions are rapidly converted to dioxygen and peroxide:

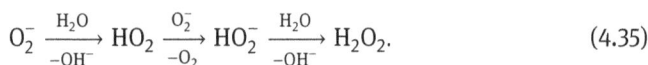

$$O_2^- \xrightarrow[-OH^-]{H_2O} HO_2 \xrightarrow[-O_2]{O_2^-} HO_2^- \xrightarrow[-OH^-]{H_2O} H_2O_2. \tag{4.35}$$

The reaction with water can be viewed as a polar-group-transfer reaction:

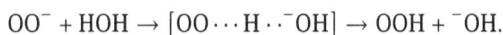

$$OO^- + HOH \rightarrow [OO \cdots H \cdot \cdot {}^-OH] \rightarrow OOH + {}^-OH.$$

This multistep process can be considered as an overall equilibrium: $k_{4.36} \approx 2.5 \cdot 10^8$ atm:

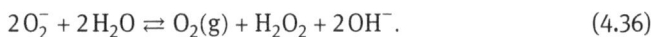

$$2\,O_2^- + 2\,H_2O \rightleftharpoons O_2(g) + H_2O_2 + 2\,OH^-. \tag{4.36}$$

Such a proton-driven disproportionation process means that O_2^- deprotonates acids much weaker than water. The first step is that O_2^- reacts with the proton source to form HO_2, which disproportionates with another O_2^-; $k_{4.37} = 1.0 \cdot 10^8 \ L\,mol^{-1}\,s^{-1}$:

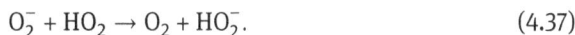

$$O_2^- + HO_2 \rightarrow O_2 + HO_2^-. \tag{4.37}$$

O_2^- undergoes competing reaction with ozone, where the ozonide anion (O_3^-) originates:

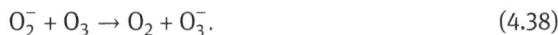

$$O_2^- + O_3 \rightarrow O_2 + O_3^-. \tag{4.38}$$

HO_2^- is the anion of the extremely weak acid H_2O_2 ($pK_a \approx 11.4$), i.e. the equilibrium lies on the right-hand side:

$$HO_2^- + H^+ \rightleftharpoons H_2O_2. \tag{4.39}$$

The two-step process is therefore $O_2^- + HO_2 \xrightarrow[-O_2]{} HO_2^- \xrightarrow{H^+} H_2O_2$. HO_2^- is also produced from HO_2 via electron transfer; transition metal ions are ubiquitous in natural waters (the reaction also proceeds with Mn^{2+} and $Cu+$):

$$HO_2 + Fe^{2+} \rightarrow HO_2^- + Fe^{3+}. \tag{4.40}$$

In competition is the deactivation (radical chain termination) of O_2^- by reaction equation (4.41):

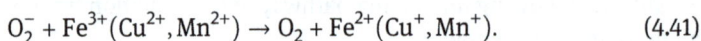

$$O_2^- + Fe^{3+}(Cu^{2+}, Mn^{2+}) \rightarrow O_2 + Fe^{2+}(Cu^+, Mn^+). \tag{4.41}$$

The HO_2 radical undergoes a rapid homolytic disproportionation, but much slower than the process with O_2^- in reaction equation (4.37); $k_{4.42} = 8.6 \cdot 10^5 \, L\,mol^{-1}\,s^{-1}$:

$$HO_2 + HO_2 \rightarrow H_2O_2 + O_2. \tag{4.42}$$

The homogeneous disproportionation of HO_2 seems to involve a 'head-to-tail' dimer intermediate that undergoes H-atom transfer:

$$2\,HOO \rightarrow [HOO \cdots HOO] \rightarrow HOOH + O_2.$$

> **i** Hydrogen dioxide (hydroperoxyl radical) undergoes protolysis ($pK = 4.8$) $HO_2 \rightleftharpoons O_2^-$ and is found in surface waters mostly as the superoxide anion O_2^-; only under acidic conditions (soils and cloud droplets) is the unprotonated HO_2 dominant. Its main role is to carry and transfer electrons:
>
> $$H_2O_2 \xleftarrow{\text{reduction}} HO_2 \rightleftharpoons O_2^- \xrightarrow{\text{oxidation}} O_2.$$

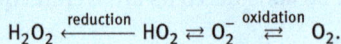

4.3.3.2 Aqueous-phase hydrogen peroxide chemistry

Before we go into OH and O_3 chemistry, we now turn to hydrogen peroxide H_2O_2 or HOOH, which is seen in equation (4.35) as the final product of oxygen reduction but is also considered an intermediate because of its reactivity in aqueous solution (in contrast to the gas phase). In the formation of oxygen from water during photosynthesis, H_2O_2 is proved as an important intermediate (see Figure 5.8). In cells, it is both a source of oxidative stress and a second messenger in signal transduction.

In his famous book, *Chemistry, Meteorology and the Function of Digestion* (1834), William Prout (1785–1850) noted the observation: "... the bleaching qualities of dew, and of the air itself; as to the large proportion of oxygen sometimes contained in snow water and in rainwater...".[56] Georg Meissner (1829–1905) provided the first evidence of H_2O_2 in the rain during a thunderstorm in 1862. Christian Friedrich Schönbein (1799–1868) confirmed this observation in 1869, and Heinrich Struve (1822–1908) detected it in the snow in 1869. Struve even proposed in 1870 that H_2O_2 is produced during all burning processes in air (Struve 1871). The German chemist Emil Schöne (1838–1896), however, was the first scientist to study atmospheric H_2O_2 systematically and in detail in rain, snow and air near Moscow in the 1870s. Since that time, in any respectable textbook of inorganic chemistry, it has been mentioned that hydrogen peroxide is found in traces in snow and rain. After the turn of the nineteenth century, evidence found for H_2O_2 in living plants, and it was proved as a primary product in respiration.

However, the role of superoxide anions in autoxidation processes only became clear after the 1950s, interestingly because of studying bleaching processes. The bleaching and germicidal properties of H_2O_2 had been found very early and opened the door to many industrial, wastewater treatment and medical applications. However, unbalancing the oxidation potential could result in oxidative stress. Thus, Möller (1989) proposed that increasing atmospheric H_2O_2 was responsible for the declining forests in Europe, which was first recognised in the second half of the 1970s (called new-type forest decline in contrast to smoke damages already caused by SO_2 in the nineteenth century).

The 'aquatic surface chemistry' (Stumm 1987) mechanism, now often termed as interfacial and photocatalytic chemistry, was not understood before the beginning of the 1960s. Since then, the reduction and oxidation of H_2O_2 has been well described in the sense of electron transfers depending on the reduction potential and pH of the aqueous solution (electron donors and/or acceptors), i. e. combining equations equation (3.183b) and equation (4.35):

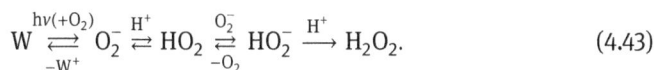

$$W \underset{-W^+}{\overset{h\nu(+O_2)}{\rightleftharpoons}} O_2^- \overset{H^+}{\rightleftharpoons} HO_2 \underset{-O_2}{\overset{O_2^-}{\rightleftharpoons}} HO_2^- \overset{H^+}{\longrightarrow} H_2O_2. \tag{4.43}$$

The high positive potential places H_2O_2 in the group of the most powerful oxidising agents known. The H_2O_2/H_2O couple has such a high potential that in many instances, the reduced species of the oxidising agent is oxidised back to its original state. The result of this behaviour is the decomposition of peroxide according to $2 H_2O_2 \rightarrow O_2 + 2 H_2O$. The OH radical oxidises H_2O_2; $k_{4.44} = 4.5 \cdot 10^7 \, L \, mol^{-1} \, s^{-1}$:

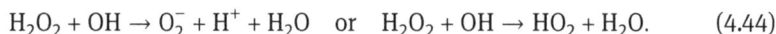

$$H_2O_2 + OH \rightarrow O_2^- + H^+ + H_2O \quad \text{or} \quad H_2O_2 + OH \rightarrow HO_2 + H_2O. \tag{4.44}$$

---

[56] The bleaching properties of dew have been known for centuries and dew has been used for the cleansing of clothes. Textiles have long been whitened by grass bleaching (spreading the cloth upon the grass for several months), a method virtually monopolised by the Dutch from the time of the Crusades to the eighteenth century.

Oxidative damage (stress) *in vivo* is often ascribed to the Fenton reaction. In 1876, Henry Fenton (1854–1929) described a coloured product obtained by mixing tartaric acid with hydrogen peroxide and a low concentration of a ferrous salt and found that iron acts catalytically. Today we know the involvement of hydroxyl radicals in the iron(II)/hydrogen peroxide system (called Fenton chemistry):

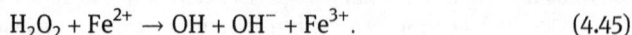

$$H_2O_2 + Fe^{2+} \rightarrow OH + OH^- + Fe^{3+}. \tag{4.45}$$

Reaction equation (4.45) also proceeds with hydrated electrons; $k_{4.46} = 1.36 \cdot 10^{10}$ L $mol^{-1} s^{-1}$:

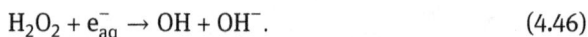

$$H_2O_2 + e_{aq}^- \rightarrow OH + OH^-. \tag{4.46}$$

Solutions, containing H_2O_2 and Fe^{2+} (ferrous ion) are called Fenton reagents and are used for the oxidation of organic compounds as well as oxidising contaminants of wastewaters. Fenton-like chemistry goes on with other TMI such as Cu and Mn:

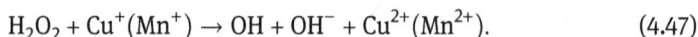

$$H_2O_2 + Cu^+(Mn^+) \rightarrow OH + OH^- + Cu^{2+}(Mn^{2+}). \tag{4.47}$$

The high importance of this pathway lies in OH generation. When H_2O_2 acts as a reducing species, the elementary step can be regarded as electron transfer onto H_2O_2, whereas $H_2O_2^-$ decays in a non-reversible manner in an acid medium (it is assumed to be in equilibrium $H_2O_2^- \rightleftharpoons OH + OH^-$) with OH generation:

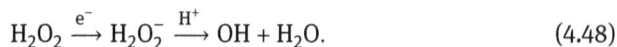

$$H_2O_2 \xrightarrow{e^-} H_2O_2^- \xrightarrow{H^+} OH + H_2O. \tag{4.48}$$

In polluted air, the main fate of aqueous H_2O_2 is the fast oxidation of dissolved SO_2 (Chapter 4.5.3), which limits the lifetime of both species. In biological systems, besides inorganic Fenton chemistry, three oxygenic base processes occur (see also Chapter 5.2.2):

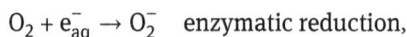

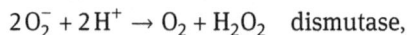

$$O_2 + e_{aq}^- \rightarrow O_2^- \quad \text{enzymatic reduction,}$$
$$2 O_2^- + 2 H^+ \rightarrow O_2 + H_2O_2 \quad \text{dismutase,}$$
$$2 H_2O_2 \rightarrow 2 H_2O + O_2 \quad \text{catalase.}$$

Hydrogen peroxide (which is between the oxygen state −2 and 0) can also act as a reducing agent and thereby turn oxidised metals back to lower oxidation states; however, at a slow rate, $k_{4.49}$ (Fe^{3+}) = $6.0 \cdot 10^2$ L $mol^{-1} s^{-1}$ and $k_{4.49}$ (Mn^{2+}) $7.3 \cdot 10^4$ L $mol^{-1} s^{-1}$:

$$H_2O_2 + Fe^{3+}(Cu^{2+}, Mn^{2+}) \rightarrow HO_2 + H^+ + Fe^{2+}(Cu^+, Mn^+). \tag{4.49}$$

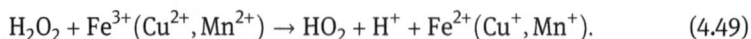

It has been suggested that superoxide reduces the iron(III) (equation (4.41)) formed on reaction equation (4.40) to explain the catalytic of the metal. All these findings emphasise the central role of H_2O_2 in radical chains during biochemical, combustion and chemical processes in natural waters in the air.

$$H_2O \xleftarrow{\text{reduced}} H_2O_2 \xrightarrow{\text{oxidised}} O_2.$$

Whereas H_2O_2 in the gas phase is more like an oxidant reservoir (the slow photolysis only plays a role in the upper troposphere), it is included in intensive cycling between OH, HO_2 and H_2O_2 in the aqueous phase. Thus, depending on the pH, the aqueous phase provides a variety of ROS, partly scavenged from the gas phase and produced in solution where oxygen photocatalysis in the presence of photosensitisers occurs in a unique way for oxidation processes. As for the gas phase, sunlight is essential, but in the absence of light transition, metal ions carry out the electron transfer processes.

4.3.3.3 Aqueous-phase ozone chemistry

Ozone is unstable in water. Ozone is one of the most powerful oxidizing agents (E° = 2.08 V), which reacts with several categories of compounds directly. Atoms with negative charge density (N, P, O, or nucleophilic carbons) or double or triple bonds are the main targets for an electrophilic attack of ozone molecules. Hydroxyl radicals (OH) produced in the indirect reaction of ozone act as a strong and non-selective oxidants.

Hence, the ozonation of drinking water has been widely used after the recognition of the germicidal properties of ozone, first shown in 1859 by Eugen Gorup-Besanez (1817–1869) and later applied to treat organically polluted water by Friedrich Emich (1860–1940) in 1885. Only after the manufacture of ozone generators was the first water treatment plant established in 1892 in Martinikenfelde[57] near Berlin by Siemens and Halske. However, before the 1980s, nothing was known about detailed chemical mechanisms (see below).

Before we summarise current knowledge about ozone decay in natural water, let us turn for a moment to early observations. It has long been known that ozone decays in an alkaline solution. Schönbein (1844) found that ozone is removed from the air after bubbling through alkaline solutions. This was proven quantitatively by Soret (1864), but Cossa (1867) found that O_3 will not be destroyed in a pure KOH solution free of any organic substances. However, a mechanism was not known (only the formation of O_2) before Weiss (1935) first proposed the following three reactions, based

---

57 This former village is now an urban region at the northwestern edge of downtown Berlin.

on the observation that ozone decay is effective only in an alkaline solution:

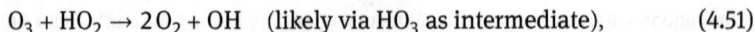

$$O_3 + OH^- \rightarrow O_2^- + HO_2, \tag{4.50}$$

$$O_3 + HO_2 \rightarrow 2O_2 + OH \quad \text{(likely via } HO_3 \text{ as intermediate)}, \tag{4.51}$$

$$O_3 + OH \rightarrow O_2 + HO_2. \tag{4.52}$$

Staehelin and Hoigné (1982), Sehested et al. (1983) and Staehelin et al. (1984) added several reactions to the Weiss mechanism:

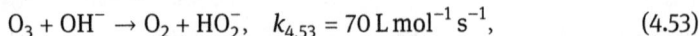

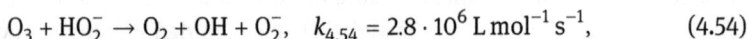

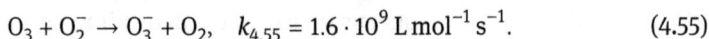

$$O_3 + OH^- \rightarrow O_2 + HO_2^-, \quad k_{4.53} = 70\,\text{L mol}^{-1}\,\text{s}^{-1}, \tag{4.53}$$

$$O_3 + HO_2^- \rightarrow O_2 + OH + O_2^-, \quad k_{4.54} = 2.8 \cdot 10^6\,\text{L mol}^{-1}\,\text{s}^{-1}, \tag{4.54}$$

$$O_3 + O_2^- \rightarrow O_3^- + O_2, \quad k_{4.55} = 1.6 \cdot 10^9\,\text{L mol}^{-1}\,\text{s}^{-1}. \tag{4.55}$$

It is clear that only reaction equation (4.55) is fast enough for further consideration; the spontaneous alkaline ozone decay (non-radical and non-photochemical) according to equation (4.50) and equation (4.53) is too slow to obtain any environmental importance. The formation of H_2O_2 in ozone decay has been controversially discussed since Schönbein where H_2O_4 ($H_2O \cdot O_3$), the hypothetical ozone acid, was first proposed by Gräfenberg (1902, 1903): H–O–O–O–O–H (also called hydrogen superoxide according to Ardon 1965) being a so-called spontaneous decay reaction:

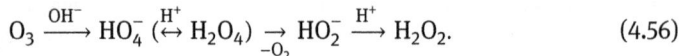

$$O_3 \xrightarrow{OH^-} HO_4^- \; (\xleftrightarrow{H^+} H_2O_4) \xrightarrow[-O_2]{} HO_2^- \xrightarrow{H^+} H_2O_2. \tag{4.56}$$

The listed ozone reactions can be expressed as electron transfer processes. The ozonide anion (systematic name: trioxide(1-)) O_3^- (not to be mixed up with the olefin-ozone adduct, Chapter 4.6.3.4) is easily produced through direct electron transfer onto dissolved ozone; the electron affinity of O_3 is several times (2.1 eV) that of O_2 (0.44 eV); $k_{4.57} = 3.6 \cdot 10^{10}\,\text{L mol}^{-1}\,\text{s}^{-1}$:

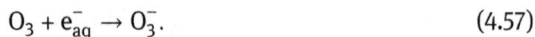

$$O_3 + e_{aq}^- \rightarrow O_3^-. \tag{4.57}$$

In summary, all the proposed reactions of O_3 decay in solution, the intermediate O_3^-, gained either via equation (4.57) or equation (4.55) quickly decays according to the following reaction sequence (Sehested et al. 1983): $k_{4.58} = 5 \cdot 10^{10}\,\text{L mol}^{-1}\,\text{s}^{-1}$ and $k_{-4.58} = 3.3 \cdot 10^2\,\text{s}^{-1}$: $k_{4.59} = 1.4 \cdot 10^5\,\text{s}^{-1}$:

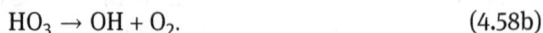

$$O_3^- + H^+ \rightleftarrows HO_3, \tag{4.58a}$$

$$HO_3 \rightarrow OH + O_2. \tag{4.58b}$$

The overall process is given by:

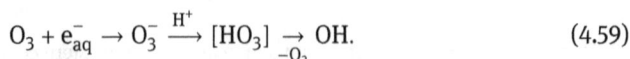

$$O_3 + e_{aq}^- \rightarrow O_3^- \xrightarrow{H^+} [HO_3] \xrightarrow[-O_2]{} OH. \tag{4.59}$$

This process is used in the so-called photocatalytic ozonation using TiO_2 as photosensitizer (see Chapter 3.3.6.4 and equation (3.184)). The direct ozonation is based on a nucleophilic attack resulting in an O-atom transfer (see equations (4.51)–(4.54). Ozone reacts rapidly with a wide range of inorganic ions in an aqueous solution, e. g. nitrite (equation (4.142)), sulphite and dissolved SO_2 (equations (4.228)–(4.230)), hydrogen sulphide, cyanide (CN^-) and halogenides. First, an adduct forms followed by oxygen-atom transfer with the direct elimination of O_2:

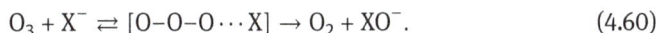

$$O_3 + X^- \rightleftarrows [O–O–O \cdots X] \rightarrow O_2 + XO^-. \tag{4.60}$$

The reaction rate is very different, for example, very slow ($k < 3 \cdot 10^{-3}\,\mathrm{L\,mol^{-1}\,s^{-1}}$) with chloride ($Cl^-$) and very fast. This mechanism for $Cl^- + O_3$ was already proposed in 1949, but in 2010 an electron transfer in analogy to equation (4.55) have been suggested, but in 2012 the oxygen atom transfer mechanism equation (4.60) was reconfirmed.

The intermediate HO_3 (also called hydrogen trioxide) has been detected. H_2O_4 (H_2O_3, produced from $OH + HO_2$, has also been suggested) have never been identified; nevertheless, the formation of H_2O_2 from O_3 in an alkaline medium has been proved to affect the direct formation of OH radicals. Thus, in the presence of O_3 and electron donors OH radicals can be produced via equation (4.60) with much higher yields than in the reaction $O_2 + e^-$ including subsequent steps finally to H_2O_2 (equation (4.35)). Once OH is gained, it regenerates ROS via degradation of DOC:[58]

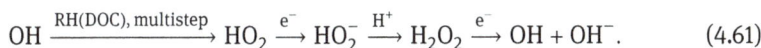

$$OH \xrightarrow{\text{RH(DOC), multistep}} HO_2 \xrightarrow{e^-} HO_2^- \xrightarrow{H^+} H_2O_2 \xrightarrow{e^-} OH + OH^-. \tag{4.61}$$

Thus, Cossa's remarkable early observations support the essential role of water-dissolved organic compounds (DOC or NOM) in reaction equation (4.62). It seems very likely that reaction equation (4.57), apart from equation (4.43), takes place on all wetted surfaces. For example, dew provides a medium where all needed reactants are available. Hence, organic matter plays a crucial role as an electron donor as well as a converter of OH into HO_2. Recently, humic-like substances (HULIS, also called macro-molecular compounds) have been found in atmospheric aerosols and cloud water, which were probably produced from primary aromatic compounds in atmospheric chemical processes and have chromophoric properties.

The most important ozone reaction in a slightly alkaline solution is $O_3 + O_2^-$ (equation (4.55)) with subsequent OH radical formation; however, in the presence of sulphurous acid (dissolved SO_2), ozone is consumed while producing sulphate (see equations (4.228)–(4.230)).

---

58 It is self-evident that this reaction provides efficient water treatment and cleaning.

4.3.3.4 Aqueous-phase OH chemistry

Hydroxyl radicals (OH) are less scavenged than HO_2 (which is an important fact when considering the gas phase ozone formation cycle; Chapter 5.3.2.2). Note that the OH yield in reaction equation (4.48) is stoichiometric to H_2O_2 and thereby provides OH concentrations that are orders of magnitudes larger than by the phase transfer of gaseous OH (remember that $[HO_2]/[OH] \sim 10$ and $[H_2O_2]$ in gas-phase is orders of magnitude higher). The OH radical has a standard reduction potential of +2.8 V in an acidic solution and is, therefore, a strong oxidant. The main sink of OH in an aqueous solution is similar to the gas phase (see Chapter 4.6.3.1), the oxidation of hydrocarbons (RH) through abstraction of the H atom from dissolved organic compounds (DOC); see equation (4.61):

$$\text{OH} + \text{RH} \xrightarrow[-\text{H}_2\text{O}]{} \text{R} \xrightarrow{\text{O}_2\text{(multistep)}} \text{HO}_2 + \text{products.} \tag{4.62}$$

If the solution contains formate (the anion of formic acid, one of the most abundant organic acids in the environment) and is saturated with O_2, then the following process produces peroxide (HO_2/O_2^-):

$$\text{HC(O)O}^- + \text{OH} \rightarrow \text{C(O)O}^- + \text{H}_2\text{O}, \tag{4.63}$$

$$\text{C(O)O}^- + \text{O}_2 \rightarrow \text{CO}_2 + \text{O}_2^-. \tag{4.64}$$

OH also reacts with (oxidises) many inorganic ions (Fe^{2+}, Mn^+, Cu^+, HSO_3^-, NO_2^- etc.) as an electron acceptor:

$$\text{OH} + \text{A}^{n+} \rightarrow \text{OH}^- + \text{A}^{(n+1)+}. \tag{4.65}$$

Another pathway is the addition with anions to radical ions (not to be confused with hypochloric acid; HOCl), which decomposes into hydroxide ions and new radicals:

$$\text{OH} + \text{Cl}^-(\text{Br}^-) \rightarrow \text{HOCl}^-(\text{HOBr}^-) \rightarrow \text{OH}^- + \text{Cl(Br).} \tag{4.66}$$

Reaction with metal ions is slower ($\leq 10^8$ L mol^{-1} s^{-1}) because of the replacement of the H_2O ligand:

$$\text{OH} + \text{Me}^{n+}(\text{H}_2\text{O})_x \xrightarrow[-\text{H}_2\text{O}]{} \text{Me}^{n+}(\text{H}_2\text{O})_{x-1}\text{OH} \xrightarrow{\text{H}_2\text{O}} \text{Me}^{(n+1)+}(\text{H}_2\text{O})_x + \text{OH}^-.$$

In a strong alkaline solution, the OH radical dissociates (it is a weak acid with $pK_a = 11.9$): $k_{4.67} = 1.2 \cdot 10^{10}$ L mol^{-1} s^{-1} and $k_{-4.67} = 9.3 \cdot 10^7$ L mol^{-1} s^{-1}:

$$\text{OH} + \text{OH}^- \rightleftharpoons \text{O}^- + \text{H}_2\text{O} \quad \text{or} \quad \text{OH} \rightleftharpoons \text{O}^- + \text{H}^+ \quad \text{or} \quad \text{OH} + \text{H}_2\text{O} \rightleftharpoons \text{O}^- + \text{H}_3\text{O}^+. \tag{4.67}$$

The standard electrode potential of OH in an alkaline solution is smaller than in an acidic solution with 1.4 V. The O^- radical reacts quickly with organic compounds under H abstraction:

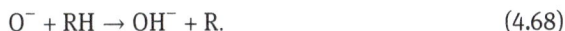

$$O^- + RH \rightarrow OH^- + R. \tag{4.68}$$

However, O^- and OH are interconvertible:

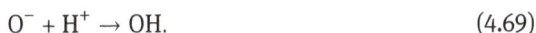

$$O^- + H^+ \rightarrow OH. \tag{4.69}$$

The dominant pathways in an alkaline solution are given by the reaction with oxygen equation (4.70); $k_{4.70} = 2.6 \cdot 10^9 \, L \, mol^{-1} \, s^{-1}$):

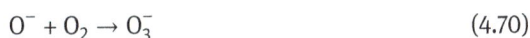

$$O^- + O_2 \rightarrow O_3^- \tag{4.70}$$

and the reaction with water ($k_{4.71} = 10^8 \, s^{-1}$); see also equation (4.48):

$$O^- + H_2O \rightarrow H_2O_2^- \rightarrow OH + OH^-. \tag{4.71}$$

O^- is produced from the reaction between e_{aq}^- and N_2O in aqueous solution (which is the most important scavenger for hydrated electrons): $k_{4.72} = 2.6 \cdot 10^9 \, L \, mol^{-1} \, s^{-1}$:

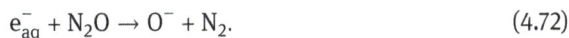

$$e_{aq}^- + N_2O \rightarrow O^- + N_2. \tag{4.72}$$

The fate of O^- (if it is ever produced in natural waters under limited conditions, corresponding to OH according to equation 4.67) is transformation into the ozonide anion O_3^- under oxic conditions (equation (4.70), which is an important intermediate in an alkaline solution with a lifetime of about 10^{-3} s. In Chapter 4.3.3.3, we have seen that it is produced through electron transfer onto ozone.

Similar to the gas phase, OH plays a central role as a key oxidant for many dissolved inorganic and organic species. Peroxides (H_2O_2 and HO_2/O_2^-) play the role of OH generation and cycling. Organic photosensitisers in the presence of sunlight produce hydrated electrons or directly O_2^- – the key role of natural water, mainly at the interface to air, is the formation of ROS, thereby providing oxidation processes, including corrosion and autoxidation.

The ultimate source of OH radicals in natural waters is the photosensitising of dissolved oxygen and the ozone decay, where hydrogen peroxide plays a role of intermediate.

4.4 Nitrogen

Nitrogen, unlike carbon, is not one of the more abundant elements on Earth – its mean mass fraction is about only 0.002 %; thus, oxygen is 23,000 times and carbon 10 times more abundant (Table 4.1). In space, however, nitrogen is the fourth most abundant element (when not considering hydrogen and the noble gases); the ratios of O/N and C/N amount to about 10 and 5, respectively. The molecular nitrogen (N_2) now present in the earth's atmosphere is considered to have remained here since the planet was first formed 4.6 billion years ago. Variation of the mass of O_2, CO_2 and H_2O, but not that of N_2, must have led to variation in total atmospheric pressure with time. The residence time of N_2 in the atmosphere, relative to exchange with and storage in crustal rocks, is estimated to be about one billion years. Even though the atmosphere is 78 % nitrogen (N_2), most biological systems are nitrogen-limited on a physiological timescale because most biota are unable to use molecular nitrogen (N_2). Two natural processes convert nonreactive N_2 to reactive N; *lightning* and *biological fixation*. The environmental importance, more exactly its role for life processes (see also Chapter 5.2.4), lies in the wide range of oxidation states from N(−3) to N(+5), hence promoting redox processes; Figure 4.7. It is the only element existing in all nine oxidation states (Table 4.11):

−3	−2	−1	±0	+1	+2	+3	+4	+5
NH_3	N_2H_4	N_2H_2	N_2	N_2O	NO	N_2O_3	NO_2	N_2O_5

Table 4.11: Inorganic nitrogen species.

ox. state	hydrides and oxides	name	oxo acids	name
−3	NH_3	ammonia		
−2	N_2H_4 ($H_2N=NH_2$)[a]	hydrazine		
−1	N_2H_2 (HN=NH)[a]	diimine[b]	$HONH_2$[a]	hydroxyl amine
	NH[a]	nitrene (or azene)		
±0	N_2 (N≡N)	nitrogen		
+1	N_2O	dinitrogen monoxide	HON[a,c]	hydrogen oxonitrate[e]
			HON=NOH[a]	hyponitrous acid
+2	NO	nitrogen monoxide	HONO	nitrous acid
+3	N_2O_3	dinitrogen trioxide	HOONO[d]	peroxonitrous acid
+4	NO_2	nitrogen dioxide		
	N_2O_4	dinitrogen tetroxide		
+5	NO_3	nitrogen trioxide	$HONO_2$	nitric acid
	N_2O_5	dinitrogen pentoxide	$HOONO_2$[c]	peroxonitric acid

[a] likely only intermediates in plants
[b] as azo group (−N=N−) in organic compounds
[c] instable (intermediates)
[e] nitroxyl radical
[d] isomer of nitric acid

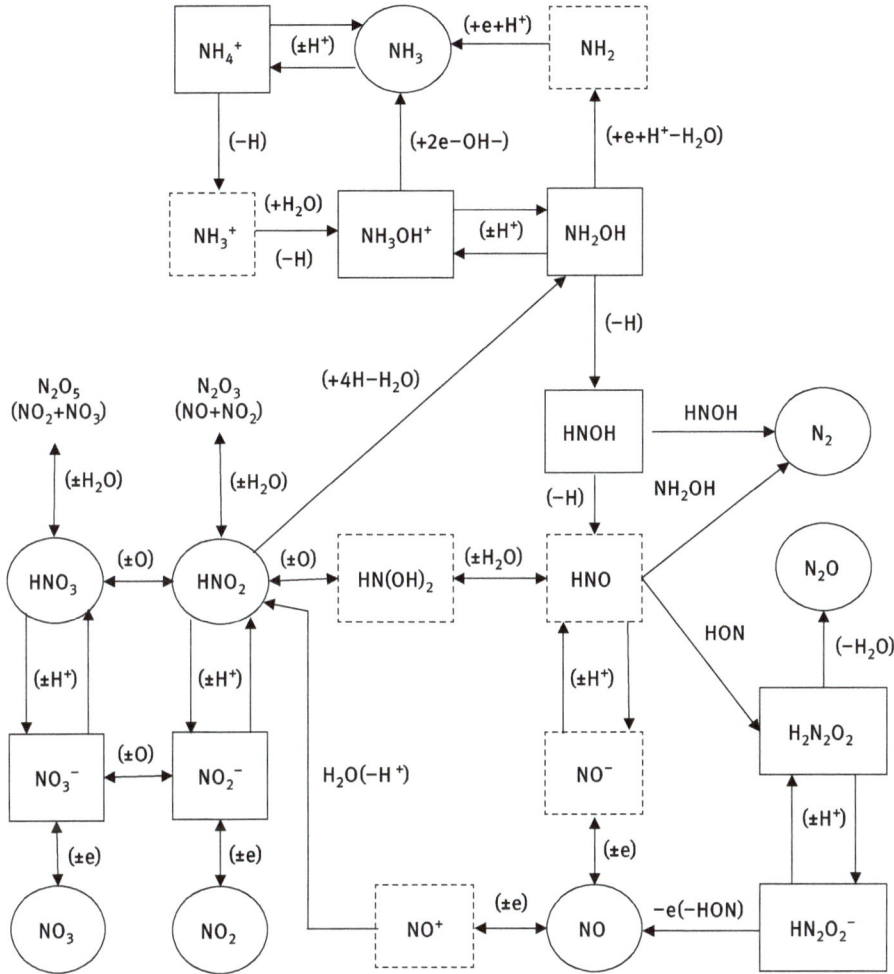

Figure 4.7: Scheme of the biochemical, soil chemical and atmospheric chemical nitrogen oxidation and reduction in the biosphere (nitrification, ammonification and denitrification). Symbols for redox processes: $-e$: red, $+e$: ox, $-H$: ox ($\equiv e + H^+$), $+H$: red ($\equiv +H^+ - e$), $-O$: red ($\equiv +2\,H^+ - H_2O$), $+O$: ox ($\equiv +H_2O - 2\,H$); circles: gases; dotted boxes: intermediates.

In animals (including humans) and some plants, nitrogen ranks in the third place (after hydrogen and carbon, excluding water as a molecule) in the formation of proteins. It is the $-NH_2$ group (together with the carboxyl group $-COOH$), forming amino acids, which build up proteins, large biological molecules that perform a vast array of functions within living organisms, including catalysing metabolic reactions, replicating DNA, and transporting molecules from one location to another.

Reactive nitrogen is defined as any single nitrogen species with the exception of N_2 and N_2O. It includes:

- NO_y ($NO + NO_2 + N_2O_3 + N_2O_4 + HNO_2 + HNO_3 + NO_3 + N_2O_5 + HNO_4 +$ organic $NO_x +$ particulate NO_2^- and NO_3^-);
- NH_x ($NH_3 + NH_4^+$), NH_2OH;
- organic bonded N (mostly NH_2^- but also SCN and other structures or functional groups with special biochemical functions).

Note that $NO_x = NO + NO_2$ and is often defined as $NO_z = NO_y - NO_x$. In addition to being important to biological systems, reactive nitrogen also affects the chemistry of the atmosphere. At very low NO concentrations, ozone (O_3) is destroyed by reactions with radicals (especially HO_2), although at higher levels of NO (greater than 10 ppt), there is a net O_3 production (because HO_2 reacts with NO to form NO_2). The photolysis of NO_2 is the only source of photochemically produced O_3 in the troposphere.

Although N_2O is not viewed as a reactive form of nitrogen in the troposphere; it adsorbs IR radiation and acts as a greenhouse gas. In the stratosphere, N_2O will be oxidised to NO_x and influences the O_3 concentration.

NH_3 is the major source of alkalinity in the atmosphere and a source of acidity in soils. A small part of atmospheric NH_3 ($\leq 5\%$) is only oxidised by OH radicals, where the main product has been estimated to be N_2O, thus contributing around 5 % to estimated global N_2O production. Deposited NH_3/NH_4^+ will be *nitrified* in soils and water to NO_3^-, where two moles of H^+ are formed for each mole of NH_3/NH_4^+. Thus, any change in the rate of formation of reactive nitrogen (and N_2O), its global distribution or its accumulation rate can have a fundamental impact on many environmental processes.

We live in an era with a surplus of ammonia (NH_3) and ammonium (NH_4^+) in many parts of the world. Following the invention of the Haber–Bosch process, patented in 1908 by Fritz Haber (1868–1934) and commercialised by Carl Bosch (1874–1940), it has been possible to produce ammonia in large quantities relatively cheaply. In particular, the widespread use of ammonia and its derivatives as agricultural nitrogen fertilisers has substantially increased emissions of ammonia to the atmosphere, leading to a wide range of different environmental problems. These include the eutrophication of semi-natural ecosystems, acidification of soils, formation of fine particulate matter in the atmosphere, and alteration of the global greenhouse balance.

Ammonia is considered to be a potential medium for hydrogen storage: Its high volumetric hydrogen density, low storage pressure and stability for long-term storage are among the beneficial characteristics. Furthermore, ammonia is also considered safe due to its high auto-ignition temperature, low condensation pressure and lower gas density than air. However, for energy harvesting, the utilization of ammonia has not been widely adopted. Intensively studies in the past have confirmed that ammonia has potential as the primary fuel for a spark-ignition engine; however, the low combustion rate of ammonia results in practical problems. In the case hydrogen needs to be released from ammonia, the decomposition of ammonia to hydrogen can be conducted via thermochemical and electrochemical routes. However, ammonia also can be utilised using a direct fuel cell, without the need of decomposition or cracking.

Ammonia is used for the abatement of NO emission from boilers by catalytic processing (see equation (4.102)).

4.4.1 Natural occurrence and sources

Biological processes in soils, waters (including oceans), and even in organisms themselves, together with inorganic conversions in the medium, produce a huge number of compounds (almost all presented in this book), found in the environmental compartments and, according to the physical conditions, exchanged and transferred between soil, water and air. The bacterial decomposition of animal excreta is the largest source of NH_3. Besides, but in much smaller quantities, many organic amines are emitted from animals. It is likely that natural ecosystems (forest, grassland) emit no or only small amounts of ammonia because normally, there is a deficit of fixed nitrogen in landscapes.

In contrast to sulphur species, there are no fundamental differences between natural and anthropogenic processes in the formation and release of reactive nitrogen species. Industrial nitrogen fixation (in separated steps: $N_2 \rightarrow NH_3, N_2 \rightarrow NO_x, NO_x \rightarrow NO_3$) proceeds via the same oxidation levels as biotic *fixation* and *nitrification*, either on purpose in chemical industries (ammonia synthesis, nitric acid production) or unintentionally in all high-temperature processes, namely combustion, as a by-product due to $N_2 + O_2 \rightarrow 2\,NO$.

In cultivated soils, as for NH_3, the primary NO source is N fertiliser application; natural soils have much less specific emissions. Denitrification (see Chapter 5.2.4) is the direct way of producing molecular nitrogen (N_2), dinitrogen monoxide (N_2O) and nitrogen monoxide (NO); Figure 4.7. This way is parallel to the formation of ammonia/ammonium (ammonification), and therefore it is assumed that all these compounds appear together but in different quantities. Soil structure and pH, oxygen content, humidity and temperature and radiation are important parameters in determining emissions. N_2O emissions from soils under natural vegetation are significantly influenced by vegetation type, soil organic C content, soil pH, bulk density and drainage, while vegetation type and soil C content are major factors for NO emissions. A soil emission similar to those of ammonia is proposed. The emission of N_2O is perhaps twice that, which is supported by the more direct chemical formation pathway during denitrification (Figure 4.7). In analogy to ammonia, microorganisms living in soils and plants assimilate both gases. As for ammonia, emission of NO and N_2O is considered to be a loss for the organisms, in contrast to the emission of N_2 by denitrification, closing the atmospheric cycle ($>100\,\text{Tg N yr}^{-1}$).

As for NO production during lightning, similar conversion processes occur in all combustion and high-temperature processes (see Chapter 4.4.2). To a minor percentage, the fuel nitrogen content contributes to NO formation, but this pathway is relatively more important for biofuels. There is no less uncertainty in estimating the anthropogenic NO emission. While the fossil-fuel source is estimated at 20–25 Tg N yr^{-1}, total NO emissions are about 44 (23–81) Tg N yr^{-1}.

4.4.2 Thermal dissociation of dinitrogen (N_2)

> **i** At high temperatures (T > 1000 °C), molecular nitrogen from air converts into NO.

This can happen during lightning (biomass combustion do not provide such high T) and in industrial combustion processes. Lightning provides – depending on the flash energy – thermal energy for molecule dissociation. Therefore, ambient air molecules are dissociated, and subsequently, new molecules are produced according to the air chemical conditions. Depending on the flash energy and the molecule dissociation energy, there is no other limit to the decomposition (and subsequently to synthesis) of any molecule in the atmosphere. Because of the large dissociation energy in N_2, the initial step is dioxygen thermal dissociation ($O_2 \rightleftharpoons O + O$) with a subsequent reaction of oxygen atoms with N_2:

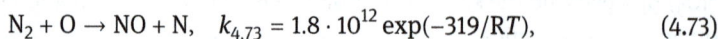

$$N_2 + O \rightarrow NO + N, \quad k_{4.73} = 1.8 \cdot 10^{12} \exp(-319/RT), \tag{4.73}$$

$$N + O_2 \rightarrow NO + O, \quad k_{4.74} = 6.4 \cdot 10^{9} \exp(-26/RT). \tag{4.74}$$

In steady state ($dN/dt = 0$), it follows that:

$$\frac{d[NO]}{dt} = 2 \cdot k_{4.73}[N_2][O]. \tag{4.75}$$

A reaction chain follows where radical reactions equation (4.76) and equation (4.77) are rapid whereas thermal molecular reactions equations (4.78)–(4.80) are slow; the radical reaction rates of equation (4.81) and equation (4.82) are slow due to less reaction probability:

$$NO + N \rightarrow N_2 + O, \tag{4.76}$$

$$NO + O \rightarrow N + O_2, \tag{4.77}$$

$$NO \rightleftharpoons N + O, \tag{4.78}$$

$$NO + NO \rightarrow N_2O + O, \tag{4.79}$$

$$NO + NO + O_2 \rightarrow 2NO, \quad k_{4.80} \sim 2 \cdot 10^{-19}. \tag{4.80}$$

Besides equation (4.79), the formation of N_2O is possible for a small percentage, which can also affect the NO formation:

$$N_2 + O + M \rightarrow N_2O + M, \tag{4.81}$$

$$N_2O + O \rightarrow NO + NO. \tag{4.82}$$

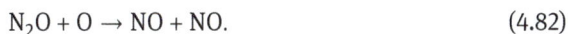

The above-described mechanism is valid for all high-temperature processes (lightning, metallurgic processes, glass production). This mechanism is known as the 'thermal NO mechanism' or the 'Zeldovich mechanism'.[59] In fuel-rich flames (hydrocarbon combustion), additionally

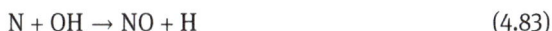

$$N + OH \rightarrow NO + H \tag{4.83}$$

occurs. During combustion of fossils fuels, CH, H, O, and OH radicals are produced, and the following reactions were proposed (Fenimore mechanism):[60]

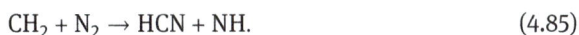

$$CH + N_2 \rightarrow HCN + N, \tag{4.84}$$
$$CH_2 + N_2 \rightarrow HCN + NH. \tag{4.85}$$

The different nitrogen products quickly oxidise to NO. After all, the primary product is NO, which then can go into further oxidation processes to NO_2.

4.4.3 Ammonia (NH_3)

All nitrogen-fixing organisms are prokaryotes (bacteria). Some of them live independent of other organisms – the so-called free-living nitrogen-fixing bacteria. Others live in intimate symbiotic associations with plants or other organisms (e. g. protozoa). Biological nitrogen fixation can be represented by the following equation, in which two moles of ammonia are produced from one mole of nitrogen gas, at the expense of 12 moles of ATP (adenosine triphosphate, empirical formula: $C_{10}H_{16}N_5O_{13}P_3$; ADP adenosine diphosphate) and a supply of electrons and protons, using an enzyme complex termed nitrogenase. This reaction is performed exclusively by prokaryotes (the bacteria and related organisms):

$$N_2 + 6\,H^+ + 6\,e^- + 12\,ATP \rightarrow 2\,NH_3 + 12\,ADP + 12\,phosphate.$$

There is stepwise hydrogenation via diimine N_2H_2, an unstable intermediate, and hydrazine N_2H_4 (diazene) to ammonia (azane):

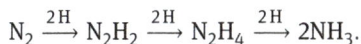

$$N_2 \xrightarrow{2H} N_2H_2 \xrightarrow{2H} N_2H_4 \xrightarrow{2H} 2NH_3.$$

---

59 Jakow Borissowitsch Zeldovich (1914–1987) was a Soviet physical chemist.
60 Charles Paine Fenimore (1913–?) was an US American chemist.

Hydrazine is found in natural waters as a pollutant from industrial manufacturing in small concentrations. The technical formation of NH_3, to provide nitrogen fertilisers (chemical nitrogen fixation), is based on the Haber–Bosch process (500 °C and 200 bar, iron-catalyst):

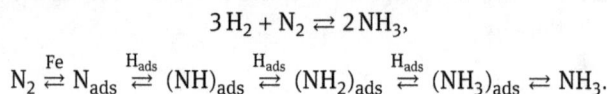

$$3\,H_2 + N_2 \rightleftharpoons 2\,NH_3,$$

$$N_2 \overset{Fe}{\rightleftharpoons} N_{ads} \overset{H_{ads}}{\rightleftharpoons} (NH)_{ads} \overset{H_{ads}}{\rightleftharpoons} (NH_2)_{ads} \overset{H_{ads}}{\rightleftharpoons} (NH_3)_{ads} \rightleftharpoons NH_3.$$

Ammonia[61] is relatively stable in air, and its importance lies in the formation of salts. The photodissociation is out of interest in the atmosphere (but was discussed in the early atmosphere) because the dissociation energy of ammonia through photolysis at 205 nm was determined to be (4.34 ± 0.07) eV:

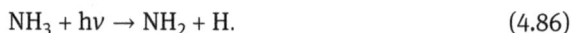

$$NH_3 + h\nu \rightarrow NH_2 + H. \tag{4.86}$$

The only gas-phase reaction of NH_3 is that with OH, forming the amidogen (amide) radical NH_2; $k_{4.87} = 1.6 \cdot 10^{-13}$ cm^3 molecules^{-1} s^{-1}:

$$NH_3 + OH \rightarrow NH_2 + H_2O. \tag{4.87}$$

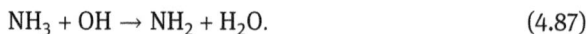

The water amidogen radical complex (H_2O-NH_2) has been detected as a reactive intermediate in atmospheric ammonia oxidation. The reaction of NH_2 with O_2 is slow $(k < 6 \cdot 10^{-21}$ cm^3 molecule^{-1} s^{-1} at 298 K), making it unimportant in the atmosphere. The products are not specified (NH_2O_2, $NO + H_2O$, $OH + HNO$). The most likely fate (Finlayson-Pitts and Pitts (2000) suggest a lifetime of about 2–3 s) is reaction with NO_x: $k_{4.88}$ ($= k_a + k_b + k_c) = 1.6 \cdot 10^{-11}$ cm^3 molecule^{-1} s^{-1} at 298 K with $k_a/k_{4.88} = 0.9$ and $(k_b + k_c)/k_{4.88} = 0.1$:

$$NH_2 + NO \rightarrow N_2 + H_2O, \tag{4.88a}$$
$$NH_2 + NO \rightarrow N_2H + OH, \tag{4.88b}$$
$$NH_2 + NO \rightarrow N_2 + H + OH. \tag{4.88c}$$

In the reaction with NO_2, channels (a) and (c) are the most probable; no evidence has been found for the occurrence of the channel (b) or the other exothermic channels

---

[61] The origin of the word ammonia is often said to relate to the classical discovery of *sal ammoniac* near the Temple of Zeus Ammon, in the Siwa Oasis of the Lybian Desert. Pliny the Elder is often cited as the first to note the existence of *hammoniacum* (i. e., *sal ammoniac*), noting its occurrence near the Temple of Ammon. There are different stories on its origin that the *sal ammoniac* (NH_4Cl) was formed by solar distillation in the sands from the urine and dung of camels at the Siwa Oasis, and because of the near sea, sea salt (NaCl) deposition. Sal ammoniac (in German *Salmiak*) was used for medical purposes and for incense in the temple ceremonies.

leading to $N_2 + 2\,OH$ and/or $2\,HNO$: $k_{4.89}$ ($= k_a + k_b + k_c$) $= 2.0 \cdot 10^{-11}$ cm^3 molecule^{-1} s^{-1} at 298 K where $k_a/k_{4.88} = 0.25$ and $k_c/k_{4.88} = 0.75$ over the temperature range 298–500 K:

$$NH_2 + NO_2 \rightarrow N_2O + H_2O, \tag{4.89a}$$

$$NH_2 + NO_2 \rightarrow N_2 + H_2O_2, \tag{4.89b}$$

$$NH_2 + NO_2 \rightarrow H_2NO + NO. \tag{4.89c}$$

In summary, ammonia oxidation is negligible (5 %) compared with the main fate of NH_3, particle formation (55 %) and deposition (40 %); the numbers in parenthesis provide the percentage of emitted NH_3. Thus, ammonia remains in its oxidation state −3, is mainly seen as ammonium (NH_4^+) in air, and returns to soils and waters as ammonium, where it moves between the amino group ($-NH_2$) in the biomass and nitrate through nitrification and ammonification. The nitrenium ions NH_2^+ have been exploited as intermediates in organic biological reactions.

In the air (and exhaust gases), gaseous NH_3 converts with gaseous HCl as well as HNO_3 to solids (called *gas-to-particle conversion*), which form a gas-solid equilibrium:

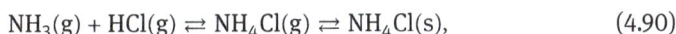

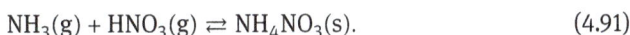

$$NH_3(g) + HCl(g) \rightleftharpoons NH_4Cl(g) \rightleftharpoons NH_4Cl(s), \tag{4.90}$$

$$NH_3(g) + HNO_3(g) \rightleftharpoons NH_4NO_3(s). \tag{4.91}$$

NH_3 is highly soluble, but it is a weak base:

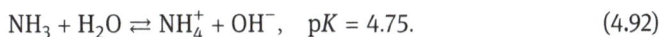

$$NH_3 + H_2O \rightleftharpoons NH_4^+ + OH^-, \quad pK = 4.75. \tag{4.92}$$

NH_3 is also a very weak acid, forming the amide anion NH_2^- (but not under environmental conditions), also called according to IUPAC azanide: $NH_3 + NH_3 \rightleftharpoons NH_2^- + NH_4^+$. After dissolution in hydrometeors (clouds, rain), ammonia neutralises acids, namely sulphuric acid (H_2SO_4) and nitric acid (HNO_3); in early time, before air pollution, the main salt in rainwater was $(NH_4)_2CO_3$ mixed with NaCl. Today, the most abundant salt in background atmospheric aerosol is ammonium sulphate $(NH_4)_2SO_4$ or hydrogen sulphate NH_4HSO_4. It is formed while homogeneous SO_2 oxidation in air to SO_3 and subsequent condensation to particulate sulphuric acids, taking up gaseous NH_3.

Let us now turn to an interesting pathway combining ammonium and nitrite chemistry. Likely in the early nineteenth century, it was already known that ammonium nitrite is a 'natural' substance found in the air.[62] Berzelius found in 1812 that aqueous solutions of NH_4NO_2 decompose, and Marcellin Berthelot (1827–1907) stated in 1875 that concentrated solutions quickly decompose under the formation of N_2

---

[62] Alchemists collected dew (see Mutus Liber) in large amounts and distilled it to find the *materia prima*; it is likely that they also mentioned the 'explosive' character of the residual salt – ammonium nitrite (Möller 2008).

where acids accelerate this. We already mentioned that dew water is very common and likely hitherto an underestimated interfacial chemical pathway linking the biosphere and atmosphere. Besides photosensitised oxidation processes under drying (in other terms evaporation) conditions, high concentrations of solutes occur. Without any doubt, ammonium (NH_4^+) and nitrite (NO_2^-) are important species in dew. It has been found that drying dew droplets, containing NH_4^+ and NO_2^-, not only evaporate HNO_2 but also N_2, NO and NO_2. The gross reactions already had been described in the 1930s:

$$NH_4^+ + NO_2^- \rightarrow N_2 + H_2O. \tag{4.93}$$

Here we propose the following hypothetical mechanisms:

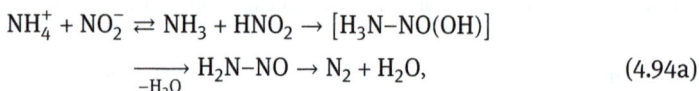

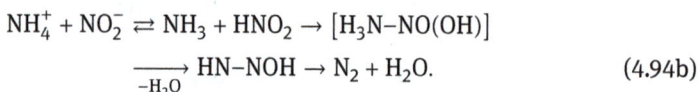

$$NH_4^+ + NO_2^- \rightleftharpoons NH_3 + HNO_2 \rightarrow [H_3N-NO(OH)]$$
$$\xrightarrow[-H_2O]{} H_2N-NO \rightarrow N_2 + H_2O, \tag{4.94a}$$

$$NH_4^+ + NO_2^- \rightleftharpoons NH_3 + HNO_2 \rightarrow [H_3N-NO(OH)]$$
$$\xrightarrow[-H_2O]{} HN-NOH \rightarrow N_2 + H_2O. \tag{4.94b}$$

The intermediates HN=NOH (hydroxyl diimide or hydroxy[1,1 or 1,2]diazene) or the tautomer H_2N-NO (nitrosamide) has been described *ab-inito* in the form of nine isomers, but none have been isolated and characterised. Several species are believed to be intermediates in processes involving the reduction of nitrogen oxides to molecular nitrogen and water, namely HN=N(O)H (diimide-N-oxide).

4.4.4 Dinitrogen monoxide (N_2O)

Dinitrogen monoxide[63] is a biologically important gas and is rather stable in the troposphere. It only undergoes either uptake by soils and vegetation or transport to the stratosphere where it photodissociates up to 90 %; however, the photolysis is effective only for $\lambda < 240$ nm:

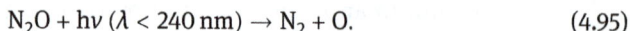

$$N_2O + h\nu \ (\lambda < 240 \text{ nm}) \rightarrow N_2 + O. \tag{4.95}$$

Another pathway is via ($k_{4.96} = 1.2 \cdot 10^{-10}$ cm^3 molecules^{-1} s^{-1}):

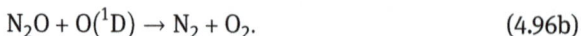

$$N_2O + O(^1D) \rightarrow 2 NO, \tag{4.96a}$$
$$N_2O + O(^1D) \rightarrow N_2 + O_2. \tag{4.96b}$$

---

63 The common names laughing gas and nitrous oxide for N_2O should not be used.

N_2O contributes to the greenhouse effect. At present, it is still negligible, but after energy conversion (end of CO_2 emission), it could become a problem because of likely further increasing emissions from agriculture (together with CH_4 emission from agriculture).

4.4.5 Nitrogen oxides (NO_x) and oxoacids (HNO_x)

Nitrogen forms oxides with the formula NO_n ($n = 1, 2, 3$) and N_2O_n ($n = 1, 2, 3, 4, 5, 6$); N_2O was presented in the previous Chapter whereas N_2O_6 and N_4O_n ($n = 1, 2$) play no role in the environment. Here we present NO (nitrogen monoxide),[64] NO_2 (nitrogen dioxide), nitrogen trioxide (NO_3), and dinitrogen pentoxide (N_2O_5). Several oxoacids exist from nitrogen; of importance for the environment are only nitrous acid HNO_2 and nitric acid HNO_3 (from which the anions nitrite NO_2^- and nitrate NO_3^- are derived), and as intermediates H_3NO (hydroxylamine NH_2OH) and nitrosyl HNO. In contrast to the equivalent oxoacids of sulphur (H_2SO_3 and H_2SO_4), both acids occur molecular gaseous and molecular in air.

It is useful to distinguish between groups (the termination of NO_x is also practical). Most *in situ* analysers based on chemiluminescence measure the sum of NO + NO_2. Only by using a two-channel technique, it is possible to detect NO and NO_x, where the difference is interpreted to be NO_2.

$$NO_x = NO + NO_2,$$
$$NO_y = NO_x + NO_3 + N_2O_5 + HNO_2 + HNO_3 + \text{organic N} + \text{particulate N}; \quad \text{and}$$
$$NO_z = NO_y - NO_x.$$

Therefore, NO_y represents the sum of all nitrogen with the exception of ammonia (and amines), N_2O and N_2.

4.4.5.1 Gas-phase chemistry

NO_2 and NO play crucial roles in the tropospheric ozone formation cycle. The former provides the source of atomic oxygen, and the latter cycles the HO_2 radical back to OH for the continuous 'burning' of the ozone precursors CO, CH_4 and NMVOC (see Chapter 5.3.2.2).

The key reaction for subsequent formation of ROS we have already mentioned (equation (4.17)), which is followed by equation (4.97), reforming NO_2:

$$NO_2 + h\nu \rightarrow O(^3P) + NO \quad (\lambda_{threshold} = 398\,nm),$$

---

64 The common name nitric oxide for NO should no longer be used.

$$NO + HO_2 \rightarrow NO_2 + OH, \quad k_{4.97} = 8.3 \cdot 10^{-12}. \tag{4.97}$$

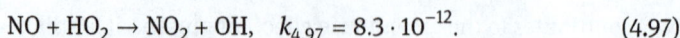

Besides reaction equation (4.97), the fast NO oxidation by O_3 occurs (and limits atmospheric O_3 concentration in polluted areas having large NO emission such as traffic sites):

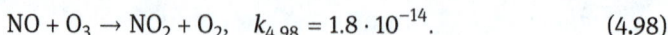

$$NO + O_3 \rightarrow NO_2 + O_2, \quad k_{4.98} = 1.8 \cdot 10^{-14}. \tag{4.98}$$

Competing with reaction equation (4.97), NO oxidises by alkyl peroxyl radicals (see also Chapter 4.6.3.1):

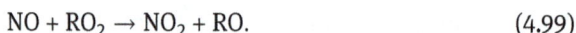

$$NO + RO_2 \rightarrow NO_2 + RO. \tag{4.99}$$

The double role of NO_x in ozone formation and destruction (or limiting) processes will be presented in Chapter 5.3.2.2. Under atmospheric conditions, the formation of NO_2 is relevant only through reactions equation (4.97) and equation (4.98). The reaction

$$NO + NO + O_2 \rightarrow 2\,NO_2, \quad k_{4.100} \approx 2 \cdot 10^{-19} \tag{4.100}$$

is too slow ($2.0 \cdot 10^{-38}$ cm^6 $molecule^{-2}$ s^{-1} at 298 K) to be considered in the atmosphere but plays a role in exhaust gases of power plants and mobile engines where large NO concentrations occur. Reaction equation (4.100) does not represent an elementary reaction; it is a multistep mechanism involving NO_3 or the dimer $(NO)_2$. This NO_3 (O=NOO) is an isomer to the nitrate radical O=N=O(O) and the first product of NO oxidation (note, the nitrate radical is not a peroxide but the isomer). It is clear that this very unstable peroxo radical will mostly decompose by quenching to NO + O_2, which results in a slow reaction probability (we will meet this reaction later in biological systems):

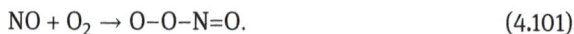

$$NO + O_2 \rightarrow O-O-N=O. \tag{4.101}$$

It is known that after the primary NO formation, according to the high-temperature mechanism (Chapter 4.4.2), 5–10 % is converted into NO_2.

> **i** Since NO_2 is the precondition for near-surface ozone formation and thereby beginning radical (oxygen) chemistry, any possible direct NO_2 emission is of large importance for changing the atmospheric ROS budget because there is no ROS consumption in the NO–NO_2 conversion.

To reduce the environmental impact of NO emissions (note that about 5–10 % is NO_2 due to reaction (equation (4.100)) from power plants and vehicles can be possibleby different catalytic processes. For large industrial boilers and process heaters, the selective catalytic reduction (SCR process) is applied using ammonia as the reducing agent at 200–450 °C and catalysts containing WO_3, V_2O_5 and TiO_2:

$$2\,NO + 2\,NH_3 + \frac{1}{2}\,O_2 \rightarrow 2\,N_2 + 2\,H_2O. \tag{4.102}$$

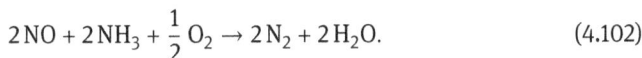

There are several secondary reactions and complications, resulting in no more than 60 % removal efficiency. For vehicle emission control, the three-way catalytic converters (TWC) have been introduced to reduce emissions of CO, NO and hydrocarbons (C_nH_m). Whereas CO and hydrocarbons are oxidised by O_2 to CO_2, NO is reduced to N_2 according to (at about 400 °C):

$$4\,NO + CH_4 \rightarrow 2\,N_2 + CO_2 + 2\,H_2O, \tag{4.103}$$

$$2\,NO + 2\,CO \rightarrow N_2 + 2\,CO_2. \tag{4.104}$$

NO and NO_2 form equilibrium with dimers:

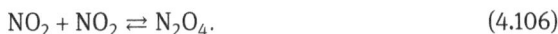

$$NO + NO_2 \rightleftarrows N_2O_3, \tag{4.105}$$

$$NO_2 + NO_2 \rightleftarrows N_2O_4. \tag{4.106}$$

Both substances (they are more soluble than NO and NO_2) have been discussed in the past as acid precursors in solution. However, compared with other pathways and because of their very low gas-phase concentrations, they have been assessed to be negligible; N_2O_3 might still play a role as an interfacial intermediate (see Chapter 4.4.5.2). N_2O_3 is the anhydride of nitrous acid ($N_2O_3 + H_2O = 2\,HNO_2$). The gas-phase equilibrium constant $K_{N_2O_3} = [NO][NO_2]/[N_2O_3] = 1.91$ atm at 298 K suggests that N_2O_3 is negligible in the air. By contrast, $K_{N_2O_4}$ strongly depends on temperature (0.0177 at 273 K and 0.863 at 323 K) but remains for the atmosphere (out of plumes) without consideration. The brown colour of some NO_x exhaust plumes is given by NO_2.

The next higher nitrogen oxide, which has only a short life but is an extremely important atmospheric intermediate, is nitrogen trioxide NO_3:[65]

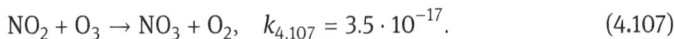

$$NO_2 + O_3 \rightarrow NO_3 + O_2, \quad k_{4.107} = 3.5 \cdot 10^{-17}. \tag{4.107}$$

The lifetime of NO_3 is very short and is much shorter in the daytime because of effective photodissociation (quantum yield 1.0 for $\lambda \leq 587$ nm); NO_3 radical dissociation is dominant to $NO_2 + O(^3P)$:

$$NO_3 + h\nu \rightarrow NO + O_2, \quad \lambda_{\text{threshold}} \geq 714\,\text{nm}, \tag{4.108a}$$

$$NO_3 + h\nu \rightarrow NO_2 + O(^3P), \quad \lambda_{\text{threshold}} = 587\,\text{nm}. \tag{4.108b}$$

[65] This reaction can proceed also when breathing polluted air containing NO_2 and O_3, hence resulting in a synergistic effect due to stronger damages by the nitrate radical, reacting with biomolecules and gaining nitric acid according to equation (4.111).

Moreover, collision with NO_x removes NO_3 quickly:

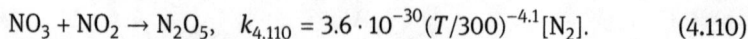

$$NO_3 + NO \rightarrow 2\,NO_2, \quad k_{4.109} = 2.6 \cdot 10^{-11}, \tag{4.109}$$

$$NO_3 + NO_2 \rightarrow N_2O_5, \quad k_{4.110} = 3.6 \cdot 10^{-30}(T/300)^{-4.1}[N_2]. \tag{4.110}$$

Hence, at night NO_3 is accumulated (daytime concentrations are negligible) and plays a role similar to OH in H abstraction, whereas stable HNO_3 is produced as the final product of the NO oxidation chain:

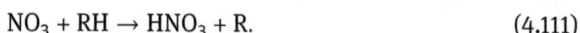

$$NO_3 + RH \rightarrow HNO_3 + R. \tag{4.111}$$

The specific reaction rates with different hydrocarbons are generally lower (about 3–4 orders of magnitude) compared with OH + RH (equation (4.277)), but the large night-time NO_3 concentration can balance it and provide absolute rates comparable with the OH pathway. Of high importance is the fast reaction of NO_3 with isoprene and α-pinene.

N_2O_5, produced by reaction equation (4.110), is the anhydride of HNO_3, but it reacts negligibly in the gas phase with H_2O:

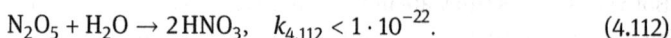

$$N_2O_5 + H_2O \rightarrow 2\,HNO_3, \quad k_{4.112} < 1 \cdot 10^{-22}. \tag{4.112}$$

In a certain sense, N_2O_5 is in (dynamic) equilibrium with NO_2 and NO_3 according to the fast reaction (the kinetic is unknown). It is remarkable that till now, no measurements of N_2O_5 but extensive measurements of NO_3 have existed, from which indirect conclusions about N_2O_5 have been drawn:

$$N_2O_5 + M \rightarrow NO_2 + NO_3 + M. \tag{4.113}$$

Clouds and precipitation will quantitatively scavenge NO_3 and N_2O_5 (see Chapter 4.4.5.2), forming HNO_3 and NO_3^-, respectively. Some reactions of N_2O_5 onto sea salt particles have been studied, producing gaseous nitryl chloride from NaCl (and similar with NaBr and NaI):

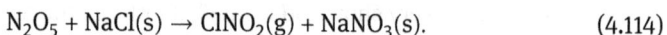

$$N_2O_5 + NaCl(s) \rightarrow ClNO_2(g) + NaNO_3(s). \tag{4.114}$$

NO_2 can react with sea salt to produce gaseous nitrosyl chloride:

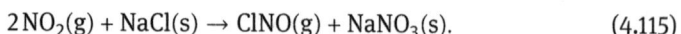

$$2\,NO_2(g) + NaCl(s) \rightarrow ClNO(g) + NaNO_3(s). \tag{4.115}$$

In both reactions, the significance consists of the possible subsequent photolytic release of Cl (or related halogen) radicals, which can go, for example, in O_3 destruction cycles:

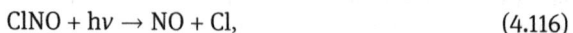

$$ClNO + h\nu \rightarrow NO + Cl, \tag{4.116}$$

$$\mathrm{ClNO_2} + h\nu \rightarrow \mathrm{NO_2} + \mathrm{Cl}. \tag{4.117}$$

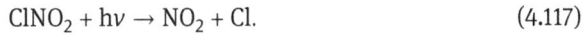

In summary, we present the NO_x–NO_y chemistry in the following line:

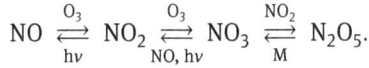

$$\mathrm{NO} \underset{h\nu}{\overset{\mathrm{O_3}}{\rightleftarrows}} \mathrm{NO_2} \underset{\mathrm{NO},\,h\nu}{\overset{\mathrm{O_3}}{\rightleftarrows}} \mathrm{NO_3} \underset{\mathrm{M}}{\overset{\mathrm{NO_2}}{\rightleftarrows}} \mathrm{N_2O_5}.$$

We now turn to the formation of oxo acids. As noted, nitric acid ($\mathrm{HNO_3}$) is the final product that is formed from $\mathrm{NO_3}$ and $\mathrm{N_2O_5}$ via phase transfer as well as from $\mathrm{NO_2}$ in a fast reaction with OH; $k_{4.118} = 3.3 \cdot 10^{-30}(T/300)^{-3.0}[\mathrm{N_2}]\,\mathrm{cm^3\,molecule^{-1}\,s^{-1}}$ over the temperature range 200–300 K (about $6.0 \cdot 10^{-11}\,\mathrm{cm^3\,molecule^{-1}\,s^{-1}}$ at 298 K):

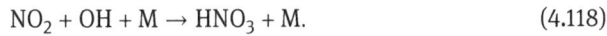

$$\mathrm{NO_2} + \mathrm{OH} + \mathrm{M} \rightarrow \mathrm{HNO_3} + \mathrm{M}. \tag{4.118}$$

This reaction is the ultimate OH sink in the air; the photodissociation of $\mathrm{HNO_3}$ is negligible and the fate of nitric acid is scavenging, dry deposition and particle formation with $\mathrm{NH_3}$, which also finally deposits. The reaction of $\mathrm{HNO_3}$ with OH is too slow to be important in the lower troposphere; $k_{4.119} = 1.5 \cdot 10^{-13}\,\mathrm{cm^3\,molecule^{-1}\,s^{-1}}$ at 298 K and 1 bar of air:

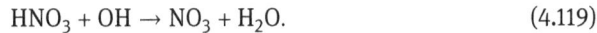

$$\mathrm{HNO_3} + \mathrm{OH} \rightarrow \mathrm{NO_3} + \mathrm{H_2O}. \tag{4.119}$$

Let us now turn to the formation of nitrous acid ($\mathrm{HNO_2}$, which often is also written as HONO). In the atmosphere, the formation by OH radicals is quickly followed by photodissociation of $\mathrm{HNO_2}$; $k_{4.120} = 7.4 \cdot 10^{-31}(T/300)^{-2.4}[\mathrm{N_2}]\,\mathrm{cm^3\,molecule^{-1}\,s^{-1}}$ over the temperature range 200–400 K and quantum yield 1.0 throughout the wavelength range 190–400 nm; $k_{4.122} = 6.0 \cdot 10^{-12}\,\mathrm{cm^3\,molecule^{-1}\,s^{-1}}$ at 298 K:

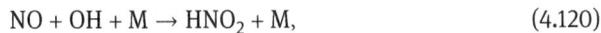

$$\mathrm{NO} + \mathrm{OH} + \mathrm{M} \rightarrow \mathrm{HNO_2} + \mathrm{M}, \tag{4.120}$$

$$\mathrm{HNO_2} + h\nu \rightarrow \mathrm{NO} + \mathrm{OH}, \quad \lambda_{\mathrm{threshold}} = 578\,\mathrm{nm}, \tag{4.121a}$$

$$\mathrm{HNO_2} + h\nu \rightarrow \mathrm{H} + \mathrm{NO_2}, \quad \lambda_{\mathrm{threshold}} = 361\,\mathrm{nm}, \tag{4.121b}$$

$$\mathrm{HNO_2} + \mathrm{OH} \rightarrow \mathrm{NO_2} + \mathrm{H_2O}. \tag{4.122}$$

Gaseous $\mathrm{HNO_2}$ in ambient air was first measured at the end of the 1970s using differential optical absorption spectroscopy (DOAS). Recent measurements indicate that $\mathrm{HNO_2}$ also plays a much larger role in the reactive nitrogen budget of rural sites than previously. The formation of nitrous acid (HONO) via heterogeneous and interfacial pathways (Chapter 4.4.5.2) provides a source (especially in the morning after sunrise) to produce OH radicals parallel to the photolysis of $\mathrm{O_3}$ (equation (4.14)) and HCHO (equation (4.294)). From measurements, it has been derived that $\mathrm{HNO_2}$ accounts for about 30–40 % of the radical production in the air close to the ground, similar to contributions from photolysis of HCHO and $\mathrm{O_3}$.

Figure 4.8 shows the significant difference between nocturnal and daytime NO_y chemistry. Note that $\mathrm{HNO_3}$ formation goes through very different pathways. The daytime

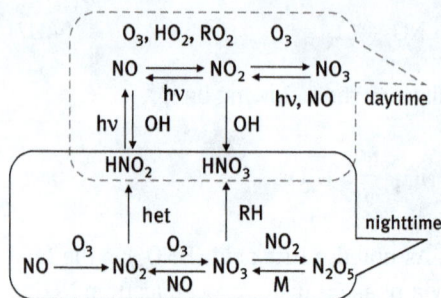

Figure 4.8: Gas-phase NO_y chemistry at daytime and night-time.

chemistry can be characterised as an interrelationship

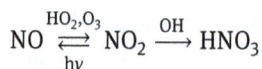

and the night-time chemistry by the chain

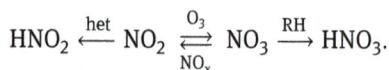

4.4.5.2 Aqueous-phase chemistry

NO_3, N_2O_5 and HNO_3 will be quantitatively scavenged by natural waters. Nitric acid is a strong acid and thereby fully dissociated, whereas nitrous acid is roughly 50 % dissociated in hydrometeors, but in other natural waters (pH ≥ 7), the equilibrium lies on the right side:

$$HNO_3 \rightarrow NO_3^- + H^+, \tag{4.123}$$

$$HNO_2 \rightleftharpoons NO_2^- + H^+, \quad pK_a = 3.3. \tag{4.124}$$

When N_2O_5 sticks to water surfaces, it is completely and quickly converted into nitrate ions ($N_2O_5 + H_2O \rightarrow 2\,NO_3^- + 2\,H^+$). The nitrate radical NO_3 can react with all electron donors according to equation (4.125) and is, therefore, a strong oxidant:

$$NO_3 + e^- \rightarrow NO_3^-. \tag{4.125}$$

It reacts with dissolved hydrocarbons according to equation (4.111). However, its removal is likely to be dominated by the reactions with chloride and sulphite; $k_{4.126} = 9.3 \cdot 10^6\,L\,mol^{-1}\,s^{-1}$ and $k_{4.127} = 1.7 \cdot 10^9\,L\,mol^{-1}\,s^{-1}$:

$$NO_3 + Cl^- \rightarrow NO_3^- + Cl, \tag{4.126}$$

$$NO_3 + HSO_3^- \rightarrow NO_3^- + HSO_3. \tag{4.127}$$

The fate of Cl radicals is described in Chapter 4.7.2 and that of sulphite radicals in Chapter 4.5.3.

Nitrate ions can be photolysed; however, in the bulk water phase, the reaction is very slow ($j \approx 10^{-7} \, \text{s}^{-1}$). The photodecomposition of NO_3^- into OH and NO_x species within and upon ice has been discussed over the past two decades and can be crucial to the chemistry of snowpacks and the composition of the overhead atmospheric boundary layer; $\lambda > 300$ nm and pH < 6:

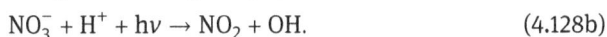

$$NO_3^- + h\nu \rightarrow NO_2^- + O(^3P), \tag{4.128a}$$

$$NO_3^- + H^+ + h\nu \rightarrow NO_2 + OH. \tag{4.128b}$$

It has been found that channels (a) and (b) contribute 10 % and 90 %, respectively, but that nitrate photolysis is generally too slow (a lifetime of about seven days) to be significant in the atmosphere. In the presence of O_2, subsequent to reaction equation (4.128a), O_3 formation follows (see equation (4.11)). Therefore, these pathways are only of local interest in the lower part of the boundary layer.

In contrast to nitrate, nitrite will be photolysed, yielding OH; thereby independent of the channels, OH is produced according to the following budget equation:

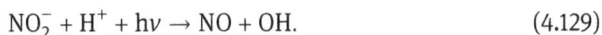

$$NO_2^- + H^+ + h\nu \rightarrow NO + OH. \tag{4.129}$$

In the absence of radical scavengers, nitrate can react with monooxygen; $k_{4.130} = 2 \cdot 10^8 \, \text{L mol}^{-1} \text{s}^{-1}$:

$$NO_3^- + O(^3P) \rightarrow NO_2^- + O_2. \tag{4.130}$$

OH radicals quickly convert nitrite back to NO_2; $k_{4.131} = 2 \cdot 10^{10} \, \text{L mol}^{-1} \text{s}^{-1}$:

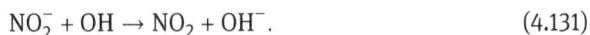

$$NO_2^- + OH \rightarrow NO_2 + OH^-. \tag{4.131}$$

Later, we will see that NO_2 will again be converted to nitrite (equation (4.135)) via electron transfer processes independent from

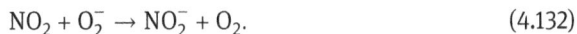

$$NO_2 + O_2^- \rightarrow NO_2^- + O_2. \tag{4.132}$$

For more than 200 years it has been known that a mixture of nitrogen oxides (NO + NO_2) in contact with water produces nitrous acid. Remember that NO + NO_2 = N_2O_3 (anhydride of HNO_2):

$$NO + NO_2 + H_2O \rightarrow 2\,HNO_2. \tag{4.133}$$

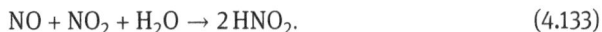

When passing NO_2 through water, nitrous, as well as nitric acid, is produced, thereby N_2O_4 is interpreted as a 'mixed' anhydride (these reactions are used for commercial production of the oxoacids):

$$NO_2 + NO_2 + H_2O \rightarrow HNO_2 + HNO_3. \tag{4.134}$$

Reaction equation (4.134) has been proposed to proceed at wetted surfaces. Near ground during the night, a continuous increase of HNO_2 in the air is found and interpreted as heterogeneous surface formation. The heterogeneous hydrolysis of NO_2 which is believed to occur with the same mechanism during the day as at night, has been investigated by numerous field and laboratory studies on many different surfaces. During the daytime, however, HNO_2 is destroyed by photolysis. Soils, buildings, roads and vegetation provide similar solid support and should hold surface water in sufficient amounts to promote heterogeneous reactions during the day. A process with dinitrogen tetroxide (N_2O_4) after the adsorption of NO_2 and steric rearrangement as a key intermediate has been proposed (Figure 4.9).

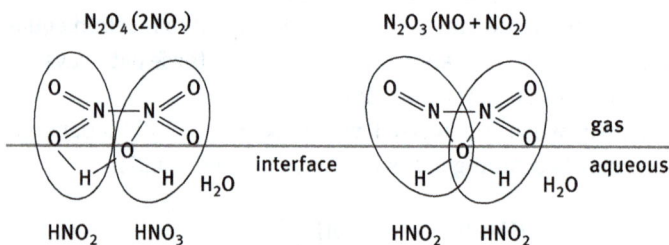

Figure 4.9: Scheme of N_2O_4 and N_2O_3 interfacial reaction to HNO_2 and HNO_3 (adapted from Möller 2019).

Whereas the interpretation of observed night-time HNO_2 formation rates is mainly based on equation (4.134), this 'classical' heterogeneous HNO_2 formation via NO_2 disproportionation is too slow to account for the observed atmospheric daytime HNO_2 mixing ratios. Hence, a daytime photoenhanced simple electron transfer onto NO_2 has recently been proposed, explaining the diurnal HNO_2 maximum at some sites:

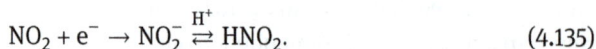

$$NO_2 + e^- \rightarrow NO_2^- \overset{H^+}{\rightleftharpoons} HNO_2. \tag{4.135}$$

Other more 'exotic' formation pathways of HNO_2 in the condensed phase include possible inorganic reactions similar to biogenic denitrification and nitrification processes, for example, from NO (which is the anhydride) via the formation of 'hydronitrous acid' (H_2NO_2):

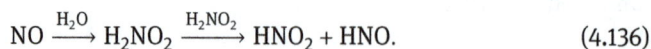

$$NO \xrightarrow{H_2O} H_2NO_2 \xrightarrow{H_2NO_2} HNO_2 + HNO. \tag{4.136}$$

The nitroxyl radical HNO can form, via dimerisation, hyponitrous acid ($H_2N_2O_2$) and this can oxidise to nitrite, likely via the not freely existing hyponitric acid (known as Angeli salt):

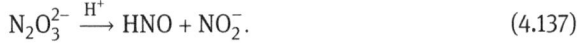

$$N_2O_3^{2-} \xrightarrow{H^+} HNO + NO_2^-. \tag{4.137}$$

Hence, it cannot be excluded that NO is directly associated with HNO_2 formation via the aqueous phase despite its very low solubility, which can be increased because of N_2O_3 formation ($NO + NO_2$); see below for a speculative transfer of NO into biological chemistry through atmospheric interfacial chemistry. The atmospheric importance would be that primary emitted NO could produce HNO_2 via the condensed phase and finally OH radicals.

> Recently, it has been shown experimentally under laboratory conditions, the reduction of NO_2 to HNO_2 when a TiO_2 aerosol was present. H_2O_2 was detected when reducing NO_2 on TiO_2 particles. Combining oxygen and nitrogen chemistry via heterogeneous (or interfacial) photochemistry would explain all experimental results. In Chapter 3.3.6.4, we have discussed that photosensitisers (organic compounds but also many transition state metals such as Wo, No, Ir, Ti, Mn, V, Ni, Co, Fe, Zn and others) seem to be ubiquitous in the environment. The specific abundance in natural waters and aerosol particles vary; thereby, the potential to provide electrons can vary, and this explains the different ratios of HNO_2/NO_x found in different regions.

Figure 4.10 shows the cycle between NO_x and HONO via the condensed phase:[66] $NO_2(g) \rightleftarrows NO_2(ads)$. In detail, the following reactions proceed in the NO_2 reduction process (Figure 4.10); g denotes the gas phase; the adsorption phase (particulate and/or aqueous) is not indexed; $k_{4.139} = 4.5 \cdot 10^9 \, L\,mol^{-1}\,s^{-1}$ and $k_{4.140} = 1.2 \cdot 10^7 \, L\,mol^{-1}\,s^{-1}$:

$$NO_2 + e^- \rightarrow NO_2^- \xrightarrow{H^+} HONO \rightleftarrows HONO(g) \xrightarrow{h\nu} NO + OH, \tag{4.138}$$
$$NO_2 + O_2^- \rightarrow NO_2^- + O_2, \tag{4.139}$$
$$NO_2 + HSO_3^- \rightarrow NO_2^- + HSO_3. \tag{4.140}$$

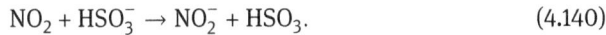

From this sequence, the budget equation $NO_2 + H^+ \xrightarrow{e^- + h\nu} NO + OH$ follows, which shows a smaller stoichiometry to OH than the budget from the pure gas-phase cycle of reactions equation (4.17), equation (4.11), equation (4.14) and equation (4.19): $NO_2 + H_2O \xrightarrow{h\nu} NO + 2\,OH$. The overall budget also depends on the interfacial conditions. Acid solutions favour HONO formation and desorption and thereby reduce nitrite oxidation (note that HONO formation is a reducing step in an aerobic environment). Therefore,

---

66 A good compilation of aqueous phase chemical reactions and equilibria can be found in Herrmann et al. (2005).

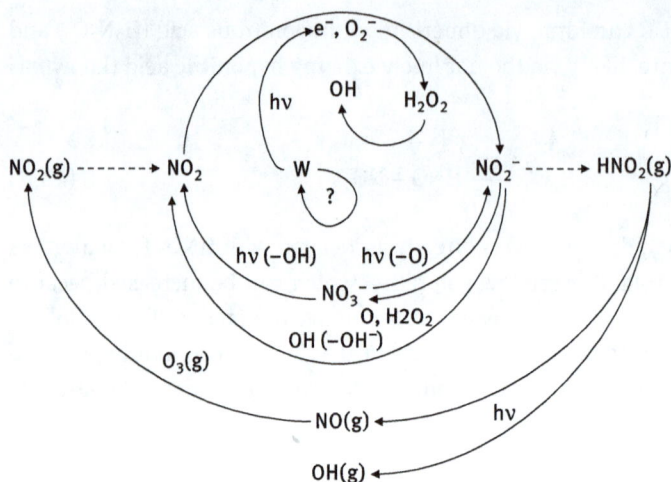

Figure 4.10: Photocatalytic NO_2 conversion at condensed phases (natural waters, hydrometeors and aerosol particles); (g) gas phase.

alkaline conditions and high oxidation potential lead to nitrate formation; $k_{4.141} = 4.6 \cdot 10^3 \, \text{L mol}^{-1} \text{s}^{-1}$ and $k_{4.142} = 5.0 \cdot 10^5 \, \text{L mol}^{-1} \text{s}^{-1}$:

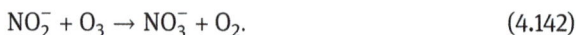

$$NO_2^- + H_2O_2 \rightarrow NO_3^- + H_2O, \tag{4.141}$$
$$NO_2^- + O_3 \rightarrow NO_3^- + O_2. \tag{4.142}$$

Both reactions are relatively slow compared with the oxidation of dissolved SO_2 (i. e. HSO_3^-), which has to be assumed to be in the presence of NO_x (Chapter 4.5.3), H_2O_2 will exclusively react with HSO_3^- and O_3 with SO_3^{2-}. Recently, the very slow autoxidation ($NO_2^- + O_2$) of nitrous acid has been studied according to:

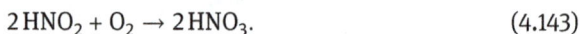

$$2\,HNO_2 + O_2 \rightarrow 2\,HNO_3. \tag{4.143}$$

It is proposed an HNO_2 dimerisation with intermediate peroxide formation and hence complicated steric arrangement (see Figure 4.9), likely explaining the very high acceleration of the reaction rate under freezing conditions:

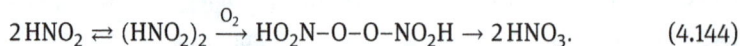

$$2\,HNO_2 \rightleftharpoons (HNO_2)_2 \xrightarrow{O_2} HO_2N-O-O-NO_2H \rightarrow 2\,HNO_3. \tag{4.144}$$

Experimental studies also show that nitrite in rainwater and cloud water samples exists for hours and even days, whereas S(IV) is oxidised within minutes. Therefore, it is very likely that, especially at interfaces, nitrite is accumulated and transferred back to the gas phase as HONO. The very fast reaction equation (4.140) is dominant in the presence of low SO_2 concentrations and has not been considered until now to budget for the NO_2 transfer to HONO. In Figure 4.10, the nitrate chemistry is included; however, it is likely that we can neglect this and conclude the following main pathway:

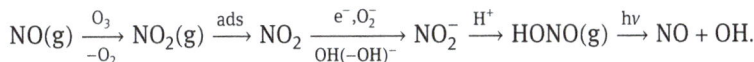

$$NO(g) \xrightarrow[-O_2]{O_3} NO_2(g) \xrightarrow{ads} NO_2 \xrightarrow[OH(-OH)^-]{e^-, O_2^-} NO_2^- \xrightarrow{H^+} HONO(g) \xrightarrow{h\nu} NO + OH.$$

The budget equation follows, which represents a water-splitting process similar to the gas phase OH production from O_3 ($O_3 + H_2O \xrightarrow{h\nu} O_2 + 2\,OH$, reactions equation (4.14) and equation (4.19)):

$$O_3 + e^- + H_2O \rightarrow O_2 + OH + OH^-. \tag{4.145}$$

It is fascinating that this overall conversion is also given by the aqueous phase ozone decay, which here is given in a slightly different reaction pathway to equation (4.60). Much experimental evidence for HONO formation from NO_2 condensed-phase photoconversion is available, but this pathway is not a net source of OH because in the first step of NO to NO_2 conversion, ozone (or HO_2) is consumed (note the budget equation in the gas phase: $2\,O_3 + H_2O \rightarrow OH + HO_2 + 2\,O_2$). This widely discussed process only shifts (because of the much faster photolysis of HONO compared with that of O_3) the photo-steady states and species reservoir distribution.

Under specific conditions (in biochemistry but also flue-gas chemistry), nitrous acid reduces:

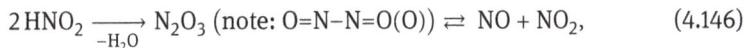

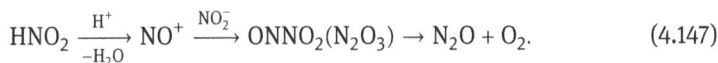

$$2\,HNO_2 \xrightarrow[-H_2O]{} N_2O_3 \text{ (note: O=N–N=O(O))} \rightleftarrows NO + NO_2, \tag{4.146}$$

$$HNO_2 \xrightarrow[-H_2O]{H^+} NO^+ \xrightarrow{NO_2^-} ONNO_2(N_2O_3) \rightarrow N_2O + O_2. \tag{4.147}$$

Note that dinitrogen monoxide N_2O is tautomer: $^-N=N^+=O$ and $N\equiv N^+-O^-$. Reaction equation (4.146) is the 'back' reaction equation (4.133); see also Figure 4.9 for steric arrangement.

Nothing is found in the modern chemistry literature on the aqueous phase chemistry of nitric oxide (NO). In textbooks of inorganic chemistry (e. g. Wiberg et al. 2001), it is noted that NO does not react with water.

However, from older literature (Gmelin 1936), we learn that NO slowly reacts with water under the formation of HNO_2, N_2 and N_2O. It was speculated that NO combines with OH^- via $H_2N_2O_2$ (hyponitrous acid, an isomer of nitramide) formation, which decays to N_2O, or that NO directly combines with H_2O to generate $H_2N_2O_3$ (oxo hyponitrous acid), which decays to HNO and HNO_2. The dimerisation of HNO to $H_2N_2O_2$ was also proposed, but detailed mechanisms were unknown (see below). The intermediate existence of HNO in HNO_2 reduction as well as NH_3 or NH_2OH (hydroxylamine) oxidation is well established (Heckner 1977).

With the findings that NO (one of the smallest and simplest molecules) is an important signalling molecule in biological chemistry and its inactivation is unique with respect to other signalling molecules because it depends solely on its non-enzymatic chemical reactivity with other molecules (Miranda et al. 2000), an explosion in NO solution chemistry began in the 1990s. Until recently, most of the biological effects of nitric oxide have been attributed to its uncharged state (NO), yet NO can also exist in a reduced state as nitroxyl (HNO and its protolytic form, the nitroxyl anion NO^-) and in an oxidised form as nitrosonium ion (NO^+):

$$NO^+[N{\equiv}O]^+ \leftrightarrow NO[N{=}O] \leftrightarrow NO^-[N{=}O^-] \leftrightarrow HNO[H{-}N{=}O].$$

Thus, unlike NO, HNO (not N–OH) can target cardiac sarcoplasmic ryanodine receptors to increase myocardial contractility, can interact directly with thiols and is resistant to both scavenging by superoxide (O_2^-) and tolerance development.

ℹ️ Nitrosonium, sometimes also termed nitrosyl cation (NO^+), is very short-lived in aqueous solutions. Nitrosonium ions react with secondary amines to generate nitrosamines, many of which cause cancer at very low doses (equation (4.151)).

NO^+ is formed in very small concentrations from nitrous acid in strong acid solution, when HNO_2 reacts as a base:

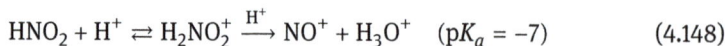

$$HNO_2 + H^+ \rightleftharpoons H_2NO_2^+ \xrightarrow{H^+} NO^+ + H_3O^+ \quad (pK_a = -7) \tag{4.148}$$

or

$$HONO \rightleftharpoons OH^- + NO^+. \tag{4.149}$$

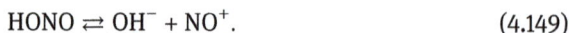

A subsequent step would be the formation of nitrosyl chloride ClNO (an important species in gas phase chemistry). The likely importance of this pathway lies in the photolysis of ClNO gaining Cl radicals:

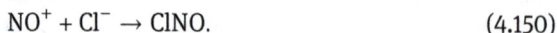

$$NO^+ + Cl^- \rightarrow ClNO. \tag{4.150}$$

The nitrosonium ion is the precursor of nitrosamines (see also equation (4.169)) reacting with secondary amines:

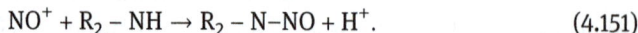

$$NO^+ + R_2 - NH \rightarrow R_2 - N{-}NO + H^+. \tag{4.151}$$

We cannot exclude that NO^+ is produced from electron holes:

$$NO + W^+(h_{vb}^+) \rightarrow NO^+. \tag{4.152}$$

See reactions equation (4.147) for another fate of NO^+ in high concentrated nitrite solutions. Nitroxyl anion NO^- is the base of nitroxyl radical HNO, which is a very weak acid; $pK_a = 11.4$, when both species are in their ground states, ^1HNO and $^3NO^-$:

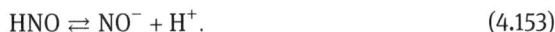

$$HNO \rightleftharpoons NO^- + H^+. \tag{4.153}$$

HNO quickly dimerises to hyponitrous acid $H_2N_2O_2$ including electronic rearrangement[67] and H shift. $H_2N_2O_2$ (HO–N=N–OH) slowly decompose in aqueous solutions into nitrous oxide and water (see Figure 4.7 and discussion in Chapter 5.2.4 with respect to nitrification):

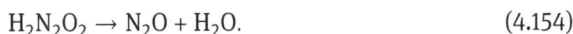

$$H_2N_2O_2 \rightarrow N_2O + H_2O. \tag{4.154}$$

Hence, N_2O can be regarded as an anhydride of hyponitrous acid. HNO also adds NO and generates the anionic radical of the NO dimer; $k_{4.155} = 5.8 \cdot 10^6 \, L \, mol^{-1} \, s^{-1}$, which further reacts with NO ($k_{4.155} = 5.4 \cdot 10^9 \, L \, mol^{-1} \, s^{-1}$) to the long-lived $N_3O_3^-$, which finally decomposes into N_2O and nitrite ($k_{4.157} = 3 \cdot 10^2 \, s^{-1}$):

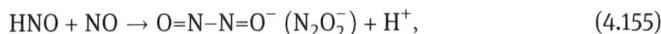

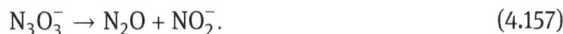

$$HNO + NO \rightarrow O=N-N=O^- \, (N_2O_2^-) + H^+, \tag{4.155}$$
$$N_2O_2^- + NO \rightarrow O=N-N(O)-N=O^- \, (N_3O_3^-), \tag{4.156}$$
$$N_3O_3^- \rightarrow N_2O + NO_2^-. \tag{4.157}$$

As for other electron receptors (such as NO_2 and NO_3), NO is easily reduced and we assume that the electron source is the hydrated electron and/or the conduction band electron:

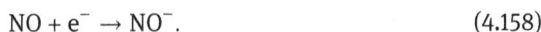

$$NO + e^- \rightarrow NO^-. \tag{4.158}$$

NO^- forms peroxonitrite (not to be confused with nitrate ion NO_3^-); $k_{4.159} = 2.7 \cdot 10^9 \, L \, mol^{-1} \, s^{-1}$, which converts with CO_2 via nitroperoxo carbonate ($ONOOCOO^-$) to nitrate; $k_{4.160} = 3 \cdot 10^4 \, L \, mol^{-1} \, s^{-1}$:

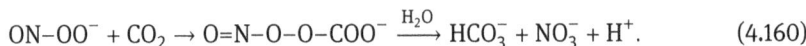

$$NO^- + O_2 \rightarrow O=N-OO^-, \tag{4.159}$$
$$ON-OO^- + CO_2 \rightarrow O=N-O-O-COO^- \xrightarrow{H_2O} HCO_3^- + NO_3^- + H^+. \tag{4.160}$$

---

67 A rearrangement reaction is a broad class of reactions where within a molecule or an activated complex (intermediate) of two molecules charges and/or atoms are changed to obtain a new structure; for example, in organic reactions where the carbon skeleton of a molecule is rearranged to give a structural isomer of the original molecule. It is separated between intermolecular reactions, covalency changes take place in two separate molecules, and in intramolecular (or internal) reactions, two or more reaction sites within the same molecule are involved.

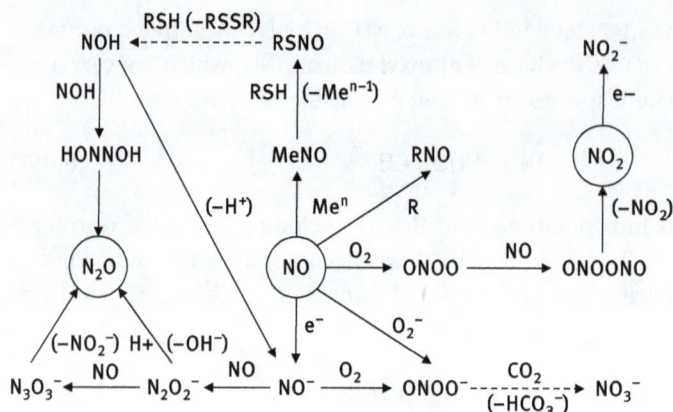

Figure 4.11: Biological NO_x chemistry; adapted and simplified from Fukuto et al. (2000). Dotted arrow: multistep process, Me transition metal ion, RSH organic sulphide (–SH thiol group), RSSH disulphide, ONOO is an isomeric form of the nitrate radical NO_3, $ONOO^-$ (ONOOH) peroxonitrous acid, HON=NOH ($H_2N_2O_2$) hyponitrous acid, $N_2O_2^-$ (O=N–N–O^-) radical anion of NO dimer, $N_3O_3^-$ (O=N–N–O–NO^-) trimer radical anion (not to be confused with Angeli salt $N_3O_4^{2-}$: $[O_2N–N–NO_2]^{2-}$).

Biochemical routes for the formation of nitroxyl ions are shown in Figure 4.11 but without considering inorganic or non-enzymatic solution chemistry. In cells, reactions of NO with haems,[68] thiols (R–SH) and metals form metal nitrosyl complexes (Me–NO) as carriers of NO^+ at physiological pH and nitrosation (formation of S-nitrosothiols) is of great biological significance. An important role of NO lies in the interchange between oxyhemoglobin Hb(Fe–O_2) and deoxyhemoglobin Hb(Fe^{III}), which is the primary mechanism by which the movement and concentration of NO are controlled *in vivo*. Because of NO lipid solubility, it can enrich high concentrations. NO can add onto iron complexes and go into internal electron transfer, thereby forming electropositive nitrosyl ligands:

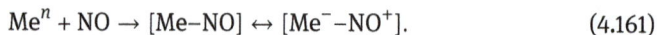

$$Me^n + NO \rightarrow [Me–NO] \leftrightarrow [Me^- – NO^+]. \tag{4.161}$$

The nitrosation of thiol compounds goes directly via N_2O_3 and indirectly through metallic nitrosyl ligands:

$$RSH + N_2O_3 \rightarrow RS–NO + HNO_2, \tag{4.162}$$

$$RSH + Me^{n+} – NO \rightarrow RS–NO + Me^{(n-1)+} + H^+. \tag{4.163}$$

---

68 Compounds of iron complexed in a porphyrin (tetrapyrrole) ring differ in side chain composition. Haems are the prosthetic groups of cytochromes and are found in most oxygen carrier proteins.

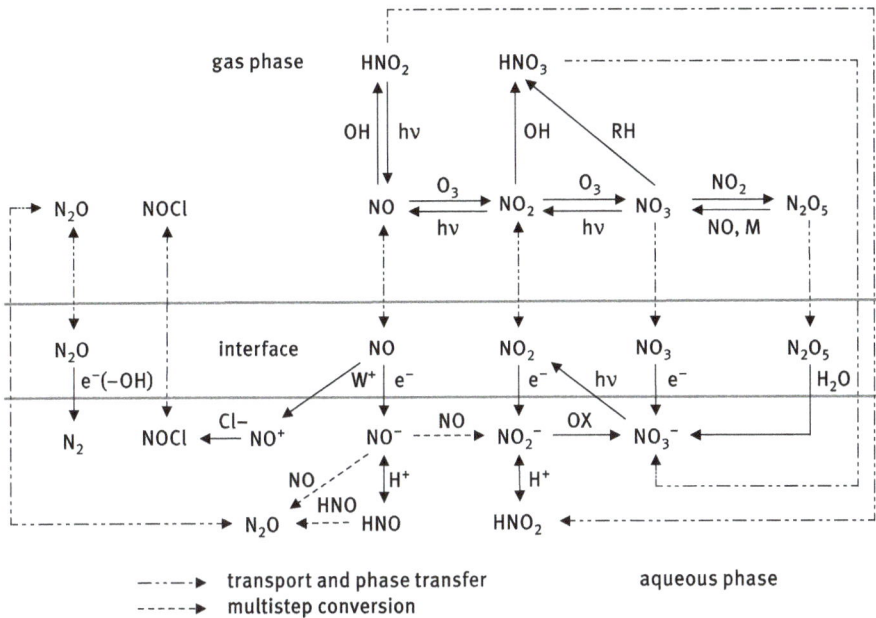

Figure 4.12: Scheme of multiphase NO_x–NO_y chemistry (main pathways); OX oxidants (OH, H_2O_2); the N_2O_3 chemistry is not included, which might be important during the night for nitrite formation in solutions and at interfaces.

The nitrosothiol is quickly decomposed by cuprous, thereby transforming nitrosyl back to nitrogen monoxide:

$$RS{-}NO + Cu^+ \rightarrow RS^- + NO + Cu^{2+} \tag{4.164}$$

Figure 4.12 summarises the main routes in the multiphase NO_x–NO_x chemistry. In the air, odd oxygen tends to accumulate in NO_x and NO_y but is in a photo-steady state between oxygen and nitrogen species. This condensed phase acts as a sink of oxygen gas-phase species, mainly in the form of NO_z (HNO_3, NO_3, N_2O_5). It is obvious that NO_x can transfer either non-photochemical (via $NO + NO_2$ reaction, not shown in Figure 4.10) or photosensitised to nitrite, which might return to the gas phase as HNO_2 and photolysed to OH and NO. N(III) oxidation in solution to nitrate is relatively slow and unimportant as an oxidant consumer. Not shown in Figure 4.12 is the decomposition of NH_4NO_2 (comproportionation of NH_3 and HNO_2) into gases (N_2, NO, NO_2, N_2O) while evaporating droplets (dew, fog, cloud); see Chapter 4.4.3.

An interesting example of complex atmospheric multiphase chemistry is the formation of Chile saltpetre ($NaNO_3$), the most important nitrogen resource, mined on a large scale from the beginning of 1830 until the early 1940s. It is found together with NaCl, NaBr and NaI, all sea salt constituents. The most accepted theory on its formation is the 'air-electrical' theory already proposed 100 years ago. To understand its formation, we also should mention geographical aspect.

The Atacama Desert in the North of Chile occupies a continuous strip for more than 1,000 km. There is no coastal plain; the mountain chain hovers around 1,500 m or so in elevation. There is frequent sea fog (cold Pacific) enriched with sea salt compounds (mainly NaCl). The fog moves from the sea to land, touching the upslope of the mountain chain (giving a unique ecosystem) and dissipating at a distance where the residual particles deposit. The Atacama Desert is commonly known as the driest place on the Earth; it may be the oldest desert and has experienced extreme hyperaridity for at least three million years, making it the oldest continuously arid region on Earth. Hence, salts cannot be washed out and accumulate over a long time. Lightning causes the formation of NO and finally NO_3, N_2O_5 and HNO_3 that is scavenged by the fog droplets. According to the following reaction, after drying, $NaNO_3$ remains. The degassing of HCl from sea salt particles by strong acids (such as HNO_3) is presented in Chapter 5.3.3.2:

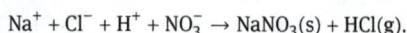

$$Na^+ + Cl^- + H^+ + NO_3^- \rightarrow NaNO_3(s) + HCl(g).$$

4.4.6 Organic nitrogen compounds

Nitrogen is an essential element in life and biogeochemical cycling (Figures 4.7, 5.12 and 5.13 and Chapter 5.2.4). In the biomass, nitrogen in organic molecules is almost in a reduced state, derived from ammonia NH_3 or ammonium NH_4^+, respectively. The main building block is the amino group NH_2, occurring in its simplest organic compounds, amines (Chapter 4.4.6.1 and Table 4.12). In the biomass, amino acids $=C(NH_2)C(O)OH$ are the building blocks of peptides (which are polymers of amino acids with the $=C-N(H)-C=$ central group), and next are proteins, which are polymers of more than 50 amino acids and have enormous diversity in structure and function. Additionally, nitrogen is found heterocyclically (mainly in nucleic acids) in biomolecules. Oxidised nitrogen (such as organic nitrites and nitrates) should be excluded in organisms and

Table 4.12: Organic nitrogen species: functional groups of environmental importance.

structure	name	formula / symbol
$-NH_2$	amino group (primary amine)	RNH_2
$=NH$	secondary amine	R_2NH
$\equiv N$	tertiary amine	R_3N
$-N=N-$	azo group	R^1NNR^2
$-C\equiv N$	cyano group (nitrile)	RCN
$-N=C<$	imine	$R^1NCR^2R^3$
$-C(=O)NH_2$	amide	$RC(O)NH_2$
$-N=O$ and $^+N=O$	nitroso group (nitrosyl)	RNO
$-O-N=O$	nitrite	RONO
$-N(=O)O$ or $-NO_2$	nitro group (nitryl)	RNO_2
$-O-NO_2$	nitrate	$RONO_2$
$=N-N=O$	nitrosamine	R_2NNO
$=N-N=(O)_2$	nitramine[a]	R_2NNO_2

[a] also called nitroamine, not to be confused with nitrosamine

expected only because of reactions between organic compounds and NO_x/NO_y (Chapter 4.4.6.2). Because of the large N-containing biomolecules, mostly non-volatile and after degradation highly water-soluble, direct emissions are less probable. Biomass burning has been identified as the source of organic nitrogen, but it was probably already in a decomposed form such as HCN or CH_3CN. The simplest amine is methylamine CH_3NH_2, whose global emissions from animal husbandry was estimated by Schade and Crutzen (1995) in 1988 to be 0.15 ± 0.06 Tg N. Almost three-quarters of these emissions consisted of trimethylamine–N. Other sources were marine coastal waters and biomass burning.

Nothing is known about the natural emissions of amines. Several anthropogenic sources (traffic, fertiliser production, paper mills, rayon manufacturing, and the food industry) have been identified. Possible amine emissions along with the scrubbed flue gas from CO_2 capture facilities (using amine-washing solutions) were recently studied. Other simple degradation products of amino acids and peptides are amides, consisting of the building block $R–C(=O)NH_2$. The simplest molecule is formamide $HC(=O)NH_2$. Amides are derivatives of carboxylic acids $(R(O)OH)$ where the hydroxyl group $(–OH)$ is replaced by the amino group $(–NH_2)$. Nitriles with the cyano group $–CN$ are potential biomass combustion products from degraded biomolecules.

4.4.6.1 Amines, nitriles and cyanides

Amines have been detected in ambient air, namely close to industrial areas, livestock homes and animal waste and wastewater treatments. For example, monomethylamine, trimethylamine, isopropylamine, ethylamine, n-butylamine, amylamine, dimethylamine and diethylamine have all been found (Schade and Crutzen 1995 and citations therein). The major amines in both air and rain were trimethylamine and methylamine, but dimethylamine, diethylamine and triethylamine were also detected and individually quantified. Amines can be found in many different matrices, from environmental samples to industrial raw materials, products, and wastes. Amides (dimethylformamide) were also detected in air, whereas other amides have been found to be emitted in industrial processes (acetamide, acrylamide). Under the aspect of carbon capture and storage (CCS technology), recently intensive studies on amine emission and degradations have been carried out. Amines until C_3 are gaseous and highly volatile. They are the Lewis bases (somewhat more basic than NH_3) and water-soluble (without longer carbon chains), thereby forming salts, the cation NR_3H^+ is called aminium:

$$NR_3(g) + HNO_3(g) \rightleftharpoons NR_3HNO_3(s). \qquad (4.165)$$

Therefore, the acidity of individual particles can significantly affect gas/particle partitioning, and the concentrations of amines, as strong bases, should be included in estimations of aerosol pH.

The amines by themselves are not very harmful at typical ambient concentrations. However, amine degradation reactions can lead to the formation of potentially carcinogenic substances such as nitrosamines and nitramines in both the gas phase and the aqueous phase. The general fate of amines is similar to that of ammonia, i. e. reactions with OH (note that similar reactions, i. e. H abstraction, occur with the radicals Cl and NO_3 under conditions of a sufficient radical concentration). They are not photolysed because the C–H bond energy is less than the N–H bond, predominantly OH abstracts H from C–H forming bifunctional compounds (e. g. amides $OC(H)NH_2$, see below for further fates). In the less probable case of N–H abstraction, the following radicals are formed:

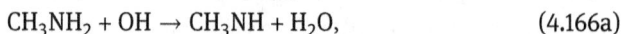

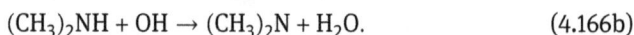

$$CH_3NH_2 + OH \rightarrow CH_3NH + H_2O, \tag{4.166a}$$

$$(CH_3)_2NH + OH \rightarrow (CH_3)_2N + H_2O. \tag{4.166b}$$

The further fate is not explicitly known, but molecule rearrangements and/or the addition of O_2, NO_x and NO_3 are possible:

$$CH_3NH + O_2 \rightarrow CH_3N(H)O_2 \xrightarrow[-NO_2]{NO} CH_3N(H)O \xrightarrow[-HO_2]{O_2} CH_3N = O \quad \text{(nitrosyl).}$$

Imines, $R^1N=CR^2R^3$, (the simplest one is methylene imine $HN=CH_2$) are reported as main products in amine gas phase photo-oxidation experiments; nothing is known on its gas-phase oxidation. They are known to hydrolyse in aqueous solutions resulting in amines and carbonyl compounds. From secondary amines, azo compounds or organic hydrazines (derived from hydrazine N_2H_4) can be given:

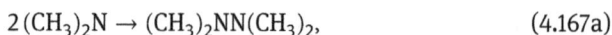

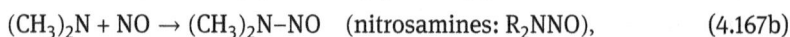

$$2\,(CH_3)_2N \rightarrow (CH_3)_2NN(CH_3)_2, \tag{4.167a}$$

$$(CH_3)_2N + NO \rightarrow (CH_3)_2N-NO \quad \text{(nitrosamines: } R_2NNO), \tag{4.167b}$$

$$(CH_3)_2N + NO_2 \rightarrow (CH_3)_2N-NO_2 \quad \text{(nitramines: } R_2NNO_2). \tag{4.167c}$$

They also quickly photolyse back to R_2N and NO_x in the gas phase (hence, they are enriched in the air during night-time). Nitrosamines are of high interest because of their cancerogenic properties. They have been found everywhere, but mostly in small concentrations.

As good electron donors, secondary amines are easily oxidized in solution to form derivates of hydroxylamine (NH_2OH):

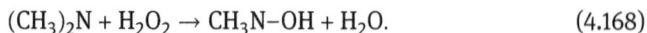

$$(CH_3)_2N + H_2O_2 \rightarrow CH_3N-OH + H_2O. \tag{4.168}$$

In water, tertiary amines form quaternary ammonium compounds, for example with methyl halogenides. In aqueous solutions, secondary (aliphatic and aromatic) amines react with nitric acid HONO (namely the nitrosonium ion NO^+) as already mentioned (reaction equation (4.151)) to generate nitrosamines:

$$(CH_3)_2NH + HONO \rightarrow (CH_3)_2N-NO + H_2O. \tag{4.169}$$

In water, amines are decomposed to a greater extent than in the gas phase and will be accumulated in the condensed phase. Aromatic nitrosamines are formed even more easily (Ar – aryl, e. g. C_6H_5):

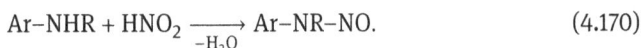

$$Ar-NHR + HNO_2 \xrightarrow[-H_2O]{} Ar-NR-NO. \tag{4.170}$$

They reform by internal rearrangement (the Fischer–Hepp rearrangement) of the N-nitroso group into p-nitrosoarylamines:

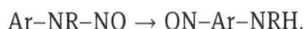

$$Ar-NR-NO \rightarrow ON-Ar-NRH.$$

The rate of all nitrosation (nucleophilic substitution) increases with the basicity of leaving group X:

$$X-NO \xrightarrow{Y^-} [X \ldots NO \ldots Y]^- \rightarrow Y-NO + X^-.$$

For amines (HY $=$ HNR_2), this reaction is adequate to equation (4.169) – there is an equilibrium $HNR_2 \rightleftarrows H^+ + R_2N^-$; however, nitrous acid (ON–OH) is not the direct nitrosation agent but the protonated form $ON-OH_2^+$ with measurable rates only for pH < 1. Other excellent nitrosation agents are nitrosamines themselves ($OH-NR_2$) and dinitrogen trioxide N_2O_3 ($ON-NO_2$):

$$NO-NO_2(N_2O_3) + RNH_2 \rightarrow R-NH-NO + HNO_2. \tag{4.171}$$

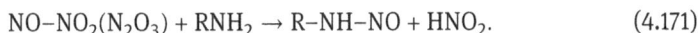

Dimethylformamide is expected to exist almost entirely in the vapour phase in ambient air and react with photochemically produced hydroxyl radicals in the atmosphere:

$$(CH_3)_2C(O)NH_2 + OH \rightarrow (CH_3)_2C(O)NH + H_2O. \tag{4.172}$$

Further dimerisation leads to imides or imido compounds ($-CONH_2$), which quickly react with OH (Barnes et al. 2010). Formamide $HC(O)NH_2$ is the simplest amide in the air. The following main and/or dominant products have been identified, being, therefore, the final product of amine oxidation:

isocyanic acid	$O=C=NH$
methyl isocyanate	$O=C=NCH_3$
N-formylformamide	$HN(C(O)H)_2$
N-formyl-N-formamid	$N-CH_3(C(O)H)_2$

In the aqueous phase, tertiary amines react with H_2O_2 to amine oxides, which act as surfactants and can be of importance in cloud droplets and surface water:

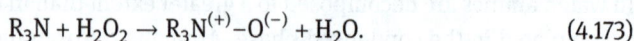

$$R_3N + H_2O_2 \rightarrow R_3N^{(+)}-O^{(-)} + H_2O. \tag{4.173}$$

Organic cyanides (nitriles) react with OH according to the general C–H abstraction, probably leading ultimately to the formation of CN radicals, which turn to HCN by reaction with hydrocarbons:

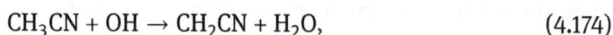

$$CH_3CN + OH \rightarrow CH_2CN + H_2O, \tag{4.174}$$

$$CN + RH \rightarrow HCN + R. \tag{4.175}$$

HCN is highly soluble in water but a weak acid ($pK_a = 9.2$), forming various salts (cyanides):

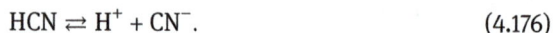

$$HCN \rightleftharpoons H^+ + CN^-. \tag{4.176}$$

Hydrogen cyanide or hydrocyanic acid (HCN) is found in the air, waters and soils (Ghosh et al. 2006). It is emitted from several thousand plant species, including many economically important food plants, synthesising cyanogenic glycosides and cyano-lipids. Upon tissue disruption, these natural products are hydrolysed, liberating the respiratory poison hydrogen cyanide. This phenomenon of cyanogenesis accounts for numerous cases of acute and chronic cyanide poisoning of animals, including humans. The $-C\equiv N$ group (as well as thiocyanate or, in older terminology, rhodanide $-S-C\equiv N$) is the simplest organic nitrogen bond that was produced in interstellar chemistry (Chapter 5.1.1) and is a building block of biomolecules. In the air, HCN is insignificantly destructed by OH ($k = 3 \cdot 10^{-14}$ cm^{-3} molecule^{-1} s^{-1}) into the CN radical. HCN is cycled back to vegetation and soils through dry and wet deposition and biologically taken up.

$\boxed{\text{i}}$ Therefore, cyanides are ubiquitously found but in concentrations far from being poisonous.

In contact with ozone, the cyanide anion is fast oxidised to cyanate (OCN^-), which is far less toxic than cyanide; it is generally present in metallurgical solutions at low concentrations and does not persist in the environment for long periods of time.

4.4.6.2 Organic NO$_x$ compounds

Organic nitrates, including nitroglycerine, represent the oldest class of NO donors and agents available to treat or prevent acute attacks of angina pectoris; nitroglycerine is still in *use* as an *industrial* explosive (dynamite). The direct natural emission of oxidised organic nitrogen compounds is unknown. Organic nitrate and nitrites are produced during the decomposition of hydrocarbons (Chapter 4.6.3.1), where alkoxyl radicals (RO) can combine with NO (leading to nitrites) and NO$_2$ (leading to nitrates):

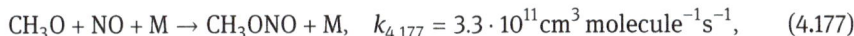

$$CH_3O + NO + M \rightarrow CH_3ONO + M, \quad k_{4.177} = 3.3 \cdot 10^{11} cm^3 \, molecule^{-1} s^{-1}, \quad (4.177)$$

$$CH_3O + NO_2 + M \rightarrow CH_3ONO_2 + M, \quad k_{4.178} = 2.1 \cdot 10^{11} cm^3 \, molecule^{-1} s^{-1}. \quad (4.178)$$

Organic nitrate chemistry is the primary control over the lifetime of nitrogen oxides (NO$_x$) in rural and remote continental locations. As NO$_x$ emissions decrease, organic nitrate chemistry becomes increasingly important to urban air quality. However, the lifetime of individual organic nitrates and the reactions that lead to their production and removal remain relatively poorly constrained.

Alkyl peroxyl radicals (RO$_2$) also add NO, where the intermediate CH$_3$OONO rearranges to CH$_3$ONO(O):

$$CH_3OO + NO + M \rightarrow CH_3ONO_2 + M, \quad k_{4.179} \leq 1.3 \cdot 10^{13} cm^3 \, molecule^{-1} s^{-1}. \quad (4.179)$$

The organic peroxynitrate (not to be confused with peroxyacyl nitrates) decomposes by collision:

$$CH_3OO + NO_2 + M \rightarrow CH_3OONO_2 + M, \quad (4.180)$$

$$CH_3OONO_2 + M \rightarrow CH_3O_2 + NO_2 + M, \quad k_{4.181} = 4.5s^{-1}. \quad (4.181)$$

NO reacts with the methyl radical to form formaldehyde and nitroxyl (remember: in air, CH$_3$ + O$_2$ is absolutely predominant):

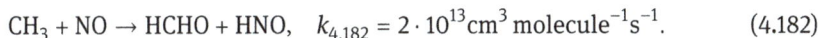

$$CH_3 + NO \rightarrow HCHO + HNO, \quad k_{4.182} = 2 \cdot 10^{13} cm^3 \, molecule^{-1} s^{-1}. \quad (4.182)$$

The methoxy radical CH$_3$O can, in contrast to equation (4.178), react to generate nitrous acid; however, both reactions proceed very slowly in the air and therefore are not interesting:

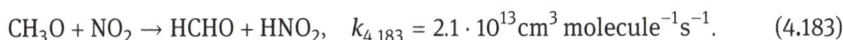

$$CH_3O + NO_2 \rightarrow HCHO + HNO_2, \quad k_{4.183} = 2.1 \cdot 10^{13} cm^3 \, molecule^{-1} s^{-1}. \quad (4.183)$$

Alkyl nitrite is photolysed to give the initial molecules in reaction equation (4.178); the photolytic lifetime is only 10–15 min. Organic nitrates also photodissociate, where equation (4.185a) is the dominant pathway; reaction equation (4.185b) includes inter-molecular rearrangement:

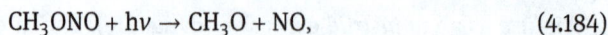

$$CH_3ONO + h\nu \rightarrow CH_3O + NO, \tag{4.184}$$

$$RONO_2 + h\nu \rightarrow RO + NO_2, \tag{4.185a}$$

$$RONO_2 + h\nu \rightarrow RCHO + HONO, \tag{4.185b}$$

$$RONO_2 + h\nu \rightarrow RONO + O. \tag{4.185c}$$

Peroxyacyl nitrates (PAN) have been intensively studied as NO_x reservoirs because they form in photochemical processes but are relatively stable and decompose thermally, thereby providing the long-range transport of nitrogen and regaining organic radicals. They are yielded from the peroxyacyl radical $RC(O)OO$, which is produced after the photolysis of aldehydes (reactions equation (4.286) and (4.287)):

$$RC(O)OO + NO_2 \rightleftarrows RC(O)OONO_2. \tag{4.186}$$

PAN can also thermally decompose as follows (this is a very rare example of a unimolecular reaction):

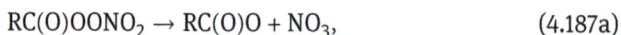

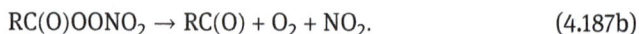

$$RC(O)OONO_2 \rightarrow RC(O)O + NO_3, \tag{4.187a}$$

$$RC(O)OONO_2 \rightarrow RC(O) + O_2 + NO_2. \tag{4.187b}$$

Many nitration reactions proceed on the surface of condensed phases (waters and particles). The nitrating properties of N_2O_3 ($NO + NO_2$) and N_2O_5 are seen from the structure, transferring $-NO_2$, $-N=O$ and $-O-N=O$:

N_2O_3 $O=N-O-N=O$
N_2O_5 $(O)O=N-O-N=O(O)$

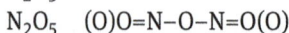

> Organic nitrates are among the extremely low volatility organic compounds that may play an important role in the nucleation and growth of atmospheric nanoparticles.

4.5 Sulphur

In the environment, sulphur ranks concerning total mass on the fourth site (Table 4.1). After oxygen and nitrogen, sulphur (and later phosphorus) is an important constituent of biomolecules with specific properties in the form of functional groups. As for the other life-essential elements, occurring dominantly in its most stable chemical compound (Table 4.2), sulphur occurs as sulphate dissolved in oceans.

Sulphur is the oldest known element (mentioned in the Bible, but known long before) as it could be found to reside close to volcanoes. The people quickly recognised its specific properties, such as burning with a penetrative odour.

In Homer's *Odyssey*, Ulysses said: "Bring my sulphur, which cleanses all pollution, and fetch fire also that I might burn it, and purify the cloisters" (translated by S. Butcher, Orange Street Press, 1998, p. 278). However, despite sulphur being the main chemical (besides mercury) of the alchemists, it was regarded until 1809 (by Humphrey Davy (1778–1829)) as a composite body even though Lavoisier in 1777 had already recognised it as an element (simple body). Gay-Lussac and Louis Jacques Thénard (1777–1857) confuted in 1809 this mistake (Kopp 1931, pp. 310–311), and from this time, sulphur has been seen as an element.

As for nitrogen, biogenic redox processes are essential to convert sulphur from its largest oxidation state +VI (sulphate) to its lowest −II (sulphide). Sulphate is the most stable compound in the atmosphere; once it has been produced, there is no abiotic reduction possible. Organisms need sulphur in the form of thiols RSH and sulphides R_2S. Thiols are (similar to alcohols ROH) weak acids but stronger then alkohols. Thiols can be easily oxidised into sulphides (this reaction is the main function in biological chemistry); thereby thiols provide hydrogen for the reduction of other molecules (the dissociated form RS^- acts as an electron donor):

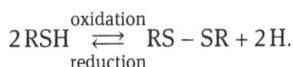

$$2\,RSH \underset{\text{reduction}}{\overset{\text{oxidation}}{\rightleftarrows}} RS - SR + 2\,H.$$

The following scheme shows the biological cysteine sulphur turnover (Gupta and Carroll 2014); see Table 4.14 for names of compounds:

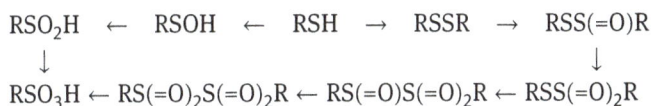

$$RSO_2H \leftarrow RSOH \leftarrow RSH \rightarrow RSSR \rightarrow RSS(=O)R$$
$$\downarrow \qquad\qquad\qquad\qquad\qquad\qquad\qquad\qquad \downarrow$$
$$RSO_3H \leftarrow RS(=O)_2S(=O)_2R \leftarrow RS(=O)S(=O)_2R \leftarrow RSS(=O)_2R$$

There are many organic sulphur compounds in nature, but all probably exist in higher oxidation states than S(−II), such as sulphides (including disulphides) and thiols, because they are oxidised after their release from organisms (see Table 4.14 for different functional sulphur groups). Carbonyl sulphide O=C=S (commonly written as COS) is the most abundant sulphur compound naturally present in the atmosphere. Before 1970, hydrogen sulphide H_2S was believed to be the major emitted species, but it is almost all oxidised before escaping from anoxic environments into the atmosphere. In the natural environment, dimethyl sulphide $(CH_3)_2S$ (DMS) is the most important sulphur species because of its large emission and contribution to climate-relevant atmospheric particulate sulphates. With industrialisation, sulphur dioxide (SO_2) became the key pollutant and the most important sulphur species in the air. Nowadays, after the introduction of flue gas desulphurisation, NO_x has replaced SO_2 in air chemical significance. However, concerning the atmospheric background submicron aerosol, sulphates remain dominant and important in climate control. It is a subtle irony of too much SO_2 abatement that cooling sulphate aerosol disappears demasking the greenhouse effect.

The five predominant sulphides (COS, CS_2, MeSH, DMS and DMDS) represent a source of some 46 Tg S yr^{-1}; a flux dominated by DMS. Natural sources dominate in the release of COS, MeSH and DMS to the atmosphere and also DMDS and polysulphides, although it would be sensitive to whether biomass burning was classified as natural or anthropogenic. However, the budget suggests that more than a third of COS arises from anthropogenic activities. Some 70 % of the CS_2 comes from human activities and almost all of the thiophenes. Table 4.13 summarizes global natural sulphur emissions.

Table 4.13: Natural sulphur emission.

source	Möller (2003)	Lee and Brimblecombe (2016)	
volcanism	–	0.03	COS, CS$_2$, DMDS
soils and plants	1–4	6.1	CS$_2$, MeSH, DMS, DMDS
wetlands	2	–	COS, CS$_2$, DMS, MeSH,
biomass burning	2–3	0.17	COS, CS$_2$, DMS, DMDS
ocean	36 (±20)	34.3	COS, CS$_2$, MeSH, DMS, DMDS
compounds	**natural**	**natural**	**anthropogenic**
DMS	35 (±20)	31.66	2.20
H$_2$S	3 (±2)	–	–
CS$_2$	1 (±1)	0.33	0.75
COS	1 (±0.5)	1.06	0.59
DMDS	–	1.08	1.10
MeSH	–	6.47	2.12
Thiophene	–	0.001	0.03
BTs	–	0	0.003
DBT	–	0	0.009
total	40 (±20)	40.6	6.8

The atmospheric chemistry of sulphur is simpler than that of nitrogen in two aspects. First, the number of stable species in air is smaller.[69] Second, the variety of interactions in the environment is less – almost all effects come from sulphates, such as acidity and radiation scattering. In the oxidation line to sulphate, oxidants are consumed:

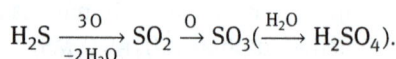

$$H_2S \xrightarrow[-2H_2O]{3\,O} SO_2 \xrightarrow{O} SO_3 (\xrightarrow{H_2O} H_2SO_4).$$

---

[69] The number of possible intermediates (often hypothetical) is huge (Tables 4.14, 4.15 and 4.16), especially in aquatic environments.

Table 4.14: Organic sulphur compounds. Sulphenes and sulphenic acid are extremely unstable and convert into disulphides and sulphinic acids. The open bond – is connected with organic rest R.

structure	formula	name of compound	formal oxidation state
$-S-H$	RSH	thiol (thio group SH)[a]	-2
$-S-$	R_2S	sulphide or sulphane	-2
$-S-S-$	R_2S_2	disulphide	$-1/-1$
$>S(=O)$	RSO	sulphine or sulphoxide (sulphinyl group SO)	±0
$-S(=O)_2$	RSO_2	sulphone (sulphonyl group SO_2)	$+3$
$=S(=O)_2$	$R_2C=SO_2$	sulphene[b]	$+6$
$-S(=O)-S-H$	$RS(O)SH$	thiosulphinic acid (thiosulphinate)	$\pm0/-1$
$-S-S(=O)_2-$	$R_2S_2O_2$	thiosulphonic acid (thiosulphanate)[c]	$-1/+3$
$-S(=O)-S(=O)-$	$R_2(SO)_2$	disulphoxide	$+4$
$-S(=O)-S(=O)_2-$	$R_2S_2O_3$	disulphide trioxide (sulphoxide-sulphone)	$+1/+3$
$-S(=O)_2-S(=O)_2-$	$R_2S_2O_4$	disulphone	$+3/+3$
$-S-OH$	$RSOH$	sulphenic acid[d]	±0
$-S(=O)-OH$	RSO_2H	sulphinic acid	$+2$
$-S(=O_2)-OH$	RSO_3H	sulphonic acid (sulpho group)	$+4$
$-O-S(=O_2)-OH$	RSO_4H	(organo)sulphate	$+4$

[a] also called mercaptan (old)
[b] note that S is in double bond with a C atom compared with to sulphone, e. g. $H_2C=SO_2$
[c] allicin is the main thiosulphanate compound in the garlic, causing the specific odour
[d] tautomer: $-S(=O)-H$

Hydrogen sulphide H_2S is the parent compound of organic thiols and sulphides; thiols R–SH (earlier name mercaptan) contain one or more SH groups in their molecules and can be considered analogues of alcohols, generated by replacing the OH group with SH. Replacing the H atom with any organic rest, we obtain sulphides R_2S. Sulphides are easily oxidised into sulphones R_2SO_2 via sulphoxides R_2SO. Sulphenic acid is the first member of the family of organosulphur oxoacid (formally derived from H_2S); it is formed by the action of reactive oxygen species on protein thiols. They are usually short-lived transient species in the oxidation of thiols to both disulphides (RSSR) and sulphinic acids R–S(=O)OH (derived from sulphurous acid H_2SO_3, replacing the OH group by organic R). Further in this oxidation line are sulphonic acids R–S(=O)$_2$OH and organic sulphates R–O–S(=O)$_2$OH. Besides, its biochemical role, the latter compounds are generated in the atmosphere in the reaction of SO_2 and SO_3 with organic compounds and play an important role in SOA formation.

4.5.1 Natural occurrence and sources

Like nitrogen, sulphur is an important element in biomolecules with specific functions. In contrast to nitrogen, for which the largest pool is the atmosphere (molecular N_2), for sulphur, the largest pool is the ocean (as dissolved sulphate); both components are chemically stable. Elemental sulphur (it forms cyclic octatomic molecules with the chemical formula S_8) was found in huge deposits in Italy (Sicily), Poland, Japan and America; nowadays, these are almost empty. Sulphur is now produced from H_2S in natural gas and still from sulphidic ores. Many hydrocarbon reservoirs contain native H_2S (and CO_2), which have been geologically generated in situ. Formation mechanisms are: (a) aquathermolysis, (b) microbial sulphate reduction and (c) thermal sulphate reduction. The C–S bond is weak in comparison to C–C bonds; therefore, reaction with high-temperature acidic H_2O normally comes at the expense of more organosulphur species when compared to non-sulphur-containing hydrocarbons.

In air, carbonyl sulphide (COS) represents the major sulphur component due to its long residence time. Similar to nitrogen and carbon, the sulphur content in the lithosphere is low because of degassing volatile sulphur compounds in the early history of the Earth. Primordial sulphides and elemental sulphur are almost all oxidised in atmospheric turnover. Moreover, humans have extracted them by mining from the lithosphere to such an extent that the remaining resources are negligible now (Figure 5.15). The great 'role' of life again is the reduction of sulphate. Volcanism is an important source of sulphur dioxide (SO_2) and promotes the formation of a strong acid (H_2SO_4) which may play an important role in weathering. Consequently, the dominant anthropogenic SO_2 emission since the Industrial Revolution has resulted in significant acidification of many parts of the world (see also Chapter 5.3.2).

More than 90 % of SO_2 emission is related to the combustion of fossils fuels (coal and oil with an approximate share of two-thirds and one-third, respectively), primarily metal smelting (Cu, Zn, Mn, and Ni) and sulphuric acid production (only in the past). Air pollution in the industrialised world has undergone drastic changes in the past 50 years. Until World War II, the most important urban compound was sulphur dioxide combined with soot from the use of fossil fuels in heat and power production. Almost everything that was known in the 1950s about the causes of smoke and its elimination had already been said by the turn of the nineteenth century. Still, hardly anything had been done to reduce the smokiness of cities. Although SO_2 concentration in the air of cities was at the level of 110 mg m^{-3} around 1900, it dropped to the upper range of μg m^{-3} (some hundreds) by the 1950s and nowadays is in the lowest range of μg m^{-3} (5–10), close to the remote background values of the order of 0.5 to 4μg m^{-3}.

Since 1980, flue gas desulphurisation (FDG) has been introduced stepwise in industrialised countries; FDG works with an efficiency of about 95 %. Global anthropogenic sulphur emissions increased until around 1980. The different estimates in the period 1990–2000 show significant variations between 65 and 85 Tg S yr^{-1}. Newer estimates show relative stability throughout the decade of the 1980s and a 25 % decline from 1990 to 2000 to a level not seen since the early 1960s. The decline is evident in North America, but most significant in Europe. The combustion of fossil fuels has increased continuously until the present. China's maximum SO_2 emission was in 2005 (16.5 Tg SO_2–S) and decreased to 14 Tg in 2010. Largely uncontrolled and rising is the SO_2 emission in India.

Biogenic sources emit so-called reduced sulphur compounds, being in the oxidation state of S^{2-} such as hydrogen sulphide (H_2S), dimethyl sulphide (DMS), dimethyl disulphide (DMDS), carbon disulphide (CS_2) and carbonyl sulphide (COS).[70] Plants emit DMS and COS (similar to other organic compounds) predominantly by respiration; see Table 4.13.

Sulphur in coal can be classified into two categories: inorganic and organic sulphur. Inorganic sulphur is present in the form of pyrite and sulphate, mainly the former. Often, organic sulphur and the macromolecular structure of coal join together in the form of covalent bonds with complex structure and are difficult to separate. It mainly includes thiols, sulphide, disulphides, thioethers, sulphoxides, sulphones, thiophene and others. In power plants, the sulphur from the fuel is oxidised almost completely, mainly to sulphur dioxide but partly to sulphur trioxide. Once SO_3 is formed, it converts quickly into particles of sulphuric acid (smaller than 10 nm) which can be much less removed by flue gas desulphurisation than SO_2. Therefore, the relative percentage of SO_3 to SO_2 is drastically increased in modern coal-fired power plants and might cause "blue plumes".

4.5.2 Reduced sulphur: H_2S, COS, CS_2, and DMS

In chemical textbooks, carbon disulphide and carbonyl sulphide are treated among carbon compounds. In the environment, the contribution of CS_2 and COS to the carbon budget is negligible, and both species play no role in carbon chemistry. Moreover, COS only decomposes in the stratosphere because of its chemical stability. Most tropospheric COS is removed by dry deposition (plant assimilation and uptake by microorganisms in soils). COS has been suggested to be a contributor to the stratospheric sulphate layer (Junge layer). However, nowadays, it is assumed that it plays a minor role in the stratospheric sulphur budget. Stratospheric chemistry is simple. Either it photodissociates into CO and S, which subsequently form CO_2 and SO_2 Or it reacts slowly with OH to CO + SO or CO_2+ SH. All sulphur species are finally oxidised to H_2SO_4 Which is important in the formation of PSCs (see Chapter 5.3.2.1). SO and SH are short-lived intermediates, which also occur in the oxidation chain of CS_2 and H_2S: $k_{4.188} = 7.6 \cdot 10^{-17}$ cm^3 molecule^{-1} s^{-1}, $k_{4.189} < 4 \cdot 10^{-19}$ cm^3 molecule^{-1} s^{-1} and $k_{4.190} = 3.7 \cdot 10^{-12}$ cm^3 molecule^{-1} s^{-1}.

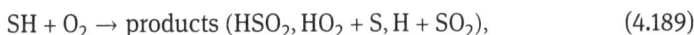

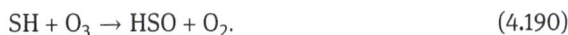

$$SO + O_2 \rightarrow SO_2 + O, \tag{4.188}$$

$$SH + O_2 \rightarrow products\ (HSO_2, HO_2 + S, H + SO_2), \tag{4.189}$$

$$SH + O_3 \rightarrow HSO + O_2. \tag{4.190}$$

---

[70] Some authors use the formula OCS, which also represents the molecule structure O=C=S. Plants emit DMS and COS (similar to other organic compounds) predominantly by respiration; see Table 4.13.

The radicals HSO and HSO$_2$ quickly decompose (with O$_3$ and O$_2$) into several other radicals (SH, SO, HSO$_2$ and OH; $k \approx 6 \cdot 10^{-14}$ cm^3 molecule^{-1} s^{-1}); finally, SO$_2$ is produced. Radical sulphur quickly oxidises; $k_{4.191} = 2.1 \cdot 10^{-12}$ cm^3 molecule^{-1} s^{-1}):

$$S + O_2 \rightarrow SO + O. \tag{4.191}$$

i | Carbonyl sulphide (COS) presents the largest sulphur pool in the atmosphere due to its long atmospheric lifetime. Its concentration is slightly increasing, likely because of anthropogenic emissions from coal combustion. It contributes (negligibly) to the greenhouse effect.

The only important (and very fast) reaction of CS$_2$ goes via OH radicals; $k_{4.192} = 2.5 \cdot 10^{-11}$ cm^3 molecule^{-1} s^{-1} and $k_{4.193} = 2.8 \cdot 10^{-14}$ cm^3 molecule^{-1} s^{-1}:

$$CS_2 + OH + M \rightarrow HOCS_2 + M, \tag{4.192}$$
$$HOCS_2 + O_2 \rightarrow products \ (CO, CS, SO, COS). \tag{4.193}$$

The CS radical reacts in two channels; $k_{4.194} = 2.9 \cdot 10^{-19}$ cm^3 molecule^{-1} s^{-1}:

$$CS + O_2 \rightarrow CO + SO, \tag{4.194a}$$
$$CS + O_2 \rightarrow COS + O. \tag{4.194b}$$

It is obvious that O reacts with O$_2$ to give O$_3$ and CO forms CO$_2$ via OH attack. Finally, the oxidation of H$_2$S is also relatively fast (2–3 days lifetime):

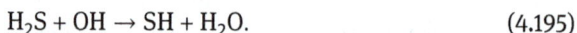

$$H_2S + OH \rightarrow SH + H_2O. \tag{4.195}$$

Analogically, OH reacts with methanethiol (old name: methyl mercaptan) CH$_3$SH ($k = 3.3 \cdot 10^{-11}$ cm^3 molecule^{-1} s^{-1}). We can summarise that H$_2$S is 100 % converted into SO$_2$ and CS$_2$ where the latter is transformed each 50 % into SO$_2$ and COS.

Organosulphur compounds are widely present in our bodies and the natural environment. There are vegetables that contain it with special properties; garlic, onion, shallot, leek, and chives are well-known representatives. Theories have emerged that explain these properties of organosulphur compounds by their correction of redox-sensing and redox-signalling properties. Biogenic sources emit so-called reduced sulphur compounds (RSCs), being in the oxidation state of S^{2-} such as hydrogen sulphide (H$_2$S), dimethyl sulphide (DMS), and dimethyl disulphide (DMDS) carbon disulphide (CS$_2$) and carbonyl sulphide (COS). Plants emit DMS and COS (similarly to other organic compounds) predominantly by respiration. It is likely that other organic sulphur compounds (in analogy to organic nitrogen), especially sulphides, are also emitted; for example, from *Allium ursinum* (a type of garlic): dipropenyldisulphide, methyl-propenyldisulphide, diallylsulphide and cis-propenyldisulphide.

The most important reduced sulphur species is DMS. Many studies have been carried out to clear the oxidation mechanisms. The fate of radical intermediates such as CH_3O, SO and CH_3 have been described elsewhere, but many other radicals can only be grouped together within the term 'product' on the right-hand side of the equation. Besides the OH attack, NO_3 (nitrate radical) has been described as a very effective oxidiser. OH initiates two channels: the H abstraction and the addition: $k_{4.196} = 4.8 \cdot 10^{-12} \, cm^3 \, molecule^{-1} \, s^{-1}$ and $k_{4.197} = 2.2 \cdot 10^{-11} \, cm^3 \, molecule^{-1} \, s^{-1}$.

$$DMS + OH \rightarrow CH_2SCH_3 + H_2O, \tag{4.196}$$

$$DMS + OH \rightarrow CH_3S(OH)CH_3. \tag{4.197}$$

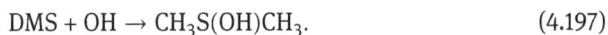

Two more channels that break down the DMS molecule are described in $CH_3 + CH_3OH$ and $CH_3 + CH_3SOH$. Because of the nocturnal NO_3 attack on DMS (H abstraction channel), its lifetime is short (only about one day). It is remarkable that via the abstraction channel, only sulphate is seen at the end, whereas in the OH addition channel, many stable organic sulphur species, such as methanesulphonic acid (MSA) and dimethyl sulphoxide (DMSO) as well as dimethyl sulphone ($DMSO_2$), can be identified; see also Table 4.14.

The further fate of these compounds is a transfer into the aqueous phase (clouds and precipitation). It is known that organic sulphide (here DMS) can also be oxidised to sulphoxide (here DMSO) and further to sulphone in solution:

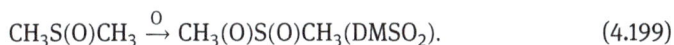

$$CH_3SCH_3 (DMS) + H_2O_2 \rightarrow CH_3S(O)CH_3 (DMSO) + H_2O, \tag{4.198}$$

$$CH_3S(O)CH_3 \xrightarrow{O} CH_3(O)S(O)CH_3 (DMSO_2). \tag{4.199}$$

Although DMSO is the major product of aqueous phase DMS oxidation (and is ubiquitously found in all aquatic environments), minor amounts of methanethiol CH_3SH, dimethyl disulphide CH_3SSCH_3 (DMDS), MSA and $DMSO_2$ have been observed.

> Because of the huge amount of maritime DMS emission (20–50 Mt S yr^{-1}), it plays a significant role in particulate sulphate formation, acting as cloud condensation nuclei above the ocean and therefore controlling the climate.

$DMSO_2$ will oxidise in droplets like MSA, which is found (together with DMSO) in clouds and precipitation water with a distinct seasonal variation. The rate of DMS oxidation has been observed to increase in the presence of humic substances as well as model substances that could act as photosensitisers. This again is a remarkable observation supporting that in aquatic environments, oxidising agents (where OH plays the key role) are photoenhanced (reaction equation (4.31)). It has been found that besides the formation of $DMSO_2$, DMSO also reacts in the following chain; MSA forms in solution methanesulphonate ($CH_3O_3S^-$):

$$DMSO \xrightarrow{OX} CH_3SO_2H \xrightarrow{OX} CH_3SO_3H(MSA).$$

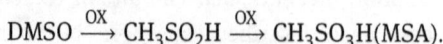

A number of studies on the oxidation of H_2S with O_2 in natural waters have been conducted in the laboratory and the field, but only one study is known to have any atmospheric relevance (Hoffmann 1977). With respect to the sediment chemistry of natural waters and the pollution treatment of wastewater, H_2S autoxidation was of interest long before. It has been found that the process is in the first order with respect to H_2S. Trace metals (especially Fe and Mn) increase the rate of oxidation, and sulphate (SO_4^{2-}), thiosulphate ($S_2O_3^{2-}$) and elemental sulphur have been detected as products. From the water bottom to the surface, the H_2S oxidation rate increases; this is attributed to a greater amount of dissolved oxygen as well as additional oxidants such as hydrogen peroxide and ozone closer to the interface with air. The oxidation of intermediate sulphite (HSO_3^-) is discussed in Chapter 4.5.3. Atmospheric H_2S concentrations are significant only near their sources. Therefore, considering the much more considerable significance of DMS and SO_2 in the sulphur cycle, the aqueous phase H_2S oxidation has never been considered in cloud and precipitation chemistry models. Thus, hydrogen sulphide as an odorous pollutant is only of interest with respect to wastewater. H_2S is only slightly soluble in water and partly dissociates; $pK_a = 7$ (the second dissociation with $pK_a \approx 13$ can be neglected):

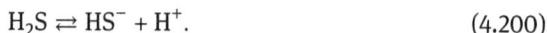

$$H_2S \rightleftharpoons HS^- + H^+. \qquad (4.200)$$

In the studies found in the literature, only gross reaction equations are given. However, with modern knowledge, we can speculate about aqueous oxygen chemistry and photocatalytic enhanced redox processes that reactive oxidants such as OH, O_3 and H_2O_2 will elementarily react with HS^- (in the autoxidation process, all these species are slowly produced from O_2), similar to the sulphite oxidation:

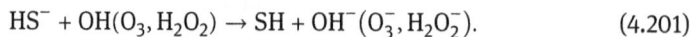

$$HS^- + OH(O_3, H_2O_2) \rightarrow SH + OH^-(O_3^-, H_2O_2^-). \qquad (4.201)$$

The SH radical (in solution it dissociates to a minimal extent into $H^+ + S^-$; thereby instable S^- could donate its electron and provide sulphur) can react in analogy to the gas phase reaction equation (4.190) with O_3 gaining HSO but also with OH forming S:

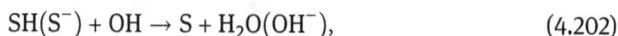

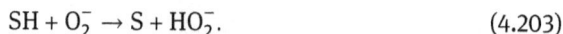

$$SH(S^-) + OH \rightarrow S + H_2O(OH^-), \qquad (4.202)$$
$$SH + O_2^- \rightarrow S + HO_2^-. \qquad (4.203)$$

The elemental sulphur becomes colloidally stable in S_8 molecules but can also add to sulphite and form thiosulphate, which also loses the sulphur via disproportionation into H_2S and sulphate:

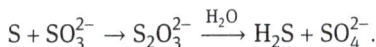

$$S + SO_3^{2-} \rightarrow S_2O_3^{2-} \xrightarrow{H_2O} H_2S + SO_4^{2-}.$$

In hydrometeors and interfacial waters, however, sulphur reacts with oxygen (reaction equation (4.191)) via SO_2 to form hydrogen sulphite HSO_3^-. Thiosulphate is an important intermediate in biological sulphur chemistry from both sulphate reduction and sulphide oxidation. Many hypothetical so-called lower sulphuric acids (Table 4.14) might appear as intermediates or in the form of radicals (such as SOH, HSO, HSO_2, HSS and SH as seen from the structure formulas) in the oxidation chain from sulphide to sulphate:

$$HS^- \rightarrow S \rightarrow HSO_3^- \rightarrow SO_4^{2-}.$$

The role of such intermediates and organosulphur compounds as antioxidants (due to radical scavenging) has been known for many years.

4.5.3 Oxides and oxoacids: SO_2, H_2SO_3, SO_3, and H_2SO_4

Sulphur forms low molecular oxides of the composition SO_m (m = 1, 2, 3, and 4). We already met sulphur monoxide (SO) as an intermediate in the oxidation of reduced sulphur compounds. Environmentally important are only sulphur dioxide (SO_2) and sulphur trioxide (SO_3). Sulphur forms three oxoacids of the composition H_2SO_n (n = 3, 4, and 5), six of the composition $H_2S_2O_n$ (n = 3, 4, 5, 6, 7, and 8), and several acids of the composition $H_2S_nO_m$ (Table 4.15).

However, in the environment, only sulphurous acid (H_2SO_3) and sulphuric acid (H_2SO_4) with the corresponding anhydrides sulphur dioxide (SO_2) and sulphur trioxide (SO_3) are of importance.

Some acids (or anions, respectively) exist only intermediary. As said, the primary emission of burning (sulphur-containing) fuels is SO_2; however, it is known that around 2–5 % of the sulphur emission occurs already as SO_3. Once SO_3 is formed, it converts quickly into particles of sulphuric acid (smaller than 10 nm) which can be much less removed by flue gas desulphurisation than SO_2. Therefore, the relative percentage of SO_3 to SO_2 drastically increased in modern coal-fired power plants and might cause 'blue plumes'. In the gas phase, SO_2 can be oxidised only by OH radicals, followed by gas-to-particle conversion to sulphuric acid and sulphate particulate matter. The main route of S(IV) oxidation goes via the aqueous phase, where cloud droplet evapora-

Table 4.15: Oxoacids of sulphur and its isomeric forms (hypothetically). Note: many are unstable intermediates or exist only as anions in solution. The –S–S– group is called sulphane, from which polysulphane is derived.

structure	formula	name (acid)	name (salt)
$(HO)_2S=O$	H_2SO_3	sulphurous acid	sulphite
$(HO)_2S(=O)_2$	H_2SO_4	sulphuric acid	sulphate
$HO-S(=O)_2-OOH$	H_2SO_5	peroxomonosulphuric acid[a]	peroxomonosulphate
$HS-S(H)=O$	H_2S_2O	thiosulphoxyl acid	thiosulphinate
$HS-S(=O)-OH$	$H_2S_2O_2$	thiosulphurous acid[b]	thiosulphite
$(HO)_2S=S(=O)$	$H_2S_2O_3$	thiosulphuric acid	thiosulphate
$HO-S(=O)-S(=O)-OH$	$H_2S_2O_4$	hyposulphurous acid[c]	hyposulphite[c]
$HO-S(=O)_2-O-S(=O)-OH$	$H_2S_2O_5$	disulphurous acid	disulphite
$HO-S(=O)_2-S(=O_2)-OH$	$H_2S_2O_6$	dithionic acid[d]	dithionate
$HO-S(=O)_2-O-S(=O_2)-OH$	$H_2S_2O_7$	disulphuric acid[e]	disulphate
$HO-S(=O_2)-OO-S(=O_2)-OH$	$H_2S_2O_8$	peroxodisulphuric acid[f]	peroxodisulphate

[a] also called Caro's acid
[b] disulphur(I)acid, anhydride: S_2O (disulphur monoxide); tautomers: HO–S–S–OH (dihydroxy disul-phane) and $S-S(OH)_2$ (thiothionyl hydroxide)
[c] also called dithionous acid (salt: dithionite and old: hydrosulphite)
[d] generic name: thionic or polythionic acids $H_2S_nO_6$ ($n = 2...5...$), also named disulphanic acid and hyposulphuric acid
[e] also called pyrosulphuric acid; generic name: polysulphuric acids $H_2S_nO_t$
[f] also called *Marshall's* acid

tion provides sub-μ aerosol particles containing sulphate as illustrated in the scheme below:

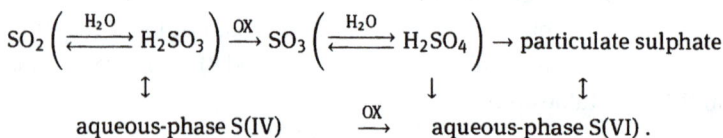

$$SO_2 \left(\xrightleftharpoons{H_2O} H_2SO_3 \right) \xrightarrow{OX} SO_3 \left(\xrightleftharpoons{H_2O} H_2SO_4 \right) \rightarrow \text{particulate sulphate}$$

$$\text{aqueous-phase S(IV)} \quad \xrightarrow{OX} \quad \text{aqueous-phase S(VI)}.$$

From all gas-phase reactions studied since the 1950s (see Möller 1980 and cita-tions therein), OH remains the sole component for SO_2 oxidation: $k_{4.204} = 1.3 \cdot 10^{-12}$ cm^3 molecule^{-1} s^{-1}: $k_{4.205} = 4.3 \cdot 10^{-13}$ cm^3 molecule^{-1} s^{-1} and $k_{4.206} = 5.7 \cdot 10^4$ s^{-1} (50 % RH). Therefore, it is seen that equation (4.204) determines the overall rate of conversion:

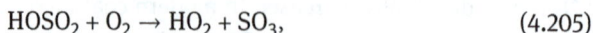

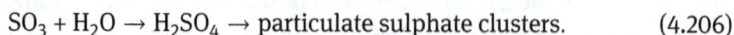

$$SO_2 + OH + M \rightarrow HOSO_2 + M, \tag{4.204}$$

$$HOSO_2 + O_2 \rightarrow HO_2 + SO_3, \tag{4.205}$$

$$SO_3 + H_2O \rightarrow H_2SO_4 \rightarrow \text{particulate sulphate clusters.} \tag{4.206}$$

Note that H_2SO_4 does not exist molecularly gaseous like HNO_3. The H_2SO_4 molecules produced *in situ* (up to 10^8 cm^{-3}) agglomerate very fast to particles in the low nanometre mode.

Assuming a mean OH radical concentration of 10^6 molecule cm^{-3}, the SO_2 lifetime is about 10 days. Therefore, dry deposition and uptake by clouds and precipitation are important removal pathways. Because of the relatively slow gas-phase SO_2 oxidation, aqueous-phase oxidation in clouds contributes to 80–90 % of sulphate formation in the northern mid-latitudes. The in-droplet S(IV) oxidation rate is related to the volume of air by using equation (2.9) and equation (2.17) in terms of:

$$\left(\frac{d[sulphate]}{dt}\right)_{air} = k[SO_2]_{gas} = RT \cdot LWC \cdot k_{aq}[S(IV)]. \tag{4.207}$$

In a cloud under normal daytime conditions of LWC and concentrations of H_2O_2 and O_3, the SO_2 lifetime is $\leq 1\,h$, and sulphate production can reach 8 ppb h^{-1}. This is by a factor of 100 higher than the maximum gas-phase production (Table 4.17). On a yearly basis, however, we need the statistical information on the occupancy of the lower atmosphere by clouds and the occurrence of clouds to calculate the mean aqueous-phase S(IV) oxidation. Any error in k_{aq} is insignificant compared with the uncertainty of cloud statistics. The atmospheric SO_2 residence time strongly depends on the event-related cloud and precipitation statistics.

SO_2 is moderately soluble and forms sulphurous acid H_2SO_3, not known as a pure substance; several isomeric forms have been spectroscopically detected. Therefore, the symbol $SO_2 \cdot aq$ is often used; the equilibrium equation (4.208) lies fully on the left-hand side ($K_{4.208} \ll 10^{-9}$). Sulphurous acid, however, is dissociated mainly with $pK_{4.209} = 2.2$ and $pK_{4.210} = 7.0$:

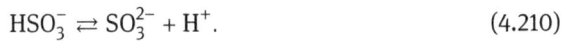

$$SO_2 + H_2O \rightleftharpoons SO_2 \cdot aq \rightleftharpoons H_2SO_3, \tag{4.208}$$

$$H_2SO_3 \rightleftharpoons HSO_3^- + H^+, \tag{4.209}$$

$$HSO_3^- \rightleftharpoons SO_3^{2-} + H^+. \tag{4.210}$$

Because of the complication with H_2SO_3, dissolved SO_2 can be directly associated with bisulphite: $k_{4.211} = 6.3 \cdot 10^4\,s^{-1}$.

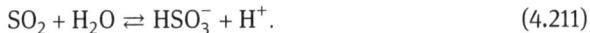

$$SO_2 + H_2O \rightleftharpoons HSO_3^- + H^+. \tag{4.211}$$

SO_2 also directly reacts with hydroxide ions: $k_{4.212} = 1.1 \cdot 10^{10}$ $L\,mol^{-1}\,s^{-1}$:

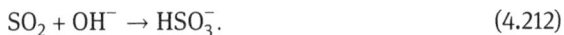

$$SO_2 + OH^- \rightarrow HSO_3^-. \tag{4.212}$$

It is useful to name with S(IV) all dissolved sulphur species in this oxidation state: $SO_2 + H_2SO_3 + HSO_3^- + SO_3^{2-}$, whereby $[H_2SO_3] \rightarrow 0$.

Sulphite forms adducts with dissolved aldehydes, of which the most important is α-hydroxymethanesulphonate (HMS), which is stable against oxidation: $pK_{4.213} = 3.4$.

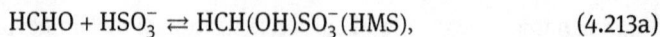

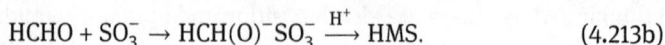

$$HCHO + HSO_3^- \rightleftharpoons HCH(OH)SO_3^- \text{(HMS)}, \tag{4.213a}$$

$$HCHO + SO_3^- \rightarrow HCH(O)^-SO_3^- \xrightarrow{H^+} HMS. \tag{4.213b}$$

However, competitive formaldehyde gives a hydrate named methanediol, also known as formaldehyde monohydrate or methylene glycol, which is unable to add onto sulphite:

$$HCHO + H_2O \rightleftharpoons CH_2(OH)_2. \tag{4.214}$$

HMS is the anion of the strong hydroxymethanesulphone acid (HMSA), fully dissociated. The second dissociation step to $HCH(O)^-SO_3^-$ corresponds to a weak acid with $pK_a \approx 10$. HMS slowly decomposes into the initial substances, thereby sometimes describing equation (4.213a), (4.213b) and equation (4.215) as equilibrium with $K_{4.213/4.215} = 6.6 \cdot 10^9$, $k_{4.215} = 7.7 \cdot 10^{-3} \, s^{-1}$:

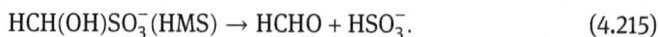

$$HCH(OH)SO_3^- \text{(HMS)} \rightarrow HCHO + HSO_3^-. \tag{4.215}$$

Often, formaldehyde hydration is also included in the equilibrium:

$$K = \frac{[CH_2(OH)SO_3^-]}{[CH_2(OH)_2][HSO_3^-]} = 3.6 \cdot 10^6 \, L \, mol^{-1}. \tag{4.216}$$

In alkaline solution, HMS decomposes: $k_{4.217} = 3.7 \cdot 10^{-3} \, L \, mol^{-1} \, s^{-1}$.

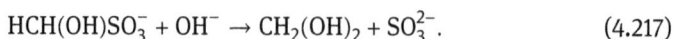

$$HCH(OH)SO_3^- + OH^- \rightarrow CH_2(OH)_2 + SO_3^{2-}. \tag{4.217}$$

The only oxidation of HMS proceeds via OH attack:

$$HCH(OH)SO_3^- + OH \xrightarrow{O_2 \text{(multistep)}} HC(O)SO_3^- + H_2O + HO_2. \tag{4.218}$$

From all three HMS sinks (oxidation, alkaline decomposition and decay), only equation (4.215) is considered to be important. Sulphite also forms adducts with other aldehydes such as benzaldehyde, methylglyoxal, acetaldehyde and hydroxyacetaldehyde.

> To protect a solution (e. g., a rainwater sample) against S(IV) oxidation, adding an excess of the expected equimolar amount of formaldehyde solution (HCHO) forms the adduct with sulphite. The remaining sulphate corresponds analytically to the real and original concentration. Afterwards, the S(IV)-formaldehyde adduct can be destroyed (or directly estimated by ion chromatography), and after the addition of a H_2O_2 solution all sulphite oxidises to sulphate, which is analysed as a total sum of S(IV) and S(VI).

Because of the importance of SO_2 and sulphate in the atmosphere, the oxidation pathways in solution have been studied extensively and typically been subdivided as follows:

- by peroxides (H_2O_2 and ROOH),
- by ozone O_3,
- by oxygen (autoxidation),
- by oxygen with the participation of TMI,
- by radicals (OH, NO_3, Cl and others),
- by other oxidants (e. g. HNO_4, HOCl).

The decades of SO_2 research have given almost only gross reaction rates such as those first studied by Mader (1958) and later recognised as being atmospherically important by Hoffmann and Edwards (1975) and Penkett et al. (1979); $k_{4.219} = (5.3 \pm 2.7) \cdot 10^7 \, L^2 \, mol^{-2} \, s^{-1}$).

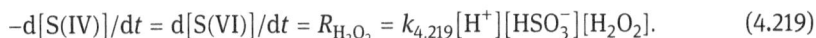

$$-d[S(IV)]/dt = d[S(VI)]/dt = R_{H_2O_2} = k_{4.219}[H^+][HSO_3^-][H_2O_2]. \qquad (4.219)$$

The rate law equation (4.219) has been confirmed by many experimental studies. Remarkably, soon after its discovery, it was known that H_2O_2 oxidises sulphurous acid into sulphuric acid without the formation of free oxygen (Gmelin 1827), a mechanism recognised to be important in air chemistry almost 150 years later (Möller 1980) as the most important pathway in the oxidation of dissolved SO_2 in hydrometeors. It is remarkable that the detailed mechanism of the S(IV)–H_2O_2 reaction is unknown. Möller (2009) proposed that a single electron transfer occurs in the sense of another Fenton-like reaction:

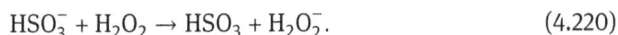

$$HSO_3^- + H_2O_2 \rightarrow HSO_3 + H_2O_2^-. \qquad (4.220)$$

The fate of $H_2O_2^-$ is well known (equation (4.48)); the enhancement of the H_2O_2 pathway in acidic solution is clearly seen:

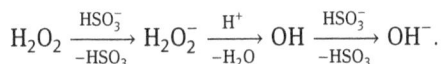

$$H_2O_2 \xrightarrow[-HSO_3]{HSO_3^-} H_2O_2^- \xrightarrow[-H_2O]{H^+} OH \xrightarrow[-HSO_3]{HSO_3^-} OH^-.$$

Figure 4.13 shows a generalised scheme of S(IV) oxidation, depending on pH, where the radical chain mechanism is most likely in S(IV) oxidation. The existence of the sulphite radical and its role in biological damage (whereby the subsequently produced peroxosulphate radical SO_5^- is a much stronger oxidant) has long been known. Many molecules, radicals and metal ions react with sulphite and bisulphite in one-electron oxidation; A – electron acceptor (see also Figure 4.13):

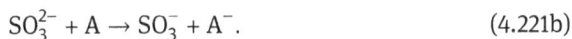

$$HSO_3^- + A \rightarrow HSO_3 + A^-, \qquad (4.221a)$$
$$SO_3^{2-} + A \rightarrow SO_3^- + A^-. \qquad (4.221b)$$

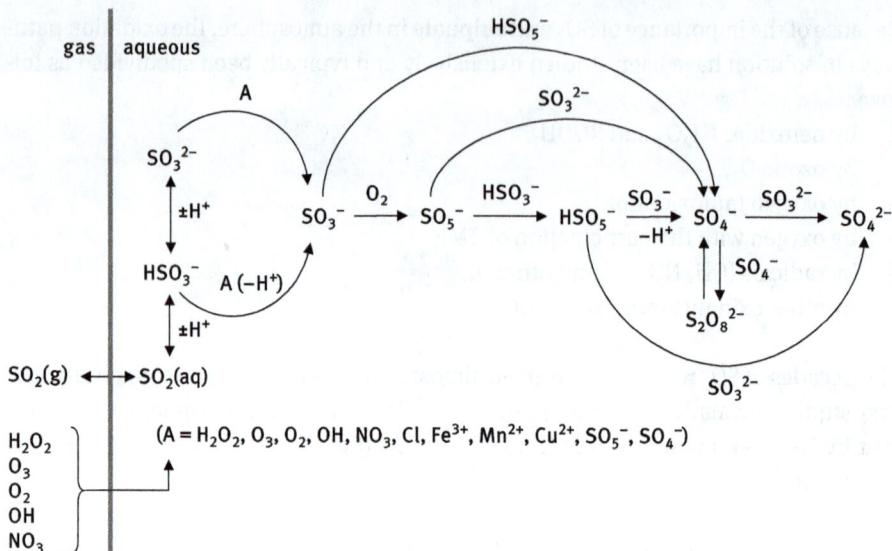

Figure 4.13: Scheme of general S(IV) oxidation in alkaline and acidic solution. Electron acceptors: H_2O_2, O_3, O_2, OH, NO_3, Cl, Fe^{3+}, Mn^{2+}, Cu^{2+}, electron donors: Fe^{2+}, Mn^+, Cu^+, HSO_3^-, SO_3^{2-}, OH^-.

The sulphite radical SO_3^- reacts quickly with oxygen to form the peroxosulphate radical; $k_{4.222} = 2.5 \cdot 10^9 \, \text{L mol}^{-1}\,\text{s}^{-1}$:

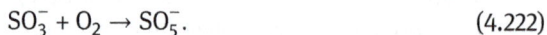

$$SO_3^- + O_2 \rightarrow SO_5^-. \tag{4.222}$$

The peroxosulphate radical SO_5^- reacts almost with sulphite (HSO_3^- and SO_3^{2-}) in a different pathway to peroxomonosulphate (HSO_5^-) and the final product sulphate (SO_4^{2-}), generating sulphur radicals (SO_3^- and SO_4^-), see Figure 4.13:

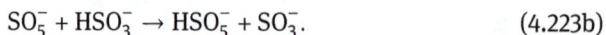

$$SO_5^- + SO_3^{2-} \rightarrow SO_4^- + SO_4^{2-}, \tag{4.223a}$$
$$SO_5^- + HSO_3^- \rightarrow HSO_5^- + SO_3^-. \tag{4.223b}$$

At this stage, the pathway splits from sulphate radical SO_4^- and peroxomonosulphate (HSO_5^-); the latter decomposes with sulphite (HSO_3^- and SO_3^{2-}) to sulphate:

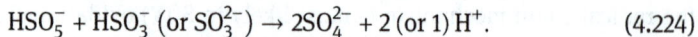

$$HSO_5^- + HSO_3^- \text{ (or } SO_3^{2-}) \rightarrow 2SO_4^{2-} + 2 \text{ (or 1) } H^+. \tag{4.224}$$

The sulphate radical reacts similarly with sulphite but regenerates sulphite radicals:

$$SO_4^- + HSO_3^- \text{ (or } SO_3^{2-}) \rightarrow SO_4^{2-} + SO_3^- + H^+. \tag{4.225}$$

Hence a radical chain has been established in which, initially, sulphite ions will also be transformed into sulphite radicals by other sulphur radicals (SO_4^- and SO_5^-). Of less importance are radical-radical reactions; we cited first dimerisation to relative stable dithionate ($S_2O_6^{2-}$) and peroxodisulphate ($S_2O_8^{2-}$):

$$SO_4^- + SO_4^- \rightarrow S_2O_8^{2-}, \tag{4.226}$$

$$SO_5^- + SO_5^- \rightarrow S_2O_8^{2-} + O_2, \tag{4.227a}$$

$$SO_5^- + SO_5^- \rightarrow 2SO_4^- + O_2. \tag{4.227b}$$

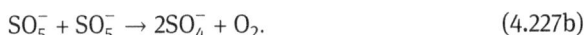

Peroxodisulphate (it is produced as a strong oxidant) decays back in an aqueous solution by homolysis into sulphate radicals (see equation (4.227b)). There are several competing reactions in this radical sulphur oxidation mechanism (not noted here) because the chain

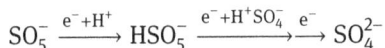

$$SO_5^- \xrightarrow{e^- + H^+} HSO_5^- \xrightarrow{e^- + H^+ SO_4^-} \xrightarrow{e^-} SO_4^{2-}$$

also proceeds by all available electron donors (such as reduced TMI, O_2^-, OH^- and Cl^-). Furthermore, there are described transfers of sulphur radicals by ROS (H_2O_2, HO_2, and OH), whereby OH acts as H abstractor (formation of H_2O) and the ionic species acts as electron donor according to (the electron acceptor is any kind of sulphur radical shown above):

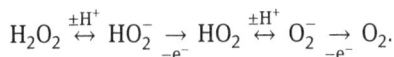

$$H_2O_2 \overset{\pm H^+}{\leftrightarrow} HO_2^- \underset{-e^-}{\rightarrow} HO_2 \overset{\pm H^+}{\leftrightarrow} O_2^- \underset{-e^-}{\rightarrow} O_2.$$

Note that there are significant differences in the reaction rate constants independent of whether sulphite or bisulphite is the reagent. This makes the overall process strongly dependent on pH. This pH dependence, however, is much more subtle because the sulphur radicals (sulphite, sulphate and peroxomonosulphate) undergo acid-base equilibrium, which likely is on the left-hand side because the hydrogenated species are strong acids:

$$SO_3^- \overset{\pm H^+}{\leftrightarrow} HSO_3, \quad SO_4^- \overset{\pm H^+}{\leftrightarrow} HSO_4, \quad \text{and} \quad SO_5^- \overset{\pm H^+}{\leftrightarrow} HSO_5.$$

Parallel to all possible starting reactions of the type equation (4.221a), (4.221b), the direct nucleophilic attack of oxygen species (O_3, OH and HO_2) might be possible

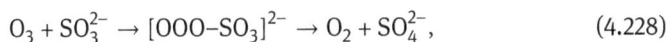

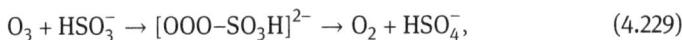

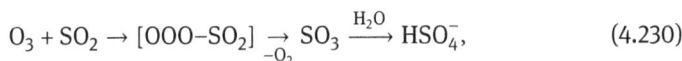

$$O_3 + SO_3^{2-} \rightarrow [OOO{-}SO_3]^{2-} \rightarrow O_2 + SO_4^{2-}, \tag{4.228}$$

$$O_3 + HSO_3^- \rightarrow [OOO{-}SO_3H]^{2-} \rightarrow O_2 + HSO_4^-, \tag{4.229}$$

$$O_3 + SO_2 \rightarrow [OOO{-}SO_2] \underset{-O_2}{\rightarrow} SO_3 \xrightarrow{H_2O} HSO_4^-, \tag{4.230}$$

$$OH + SO_3^{2-} \rightarrow [HO\text{--}SO_3]^{2-} \xrightarrow{O_2} SO_4^{2-}, \tag{4.231}$$

$$HO_2 + SO_3^{2-} \rightarrow [HOO\text{--}SO_3]^{2-} \xrightarrow[-OH]{} SO_4^{2-}. \tag{4.232}$$

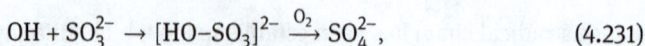

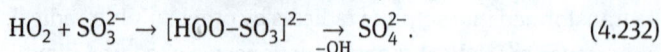

The reaction of ozone with S(IV) is assumed as a nucleophilic attack onto all S(IV), see equation (4.228), equation (4.229) and equation (4.230):

$$R_{O_3} = (k_a[SO_2] + k_b[HSO_3^-] + k_c[SO_3^{2-}])[O_3]. \tag{4.233}$$

The still best accepted k-values are (in $L\ mol^{-1}s^{-1}$): $k_a = 2.4 \cdot 10^4$, $k_b = 3.7 \cdot 10^5$ and $k_c = 1.5 \cdot 10^9$. Using the expressions for the dissociation equilibriums of different S(IV) species and simplification for pH > 3, we get:

$$R_{O_3} = (k_b + k_c K_b[H^+]^{-1})[S(IV)][O_3], \tag{4.234}$$

where K_b is the equilibrium constant of the second dissociation ($HSO_3^- \rightleftharpoons SO_3^{2-}$). A general rate law can be derived from the studies suggesting the radical mechanism:

$$R_{O_3} = k[H^+]^{-1/2}[S(IV)][O_3], \quad (k = 1.3 \cdot 10^{-4}\ L^{1/2}\ mol^{-1/2}\ s^{-1}). \tag{4.235}$$

Figure 4.14 shows the strong influence of pH on both pathways, where H_2O_2 is dominant in acidic solution and O_3 in alkaline solution.

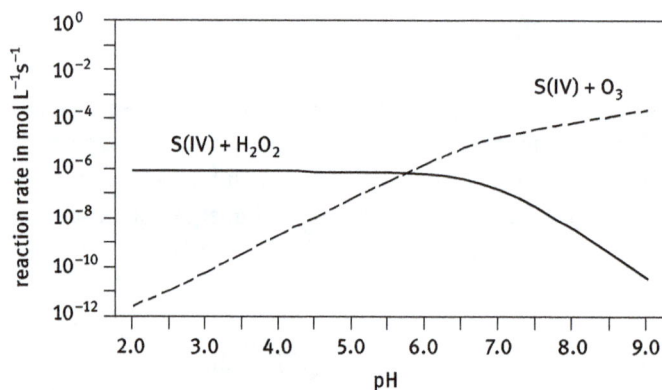

Figure 4.14: Dependence of S(IV) by H_2O_2 and O_3 from pH.

Accepting the radical mechanism theory, the reaction rate of the sulphite radical formation according to equation (4.221a), (4.221b) determines the overall rate. Amplifying the S(IV) oxidation is given by the subsequent formation of radicals (OH, O_2^-, SO_4^-, SO_5^-) that further react with sulphite or bisulphite (Figure 4.13). It is impossible to study the S(IV) oxidation under natural conditions (i. e. outside laboratory conditions) in the sense of a definite mechanism because all reactive species (providing A to E in the following scheme) are available and participate in the oxidation. The species are various and mostly unknown concentrations and superposed, making sulphate formation very complex and dependent on the redox state and the pH of the solution as well as the radiation and photosensitisers:

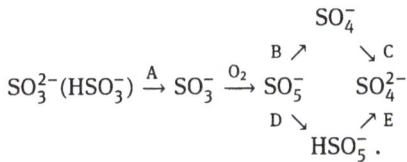

$$SO_3^{2-}(HSO_3^-) \xrightarrow{A} SO_3^- \xrightarrow{O_2} SO_5^- \quad\quad SO_4^{2-}$$

with branches: SO_4^- (B↗ ↘C), SO_4^{2-}, HSO_5^- (D↘ ↗E).

All conversions through species A to E can proceed through sulphur radicals (sulphite, peroxomonosulphate and sulphate) as well as via many other species. In contrast, some of them are permanently abundant and others are produced via photochemical processes. Only coupled gas-aqueous phase models, comprising the whole chemistry, can describe the complex S(IV) oxidation in hydrometeors.

Dithionate ($S_2O_6^{2-}$) and peroxodisulphate ($S_2O_8^{2-}$) have never been detected in cloud and rainwater. Dithionate slowly disproportionates (into $SO_4^{2-} + SO_2$) and peroxodisulphate hydrolyses in acid solution (into $HSO_5^- + SO_4^{2-}$). Table 4.16 lists all higher oxidised sulphur species of interest in the environment.

Table 4.16: Ions and radicals of oxoacids of sulphur (instead of hydrogen, bi– is also used to term protonated forms).

ion	radical	name	acid	name
HSO_3^-	HSO_3	hydrogensulphite	H_2SO_3	sulphurous acid
SO_3^{2-}	SO_3^-	sulphite		
HSO_4^-	HSO_4	hydrogensulphate	H_2SO_4	sulphuric acid
SO_4^{2-}	SO_4^-	sulphate		
HSO_5^-	HSO_5	hydrogenperoxomonosulphate	H_2SO_5	sulphuric acid[a]
SO_5^{2-}	SO_5^-	peroxomonosulphate		

[a] $HO_3S–O–OH$

Table 4.17 shows mean conversion rates and percentages of daytime and nighttime conversion as well as the significance of the hydrogen peroxide and ozone pathway depending on the season; these results come from complex modelling.

The rate of S(IV) oxidation increases dramatically with decreasing acidity (or increasing pH, resp.) of the aqueous phase. In many parts of the world, an increase in cloud water pH has been observed due to air pollution abatement, reduction of SO_2 and NO_x emissions. Thus, the dominant S(IV) oxidation pathway changed from H_2O_2

Table 4.17: Atmospheric SO_2 oxidation (percentage of pathwaya) and sulphate formation rates (in % h^{-1}) for central European conditions; data from Möller (2019).

pathway	summer day	summer night	winter day	winter day
liquid phase	75.8	8.1	11.3	4.4
gas phase	1.4	0.0013	0.22	0.0002
O_3	0.3	2.2	65.5	24.1
H_2O_2	99.6	87.3	26.7	34.7

[a]the difference to 100 % is given by other not listed pathways such as radicals and TMI catalytic

pathway in a previous time to O_3 pathway nowadays. This change results further in an increasing ozone removal capacity of clouds.

> **i** Sulphur dioxide is the most important consumer of atmospheric hydrogen peroxide (H_2O_2) and ozone (O_3) via the aqueous phase; therefore, it limits the oxidation capacity of the atmosphere. However, it is (or was) the main cause of environmental acidification ('acid rain'). Acid formation (acidifying capacity) is inextricably linked with oxidation capacity.

4.6 Carbon

Carbon is within the main group 4 (tetrele) of the periodic table of the elements. Carbon is very central in this group of elements. No group shows more differences: carbon is a non-metal, and lead, the final item in this group is a typical metal showing no similarity with carbon. The similarities between the elements increase from the middle group to the beginning groups (alkali and alkaline earth metals) and ending group (halogens). It is the harmonically balanced affinity of carbon to electropositive and electronegative elements, which additionally provides the largest quantity of different chemical compounds. This and the huge reservoir of CO_2 (including water-dissolved bicarbonate) supports the maintenance of life in the form of omnipresent plants and animals and carbon cycling. By contrast, elemental silicon with similar properties provides non-volatile SiO_2 with the tendency to form polymers and cannot provide global turnover rates compared with carbon but is the foundation of inorganic 'life': the rocky world.

Without life, there would be no carbon cycle on Earth. However, this is also true for all other elements: geochemical processes cannot reduce carbonate, nitrate, sulphate and phosphate in the environment – this is done only by biochemical processes. With regard to climate change due to artificial CO_2 emissions, many people (almost politicians) propose carbon neutrality (which is a state of net-zero carbon dioxide emissions) or even a carbon-free world for the future. However, understanding humans as part of nature, we have to learn to live (including generation of energy and materials) within biogeochemical cycling: I called it carbon economy in a solar era (see Chapter 6). Thus 'carbon neutrality' is the right approach in future, not by compensating CO_2 emissions but rather by balancing CO_2 emissions through carbon removal technologies.[71]

Deep in the Earth, however, we cannot exclude – even hypothesising the existence of elemental carbon – reducing chemical regimes, turning elements on geological timescales. The separation between inorganic and organic carbon chemistry (and compounds) is not strongly fixed. In nature, the synthesis of organic compounds only occurs in living cells of plants and animals, where only plants are able, through photosynthesis, to link the organic world with the inorganic, i. e. to use CO_2 as feedstock. Nature provides organic matter in a large variety for food, materials and energy carriers. So far, the extraction of such compounds has been limited to the carbon cycle (i. e. limited to renewable sources), and problems have only arisen because of local limits of carbon supply and local waste loadings. Only because of the exhaustion of fossil fuels, people again face the same general problems, interrupting biogeochemical cycles, but now on a global scale. Thousands of organic compounds used as chemicals by man are described in relation to properties and environmental fates in air, soil and water. However, the detailed chemistry is almost unknown and only studied for a few hundreds of substances.

Table 2.3 shows that CO_2, CH_4 and CO are the main carbon compounds in the air, roughly in a ratio of 1,000 : 10 : 1. It is noteworthy that these ratios also express approximately the ratios of the chemical lifetime of these species in the atmosphere ($\tau_{CH4} \approx$ 10 years). The exceptional physical and chemical characteristics of carbon make it not only unique but also fundamental in the environment as a carrier of specific properties, such as organic life and the ubiquity of gaseous carbon dioxide and aqueous carbonate. In the following chapters, we will summarise the basic principles of carbon chemistry; the interested reader is recommended to refer also to textbooks on organic chemistry, biogeochemistry and biochemistry.

4.6.1 Elemental carbon

Elemental carbon exists naturally as graphite (hexagonal C structure) and diamond (tetrahedral C structure). In graphite, very small amounts of fullerenes, where C_{60} molecules are most known, have been detected. We also have clear evidence that unoxidised carbon exists at depths between 150 km and 300 km in the form of diamonds, which moves under certain conditions up to the earth's surface (see Chapter 5.1.3).

---

71 Today, however, the drastic reduction of CO_2 emissions is the ultimate way to control global warming. The knowledge how to organize a carbon-balanced word exists for two decades but was not recognized by a broader community; an implementation needs further decades being too late for actual climate control but the only way for long-term survival in a sustainable word – by sustainable (or green) chemical processing.

Elemental carbon is chemically extremely stable. Only at high temperatures, carbon reacts with other elements and burns in O_2 (known for centuries as a coal dust explosion). Another phenomenon is the self-ignition of coal, but locally the required increased temperature must rise, and several processes have been suggested. This process of self-oxidation until self-ignition needs time and is only possible in condensed coal stocks – when burning in deposits, it can occur over centuries. However, this ignition was always initiated by humans who interrupted the chemical regime of the deposit through contact with atmospheric oxygen. In the environment, soot is a phenomenon that is as old as the culture of fire. People dominantly cause biomass burning, and thereby 'natural' burning caused by lightning strikes has always been negligible. Therefore, soot is primarily an artefact of nature (Table 4.18).

Table 4.18: Soot types.

soot origin	characteristics
wood combustion soot	large OC fraction, usually only 20 % BC; lignin-derived substances with OC
biomass burning soot	similar to wood soot but OC fraction larger up to 90 %
coal combustion soot[a]	different from biomass burning soot; large BC and EC fraction, mainly in the coarse mode
diesel soot	OC may approache 50 %, the remainder is BC and EC; smallest size fraction 3–20 nm consists of oil nanodroplets, accumulation mode (50–250 nm) contains EC and OC, and the coarse mode is EC due to coagulation
aviation soot	includes undefined OC fraction up to 300 nm

[a] This soot was important in the past from household coal heating and steam locomotives; coal-fired power plants with low efficiency, such as largely still in use in India and China, may also produce large soot emissions.

A large fraction of particulate matter (PM) is soot, the historical *symbol* of air pollution. There has been a long dispute in the literature on the definition of soot, which is also called elemental carbon (EC), black carbon (BC) and graphitic carbon. The term elemental matter (EM) is also found in literature and might 'integrate' EC and BC. Surely, soot is the best generic term that refers to impure carbon particles resulting from the incomplete combustion of a hydrocarbon. The formation of soot depends strongly on the fuel composition. It spans carbon from graphitic through BC to organic carbon fragments (OC). Each of the available methods (optical, thermal, and thermo-optical) refers to a different figure; it remains a simple question of definition. Hence, the comparison of different soot methods is senseless. The atmospheric implications of soot are:
- provide the highest surface-to-volume ratio for heterogeneous processes;
- form most complex structural nanoparticles;

– carry (toxic) organic substances; and
– warm the atmosphere through light absorption.

The answers to the question "What is soot?" are as different as different people will ask this question. A general definition was given by Popovicheva et al. (2007): "Soot is a carbon-containing aerosol resulting from incomplete combustion of hydrocarbon fuel of varying stoichiometry, defined by the ratio of fuel to oxygen". Soot not only addresses the properties of BC and EC fractions commonly associated with *soot* but also includes the organic fraction (OC). The chosen combustion conditions largely control the soot properties.

Soot aerosol consists of harmful substances, such as adsorbed PAHs as well as their hydroxylated and nitro-substituted congeners, which have significant carcinogenic and mutagenic potential. New research has found that the sooty *brown clouds*, caused primarily by the burning of coal and other organic materials in India, China and other parts of South Asia, might be responsible for some of the atmospheric warming that had been attributed to greenhouse gases.

> For centuries, until the end of the twentieth century, when the air pollution problems associated with the combustion of fossil fuels, sulphur dioxide and soot were the key air pollutants, termed as the so-called smoke plagues. Coal has been used in cities on a large scale since the beginning of the Middle Ages; this 'coal era' has not yet ended. In most parts of the world until the 1960s, coal was the primary source of energy for electricity generation, railway traffic using steam locomotives, industry and domestic heating. Air purification devices were practically non-existent. This led to very high levels of air pollution, particularly in cities, with soot, dust, sulphur dioxide and nitrogen oxides. Winter smog, particularly the notorious episodes in London during the 1950s, had serious effects on health as well as on building materials and historical monuments. The soot problem (and mostly that of sulphur dioxide) seems to have been solved nowadays; however, the problem of climate change due to carbon dioxide remains still unsolved.

A lot is known about the direct and indirect climate impacts of atmospheric soot, the absorption of gases, possible heterogeneous processes and water-soot interactions, but nothing is known about the fate of soot, especially elemental carbon. Studies on the chemistry of NO_y, O_3, SO_2 and many other species on soot have been carried out over recent decades, showing that soot might provide a reactive surface in air. However, such surface chemistry has been assessed to be insignificant in the budget of chemical species compared with gas phase and liquid phase processes, mainly because of the limited PM surface-to-air volume ratio. Many studies suggest that direct ozone loss on soot aerosol is unlikely under ambient conditions in the troposphere. Without a doubt, the large OC fraction in 'soot' will undergo 'ageing' by oxidation.[72] However, we can only speculate that reactive oxygen species such as O_3 and OH can

---

72 Decesari et al. (2002) showed that the WSOC produced from the oxidation of soot particles increased rapidly with ozone exposure and consisted primarily of aromatic polyacids found widely in

react with carbon similar to CO:

$$C + OH \rightarrow CO + H, \tag{4.236}$$

$$C + O_3 \rightarrow CO + O_2. \tag{4.237}$$

This process is extremely slow and can result in a chemical lifetime of hundreds or more years under environmental conditions. It is known that surfaces covered with photocatalytic active TiO_2 obviously remain 'clean' with respect to soot pollution, whereas reference surfaces become black. As discussed before, under such photocatalytic conditions, high OH concentrations might locally be produced, oxidising EC and OC.

The fate of the about 8 Tg BC yr^{-1} widely dispersed on the globe is deposition to oceans, soils and other surfaces. It is known that coal can survive in soils for hundreds of years and improve soil structure and water budget. The survival of atmospheric soot from coal combustion in the Middle Ages can still be seen at old churches and palaces – it is a cultural question whether to regard it as patina with respect or simply dirty pollution.

4.6.2 Inorganic C_1 chemistry: CO, CO_2, and H_2CO_3

As said at the very beginning, it does not make much sense to separate inorganic and organic chemistry. However, we follow a convention here. Sulphur, nitrogen and halogen compounds of carbon are treated elsewhere (Table 4.19). Here we present the two oxides and the carbonic acid. The Scottish physician and chemist *William Cruickshank* (?–1811) prepared in 1800 carbon monoxide (CO) by passing carbon dioxide over heated iron. CO is the product of incomplete combustion of carbon and hydrocarbon. CO was an important constituent of coal gas (or town gas); his poisoning properties

Table 4.19: Inorganic carbon compounds in the environment.

name	formula	comment
oxides	CO, CO_2	carbon monoxide, carbon dioxide
oxoacids	H_2CO_3	carbonic acid
sulphur compounds	COS, CS_2	see Chapter 4.5.2
nitrogen compounds	HCN	see Chapter 4.4.6.1
halogen compounds	CCl_4, CF_4	see Chapter 4.7.2 and 5.3.2.1

atmospheric aerosols and which are frequently referred to as macromolecular humic-like substances (HULIS).

have been examined in the mid-nineteenth century.[73] Since the 1940s, CO was observed spectroscopically in the earth's atmosphere; but until the end of the 1960s, CO was believed to be inert in the troposphere. Carbon monoxide (as a product of incomplete biomass and fossil fuel burning processes) oxidises by OH radicals direct to CO_2:

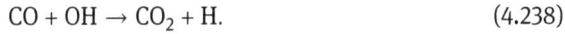

$$CO + OH \rightarrow CO_2 + H. \tag{4.238}$$

This is the only reaction where OH dissociates. The fate of H is known; hence, CO is an important converter OH $\rightarrow$ HO_2 and plays acritual role in photochemical smog chemistry through the subsequent radical reactions H + O_2 $\rightarrow$ HO_2 and HO_2 + NO $\rightarrow$ OH + NO_2, providing radical chains and tropospheric net ozone formation (see Chapter 5.3.2.2):

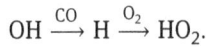

$$OH \xrightarrow{CO} H \xrightarrow{O_2} HO_2.$$

CO_2 is slightly soluble in water; the ratio between atmospheric and water dissolved CO_2 is described by the Henry equilibrium:

$$CO_2(g) \rightleftharpoons CO_2(aq) \quad \text{true } Henry \text{ constant } H_{CO_2}. \tag{4.239}$$

This 'physical' equilibrium depends only on the temperature. The effective equilibrium is given through subsequent chemical reactions, leading to the higher solubility of CO_2 in water. Dissolved carbon dioxide forms bicarbonate via different steps:

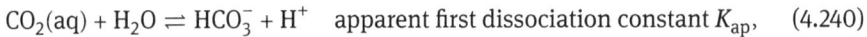

$$CO_2(aq) + H_2O \rightleftharpoons HCO_3^- + H^+ \quad \text{apparent first dissociation constant } K_{ap}, \tag{4.240}$$

$$CO_2(aq) + H_2O \overset{k_{4.241}}{\underset{-k_{4.241}}{\rightleftharpoons}} H_2CO_3 \quad \text{hydration constant } K_h, \tag{4.241}$$

$$H_2CO_3 \rightleftharpoons H^+ + HCO_3^- \quad \text{true first dissociation constant } K_1, \tag{4.242}$$

$$HCO_3^- \rightleftharpoons H^+ + CO_3^{2-} \quad \text{second dissociation constant } K_2. \tag{4.243}$$

CO_2 hydration (equation (4.241)) is relatively slow and in comparison to total dissolved CO_2 the concentration of H_2CO_3 is very low (negligible). Reaction equation (4.241) occurs for pH < 8; for pH > 10 reaction equation (4.244) is dominant, in the pH

---

73 This gas (also illuminating gas called) is a mixture of very variable composition (in parenthesis %) of H_2 (20–50), CO (10–60), CH_4 (2–20), N_2 (5–15) and other volatile hydrocarbons (<5), produced when coal is heated strongly in the absence of air. Originally it was created as by-product of the coking process since the early nineteenth century only for street lightning and in the late nineteenth century also for heating and cooking; replaced by natural gas in the early 1970s.

region 8–10 both reactions are parallel, and it is complicated to study the kinetics. Hence, reliable kinetic data are valid only for pH < 8 and pH > 10, respectively.

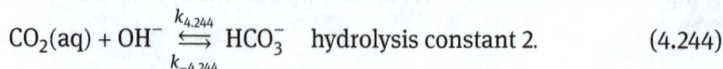

$$CO_2(aq) + OH^- \overset{k_{4.244}}{\underset{k_{-4.244}}{\rightleftharpoons}} HCO_3^- \quad \text{hydrolysis constant 2.} \tag{4.244}$$

However, reaction equation (4.244) plays no role in natural waters with the exception of the initial state of cloud/fog droplet formation from alkaline CCN (e. g. flue ash and soil dust particles). The reaction rate constants (minus prefix means the inverse reaction) have been 'best' estimated to be $k_{4.241} = 0.03\,s^{-1}$ (25 °C) and $k_{-4.241} = 20\,s^{-1}$ (25 °C) as pseudo-first order rates and $k_{4.244} = 8,400\,L\,mol^{-1}\,s^{-1}$ and $k_{-4.244} = 2 \cdot 10^{-4}\,s^{-1}$.

> In water, the following chemical carbon(IV) species exist in equilibrium: carbon dioxide (CO_2), carbonic acid (H_2CO_3), bicarbonate (HCO_3^-) and carbonate (CO_3^{2-}). Additionally, the phase equilibrium with gaseous CO_2 and a possible solid bodies, such as $CaCO_3$ and $MgCO_3$, has to be considered.

Free carbonic acid is not isolated, but the structure $O=C(OH)_2$ in aqueous solution has been confirmed. Often the expression $CO_2 \cdot H_2O$ is also used for carbonic acid. The sum of the dissolved carbonate species is denoted as total DIC (dissolved inorganic carbon) and is equivalent to other terms used in literature (T stands for 'total'):

$$DIC \equiv \sum CO_2 \equiv TCO_2 \equiv C_T = [CO_2] + [H_2CO_3] + [HCO_3^-] + [CO_3^{2-}].$$

The carbon dioxide (physically) dissolved in water – we denote it as $CO_2(aq)$ – is in equilibrium with gaseous atmospheric carbon dioxide $CO_2(g)$. There is no way to separate non-ionic dissolved $CO_2(aq)$ and H_2CO_3; therefore, it is often lumped into $CO_2^*(aq)$. Analytically, DIC can be measured by acidifying the water sample, extracting the CO_2 gas produced and measuring it. The marine carbonate system represents the largest carbon pool in the environment, and it is of primary importance for the partitioning of atmospheric excess carbon dioxide produced by human activity.

The equilibrium constant of the apparent first dissociation equation (4.240) is given by:

$$K_{ap} = \frac{[HCO_3^-][H^+]}{[CO_2(aq)]} = K_1 K_h, \tag{4.245}$$

where K_{ap} can be relatively easily estimated from equilibrium concentration measurements and the hydration constant K_h is calculated according to equation (4.244); Table 4.20. The adjustment of equilibriums equation (4.242) and equation (4.243) is fast;

Table 4.20: Equilibrium constants in the aqueous CO_2 – carbonate system.

T (in °C)	0	5	10	15	20	25
H (in 10^{-2} atm^{-1} mol L^{-1})	7.70	–	5.36	–	3.93	3.45
K_{ap} (in 10^{-7} mol L^{-1})	2.64	3.04	3.44	3.81	4.16	4.45
K_1 (in 10^{-4} mol L^{-1})	–	1.56	–	1.76	1.75	1.72
$K_h \cdot 10^3$ ($K_h = [H_2CO_3]/[CO_2(aq)]$)	–	1.96	–	2.16	2.52	2.59
K_2 (in 10^{-11} mol L^{-1})	2.36	2.77	3.24	3.71	4.20	4.29
$K_{ap} = K_s \cdot K_h$ (in 10^{-7} mol L^{-1})	–	3.06	–	3.80	4.41	4.45

the direct estimation of the dissociation constants K_1 and K_2 is not possible from concentration measurements (only indirectly through potentiometric and/or conducto-metric measurements). The true first dissociation constant K_1 (pK = 3.8) is three orders of magnitude larger than the apparent dissociation constant K_{ap}. Hence, carbonic acid is 10 times stronger than acetic acid, but acetic acid can degas CO_2 from carbonised solutions because H_2CO_3 is decomposed to about 99 % into CO_2 (which escapes from the water body) and H_2O as it follows from K_h. This makes the aqueous carbonic system unique (Figure 4.15): carbonic acid exists as well (but in very low concentrations) as H_2CO_3 (in kinetic-inhibited equilibriums) and largely as $CO_2 \cdot H_2O$ where CO_2 degassing is also inhibited. The second dissociation constant characterises bicarbonate as a very weak acid (pK_2 = 10.4). The aqueous-phase concentrations of different DIC species can be calculated from the equilibrium expressions:

$$[CO_2(aq)] = H \cdot [CO_2(g)], \tag{4.246}$$
$$[H_2CO_3] = H \cdot K_h[CO_2(g)], \tag{4.247}$$
$$[HCO_3^-] = H \cdot K_1 K_h[CO_2(g)][H^+]^{-1}, \tag{4.248}$$
$$[CO_3^{2-}] = H \cdot K_1 K_2 K_h[CO_2(g)][H^+]^{-2}. \tag{4.249}$$

Seawater is slightly alkaline (pH ≈ 8.2) because of the equilibrium between solid suspended $CaCO_3$ and dissolved carbonate. At a typical surface seawater pH of 8.2, the speciation between $[CO_2]$, $[HCO_3^-]$ and $[CO_3^{2-}]$ is 0.5 %, 89 % and 10.5 %, showing that most of the dissolved CO_2 is in the form of HCO_3^- and not CO_2:

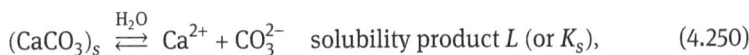

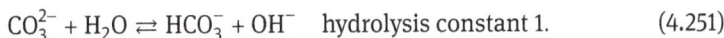

$$(CaCO_3)_s \overset{H_2O}{\rightleftharpoons} Ca^{2+} + CO_3^{2-} \quad \text{solubility product } L \text{ (or } K_s\text{)}, \tag{4.250}$$
$$CO_3^{2-} + H_2O \rightleftharpoons HCO_3^- + OH^- \quad \text{hydrolysis constant 1.} \tag{4.251}$$

Carbonate acts as a base (Chapter 3.2.2.3). The solubility of $CaCO_3$ at 20 °C in water is only about 0.007 g L^{-1} as carbon. $CaCO_3$ water solubility (in mg L^{-1}) decreases linearly with increasing temperatures – $[CaCO_3] = 80.3 - T$ ($r^2 = 0.997$), T in °C – and increases slightly with increasing CO_2 partial pressure – $[CaCO_3] = 56.5 + 0.0219 \cdot [CO_2(g)]$ ($r_2 = 0.986$; 20 °C), valid in the range 20–1,000 ppm CO_2.

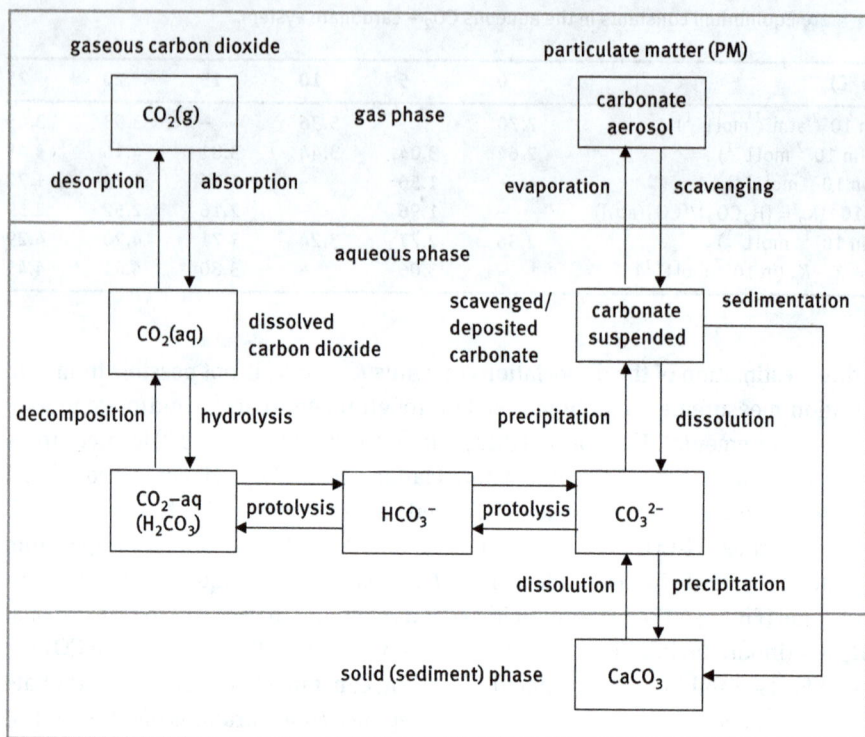

Figure 4.15: Scheme of the multiphase CO_2–carbonate system.

Processes according to equation (4.250) and equation (4.251) describe the solid-liquid equilibrium at the water bottom (sediment–seawater interface) with the suspended matter in seawater (calcareous organisms, Figure 4.15). The ocean is saturated with $CaCO_3$, which represents the largest carbon reservoir in sediments in the form of calcite and aragonite. An increase in carbonate (in terms of DIC) solubility is because of dissolved CO_2, which converts carbonate (CO_3^{2-}) into higher soluble bicarbonate (HCO_3^-). It follows that the capacity of the ocean for CO_2 uptake is still very large – the system is far from saturation in DIC (or total carbonate, respectively) but rather in equilibrium. With increasing atmospheric CO_2, the seawater CO_2/carbonate concentration increases, and vice versa, i. e. in the case of decreasing atmospheric CO_2 concentrations the ocean will degas CO_2 thereby leading to a new equilibrium.

However, the relationship between atmospheric CO_2 is more complicated because of the buffer capacity of seawater (besides carbonate in a more exact treatment, all buffering chemical species – for example, borate – have to be considered). The buffer capacity of carbonised water (here seawater) is given to complete the acid-based reaction:

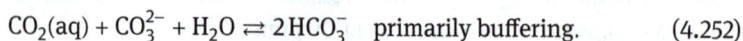

$$CO_2(aq) + CO_3^{2-} + H_2O \rightleftharpoons 2\,HCO_3^- \quad \text{primarily buffering.} \tag{4.252}$$

Anthropogenic CO_2 dissolves in seawater and produces hydrogen ions (called oceanic acidification), turns carbonate into bicarbonate ions and shifts the solid-aqueous carbonate equilibrium
more to the aqueous site (dissolution of carbonaceous species such as corals).

Hence, H^+ concentration (and pH) will not change in small ranges depending on the CO_2 partial pressure increase and the available carbonate in seawater. However, when seawater pH declines because of rising CO_2 concentrations, the concentration of CO_3^{2-} will also fall (see reaction equation (4.252)), reducing the calcium carbonate saturation state. Marine carbonates also react with dissolved CO_2 through the reaction:

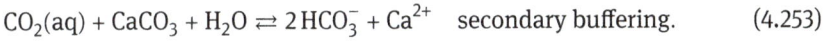

$$CO_2(aq) + CaCO_3 + H_2O \rightleftharpoons 2\,HCO_3^- + Ca^{2+} \quad \text{secondary buffering.} \qquad (4.253)$$

In aqueous solutions, especially in cellular environments, the carbonate radical an-
ion (CO_3^-) is produced by the reaction between the ubiquitous carbon dioxide and peroxonitrite (ONOO−), which is an unstable intermediate (Figure 4.11) in biological NO reduction and first forms as a CO_2 adduct nitrosoperoxocarboxylate, which then decomposes:

$$CO_2 + ONOO^- \rightarrow ONOOCO_2^- \rightarrow CO_3^- + NO_2. \qquad (4.254)$$

Carbonate radicals react with many organic compounds in the general H abstraction reaction (competing with OH); $k_{4.255} = 10^4 \ldots 10^7 \, \text{L mol}^{-1}\text{s}^{-1}$ depending on RH:

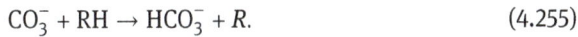

$$CO_3^- + RH \rightarrow HCO_3^- + R. \qquad (4.255)$$

It is a strong acid (i. e. the radical HCO_3 fully dissociated) with $pK_a = -4.1$ and a strong oxidising agent with $E°\,(CO_3^-/CO_3^{2-}) = 1.23 \pm 0.15\,\text{V}$, which likely exists as a dimer $H(CO_3)_2^-$. The importance in natural waters is likely limited because it is produced in radical reactions such as listed in the following: ($k_{4.256a} = 3.9 \cdot 10^8 \, \text{L mol}^{-1}\text{s}^{-1}$, $k_{4.256b} = 1.7 \cdot 10^7 \, \text{L mol}^{-1}\text{s}^{-1}$, $k_{4.257} = 4.1 \cdot 10^7 \, \text{L mol}^{-1}\text{s}^{-1}$, $k_{4.258} = 2.6 \cdot 10^6 \, \text{L mol}^{-1} \cdot \text{s}^{-1}$) and reacts back to carbonate:

$$CO_3^{2-} + OH \rightarrow CO_3^- + OH^-, \qquad (4.256a)$$
$$HCO_3^- + OH \rightarrow CO_3^- + H_2O, \qquad (4.256b)$$
$$HCO_3^- + NO_3 \rightarrow CO_3^- + H^+ + NO_3^-, \qquad (4.257)$$
$$CO_3^{2-}(HCO_3^-) + Cl_2^- \rightarrow CO_3^- + 2\,Cl^- \; (+H^+), \qquad (4.258)$$
$$CO_3^{2-} + SO_4^- \rightarrow CO_3^- + SO_4^{2-}. \qquad (4.259)$$

It reacts with TMI ($k_{4.260} = 2 \cdot 10^7 \, \text{L mol}^{-1}\text{s}^{-1}$) and with peroxides ($k_{4.261} = 6.5 \cdot 10^8 \, \text{L mol}^{-1}\text{s}^{-1}$), which represents radical termination in one-electron transfers:

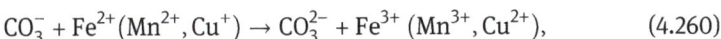

$$CO_3^- + Fe^{2+}(Mn^{2+}, Cu^+) \rightarrow CO_3^{2-} + Fe^{3+}\,(Mn^{3+}, Cu^{2+}), \qquad (4.260)$$

$$CO_3^- + HO_2 \rightarrow HCO_3^- + O_2, \tag{4.261a}$$

$$CO_3^- + O_2^- \rightarrow CO_3^{2-} + O_2, \tag{4.261b}$$

$$CO_3^- + H_2O_2 \rightarrow HCO_3^- + HO_2. \tag{4.262}$$

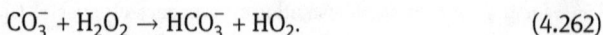

The following fast reaction obviously transfers O^- (adequate reactions concerning $NO \rightarrow NO_2^-$ and $O_2 \rightarrow O_3^-$ are not described in the literature): $k_{4.263} = 1 \cdot 10^9 \, L \, mol^{-1} \, s^{-1}$.

$$CO_3^- + NO_2 \rightarrow CO_2 + NO_3^-. \tag{4.263}$$

A reaction with ozone is slow and implies the intermediate $O_4^- (\overset{H^+}{\leftrightarrow} HO_4)$: $k_{4.264} = 1 \cdot 10^5 \, L \, mol^{-1} \, s^{-1}$ (if so, then also O^- transfer occurs similar to equation (4.263)):

$$CO_3^- + O_3 \rightarrow CO_2 + O_2 + O_2^-. \tag{4.264}$$

The carbon dioxide anion radical CO_2^-, produced by electron transfer onto CO_2, represents another interesting species, which is an efficient reducing agent in two ways, providing electron transfer and radical addition; $k_{4.265} = 4 \cdot 10^9 \, L \, mol^{-1} \, s^{-1}$:

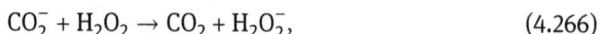

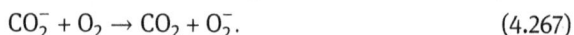

$$CO_2 + e_{aq}^- \rightarrow CO_2^-, \tag{4.265}$$

$$CO_2^- + H_2O_2 \rightarrow CO_2 + H_2O_2^-, \tag{4.266}$$

$$CO_2^- + O_2 \rightarrow CO_2 + O_2^-. \tag{4.267}$$

The CO_2^- radical is also gained by oxidation of formate ions:

$$HCOO^- + OH \rightarrow CO_2^- + H_2O. \tag{4.268}$$

It adds onto organic radicals and double bonds:

$$CO_2^- + RCH(OH) \rightarrow RCH(OH)COO^-, \tag{4.269}$$

$$CO_2^- + R-CH=CH-R \rightarrow RCH(COO^-)-CH-R. \tag{4.270}$$

It disproportionates and dimerises to oxalate:

$$CO_2^- + CO_2^- \rightarrow CO_3^{2-} + CO, \tag{4.271}$$

$$CO_2^- + CO_2^- \rightarrow {}^-O(O)C-C(O)O^-. \tag{4.272}$$

These processes represent a way of sustainable chemistry for future CO_2 air capture and subsequent CO_2 reduction to fuels. CO_2 removal from atmospheric air is a technology already tested in the laboratory and the only way for climate sanitation and long-term carbon fuel use with respect to climate control (see Chapter 6). Two-electron steps onto adsorbed CO_2 are favoured compared to equation (4.265), whose reduction potential amounts $-1.0\,V$:

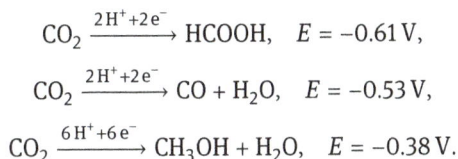

$$CO_2 \xrightarrow{2H^+ + 2e^-} HCOOH, \quad E = -0.61\,V,$$

$$CO_2 \xrightarrow{2H^+ + 2e^-} CO + H_2O, \quad E = -0.53\,V,$$

$$CO_2 \xrightarrow{6H^+ + 6e^-} CH_3OH + H_2O, \quad E = -0.38\,V.$$

In aqueous solutions, some metal-ligand complexes form CO_2 adducts, which internally undergo a two-electron step:

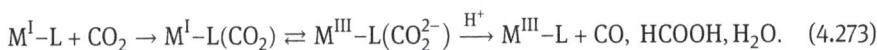

$$M^I-L + CO_2 \rightarrow M^I-L(CO_2) \rightleftharpoons M^{III}-L(CO_3^{2-}) \xrightarrow{H^+} M^{III}-L + CO, HCOOH, H_2O. \quad (4.273)$$

Thus, one of the best routes to remedy the CO_2 problem is to convert it into valuable hydrocarbons using solar energy and recycle it by different capture technologies.

Finally, another radical is given from CO, forming the carbon monoxide anion CO^-, which very quickly reacts with water to give the formyl radical HCO (see further fate with equation (4.296)):

$$CO + e_{aq}^- \rightarrow CO^-, \tag{4.274}$$

$$CO^- + H_2O \rightarrow HCO + OH^-. \tag{4.275}$$

4.6.3 Organic carbon

In Chapter 1.2, we have defined what 'organic chemistry' means. The simplest organic compounds are hydrocarbons, consisting of only carbon and hydrogen, often symbolised as HC; the simplest molecule is methane CH_4. The specific chemical role of organic molecules, however, is given by functional groups; the most important are listed in Table 4.21. The functional groups determine the chemical reactivities – the reactive influence of the carbon rest is much less significant. Here we deal only with HC and HCO compounds; nitrogen, sulphur and halogen organic compounds are treated in the relevant element chapters.

Classification of organic chemistry can be done according to the functional groups (Table 4.21) or via compounds classes:

Table 4.21: Most important functional groups in the environment relevant for organic compounds.

symbol	formula	name	examples of compounds
–OH	–O–H	hydroxy	alcohols, phenols, sugars
>CO	>C=O	carbonyl	aldehydes, ketones
–COOH	–C(=O)(–OH)	carboxy	carboxylic acids
–NH$_2$	–N(–H)$_2$	amino	amines, amino acids (together with –COOH)
–NO$_2$	–N(=O)$_2$	nitro	nitrophenol
–CN	–C≡N	cyano	nitriles
–SO$_3$H	–S(=O)$_2$(–OH)	sulpho	sulphonamides

aliphatic compounds — acyclic and cyclic, but not aromatic; can be saturated, joined by single bonds (alkanes), or unsaturated, with double bonds (alkenes) or triple bonds (alkynes). Besides hydrogen, other elements can be bound to the carbon chain, the most common being oxygen, nitrogen, sulphur, and chlorine

alicyclic compounds — three or more atoms of the element carbon are linked together in a ring; the bonds between pairs of adjacent atoms may all be of the type designated single bonds, or some of them may be double or triple bonds, but not aromatic

aromatic compounds — compounds based on the benzene ring, including polycyclic aromatic compounds

heterocyclic compounds — characterised by the fact that some or all of the atoms in their molecules are joined in rings containing at least one atom of an element other than carbon (C)

The environmentally significant group of heterocyclic compounds are *natural products*: a chemical compound or substance produced by a living organism – that is, found in nature. Heterocyclic compounds include many of the biochemical materials essential to life. For example, nucleic acids, the chemical substances that carry the genetic information controlling inheritance, consist of long chains of heterocyclic units held together by other types of materials. Many naturally occurring pigments, vitamins, and antibiotics are heterocyclic compounds, like most hallucinogens. Modern society is dependent on synthetic heterocycles for use as drugs, pesticides, dyes, and plastics. Another important class of natural products are terpenes (which are emitted in huge amounts by vegetation into the atmosphere; Table 4.22) and steroids. The interested reader should refer to textbooks on organic chemistry. In the following chapters, we only deal with aliphatic and aromatic compounds. To get an impression of how much the biosphere emits, see Table 4.22 (note that secondary formation means chemical formation in the air from other NMVOC).

Table 4.22: Average global annual emission of organic substances in Tg yr^{-1}. – no date available, 0 no emission; you should note the large uncertainty of data.

substance	vegetation	ocean	natural emission biomass burning[c]	secondary formation	anthropogenic emission	total[a]
isoprene	500	5	–	0	0	–
monoterpenes	120–200	0	–	0	0	–
methanol	150–200	10–27	3–9	30–40	4	240
acetaldehyde	37	–	–	0	–	–
formaldehyde	34	–	3–10	1600	–	–
acetone	30–40	–	–	40	–	95
glyoxal	0	–	–	40–100	0	–
carboxylic acids	–	–	5	–	0	–
propene	15	–	–	0	–	–
propane	12	–	–	0	–	–
i-pentane	5	–	–	0	–	–
methane	0	1–10	15–30	0	350	500–600[b]
ethane	4	–	–	0	–	–
ethene	4	–	–	0	–	–
i-butane	4	–	–	0	–	–
alkanes	–	1	7–30	0	15–60	–
alkenes	–	3–12	10–30	0	5–25	–
aromatics	–	0	2–19	0	10–30	–
total[a]	750–1150	–	95	–	50–200	450–4800[d]

[a] not the sum – independent estimates
[b] including wetlands
[c] biomass burning is almost anthropogenically caused
[d] total anthropogenic 100 (50–200) and total biogenic 1,200 (400–4,600)

The evolution of human culture is closely connected with hydrocarbon (Hall et al. 2003). Nature has favoured the storage of solar energy in the hydrocarbon bonds of plants and animals, and human cultural evolution has exploited this hydrocarbon energy profitably. The beginning was the harnessing of the energy in the hydrocarbon bonds of wood using fire. About 300 years ago, the industrial revolution began with stationary wind-powered and water-powered technologies, which were essentially replaced by fossil hydrocarbons: coal in the nineteenth century, oil since the twentieth century, and now, increasingly, natural gas. However, the use of hydrocarbons to meet economic and social needs is a major driver of our most important environmental changes, including global climate change, acid deposition, urban smog and the release of many toxic materials. Now, society has a great opportunity to make investments in a different source of energy, one freeing us for the first time from our dependence on hydrocarbons.

> **i** The role of hydrocarbons as a source of energy will be replaced by their role as an energy carrier, and their role as a source of materials increases both through natural renewable sources (plants) and carbon dioxide methanation.

Crude oil contains volatile compounds (1–3 %), 20–60 % light liquids (petrol and kerosene, boiling point 40–140), 10–20 % heavy naphthas and diesel (boiling point 140–350), and about 50 % oils (boiling point 350–500) as well as residues such as tar, asphalt (boiling point >500). 80–90 % are alkanes, aromatics and naphthenes, and 10–20 % sulphur, nitrogen and oxygenated organic compounds.

The only renewable hydrocarbon exists in plants, mainly wood and crop, but also in animals, used by humans as a source of food, energy and materials. Wood of all kinds has the same chemical composition; basic substances are hydrocarbons with about 50 % carbon, 43 % oxygen, 6 % hydrogen and 1 % nitrogen. The main materials of wood are cellulose (45 %), wood polyoses (cellulose-like constituents like pentosans and pectin) (25 %) and lignin (25 %); further minor constituents such as resins, tannins, dyes and others.

Biofuel, in a technical sense, is defined as a solid, liquid or gaseous fuel obtained from living organisms or from metabolic by-products (organic or food waste products). Global biofuel production consists of ethanol (90 %) and diesel (10 %) but still accounts for less than 3 % of the global transportation fuel supply.

The biomass combustion process is generally divided into four basic combustion phases; ignition, flaming, smouldering and glowing. During the initial heating period, large quantities of VOCs are released. Once the fuel is sufficiently dry, combustion proceeds from the ignition phase to the flaming phase (325–350 °C). During the flaming process, hydrocarbons are volatilized from the thermally decomposing biomass and are rapidly oxidised in a flame. Products of complete combustion are CO_2 and H_2O. Incomplete combustion leads to the emission of CO and a large variety of organic compounds.

In the atmosphere, five categories of VOCs can be detected (Table 4.23). The largest two classes are the aliphatic (alkanes, cycloalkanes, alkenes) and aromatic hydrocarbons. The third class of compounds is terpenes, emitted from plants (isoprene is dominant). Chlorinated hydrocarbons made up the fourth class, and oxygenated compounds (aldehydes, ketones, alcohols, acids) comprise the remaining class (not quantified by the measurements presented in Table 4.23). From aircraft measurements, we carried out in summer 1994 in Saxony-Anhalt (Germany), the following five compounds comprise 65 % of all measured (35) VOCs: ethane (25–35 %), ethane (19–15 %), ethene (8–10 %), propane (7–10 %), and benzene (4–5 %), emphasising the importance of C_2 compounds.

Table 4.23: Concentrations of organic compounds (52 C_3–C_{10} species were detected) in the atmosphere of Berlin (suburb, July 1998); in ppb (only n-hexane, toluene and benzene were in the ppb range; the other listed compounds were in the range 100–800 ppt, and not listed compounds 10–100 ppt).

class of compounds	c
aromatics (benzene, toluene, xylene, and others)[b]	8.2
alkanes (n-hexane, n-butane, propane, n-heptane, n-nonane, cyclohexane, and others)[a]	4.5
alkenes (butenes, pentenes, propene, hexenes)	4.0
isoprene	0.8
terpenes (α-pinene, β-pinene)	0.2

[a] ethane, ethene and ethyne were not measured
[b] compounds in order of decreasing concentration

Table 4.24: List of organic radicals. Note that in some cases, R = H (for C_1 species).

shorthand symbol	formula	name
R	R–C$^\bullet$H$_2$	alkyl[a]
–	R–CH=C$^\bullet$	alkenyl[b]
–	R–CH=C–O–O$^\bullet$	alkenyl peroxyl
RO$_2$	R–C(H$_2$)–O–O$^\bullet$	alkyl peroxyl
RO	R–C(H$_2$)–O$^\bullet$	alkoxyl[c]
RCO	R–C(=O)	acyl[d]
RCO$_3$	R–(O=)C–O–O$^\bullet$	acyl peroxy (peroxyacyl)
RCO$_2$	R–(O=)C–O$^\bullet$	acyloxy[e]
	R–(H)C$^\bullet$–O–O$^\bullet$	Criegee radical

[a] if R = H, CH$_3$ (methyl)
[b] H$_2$C=CH vinyl, H$_2$C=CH–CH$_2$ allyl, H$_3$C–CH$_2$–CH=CH 1-bytenyl
[c] derived from alcohol ROH
[d] if R = H, HCO (formyl), if R = CH$_3$, acetyl; derived from carbonyl group –CHO (generally >C=O)
[e] this radical is directly derived from carboxylic acid RCOOH

4.6.3.1 Hydrocarbon oxidation and organic ROS

Here we treat the general radical oxidation of hydrocarbons and will meet important *organic* reactive oxygen species (ROS), see Table 4.24 and Figure 4.16; the inorganic ROS (OH, HO$_2$, and H$_2$O$_2$) we already met in the oxygen Chapter 4.3.2.2. The C–H bond is distinguished in four kinds, having slightly different bond energies:

R–CH$_3$ alkylic (terminal C atom)
R–CH$_2$–R   allylic (midsised C atom)
R–CH=CH–R   vinylic (C atom is linked with a double bond
(C_2H_5)–CH$_3$ benzylic (C atom is bonded with aromatic ring).

Alkanes (CH_4 is the basic compound) are chains of carbons bound with a single atomic bond; the C–H bond (the line symbolises a σ electron pair) can be destroyed by several radicals (but not photodissociation under environmental conditions) – it remains an alkyl group RCH_2:

$$RCH_3 + OH(NO_3, Cl) \rightarrow RCH_2 + H_2O(HNO_3, HCl). \qquad (4.276)$$

Note, however, that R can be any kind of organic rest, not only C–H groups (only in the case of alkanes, R is termed alkyl), hence a more generic chemical equation of oxidation of organic compounds is

$$RH + OH \rightarrow R + H_2O. \qquad (4.277)$$

In the atmosphere, the OH radical is dominant; during the night, the NO_3 radical might be important and also in soils and waters, where chlorine is also important. The terminal CH_3 group is favoured for any radical attack and less likely is the attack on mid-sized CH_2 groups:

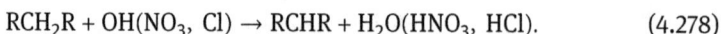

$$RCH_2R + OH(NO_3, Cl) \rightarrow RCHR + H_2O(HNO_3, HCl). \qquad (4.278)$$

The generic formula for alkanes is C_nH_{2n+2} (n number of carbon atoms). Ethane C_2H_6 ($H_3C–CH_3$) consists of two methyl groups; for propane C_3H_6 the structure formula $H_3C(CH_2)_{n-2}CH_3$ is valid. Alkenes contain one (or more) double bonds; the position of the double bond is different from butene-forming isomers. The position of the double bond is named by numbers: (1)-butene and (2)-butene.

alkane		alkene				
methane	CH_4	–				
ethane	C_2H_6	ethene	C_2H_4	C=C		
propane	C_3H_8	propene	C_3H_6	C–C=C		
butane	C_4H_{10}	butene	C_4H_8	C–C–C=C	C–C=C–C	
pentane	C_5H_{11}	pentene	C_5H_{10}	C–C–C–C=C	C–C–C=C–C	
hexane	C_6H_{14}	hexene	C_6H_{12}	C–C–C–C–C=C	C–C–C–C=C–C	C–C-C=C–C–C

For the three first alkanes and C_2–C_3 alkenes, only a linear structure is possible but from butane and butene branched structures called isomeres occur. The linear molecules have the prefix n- (n-butane) and the branched molecules i- (i-butane). All alkanes and alkenes are insoluble in water. C_1–C_4 alkanes (methane, ethane, propane and butane) are gases, C_5–C_{17} (from n-pentane to n-heptadecane) are liquids, and the higher alkanes are solids. C_1–C_4 alkenes are gases, C_5–C_{15} are liquids, and the higher alkenes are solids. The liquid alkanes and alkenes are volatile and can be found in the air. The chemistry of the double bond is described in Chapter 3.3.1. Natural gas mainly consists of CH_4 but C_2–C_6 are also found (Table 4.25).

Table 4.25: Chemical composition of natural gas (in vol %).

compound		mean	range
CH_4	(methane)	~95	62–97
C_2H_6	(ethane)	~2.5	1–15
C_3H_8	(propane)	~0.2	0–7
C_4H_{10}	(butane)	~0.2	0–3
C_5H_{12}	(pentane)	~0.03	<0.2
C_6H_{14}	(hexane)	~0.01	<0.1
N_2		~ 1.3	1–25
CO_2		~0.02	1–9
H_2S		~0.2	<3
He		~0.1	<2
H_2		~0.01	<0.02

The alkyl radical RCH_2 (often termed also simply as R) adds O_2 in aerobic environments (but any addition of other molecular entities such as NO, NO_2, Cl is possible) forming the alkyl peroxyl radical RO_2, which is similar in its chemical reactivity to the hydroperoxyl radical HO_2:

$$RCH_2(R) + O_2 \rightarrow RCH_2O_2(R\text{–}O\text{–}O) \quad \text{(terminal C atom)}, \tag{4.279}$$

$$RCHR + O_2 \rightarrow RCH(O_2)R \quad \text{(midsised C atom)}. \tag{4.280}$$

Both peroxyl radicals react in air dominantly with NO, forming the simple alkoxyl radical RO and oxidising NO into NO_2 (remember, an important step in NO_x chemistry):

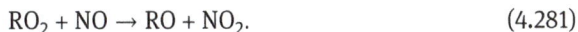

$$RO_2 + NO \rightarrow RO + NO_2. \tag{4.281}$$

There are a number of reactions competing with reaction equation (4.281), mostly with other peroxy radicals. Combining RO_2 with HO_2 leads to organic peroxides (derivatives of hydrogen peroxide H_2O_2, where an organic rest R exchanges the H atom); the simplest is methylhydroperoxide CH_3OOH:

$$RCH_2O_2(RO_2) + HO_2 \rightarrow RCH_2OOH(ROOH) + O_2. \tag{4.282}$$

Two organic peroxy radicals combine to give alcohol and aldehyde:

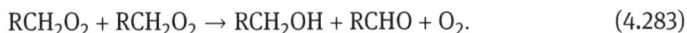

$$RCH_2O_2 + RCH_2O_2 \rightarrow RCH_2OH + RCHO + O_2. \tag{4.283}$$

However, it should be noted that $[NO] \gg [HO_2] > [RO_2]$ in the air (it can be different in other media such as biota) when discussing the percentages of different pathways. Organic peroxides (the most important are CH_3OOH and C_2H_5OOH) are permanently found in air besides H_2O_2 but in smaller concentrations. In the case of a midsized radical attack (equation (4.280)), the RO intermediate rearranges, forming a ketone:

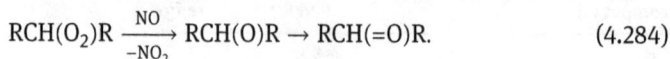

$$RCH(O_2)R \xrightarrow[-NO_2]{NO} RCH(O)R \rightarrow RCH(=O)R. \qquad (4.284)$$

In the case of terminal attack (equation (4.279)), the RO intermediate reacts with O_2, forming an aldehyde and the hydroperoxyl radical (the atmospheric importance of this reaction lies simply in the transfer $OH \rightarrow HO_2$):

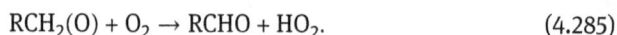

$$RCH_2(O) + O_2 \rightarrow RCHO + HO_2. \qquad (4.285)$$

In an urban environment, the addition of NO onto RO forms harmful alkyl nitrites (see Chapter 4.4.6.2, equation (4.177)): $RO + NO \rightarrow RONO$.

Aldehydes are very reactive substances. They react with OH by abstraction of H (almost from the chain) to produce finally bicarbonyls, for example, dialdehydes and ketoaldehydes (the carbonyl group is denoted by >C=O). Nevertheless, more important is the photolysis of aldehydes and ketones that initiates radical chains. We see that the simplest aldehyde HCHO is photolysed (see Tables 3.15 and 4.8 and equation (4.294)), gaining H (which turns into HO_2) and the formyl radical HCO (a carbon radical), which reacts with O_2 to HO_2 and CO. Therefore, HCHO is an efficient radical source. In analogy, higher aldehydes are photolysed and produce H atoms and acyl radicals RCO (R–C$^\cdot$ =O), for example, acetyl CH_3CO:

$$RCHO + h\nu \rightarrow RCO + H. \qquad (4.286)$$

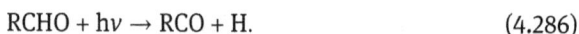

The fate of the acyl radical (which is chemically similar to the formyl radical HCO) is not decomposition but O_2 addition onto the carbon radical, giving the peroxyacyl radical $RC(O)OO$:

$$RCO + O_2 \rightarrow RC(O)OO. \qquad (4.287)$$

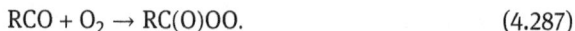

The peroxyacyl radical reacts in analogy to the RO_2 radical, thereby oxidising NO to NO_2 and giving acyloxy radicals ($RC(O)O$), which can react with HO_2 to alkyl peroxides (ROOH and ROOR), or adding NO_x (Chapter 4.4.6.2) give peroxyacyl nitrates (PAN). Many of these species are radical reservoirs (especially peroxides but also nitro and nitroso compounds), which can be transported from polluted areas away and release radicals after photolysis, starting new radical chains, oxidising trace species and decomposing O_3. Acyl radicals (not to mix with alkoxyl radical RO, see Table 4.24) are

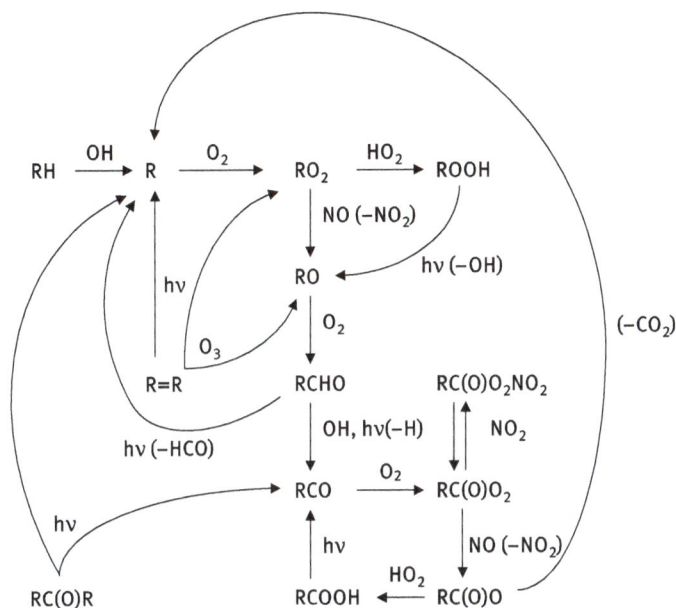

Figure 4.16: Scheme of organic radical chemistry and fate of characteristic organic groups. RH hydrocarbon, R alkyl radical, RO_2 alkyl peroxyl radical, ROOH organic peroxide, RO alkoxyl radical, RCHO aldehyde, RCOOH carboxylic acid, RC(O)R ketone, R=R olefine. Reactions between RO or RO_2 and NO or NO_2 are not included in this scheme.

also gained by the photolysis of ketones:

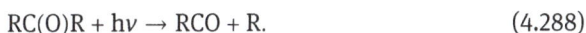

$$RC(O)R + h\nu \rightarrow RCO + R. \tag{4.288}$$

For HO_2 recycling, the photolysis of olefins (R=R) and aldehydes (RCHO) is very important. Aldehydes produce a variety of primary radicals (HCO, RCO and H) that retransform in multistep back to HO_2 (and to OH; Figure 4.16). As seen, low reactive hydrocarbons (RH), after oxidation by OH, produce many more reactive intermediates that amplify net ozone formation (see Figure 5.21). All organic oxy and peroxy radicals can also add NO and NO_2 and form organic nitroso (−N=O) and nitro (−NO_2) compounds (Chapter 4.4.6.2). These compounds present a radical reservoir and can be photolysed back to the original compounds.

The C−H bond in methane (CH_4), alkyl (−CH_3) and allylic (>CH_2) groups of any organic compounds cannot be photodissociated under environmental conditions, but the H atom is abstracted by the radical attack, almost OH. The methyl (CH_3) and alkyl (R) radicals undergo several reaction pathways, but in the presence of oxygen, aldehyde and ketones are formed. This oxygenation increases the reactivity and the water solubility of the products.

4.6.3.2 C$_1$ chemistry: CH$_4$, HCHO, CH$_3$OH and HCOOH

Let us now consider in more detail reaction equation (4.278), beginning with the simplest but also most abundant hydrocarbon methane CH$_4$. Reaction equation (4.289) is very slow ($k_{4.289}$ = 6.4 · 10^{-15} cm^3 molecule^{-1} s^{-1} at 298 K), giving a residence time of about 10 years:

$$CH_4 + OH \rightarrow CH_3 + H_2O. \tag{4.289}$$

A low specific rate (large residence time), however, does not mean that this pathway is unimportant. On the contrary, because of the large CH$_4$ concentration in the air, nearly homogeneously distributed in the whole troposphere, CH$_4$ controls to a great extent the background OH concentration and tropospheric net O$_3$ formation (see Chapter 5.3.2.2). The methyl radical CH$_3$ rapidly reacts with O$_2$, producing the methylperoxyl radical; $k_{4.290}$ = 1.2 · 10^{-12} cm^3 molecule^{-1} s^{-1} at 298 K):

$$CH_3 + O_2 \rightarrow CH_3O_2. \tag{4.290}$$

Similar to HO$_2$ (reaction equation (4.97)), CH$_3$O$_2$ oxidises with NO; $k_{4.291}$ = 7.7 · 10^{-12} cm^3 molecule^{-1} s^{-1} at 298 K):

$$CH_3O_2 + NO \rightarrow CH_3O + NO_2. \tag{4.291}$$

There are some (slow) competing reactions with equation (4.291), which we discussed in the previous Chapter 4.6.3.1 and which are relevant under very low NO concentrations and giving a secondary atmospheric source of methanol (CH$_3$OH):

$$CH_3O_2 + CH_3O_2 \rightarrow CH_3OH + HCHO + O_2. \tag{4.292}$$

However, when the methoxy radical CH$_3$O is produced, reaction equation (4.293) rapidly proceeds and gives formaldehyde, HCHO and HO$_2$, closing the HO$_y$ cycle; $k_{4.293}$ = 1.9 · 10^{-15} cm^3 molecule^{-1} s^{-1} at 298 K):

$$CH_3O + O_2 \rightarrow HCHO + HO_2. \tag{4.293}$$

Formaldehyde is the first intermediate in the CH$_4$ oxidation chain with a lifetime longer than a few seconds (Table 4.22 shows the huge amount of secondary HCHO). Formaldehyde is removed by photolysis equation (4.294) and reacts with OH (equation (4.295)); $k_{4.295}$ = 8.5 · 10^{-12} cm^3 molecule^{-1} s^{-1} at 298 K):

$$HCHO + h\nu \rightarrow H + HCO, \tag{4.294}$$

$$HCHO + OH \rightarrow H_2O + HCO. \tag{4.295}$$

The formyl radical HCO rapidly reacts with O_2 (HCO is a carbon radical and not oxygen radical):

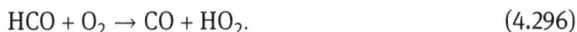

$$HCO + O_2 \rightarrow CO + HO_2. \tag{4.296}$$

We see that CH_4 oxidation results in a net gain of radicals when HCHO is photolysed (equation (4.297)) and turns equivalent OH into HO_2 when HCHO oxidises by OH (equation (4.298)). The gross budgets are:

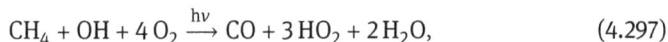

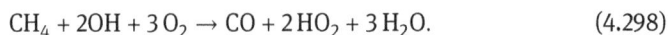

$$CH_4 + OH + 4O_2 \xrightarrow{h\nu} CO + 3HO_2 + 2H_2O, \tag{4.297}$$
$$CH_4 + 2OH + 3O_2 \rightarrow CO + 2HO_2 + 3H_2O. \tag{4.298}$$

Moreover, CO reacts much faster with OH than CH_4 to provide the first cycle in Figure 5.17; the overall "gross reaction" is:

$$CH_4 + 2OH + 3O_2 \xrightarrow{h\nu} CO_2 + 2HO_2 + 2H_2O. \tag{4.299}$$

Because C–O and O–H bonds are much stronger than the C–H bond, OH attack goes preferable onto C–H (at higher carbon chains preferably at the C–H); $k_{4.300} = 7.7 \cdot 10^{-13}$ cm^3 $molecule^{-1} s^{-1}$ and $k_{4.301} = 1.3 \cdot 10^{-13}$ cm^3 $molecule^{-1} s^{-1}$, i.e. about 85 % of methanol goes via equation (4.300):

$$CH_3OH + OH \rightarrow CH_2OH + H_2O, \tag{4.300}$$
$$CH_3OH + OH \rightarrow CH_3O + H_2O \tag{4.301}$$

The fate of the methoxy radical CH_3O is known (equation (4.293)), and CH_2OH gives the same products: $k_{4.302} = 9.7 \cdot 10^{-12}$ cm^3 $molecule^{-1} s^{-1}$:

$$CH_2OH + O_2 \rightarrow HCHO + HO_2. \tag{4.302}$$

Hence, the methanol oxidation yields formaldehyde according to the budget:

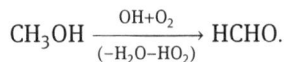

$$CH_3OH \xrightarrow[(-H_2O-HO_2)]{OH+O_2} HCHO.$$

Formaldehyde (IUPAC name: methanal) quickly converts at daytime to CO (sequence equation (4.294) to equation (4.296)) but also transfers into the aqueous phase where it hydrates (equation (4.214)) to methanediol or reacts with S(IV) (reaction equation (4.213a), (4.213b)). In aqueous solutions (such as cloud water), methanol quickly oxidises similar to the gas phase mechanisms to formaldehyde, which (together with scavenged HCHO) further oxidises via rapid oxidation of its monohydrated form, methanediol ($HOCH_2OH$) to formic acid (methane acid) by OH attack (in gas and aqueous phase); the intermolecular rearrangement is fast:

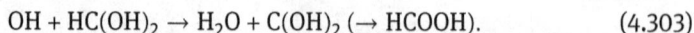

$$OH + HC(OH)_2 \rightarrow H_2O + C(OH)_2 (\rightarrow HCOOH). \tag{4.303}$$

Newest research results (Franco et al. 2021) show that formaldehyde is efficiently converted to gaseous formic acid via a multiphase pathway that involves its hydrated form, methanediol. In warm cloud droplets, methanediol undergoes fast outgassing but slow dehydration. The gas-phase oxidation of methanediol produces up to four times more formic acid than all other known chemical sources combined. The additional formic acid burden increases atmospheric acidity by reducing the pH of clouds and rainwater. The diol mechanism presented here probably applies to other aldehydes and may help to explain the high atmospheric levels of other organic acids that affect aerosol growth and cloud evolution.

Formic and acetic acids are the most abundant low molecular weight carboxylic acids in the global troposphere. They can either be emitted by direct sources, such as vehicular exhaust emissions, biomass burning, biofuel, fossil fuel and vegetation, or formed in the atmosphere by photochemical reactions.

Numerous measurements show that in the gas phase $[HCHO] > [HCOOH]$ and in hydrometeors $[HCHO] < [HCOO)]$. Formic acid was first isolated in 1671 by the English researcher *John Ray* (1627–1795) from red ants (its name comes from the Latin word for ant, *formica*). Considering methane acid as a transient between inorganic and organic carbon, the carboxyl group $-C(O)OH$ provides the huge class of organic acids RCOOH and the formate $HCOO^-$ gives the class of esters HCOOR and RCOOR.

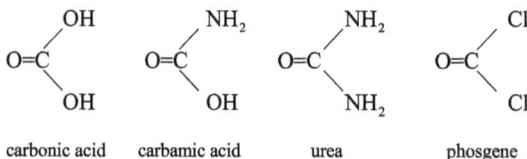

carbonic acid   carbamic acid   urea   phosgene

From carbonic acid ($CO_2 + H_2O$), important derivates can be obtained: carbamic acid (or carbamates,[74] which also provides a class of organic carbamines substituting H for organic R), urea and halogenated substitutes (such as phosgene). Urea is the "symbol" linking inorganic with organic chemistry; thereby, here are two IUPAC names: diaminomethanal (as organic compound) and carbonyl diamide (as inorganic compound).

The formyl radical HCO, which undergoes very rapid hydration in aqueous solution (it is, for example, formed from CO by electron transfer, see equation (4.274) and

---

74 They are formed while CO_2 capture from (flue) gases using aqueous amine solutions.

equation (4.275)) to $HC(OH)_2$

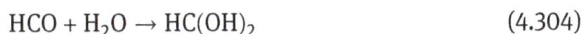

$$HCO + H_2O \rightarrow HC(OH)_2 \tag{4.304}$$

and further then dimerises to glyoxal $(HCO)_2$; see equation (4.318). The formyl radical HCO in the solution can be considered as the conjugated acid of CO^-. In the next Chapter 4.6.3.3, we will see that C_2 species are produced in the aqueous phase by carbonylation (reactions with HCO) and carboxylation (reactions with CO_2^-). It is found that directly from equation (4.274) and equation (4.275) the hydrated form is given:

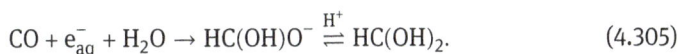

$$CO + e_{aq}^- + H_2O \rightarrow HC(OH)O^- \overset{H^+}{\rightleftharpoons} HC(OH)_2. \tag{4.305}$$

The hydrated formyl radical $HC(OH)_2$ is a strong reducing species. It has been detected as an intermediate in the reaction of hydrated formaldehyde $(H_2C(OH)_2)$ with OH, which is also oxidised by H_2O_2 to form formic acid (or the formate anion, respectively):

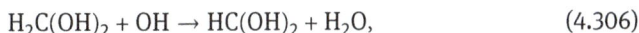

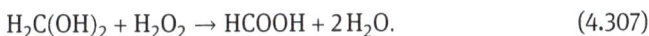

$$H_2C(OH)_2 + OH \rightarrow HC(OH)_2 + H_2O, \tag{4.306}$$

$$H_2C(OH)_2 + H_2O_2 \rightarrow HCOOH + 2H_2O. \tag{4.307}$$

A possible propagation mechanism involves:

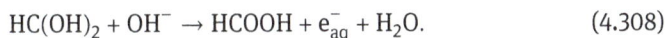

$$HC(OH)_2 + OH^- \rightarrow HCOOH + e_{aq}^- + H_2O. \tag{4.308}$$

Figures 4.17 and 4.18 summarise the C_1 chemistry in the gas and aqueous phase. We

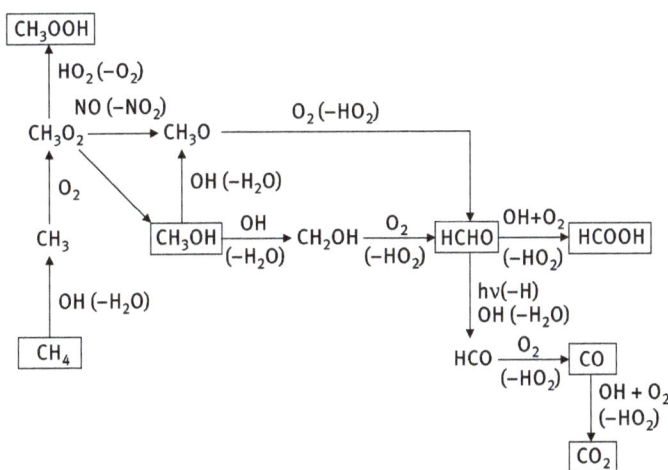

Figure 4.17: Scheme of C_1 gas-phase chemistry.

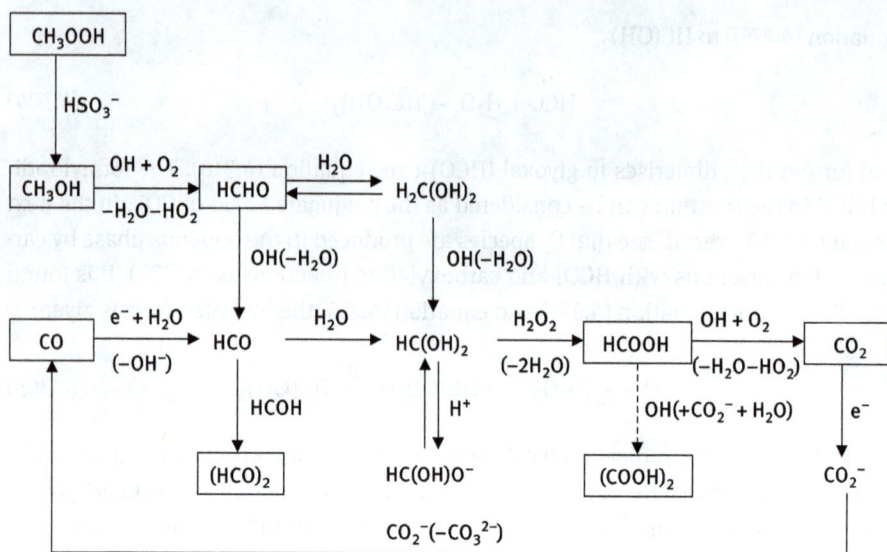

Figure 4.18: Scheme of C_1 aqueous-phase chemistry. $(HCO)_2$ – glyoxal, $(COOH)_2$ – oxalic acid; dotted line – multistep process.

again see that the aqueous phase provides more specific conditions (such as the formation of ions, hydrated electrons, interfacial arrangements) for chemical reactions and producing compounds than the gas phase. Almost all gas-phase reactions also proceed in the aqueous phase.

> Whereas CH_4 is very slowly oxidised in air (but provides methyl peroxide as a ubiquitous compound), methanol and particularly formaldehyde quickly oxidises (the latter also photodissociates) to inorganic CO and finally, CO_2. HCHO provides a net source of radicals. The formation of formic acid is likely to be negligible in the gas phase. But methanol, formaldehyde and formic acid (all produced and/or emitted in huge quantities from biogenic sources) will be scavenged and provide more efficient oxidation, finally with an accumulation of formic acid, but partly until its mineralisation to CO_2 and, most interestingly, a pathway in the formation of highly reactive bicarbonyls such as glyoxal and oxalic acid.

4.6.3.3 C_2 chemistry: C_2H_6, CH_3CHO, C_2H_5OH, CH_3COOH and $(COOH)_2$

Higher alkanes generally produce first an aldehyde by the OH oxidation of the terminal carbon atom and then a bicarbonyl (here glyoxal from ethane) at the other end of the chain (here C_2):

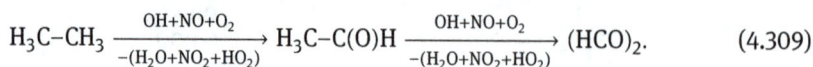

$$H_3C\text{–}CH_3 \xrightarrow[-(H_2O+NO_2+HO_2)]{OH+NO+O_2} H_3C\text{–}C(O)H \xrightarrow[-(H_2O+NO_2+HO_2)]{OH+NO+O_2} (HCO)_2. \qquad (4.309)$$

Acetaldehyde (CH_3CHO) is similar to formaldehyde photolysed and oxidised through the OH attack. Other photolysis pathways (to $CH_4 + CO$ or $CH_3CO + H$) are important under the irradiation above 300 nm in the troposphere.

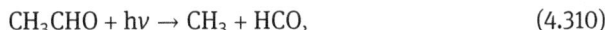

$$CH_3CHO + h\nu \rightarrow CH_3 + HCO, \tag{4.310}$$

$$CH_3CHO + OH \rightarrow CH_3CO + H_2O. \tag{4.311}$$

The acetyl radical CH_3CO adds O_2 (see equation (4.287)) and produces PAN (equation (4.186)) but in a competitive reaction as RO_2 radical, it oxidises NO and afterwards decomposes (from the methylperoxyl radical CH_3O_2 finally formaldehyde is given).

$$CH_3C(O)O_2 \xrightarrow[-NO_2]{NO} CH_3C(O)O, \tag{4.312}$$

$$CH_3C(O)O \xrightarrow[-CO_2]{O_2} CH_3O_2 \xrightarrow[-NO_2]{NO} CH_3O \xrightarrow[-HO_2]{O_2} HCHO. \tag{4.313}$$

Alcohols give acids; $k_{4.313} = 2.8 \cdot 10^{-12}$ cm^3 molecule^{-1} s^{-1}. Acetic acid CH_3COOH is likely stable in the gas phase and transferred into hydrometeors.

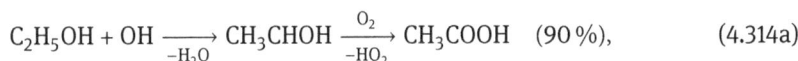

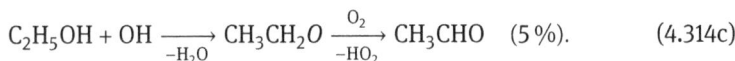

$$C_2H_5OH + OH \xrightarrow[-H_2O]{} CH_3CHOH \xrightarrow[-HO_2]{O_2} CH_3COOH \quad (90\,\%), \tag{4.314a}$$

$$C_2H_5OH + OH \xrightarrow[-H_2O]{} CH_2CH_2OH \quad (5\,\%), \tag{4.314b}$$

$$C_2H_5OH + OH \xrightarrow[-H_2O]{} CH_3CH_2O \xrightarrow[-HO_2]{O_2} CH_3CHO \quad (5\,\%). \tag{4.314c}$$

The radical adduct CH_2CH_2OH reacts the same as in the ethene + OH reaction (see equation (4.328)) and forms glycolaldehyde (IUPAC name: 2-hydroxyethanal):

$$CH_2CH_2OH \xrightarrow[-(H_2O+NO_2+HO_2)]{OH+NO+O_2} HOCH_2CHO. \tag{4.315}$$

The lifetime of glycolaldehyde in the atmosphere is about one day for reaction with OH, and >2.5 days for photolysis, although both wet and dry deposition are other important removal pathways. The primary products of OH attack and photolysis are mainly HCO (for further fate, see equation (4.296)) and CH_2OH (equation (4.302)). We will later see that glycolaldehyde is the main product of the OH + C_2H_2 (ethyne) reaction (equation (4.328)). However, the hydroxyalkoxy intermediate might also decompose (the oxidation of glycolaldehyde gives HCHO and CO as final products).

$$HOCH_2CH_2O \rightarrow HCHO + CH_2OH. \tag{4.316}$$

Finally, the CH_2OH radicals react with O_2 to give HCHO and HO_2 (equation (4.302)). Thus, C_2 is broken down into C_1 species. Figure 4.19 shows schematically the C_2 gas

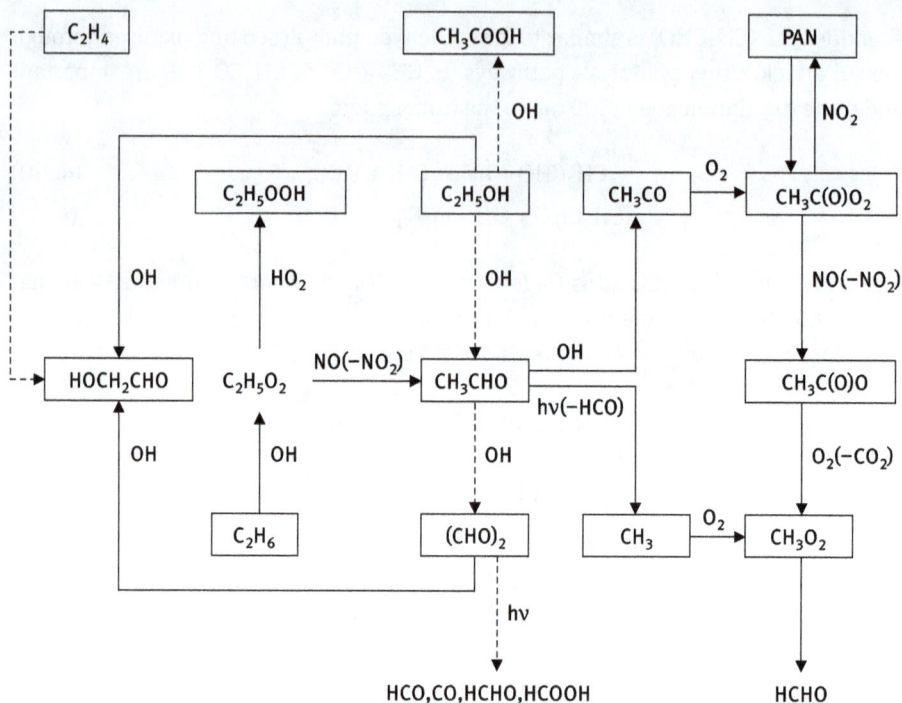

Figure 4.19: Scheme of C_2 gas-phase chemistry. C_2H_5OOH – acetyl peroxide, $HOCH_2CHO$ – glyco-laldehyde, C_2H_6 – ethane, C_2H_4 – ethene; dotted line – multistep process.

phase chemistry. It is obvious that there is no ethanol formation and acetic acid decomposition, whereas acetaldehyde provides many pathways back to C_1 chemistry. Glycolaldehyde is a highly water-soluble product from several C_2 species (ethene, acetaldehyde and ethanol); other bicarbonyls, however, are likely to be produced preferably in solution.

The aqueous phase produces other C_2 species but also decomposes them (Figure 4.20). By contrast, in an aqueous solution from C_1, C_2 species can be given as shown by the formation of glyoxal from the formyl radicals (equation (4.318)); the latter is an often-found species in solution. From methanol and formaldehyde, dicarbonyls are produced via carbonylation in the aqueous phase:

$$CH_3OH \xrightarrow[-H_2O]{OH} CH_2OH \xrightarrow{HCO} OHC-CH_2OH \quad \text{(glycolaldehyde),} \qquad (4.317)$$

$$HCHO \xrightarrow[-H_2O]{OH} HCO \xrightarrow{HCO} OHC-CHO \text{(glyoxal).} \qquad (4.318)$$

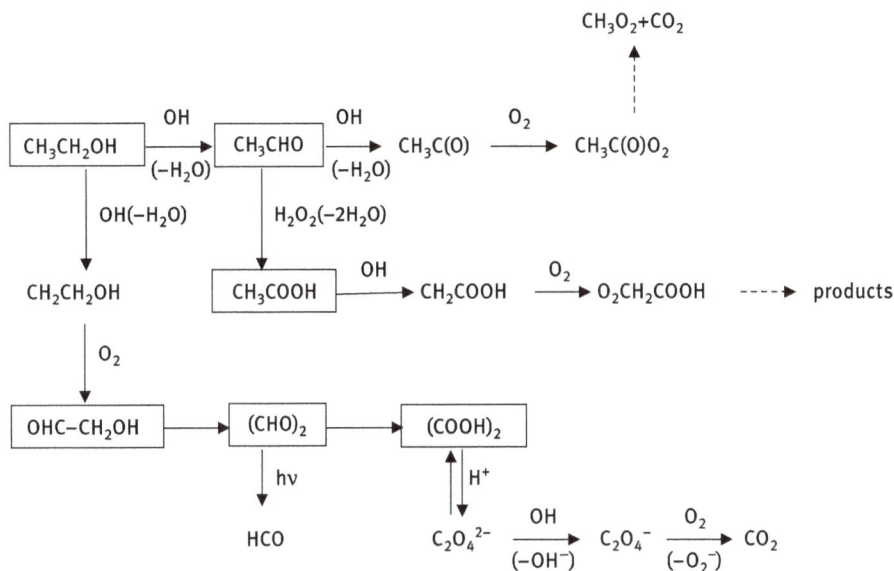

Figure 4.20: Scheme of C_2 aqueous-phase chemistry.

Furthermore, from methanol and formic acid, the corresponding acids gained through carboxylation (in reaction with the CO_2^- radical) are:

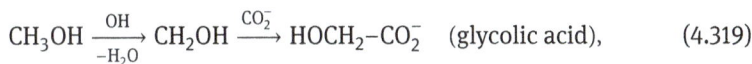

$$CH_3OH \xrightarrow[-H_2O]{OH} CH_2OH \xrightarrow{CO_2^-} HOCH_2-CO_2^- \text{ (glycolic acid)}, \qquad (4.319)$$

$$HCOOH \xrightarrow[(-H_2O)]{OH} COOH \xrightarrow{CO_2^-} COOH-CO_2^- \text{ (oxalic acid)}. \qquad (4.320)$$

Ethanol (C_2H_5OH) oxidises in a multistep process in solution to acetaldehyde.

$$CH_3CH_2OH \xrightarrow[-(H_2O+HO_2)]{OH+O_2 \text{ (multistep)}} CH_3CHO. \qquad (4.321)$$

Such starting H abstraction can also go with other H acceptors such as SO_4^-, NO_3, Cl_2^-, Br_2^- and CO_3^-. Further oxidation of the aldehyde is similar in elementary multisteps to acetic acid, likely via ethanediol $CH_3C(OH)_2$:

$$CH_3CHO \xrightarrow[-(H_2O+HO_2)]{OH+O_2 \text{ (multistep)}} CH_3COOH. \qquad (4.322)$$

Acetic acid further oxidises in a first step to give the CH_2COOH radical; in the absence of O_2, it can dimerise to succinic acid $COOH(CH_2)_2COOH$.

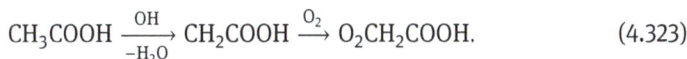

$$CH_3COOH \xrightarrow[-H_2O]{OH} CH_2COOH \xrightarrow{O_2} O_2CH_2COOH. \qquad (4.323)$$

This peroxy radical RO_2 (R = CH_2COOH) can then give glycolic acid, glyoxylic acid, oxalic acid, HCHO and CO_2. The acetylperoxy radical (ACO_3) – the precursor of PAN – is given from the oxidation of acetaldehyde. It forms with HO_2 peroxoacetic acid, also detected in air in small concentrations:

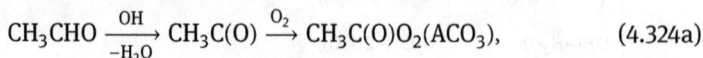

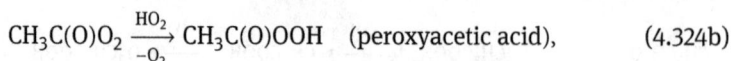

$$CH_3CHO \xrightarrow[-H_2O]{OH} CH_3C(O) \xrightarrow{O_2} CH_3C(O)O_2(ACO_3), \qquad (4.324a)$$

$$CH_3C(O)O_2 \xrightarrow[-O_2]{HO_2} CH_3C(O)OOH \quad \text{(peroxyacetic acid)}, \qquad (4.324b)$$

In anaqueous solution, peroxy radicals can also react with bisulphite, gaining the sulphite radical and organic peroxide ROOH:

$$RO_2 \xrightarrow[-SO_3^-]{HSO_3^-} ROOH. \qquad (4.325)$$

As for other peroxy radicals RO_2, the acetylperoxy radical $CH_3C(O)O_2$ can transfer O onto other dissolved species (NO → NO_2) and then decompose to the methylperoxy radical CH_3O_2, which finally forms HCHO:

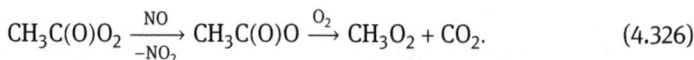

$$CH_3C(O)O_2 \xrightarrow[-NO_2]{NO} CH_3C(O)O \xrightarrow{O_2} CH_3O_2 + CO_2. \qquad (4.326)$$

Glyoxal (CHOCHO or $(CHO)_2$), the simplest dialdehyde, is one of the simplest multifunctional compounds found in the atmosphere and is produced by a wide variety of biogenically and anthropogenically emitted organic compounds. One current model estimates global glyoxal production to be 45 Tg yr^{-1}, with roughly 50 % due to isoprene photooxidation, whereas another estimates 56 Tg yr^{-1} with 70 % produced from biogenic precursors. CHOCHO is destroyed in the troposphere primarily by reaction with OH radicals (23 %) and photolysis (63 %), but it is also removed from the atmosphere through wet (8 %) and dry deposition (6 %). However, sources and sinks of glyoxal remain largely uncertain. The gas-phase photolysis of glyoxal produces two HCO radicals as the most important pathway under atmospheric conditions. Glyoxal sulphate has also been detected in filter samples:

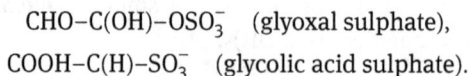

$$CHO-C(OH)-OSO_3^- \quad \text{(glyoxal sulphate)},$$
$$COOH-C(H)-SO_3^- \quad \text{(glycolic acid sulphate)}.$$

In diluted aqueous solution, glyoxal exists as a dihydrate $CH(OH)_2CH(OH)_2$, which is fast and reversibly formed. Aqueous phase photooxidation of glyoxal is a potentially important global and regional source of oxalic acid and secondary organic aerosol (SOA).

Oxalic acid is the most abundant dicarboxylic acid found in the troposphere, yet there is still no scientific consensus concerning its origins or formation process. Concentrations of oxalic acid gas at remote and rural sites range from about 0.2 ppb to 1.2 ppb with a very strong annual cycle, where high concentrations are found during the summer period. Oxalate was observed in the clouds at air-equivalent concentrations of $0.21 \pm 0.04 \ \mu g \ m^{-3}$ to below-cloud concentrations of $0.14 \ \mu g \ m^{-3}$, suggesting an in-cloud production as well. Oxalic acid is the dominant dicarboxylic acid (DCA), and it constitutes up to 50 % of total atmospheric DCAs, especially in non-urban and marine atmospheres. The large occurrence in the condensed phase led Warneck (2003) to suggest that oxalate not only originates in the gas phase but also condenses into particles (Figure 4.21). Hence, hydroxyl radicals might be responsible for the aqueous phase formation of oxalic acid from alkenes. Among different dicarboxylic acids (oxalic, adipic, succinic, phthalic and fumaric), only the dihydrate of oxalic acid, enriched in particles in the upper troposphere, acts as a heterogeneous ice nucleus. Ubiquitous organic aerosol layers above clouds with enhanced organic acid levels have

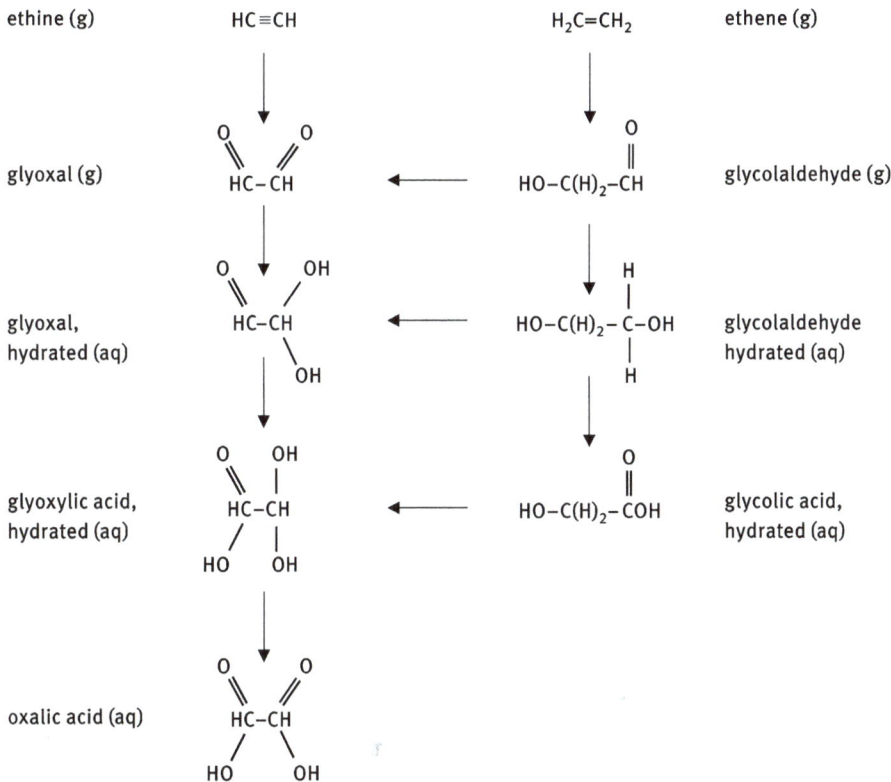

Figure 4.21: The proposed reaction pathway for the formation of oxalic acid in cloud water; after Warneck (2003).

Table 4.26: C_1 and C_2 carboxylic acids.

formula	structure	name of acid	name of salt
monocarbon acids			
H_4CO	H_3COH	methanol[a]	methanolate
H_2CO_2	$HC(O)OH$	formic acid	formate
H_2CO_3	$C(O)(OH)_2$	carbonic acid	carbonate
H_2CO_4	$C(O)(OH)OOH$[b]	peroxocarbonic acid	peroxocarbonate
dicarbon acids			
$H_4C_2O_2$	$CH_3C(O)OH$	acetic acid[d]	acetate
$H_4C_2O_3$	$HOCH_2C(O)OH$	glycolic acid[e]	glycolate
$H_2C_2O_2$	$HOC{\equiv}COH$[b,c]	dihydroxyacetylene	dihydroxyacetylate
$H_2C_2O_3$	$HOC-C(O)OH$	glyoxylic acid[f]	glyoxalate
$H_2C_2O_4$	$HO(O)C-C(O)OH$	oxalic acid	oxalate
$H_2C_2O_5$	$HO(O)C-O-C(O)OH$[b]	dicarbonic acid	dicarbonate
$H_2C_2O_6$	$HO(O)C-O-O-C(O)OH$b	peroxodicarbonic acid	peroxodicarbonate

[a] very weak acid is forming
[b] only as salts
[c] isomer with the non-acidic glyoxal $O=CH-CH=O$ (ethandial)
[d] IUPAC name: ethanoic acid; other names: methanecarboxylic acid, acetyl hydroxide
[e] IUPAC name: 2-hydroxyethanoic acid, another name: hydroxoacetic acid
[f] IUPAC name: oxoethanoic acid; other names: oxoacetic acid, formylformic acid (ubiquitous in nature in berries)

been observed, and field data suggest that aqueous-phase reactions to produce organic acids, mainly oxalic acid, followed by droplet evaporation are the source. Concentration variations of organic acids in the gas and aqueous phases have been attributed to seasonal variations in biogenic emissions.

Organic acids (Table 4.26) are ubiquitous components of the troposphere in urban and remote regions of the world. Organic acids contribute significantly to rainwater acidity in urban areas and account for as much as 80–90 % of the acidity in remote areas.

4.6.3.4 Alkenes, alkynes and ketones

Besides alkanes, C_2 carbon comprises two other classes: alkenes and alkynes. The double bond C=C is stronger than the simple C–C but paradoxically is more reactive. The simplest is ethylene (ethene) $CH_2=CH_2$. Alkenes occur widely in nature. Ethene is essential in plant physiology and phenology, functioning as a plant hormone that regulates a myriad of plant processes, including seed germination, root initiation, root hair development, flower development, sex determination, fruit ripening, senescence and response to biotic and abiotic stresses. All plants and all plant parts produce ethene, a discovery first made in the 1930s from ripe apples. Consequently, ethene is widely used as a ripening agent for plants and plays an important role in the storage and preparation of agricultural commodities. As a plant hormone that responds to various stresses, the ethene source is likely to respond to land and climate modifications. Because of its agricultural importance, the biochemistry of ethene has been well studied by plant physiologists, while the biochemistry of the other light alkenes, such as propene and butene, remains unknown.[75]

Light alkenes in the atmosphere originate from both anthropogenic and biogenic sources. Ethene, propene and butene are produced industrially by cracking petroleum hydrocarbons. In the troposphere, alkenes contribute to the photochemical production of tropospheric ozone. The "light alkenes", defined here as the C_2–C_4 alkenes, include C_2H_4 (ethene), C_3H_6 (propene) and C_4H_8 (1-butene, trans-2-butene, cis-2- butene, and 2-methylpropene). Ethene and propene have the highest ozone production rates per carbon, followed by isoprene. However, the spatial and temporal distributions of light alkene emissions are mostly unknown.

Most of the hydrocarbon flux from the biosphere to the atmosphere is just one compound, isoprene. Isoprene ($HC=C(CH_3)HC=CH_2$) is the building block of monoterpenes that was found in all structures known from plant organic molecules.[76] The general view on isoprene emission is that it results from regulated conversions of carbon and free energy in a series of photosynthetic reactions under stressful conditions caused by CO_2 deficit inside illuminated autotrophic cells. This stress generates an energy overflow far in excess of the energy-consuming capacity. The necessity of discharging this energy excess is dictated by the fact that the living cell is a dissipative structure.

Alkenes are produced in so-called elimination reactions, mainly by dehydration ($-H_2O$) of alcohols and dehydrohalogenation ($-HX$) of alkyl halides:

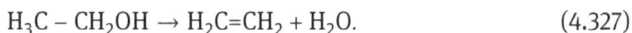

$$H_3C - CH_2OH \rightarrow H_2C=CH_2 + H_2O. \tag{4.327}$$

Alkenes add OH and provide a radical adduct (which reacts further as shown in reaction equation (4.315)) to glycolaldehyde

$$H_2C=CH_2 + OH \rightarrow CH_2CH_2OH, \tag{4.328}$$

which is quickly further oxidised to formaldehyde.

---

75 Other alkenes that occur in nature include 1-octene, a constituent of lemon oil, and octadecene ($C_{18}H_{36}$) found in fish liver. Dienes (two double bonds) and polyenes (three or more double bonds) are also common. Butadiene ($CH_2=CHCH=CH_2$) is found in coffee. Lycopene and the carotenes are isomeric polyenes ($C_{40}H_{56}$) that give the attractive red, orange, and yellow colours to watermelons, tomatoes, carrots, and other fruits and vegetables. Vitamin A, essential to good vision, is derived from a carotene. The world would be a much less colourful place without alkenes. Further, alkenes are found in insect pheromones.

76 Recent estimates of the global BVOC emission (769 Tg C yr^{-1}) include following contributions: 70 % isoprene, 11 % monoterpenes, 6 % methanol, 3 % acetone, 2.5 % sesquiterpenes, and < 2 % other BVOC.

In contrast to alkenes (also known as olefins), which are important compounds in nature, alkynes ($-C{\equiv}C-$) are relatively rare in nature but highly bioactive in some plants. The triple bond is very strong, with a bond strength of 839 kJ mol^{-1}. Alkynes are believed to be emitted almost entirely from two major anthropogenic sources: biomass burning processes and automobile tailpipe emissions. Besides ethyne, propyne, 1-butyne, and 2-butyne have also been detected in the atmosphere. Ethyne C_2H_2 (commonly known as acetylene) is generally considered to be produced only by human activities with an average tropospheric lifetime of the order of two months, allowing this compound to reach remote areas as well as the upper troposphere. Thus, the presence of C_2H_2 in open oceanic atmosphere is commonly explained by its long-range transport from continental sources together with CO originating from combustion. Both species are strongly correlated in atmospheric observations, offering constraints on atmospheric dilution and chemical ageing. In effect, its mixing ratio is typically in the range 500–3,000 ppt in inhabited countries compared with 50–100 ppt in remote oceanic areas.

The traditional view of the OH-initiated oxidation mechanism of alkynes is that it should be similar to that of the alkenes. The destruction of acetylene in the atmosphere occurs only by reaction with OH radicals via an adduct that further adds oxygen

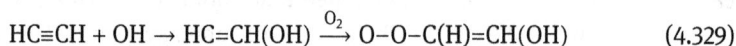

$$HC{\equiv}CH + OH \rightarrow HC{=}CH(OH) \xrightarrow{O_2} O{-}O{-}C(H){=}CH(OH) \qquad (4.329)$$

and finally (after intermolecular rearrangement) decays into glyoxal (CHO)$_2$ with OH regeneration or – more probable – into formic acid HCOOH and the formyl radical HCO. Another, earlier proposed pathway goes via the vinoxy radical CH_2CHO[77] as an intermediate to which O_2 is added, and it further reacts as described above:

$$HC{\equiv}CH + OH \rightleftharpoons (HC{=}CHOH)^* \xrightarrow{M\ (multistep)} CH_2CHO \xrightarrow{O_2} products. \qquad (4.330)$$

Alkenes are the only class of organic compounds that react in the gas phase with ozone. This occurs by the addition of a reaction called *ozonolysis* and has been known for more than 100 years:

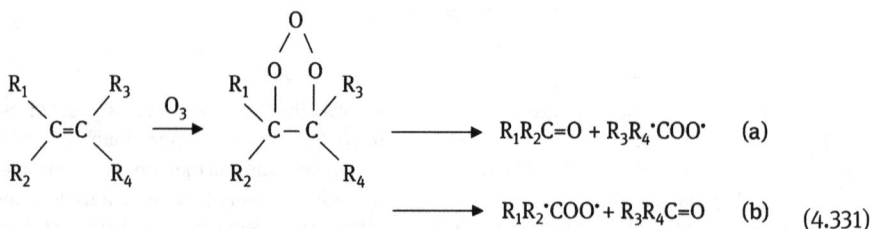

$$\qquad (4.331)$$

---

77 This alkyl radical would be formed if OH does not attack the C–H of the carbonyl group (see equation (4.321)) but the CH$_3$ group (which is much less probable).

The reaction rate increases with increasing carbon numbers, ranging between 10^{-18} and 10^{-14} cm^3 molecule^{-1} s^{-1}. Considering the O_3 concentration is larger by a factor of $>10^5$ than that of OH, the absolute rate of ozonolysis, even for lower alkenes (C_1–C_4), is about 10 % of the OH addition. Hence, at night, ozonolysis is an important pathway. For higher alkenes such as isoprene and terpene, the atmospheric lifetime is only in the range of minutes. Because of the steric consideration, the probability of each pathway (a) and (b) amounts to 50 %. The ozonide intermediate decomposes with the formation of ketone and biradicals (RRCOO), called Criegee radicals, which then stabilise and decompose. For the example of propene, the following products are given:

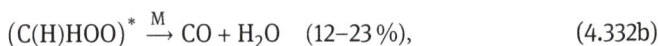

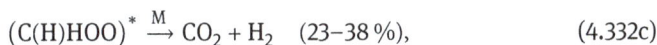

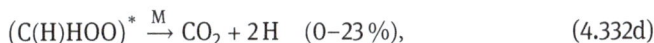

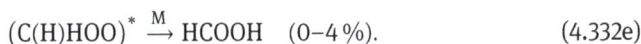

$$(C(H)HOO)^* \xrightarrow{M} HCO + OH \quad (37\text{–}50\,\%), \tag{4.332a}$$

$$(C(H)HOO)^* \xrightarrow{M} CO + H_2O \quad (12\text{–}23\,\%), \tag{4.332b}$$

$$(C(H)HOO)^* \xrightarrow{M} CO_2 + H_2 \quad (23\text{–}38\,\%), \tag{4.332c}$$

$$(C(H)HOO)^* \xrightarrow{M} CO_2 + 2\,H \quad (0\text{–}23\,\%), \tag{4.332d}$$

$$(C(H)HOO)^* \xrightarrow{M} HCOOH \quad (0\text{–}4\,\%). \tag{4.332e}$$

The produced OH and HO_2 (latter as a subsequent product from primary H) react additionally with the alkenes and provide a huge spectrum of products. Most important is SOA formation (known as a *blue haze* from biogenic emissions). The stabilised Criegee radical reacts with major species (H_2O, SO_2, NO, NO_2, CO, RCHO and ketones). In the reaction with water vapour, direct H_2O_2 can also be formed:

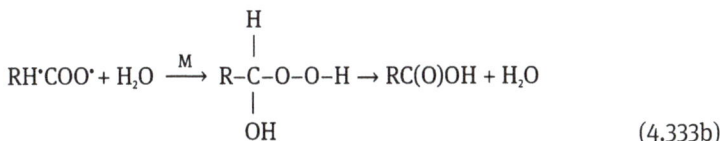

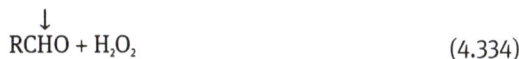

$$RH\!^{\bullet}COO^{\bullet} + H_2O \rightarrow RC(O)OH \quad \text{(carboxylic acid)} + H_2O, \tag{4.333a}$$

$$RH\!^{\bullet}COO^{\bullet} + H_2O \xrightarrow{M} R\text{–}\!\underset{\underset{OH}{|}}{\overset{\overset{H}{|}}{C}}\!\text{–O–O–H} \rightarrow RC(O)OH + H_2O \tag{4.333b}$$

$$\downarrow$$

$$RCHO + H_2O_2 \tag{4.334}$$

This is an important source of secondary organic acids; reaction equation (4.333a) is an intramolecular rearrangement via H_2O-collision intermediate.

Nucleation events over a boreal forest were driven by the condensation of terpene oxidation products. Formaldehyde (HCHO) is a high-yield product of isoprene oxidation. Isoprene emissions declined substantially in regions with large and rapid land cover changes. In addition, isoprene emission sensitivity to drought proved to have significant short-term global effects.

The hydroxyperoxide has been identified as an intermediate. Competing are the following conversions to aldehyde:

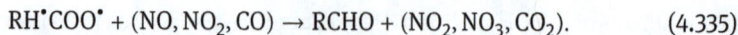

$$RH^{\bullet}COO^{\bullet} + (NO, NO_2, CO) \rightarrow RCHO + (NO_2, NO_3, CO_2). \tag{4.335}$$

The reaction with SO_2 goes via an adduct, which decays to sulphuric acid:

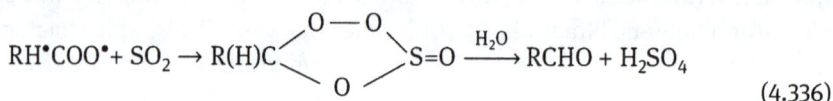

$$RH^{\bullet}COO^{\bullet} + SO_2 \rightarrow R(H)C \underset{O}{\overset{O-O}{<}} S=O \xrightarrow{H_2O} RCHO + H_2SO_4 \tag{4.336}$$

Nitrate radicals also react with alkenes by addition, which results in a variety of different compounds such as hydroxynitrates, nitrohydroperoxides and hydrocarbonyls.

> ℹ️ Another class of atmospherically important compounds is ketones (alkanone) predominantly of biogenic origin.

The simplest representative is acetone, which is ubiquitous in air. With the exception of acetone, higher ketones react preferably with OH at any carbon atom by H abstraction. With subsequent O_2 addition the fate of RO_2 is known, either forming aldehyde (RCHO) at terminal carbon or ketone (>C=O) in the middle of the chain. The photolysis of acetone (see equation (4.337)) provides the acetyl radical CH_3CO (as in the case of photolysis of acetaldehyde, see equation (4.286), whose fate is described after reaction equation (4.287); for the fate of methyl CH_3, see equation (4.290):

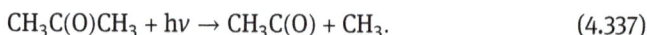

$$CH_3C(O)CH_3 + h\nu \rightarrow CH_3C(O) + CH_3. \tag{4.337}$$

> ℹ️ Oxygenated hydrocarbons, namely carboxylic acids, due to their hygroscopic properties, may play an active role in affecting the cloud condensation nucleus characteristics of aerosols and the global radiative forcing budget thus influencing the global climate.

4.6.3.5 Aromatic compounds

> ℹ️ Aromatic compounds play key roles in the biochemistry of all living organisms. The ring structure forms easily and remains stable, holding many biochemically important functions.

Aromatic hydrocarbons (arenes) can be monocyclic or polycyclic (PAH). The simplest aromatic compound is benzene C_6H_6, which is similar to the simplest alkane, methane CH_4. In contrast to alkenes, benzene is very stable (see Chapter 3.3.1 concerning bonding); namely, it is very stable against OH attack. As they are so stable, such rings tend to form easily, and once formed, they tend to be difficult to break in chemical reactions. The basic benzene structure forms many derivatives where the functional groups hold specific 'functions' in biochemistry (acidity, odour, smell, reactivity etc.). The four aromatic amino acids histidine, phenylalanine, tryptophan, and tyrosine each serve as one of the 20 basic building blocks of proteins. Furthermore, all five nucleotides (adenine, thymine, cytosine, guanine, and uracil) that make up the sequence of the genetic code in DNA and RNA are aromatic purines or pyrimidines. The haems (biomolecules) contain an aromatic system with 22 π-electrons. Chlorophyll also has a similar aromatic system.

From plants,direct emission of aromatic compounds is unknown; biomass burning is an important source. From litter, aromatic hydrocarbons are important common contaminants of soils and groundwater. Figure 4.22 shows some simple arenes.

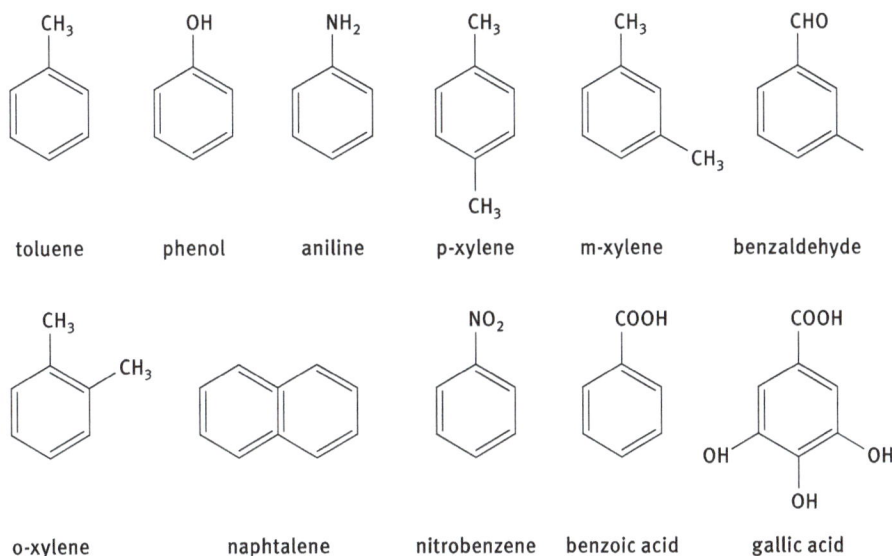

Figure 4.22: Some important arenes.

PAHs (polycyclic aromatic hydrocarbons) denominate a class of several hundreds of compounds, which share a common structure of at least three condensed aromatic rings. PAHs may be formed during natural processes such as incomplete combustion of organic materials such as coal and wood or during forest fires. PAHs are released during industrial activities such as aluminium, iron and steel production in plants and foundries, waste incineration, mining or oil refining. PAHs have also been detected at low levels in cigarette smoke and motor vehicle emissions. They are persistent organic pollutants and are slow to degrade in the environment. Benzo[a]pyrene (BaP) is commonly used as an indicator species for PAH contamination, and most of the available data refer to this compound.

Substituted arene compounds react faster with OH, for example, methylbenzenes such as toluene, p-xylene and 1,3,5-trimethylbenzene. The atmospheric aromatic hydrocarbons are emitted mainly from anthropogenic sources and are oxidized to phenolic compounds and α-dicarbonyl compounds, which contribute significantly to SOA formation. There has been a wealth of studies on the oxidation mechanism of aromatic hydrocarbons, but the degradation chemistry of aromatic VOC remains an area of particular uncertainty.

What is clear is the attack on the methyl group because it turns into the aldehyde (toluene → benzaldehyde). However, about 90 % of the OH attack goes to the ring, generating an adduct called the hydroxycyclohexadienyl radical.

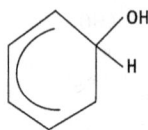

This can further react in different channels; first, it can give benzene oxide C_6H_5O, which is in equilibrium with oxepine – from which through photolysis phenols are gained, for example, α-cresol from toluene.

benzene oxide   oxepin

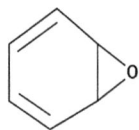

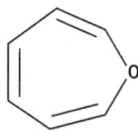

As benzene oxide has a conjugated but not localised radical carbon atom, the O_2 addition is another pathway; toluene gives the following peroxide, which is linked within the ring.

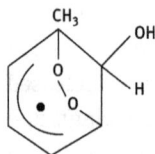

This intermediate peroxide decays while opening the ring into butenedial and methyl-glyoxal. Methylglyoxal and glyoxal have been identified to be the main components (up to 3 % only).

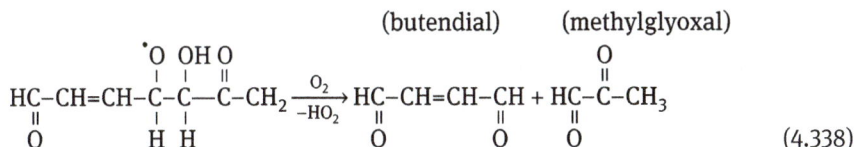

$$\underset{\text{(butendial)}}{} \quad \underset{\text{(methylglyoxal)}}{}$$

$$HC\text{-}CH\text{=}CH\text{-}\overset{\overset{\bullet}{O}}{\underset{\parallel}{C}}\text{-}\overset{\overset{OH}{}}{\underset{H}{C}}\text{-}\overset{\overset{O}{}}{\underset{H}{C}}\text{-}CH_2 \xrightarrow[-HO_2]{O_2} HC\text{-}CH\text{=}CH\text{-}\underset{\parallel}{\overset{}{C}}H + HC\text{-}\overset{\overset{O}{\parallel}}{C}\text{-}CH_3$$

(4.338)

The variety of products is large; often, bi- and polycarbonyls are found, which have been suggested to produce SOA and which transfer to the aqueous phase. However, it has also been found that during new particle formation in forested areas, OM mass fraction is significantly increasing but that the CCN efficiency is reduced by the low hygroscopicity of the condensing material.

4.7 Halogens

Among the halogens (main group 7 of the periodic table of the elements) are fluorine (F), chlorine (Cl), bromine (Br), iodine (I), and astatine (At). The radioactive At is the rarest element on Earth. The name halogens (Greek) means 'salt formers' (halides such as fluorides, chlorides etc.). Due to their large electronegativity, halogens are the only elements that form stable anions, F^-, Cl^-, Br^- and I^-; iodine also forms polyhalides such as I_3^-. Chlorine, bromine and iodine occur as elements (atomic and molecular) only in the atmosphere in very low concentrations and are very reactive. Fluorine is the most reactive and the most electronegative of all elements, so it is not found in the free elemental state in nature.[78] Despite the fact that fluorine is the 13th most common element in the earth's crust (0.027 % by weight), it is virtually absent from the biosphere.

Chlorine is essential for life. Specifically, the chloride ion is key to metabolism. In humans, the ion is obtained mainly from common salt (NaCl). It controls water-salt exchange, osmotic pressure, blood and digestive juices composition of the organisms. Its role in the synthesis of biomolecules is overwhelming (see also next Chapter 4.7.1). The chlorine atom, as one of many possible substituents used in synthetic organic chemistry, will remain in the future one of the important tools for probing structure-activity relationships in life science research and as a molecular component in commercialised compounds in order to provide safer, more selective and more environmentally compatible products with higher activity for medicine and agriculture.

---

78 F_2 reacts (such as the other halogens, but vigorously) with water, known since a century, to the hypohalogenic acid (HOF) and hydrogen halide (HF); however, only HOF further reacts with H_2O to H_2O_2 and HF.

Despite its ubiquitous yet trace presence within animals, essential functions of Br^- for tissue development and architecture for all animals have been known only very recently. Interestingly that bromide is first converted to hypobromous acid (HOBr), a compound to be found also playing a key role in environmental chemistry; HOBr further forms a bromosulphonium-ion intermediate that energetically selects for sulphilimine formation (a compound containing a sulphur-to-nitrogen double bond in biomolecules). Iodine is an essential component of the hormones produced by the thyroid gland for mammalian life. Although fluoride plays no biological role, it is an essential ion for animals, strengthening teeth and bones.

Iodine is a critical trace element involved in many diverse and important processes in the earth's system. The importance of iodine for human health has been known for over a century, with low iodine in the diet being linked to goitre, cretinism and neonatal death. Research over the last few decades has shown that iodine has significant impacts on tropospheric photochemistry, ultimately impacting climate by reducing the radiative forcing of ozone and air quality by reducing extreme O_3 concentrations in polluted regions. Iodine is naturally present in the ocean, predominantly as aqueous iodide and iodate. The rapid reaction of sea-surface iodide with O_3 is believed to be the largest single source of gaseous iodine to the atmosphere.[79] Due to increased anthropogenic O_3, this release of iodine is believed to have increased dramatically over the twentieth century by as much as a factor of three.

> **i** Without chlorine, bromine and iodine, animals will not exist.

Table 4.27 lists the halogen compounds that have been detected in nature, but not all exist in all phases (natural waters, hydrometeors and air). In the gas phase, acids such as HCl, HBr, HI, HClO, HBrO, HIO_3, $HClO_4$, $HBrO_4$ and HIO_4 exist. In the aqueous phase, perhalogenic acid, where O is exchanged by the peroxo group $-O-O-$, also exist as an intermediate in redox processes. Several interhalogens are known and might exist in air (such as iodine chloride). HF exists in air only in the condensed phase (particulate matter) despite being primarily emitted as a gas. SF_6 is exclusively from anthropogenic sources (it is used for air dispersion tracer experiments). It is extremely stable, with the longest known lifetime of 3,200 years.

The most spectacular attention on halogens was given by the implications for the stratospheric ozone layer (Chapter 5.3.2.1) and very recently for ozone depletion in the marine troposphere (Chapter 5.3.3.1).

---

79 See equation (5.96) in Chapter 5.3.3.1. Over the last decade or so, evidence has emerged that oceanic emissions of iodinated organic compounds such as CH_3I, CH_2ICl and CH_2I_2 are likely not the primary source of atmospheric iodine, as was originally thought, but may comprise only around 20 % of the total iodine flux to the atmosphere globally.

Table 4.27: Important halogen compounds in the environment.

formula[a]	names[b]	generic name
F^-, Cl^-, Br^-, I^-	chloride	halides
HF, HCl, HBr, HI	hydrogen chloride[c]	hydrogen halide[g]
Cl, Br, I	chlorine atom	halogen radical or atom
Cl_2, Br_2, I_2	dichlorine	halogen molecule
Cl_2^-, Br_2^-, I_2^-	dichlorine anion radical	dihalogen anion radical
ClO, HOBr, IO	chlorine monooxide	halogen monoxide
HOCl, HOBr, HOI	hypochlorous acid	hypohalogenic acid
OCl^-, OBr^-, OI^-	hypochlorite	hypohalogenite
ClO_2, BrO_2, IO_2	chlorine dioxide	halogen dioxide
$HClO_2$, $HBrO_2$	chlorous acid	halogenic acidg
ClO_2^-, BrO_2^-	chlorite	halogenite
$HClO_3$, $HBrO_3$, HIO_3	chloric acid	halogen acid[g]
ClO_3^-, BrO_3^-, IO_3^-	chlorate	halogenate
ClO_4^-	perchlorate	perhalogenate
$ClONO_2$, $BrONO_2$	chlorine nitrate	halogen nitrate
ClNO, BrNO	nitrosyl chloride	nitrosyl halogenide
$ClNO_2$, $BrNO_2$	nitryl chloride	nitryl halogenide
CCl_4, CF_4	carbon tetrachloride	carbon tetrahalogenide
$CHCl_3$, $CHBr_3$	trichloromethane[d] (TCM)	trihalogen methane
CH_2Cl_2, CH_2Br_2	dichloromethane[e] (DCM)	dihalogen methane
CH_3Cl, CH_3Br, CH_3I	chloromethane[f]	methyl halide
SF_6	sulphur hexafluoride	–

[a] listed are only substances found to be significant in nature
[b] given as an example for Cl compounds
[c] common name: hydrochloric acid (gas); archaic name: muriatic acid
[d] common names: chloroform, bromoform
[e] common name: methylene chloride
[f] common name: methyl chloride
[g] in English (in contrast to German where generic specific meanings exist and here proposed in English terms) halogen or halogenic acid is the general name for all acids (HCl, HOCl, $HClO_2$, $HClO_3$)

4.7.1 Halogens in the environment

There is a reservoir of chlorine (and other halogens) in the ocean and atmospheric sea salt in the form of NaCl from which HCl (and other halogens) can be released. The ratio between chloride, bromide and iodide in seawater is roughly (see Table 2.5) 300,000 : 1000 : 1. The concentration of fluoride in seawater amounts 1.2–1.4 mg L^{-1}, i. e., lies between bromide (50 times higher) and iodide (20 times smaller). The ocean has been identified as a source of many organic halogen compounds (Table 4.28), and newer insights provide evidence for halogenated compounds in soils. The ocean acts as both a source and a sink for methyl halides, where algae are responsible for the production of halogenated organic compounds, and surface photochemical pro-

Table 4.28: Emission of chlorine compounds (in Tg Cl yr^{-1}). Note, C_2 halogens are derived from ethene (>C=C<).

substance	ocean	soils	biomass burning	fossil fuel burning	other man-made emission
CH_3Cl	0.46	0.0001	0.640	0.075	0.035
$CHCl_3$	0.32	0.0002	0.0018	–	0.062
CH_3CCl_3	–	–	0.013	–	0.572
C_2Cl_4	0.016	–	–	0.002	0.313
C_2HCl_3	0.020	–	–	0.003	0.195
CH_2Cl_2	0.16	0.0003	–	–	0.487
$CHClF_2$	–	–	–	–	0.080
total organic	1.0	0.0006	0.65	0.08	1.7
total inorganic	1785[a]	15[b]	6.3[c]	4.6[d]	2[d]

[a] sea salt including the release of HCl (200–400) and $ClNO_2$ (0.06)
[b] soil dust chloride
[c] HCl and particulate chloride
[d] HCl

cesses most likely break them down to simple species (similar to the DMS production – thereby correlations have often been found). The natural emission of volatile organic chlorine (CH_3Cl is dominant) lies in the Tg range per year (Table 4.28) and those of CH_3Br only in the kt range, but the net CH_3I oceanic emission to the atmosphere is 0.2 Tg, whereas rice paddies, wetlands and biomass burning only contribute small amounts. Unexpectedly, during Saharan dust events, methyl iodide mixing ratios have been observed to be high relative to other times, suggesting that dust-stimulated emission of methyl iodide has occurred.[80]

The idea that the chloride content of continental water can be explained by airborn sea salt is often credited to František Pošepný (1836–1895), a world-renowned Czech geologist, who presented in 1877 his hypothesis on the origin of continental chloride, clearly expressed in 1878 the "salt cycle" and the formation of sea salt from seawater. Nevertheless, Wilhelm August Lampadius (1772–1842) was the first who expressed in 1820 the origin of chlorine in rainwater from the sea.[81] Various physical processes generate sea salt aerosols, especially the bursting of entrained air bubbles during whitecap formation, resulting in a strong dependence on wind speed. Sea salt

---

80 Of course, experiments with adding collected dust to seawater as well as adding H_2O_2 rapidly produced CH_3I; another example in line with photoenhanced radical aqueous chemistry.

81 Hydrochloric acid (HCl), namely from "decomposing" sea salt seems to be known since ancient time, and named by alchemists marine air. The gaseous element itself was first produced in 1774 by Carl Wilhelm Scheele at Uppsala, Sweden, by heating hydrochloric acid with the mineral pyrolusite which is naturally occurring manganese dioxide, MnO_2. Humphry Davy investigated it in 1807 and eventually concluded not only that it was a simple substance, but that it was truly an element. He announced this in 1810.

particles cover a wide size range (about 0.05 to 10 mm diameter) and have a correspondingly wide range of atmospheric lifetimes. The circulating amounts of sea salt are immense and provide chloride, bromide and iodide (together with sodium) to all parts of the world. Globally, large amounts of fine sea salt (i. e., not removed by sedimentation) are to be assumed to be emitted into the free troposphere and available for degassing processes between 100 and 800 Tg yr^{-1} as Cl. These values are based on total sea salt chloride emission in the order of 5,000 Tg yr^{-1}. Due to its ready solubility in water, Cl$^-$ is found in all natural waters and is finally transported back to the oceans.

Chlorine is one of the most abundant elements on the surface of the Earth. Until recently, it was widely believed that all chlorinated organic compounds were xenobiotic, that chlorine does not participate in biological processes and that it is present in the environment only as chloride. However, over the years, research has revealed that chlorine takes part in a complex biogeochemical cycle, that it is one of the major elements of soil organic matter and that the amount of naturally formed organic chlorine present in the environment can be counted in tonnes per km^2. More than 4,000 organohalogen compounds, mainly containing chlorine or bromine but a few with iodine and fluorine, are produced by living organisms or are formed during natural abiogenic processes, such as in volcanoes, forest fires, and other geothermal processes. The oceans are the single largest source of biogenic organohalogens, which are biosynthesised by myriad seaweeds, sponges, corals, tunicates, bacteria, and other marine life. Terrestrial plants, fungi, lichen, bacteria, insects, some higher animals, and even humans also account for a diverse collection of organohalogens.

The abundance of chlorine (as chloride) in the lithosphere seems not to be well established – in the literature, it is found in a range from 0.013 % to 0.11 %. With the last value,[82] we obtain a total mass of chlorine (using the mass of the lithosphere) of $2.2 \cdot 10^{22}$ g, which is comparable with the chloride dissolved in the oceans ($2.6 \cdot 10^{22}$ g). Elemental chlorine is one of the most reactive species and therefore is not found in nature with the exception of small volcanic emissions. Inorganic chlorine in the form of chloride (and similar to other halogens such as fluorine, bromine and iodine) is therefore stored in seawater and salt stocks from former oceans. Hence, the formation of sea salt is the dominant source of particulate chloride and gaseous HCl due to subsequent heterogeneous reactions (see Chapter 4.7.2).

Reactive halogen compounds (X, XO, X$_2$, XY, OXO, HOX, XONO$_2$, XNO$_2$, where X, Y = Cl, Br, I) – in particular halogen oxides – are present in various domains throughout the troposphere. The principal precursors for reactive Cl and Br are Cl$^-$ and Br$^-$

---

82 The Cl levels in organisms are near to (those) in seawater. Estimations of Cl average contents in plants are lower (~0.33 %) than in animals (~3.3 %), and are higher in marine organisms (plants 0.47 %, animals up to 9 %) than in terrestrial (respectively, 0.2 and 0.28 %).

in sea salt aerosol. Reactive I, in contrast, is derived mainly from organic iodine compounds produced in and emitted from the ocean and possibly from I_2.

Other secondary chlorine species (atomic Cl, ClO, ClOOCl etc.) have been responsible for Arctic ozone depletion, whereas the sources of the chlorine atoms are poorly understood. The Cl atom reacts similarly to OH (e. g. in the oxidation of volatile organic compounds). However, the photolysis of HCl is too slow (even in the stratosphere) to provide atomic Cl. Thus, the only direct Cl source from HCl is due to its reaction with OH but with a low reaction rate constant. There are several chemical means of production of elemental Cl (and other halogens) from heterogeneous chemistry (see next Chapter 4.7.2); in the troposphere, the photolysis of chloroorganic compounds is not very important, with a few exceptions. Global emission of HCl from coal combustion and from waste incineration was estimated to be 4.6 Tg Cl yr^{-1} and 2 Tg Cl yr^{-1}, respectively, whereas HCl emission from sea salt and biomass burning was estimated to be 50 Tg Cl yr^{-1} and 6 Tg Cl yr^{-1}, respectively. By measurements and modelling, the global Cl source from photolysis (see equation (4.117)) of nitryl chloride ($ClNO_2$) has been estimated to be 8–22 Tg Cl yr^{-1} whereas $ClNO_2$ is produced nocturnally, especially in polluted areas, according to equation (4.114). The global methane sink due to reaction with Cl atoms (see equation (4.349)) in the marine boundary layer could be as large as 19 Tg yr^{-1}; see Chapter 5.3.3.1 for the Cl formation process via bromine autocatalysis.

Besides HCl, volcanic plumes contain small amounts of ClO and OClO. In the air over salt lakes (the Dead Sea, Great Salt Lake in the USA), ClO, IO, OIO, I_2 and BrO were detected, likely released from salt stocks due to "bromine explosion" mechanism (see Chapter 5.3.3.1). Perchlorate (ClO_4^-) was detected on Mars (NASA's Phoenix Lander Mission) in soil samples in a range of 0.5–0.6 % ClO_4^- in an order of one magnitude higher than Cl$^-$. On Earth, natural perchlorate is rare[83] and is found only in the driest places, such as unusually arid deserts and the stratosphere. The highest concentrations – comparable with those on Mars – were found in the Atacama in Chile (0.04–0.57 %) together with nitrate ores. Perchlorate in the Atacama can be definitively attributed to an atmospheric source such as nitrate. Together with ClO_4^-, high concentrations (0.04–0.14 %) of iodate (IO_3^-) also were found in Atacama saltpetre.[84] The most important reservoirs of inorganic iodine are HOI and $IONO_2$, and for bromine are BrONO, HOBr and BrCl in the atmosphere.

As seen from Table 4.27, some volatile organochlorine compounds are emitted both naturally and anthropogenically. More than 200 chlorinated gases have been identified in the air, and more than 1,000 organic chlorine compounds have been iden-

---

83 In small concentrations, perchlorate is found everywhere in nature, in precipitation ($<0.1\,\mu g\,L^{-1}$), in soils ($<10\,\mu g\,kg^{-1}$), in groundwater (0.01 to $>100\,\mu g\,L^{-1}$) and in drinking water as well as food.

84 Caliche is the name for the deposits of natural saltpetre containing minerals in the Atacama Desert of northern Chile and west of the Andes Mountains.

tified in nature. As mentioned above, today, more than 4,000 compounds are known from natural processes. It is now an established fact that natural organohalogens are a normal part of the chlorine cycle in the environment. The group of reactive gases consists of chloromethane or methyl chloride (CH_3Cl), chloroform ($CHCl_3$), phosgene ($COCl_2$), dichloromethane (CH_2Cl_2), chlorinated ethylenes (C_2HCl_3, C_2Cl_4), chlorinated ethanes (CH_4Cl_2, $C_2H_2Cl_4$) from natural and artificial sources (see Table 4.27). The long-lived or unreactive gases are the chlorofluorocarbons (CCl_2F_2, CCl_3F, $C_2Cl_3F_3$, $C_2Cl_2F_4$, C_2ClF_5, CCl_3F) and carbon tetrachloride (CCl_4). These are all of artificial origin and will no longer be produced because of the Montreal Protocol and its amendments (international agreements to phase out global production of compounds that can deplete the stratospheric ozone layer); see also Chapter 5.3.2. Synthesised organic chlorine compounds have been widely used as solvents, cleaning materials, pesticides, pharmaceuticals and plastics. In organic chemical synthesis, chlorinated compounds (mostly via radical attack of elemental chlorine) are used for other synthesis because Cl is easily exchangeable with other functional groups. Many compounds belong to the category of persistent organic pollutants (POPs). Because chloroorganic compounds are lipophilic, they accumulate in the organs of animals and can have effects when they exceed a certain toxic threshold. Pesticides such as DDT (dichloro-diphenyl-trichloroethane), which is the best known, are banned in many countries. It seems that chlorinated (or generally halogenated) organic compounds play very special roles as biomolecules. Halogenated natural products are medically valuable and include antibiotics (chlorotetracycline and vancomycin), antitumour agents (rebeccamycin and calichemycin), and human thyroid hormone (thyroxine). Halogenation is essential to the biological activity and chemical reactivity of such compounds and often generates versatile molecular building blocks for chemists working on synthetic organic molecules.

The organic chlorine in soil was originally suggested to be of anthropogenic origin, resulting from the atmospheric transport and deposition of artificial chlorinated compounds. However, the total atmospheric deposition of organic chlorine in remote areas can only explain a small fraction of the organic chlorine found in soil. Furthermore, it has been shown that soil constituents, which originate from the period before industrialisation, also contain organic chlorine. Very little is known about the biogeochemical cycling (formation, mineralisation, leaching, etc.) of chlorinated organic matter in the soil. For example, the net formation of organic chlorine in spruce forest soil is closely related to the degradation of organic matter. The ecological role of this formation is so far unknown, but recent findings suggest that the amount of organically bound halogens in the soil increases with decreasing pH, and that production seems to be related to lignin degradation, in combination with studies that suggest that the production of organochlorine is a common feature among white-rot fungi. This makes it tempting to suggest a relationship between lignin degradation and the production of organohalogens. Such a relationship may result from an enzymatically catalysed formation of reactive halogen species as outlined below.

It has been enlightened four paradoxes that spring up when some persistent tacit understandings are viewed in the light of recent work as well as earlier findings in other areas. The paradoxes are that it is generally agreed that: (1) chlorinated organic compounds are xenobiotic even though more than 1,000 naturally produced chlorinated compounds have been identified; (2) only a few rather specialised organisms are able to convert chloride to organic chlorine even though it appears the ability among organisms to transform chloride to organic chlorine is more the rule than the exception; (3) all chlorinated organic compounds are persistent and toxic even though the vast majority of naturally produced organic chlorine are neither persistent nor toxic; and (4) chlorine is mainly found in its ionic form in the environment even though organic chlorine is as abundant or even more abundant than chloride in soil.

Considering the important role of chlorine (and other more reactive halogens; but Cl is only industrially used due to its cheap production from electrolysis) in organic synthesis, it is a small step to assume that the evolution of the metabolisms of organisms (especially animals) results in the use of chloride, which is transformed into Cl atoms used in specific organosynthesis reactions and also provides functional molecules. The ubiquitous role of chloride (as dissolved sodium chloride) in animal and human cells and blood plasma is manifold. As just discussed, it provides Cl for organosynthesis and to control the electrolytic properties such as osmosis (a process where water molecules move through a semipermeable membrane from a dilute solution into a more concentrated solution), for nutrient and waste transport, as well as providing electrical gradients (based on conductivity) for information transfer through neurons.

Elemental bromine was discovered[85] in 1826 by the French chemist Antoine-Jérôme Balard (1802–1876) in the residues from the manufacture of sea salt at Montpellier. Eugène Marchand (1816–1895), a pharmacist from Fécamp in Normandy, very clearly expresses in 1850 the origin of iodine and bromine from seawater and their removal by atmospheric waters.

Bromine-containing organic compounds (bromocarbons) are ubiquitous in the oceans, and they are mainly formed by macro- and microalgae. Of these naturally produced substances, bromoform ($CHBr_3$) is the best known since it is the most abundant brominated organic. It is formed in enzymatic processes developed to protect algal cells of reactive oxygen species, such as hydrogen peroxide, formed

---

[85] Justus von Liebig gained unwittingly bromine in 1824 while analysing brines at Bad Salzhausen. Sometimes it is written that Carl Jacob Löwig (1803–1890), German pharmacist in Kreuznach, discovered bromine independently of Balard, however, in his publication (1827), Löwig refers Liebig who soon before (1826) gained bromine form the salts of Kreuznach saline using Balard's method and thus confirmed Balard's discovery. Liebig was very unsatisfied not to discovered Br before Balard, and not to paid more attention on his investigation in 1824 (he first believed the substance to be iodine monochloride ICl).

during photosynthesis. In addition, several other bromocarbons are formed either directly through the same enzymatic pathway or by nucleophilic substitution of bromoform. In the atmosphere, bromocarbons are photochemically degraded to reactive bromine.

Iodine was first discovered in 1811 by Bernard Courtois (1777–1838),[86] a French nitrate maker and chemist while extracting potassium and sodium from seaweed ash. Iodine was first found naturally in air, water and soil by the French pharmacist and botanist Gaspard-Adolphe Chatin (1813–1901) in 1850.[87]

> The most important sources of natural iodine are the oceans, produced by marine organisms. Micro- and macro-algae turn iodine into more volatile species (e. g. CH_3I, CH_2I_2) in seawater, which easily enter the atmosphere. There are also abiotic production routes in seawater producing HOI and I_2. This emitted iodine then undergoes gas- and aerosol- phase chemistry and is mostly returned back to the ocean through the deposition. Some of this iodine is deposited on land, where it is an important bio-nutrient, and vital to global populations, which have both historically and recently suffered too low an iodine content in their diets, leading to thyroid-related medical conditions such as Goitre.

Iodine is found in appreciable concentrations in contaminated soils, with iodine concentrations reported up to 5 ppm, and in anoxic marine basins, where iodine concentrations approach 1 mM. Iodate IO_3^- (+5 oxidation state) and iodide I^- (–1 oxidation state) represent the dominant iodine redox species in the environment. In the ocean, aqueous iodide and iodate are the dominant iodine species, with a total concentration of generally 400–500 nM. Thermodynamically, iodate is the favoured form of iodine (except in very oxygen-depleted waters), and it is the overwhelmingly dominant form below the oceanic mixed layer in oxygenated seawater. The biogeochemical iodine cycle consists of coupled abiotic (purely chemical) and biotic (enzymatic) reactions. In marine environments, for example, IO_3^- is reduced to I^- by iodate-reducing microorganisms. Recent results support a link between nitrification and the oxidation of iodide to iodate. The produced iodide is subsequently volatilized from marine surface waters via transformation to a variety of volatile organic iodine compounds, including methyl iodide (CH_3I), iodomethane (CH_2I_2), iodoethane (C_2H_5I), and iodopropane

---

86 Courtis reported in 1812 his discovery to his friend, the French chemist Nicolas Clément (1779–1841) who made first investigations, published in 1813. Gay-Lussac continued the examinations and gave the substance the name "Iode". Humphry Davy gave it the name "Iodine" in 1814. Although it was assumed that iodine must occur in sea water, the German chemist Christoph Heinrich Pfaff (1773–1852) was the first to proof in 1825 its presence in water of the Baltic Sea. German atmospheric chemist Hans Cauer (1898–1962) first found in 1938 that iodine in rainwater is enhanced against chlorine comparing to its ratio in sea water, due to the emission by seaweed burning.

87 Christian Friedrich Schönbein (1799–1868), the discoverer of ozone, used starch-iodide paper (later simple named Schönbein paper) for detection of O_3 due to the coloration of the paper through precipitation of iodine from potassium iodide. Other researchers developed several iodometric papers beside potassium iodide on starch paper, such as litmus paper and thallium paper.

(C_3H_7I). I^- methylation activity is displayed by algae, phytoplankton, and bacteria. The presence and distribution of iodide in the ocean surface are essentially determined by its biologically mediated interconversion with iodate and the processes of physical mixing and advection.

> Rates and spatial distribution of nitrification in the oceans are influenced by environmental factors such as oxygen level, temperature and pH, all of which are currently changing. Recent modelling shows a global increase in surface iodide with a decrease in nitrification. An increase in oceanic iodide will lead to regional-scale decreases in O_3 concentrations through both greater O_3 deposition to the sea surface and the resulting iodine-initiated catalytic O_3-destroying cycles in the atmosphere.

Other iodine deposits, such as some minerals, natural brines and oil brines, have their origin, without any doubt, from oceanic emissions via atmospheric deposition. Whilst sources of iodine in the atmosphere were previously thought to be almost entirely organic (e. g. CH_3I, CH_2I_2), laboratory work has shown one large source is abiotic sea-surface reactions involving ozone. This means that iodine emissions are linked to tropospheric ozone.

> In recent years atmospheric chemists have detected by combining measurements of iodine and sodium in the upper 130 m of an ice-core in Greenland by modelling and laboratory studies about the chemistry between ozone, iodine and iodide that the human-driven increase in tropospheric ozone has led to an amplification of the natural cycle of oceanic iodine emissions that has consequently decreased the lifetime of ozone in the marine atmosphere, thus closing a negative feedback loop. It was concluded that the total tropospheric ozone burden probably increased by less than 40 % between 1850 and 2005, primarily near the surface, with the majority of the increase occurring between 1950 and 1980. A synchronous increase in oceanic emissions of iodine – the product of ozone-iodide reactions at the air-sea interface – seems to corroborate this timing, as well as the importance of halogen chemistry in late twentieth-century ozone budgets. Moreover, historical global emission estimates of ozone precursors, combined with current model chemical schemes, appear to capture the main features of the tropospheric ozone increase since 1850.

Bromine and iodine critically affect atmospheric ozone at all altitudes, playing a crucial role in the ozone depletion events observed in the polar lower troposphere during early spring. Despite its lower abundance, bromine is up to 45–70 times more efficient than chlorine as a catalyst of stratospheric ozone depletion. Particularly in the Arctic near-surface troposphere, bromine and iodine atoms play a central role in atmospheric reactive halogen chemistry, depleting ozone and elemental mercury, thereby enhancing the deposition of toxic mercury. A few bromine molecules per trillion (ppt) causes the complete destruction of ozone in the lower troposphere during polar spring and about half of the losses associated with the "ozone hole" in the stratosphere. Direct bromine atom measurements have been conducted very recently in the springtime Arctic where levels reached 14 ppt.

There are several chemical reasons that fluorine is practical out of the biological realm. The three richest natural sources of fluorine, the minerals fluorspar (CaF_2), fluorapatite ($Ca_5(PO_4)_3F$) and cryolite (Na_3AlF_6), are practically insoluble in water. Furthermore, the high oxidation potential of fluorine (−3.06 V, much higher than the rest of halogens) makes it impossible to form the corresponding hypohalous intermediates necessary for the known enzymatic halogenation. Thus, life evolved without fluorine, and it has no biological role. Fluorine compounds have many industrial applications.

> Mt. Etna is the largest known point atmospheric source of fluorine, even stronger than today estimated anthropogenic release over whole Europe. Average HF emission rates from Mt. Etna can be estimated at about 75 Gg/a. Mt. Etna releases through open conduit degassing on average about 200 Mg HF each day, but highly variable ranging from about 20 to 1000 Mg/d.

Fluoride enters the atmosphere mainly from anthropogenic sources. Natural sources like volcanic eruptions, rock dust, or the marine environment make only a small contribution to the global atmospheric emission of this compound. The principal anthropogenic sources include aluminium smelters, fertilizer factories, industrial activities such as brick, tile, pottery and cement works, ceramic industries, and glass manufacture. Anthropogenic fluorine emitted as HF into the atmosphere is highly reactive and reacts with many materials (in both vapour phase and aerosols), forming typically non-volatile stable fluorides. Although fluorine-containing chlorofluorocarbons (CFCs) and halons are considered responsible for the destruction of the ozone layer in the polar regions, fluorine by itself does not contribute to ozone depletion. Fluorine atoms released from the photodissociation of fluorine-bearing sources are quickly sequestered into carbonyl compounds and subsequently into the ultimate hydrogen fluoride (hydrofluoric acid HF), which is very stable in the stratosphere, but entering the troposphere, it dissolves mainly in cloud water or reacts with dust particles, forming silicon tetrafluoride (or tetrafluorosilane) SiF_4. Thus, wet and dry deposition are the atmospheric removal processes. The concentration of fluoride in precipitation and surface waters is less than 0.1 mg L^{-1}.

Fluorinated halocarbons and other gases such as perfluorocarbons or sulphur hexafluoride (SF_6) are also extremely potent greenhouse gases for which the current trends must be monitored, and future scenarios of growth must be evaluated.

> Fluorine plays a critical role in the development of modern pharmaceuticals; the reason is that introduction of fluorine usually enhances bio-activity and metabolic stability. Fluorine tends to make drugs more hydrophobic. Generally, only aromatic fluorine is used because aliphatic fluorine can create toxic metabolites. Because the C–F bond is stronger than C–H, it can also be placed at specific sites in a molecule that would otherwise be metabolised to promote a longer duration of the medication. Fluorination of drugs and agrochemicals is bio-mechanistically well-rationalised and, most definitely, very beneficial, providing more potent, selective, life-saving medicines and crop-protection agents. The increase in consumption of fluorine-containing drugs leads to higher-

than-normal concentrations of fluoride in body fluids and tissues. There are clear indications that fluoride overload is responsible for dental and skeleton fluorosis, impaired thyroid, brain and endocrine system function, as well as numerous other acute health problems. Consumption of water with fluoride concentration above 1.5 mg/l results in acute to chronic dental fluorosis. Fluoride damages plants at concentrations about 1,000 times lower than those causing detectable human health effects.

4.7.2 Halogen chemistry

As mentioned, fluorine compounds play only a limited role in our environment. Fluorinated hydrocarbons are widely in use in different products, but because the fluorine atom does not react with ozone, they are not relevant for stratospheric ozone depletion (but other halogens associated with such fluorohydrocarbons). Inorganic emissions concern hydrogen fluoride HF (also called hydrofluoric acid and fluoric acid) and silicon tetrafluoride SiF_4. The rather inert gaseous SiF_4, which is also gained in the reaction of HF with silica SiO_2, i. e., any soil-derived dust particle, is very hygroscopic and would nucleate in the presence of humidity to hexafluorosilicic acid H_2SiF_6, a strong acid, fully dissociated into stable fluorosilicate

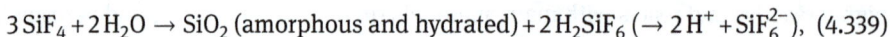

$$3\,SiF_4 + 2\,H_2O \rightarrow SiO_2 \text{ (amorphous and hydrated)} + 2\,H_2SiF_6\ (\rightarrow 2\,H^+ + SiF_6^{2-}), \quad (4.339)$$

the most commonly used agents in drinking water fluoridation. HF is a week acid ($pK = 3.19$); the acidity of halogenic acids decreases in the order HI ($pK = -9.3$) > HBr ($pK = -8.9$) > HCl ($pK = -6.1$) > HF. Among these acids, only HF would be liquid under environmental conditions (boiling point about 20 °C). Thus HI, HBr and HCl are fully dissociated in an aqueous solution, but also HF, which, however, forms ionic pairs $H_3O^+ \cdots F^-$, being "less acidic" than $H_3O^+ \cdot 3\,H_2O$.

Halogens and their compounds play an important role in atmospheric chemistry (see Chapter 5.3.3); in the following, mostly the chemistry of chlorine is treated – that of bromine and iodine is very similar. Hydrochloric acid HCl was known a long time ago to be freed in the atmosphere from the combustion of coal, waste incineration, biomass burning, potash salt production from the mineral carnallite ($KCl \cdot MgCl_2 \cdot 6\,H_2O$) and sea salt produced by the evaporation of seawater.

World-wide the largest source of HCl to the air is its acid displacement from sea salt aerosol (SSA). The depletion of chloride (or in other words, the enrichment of sodium) in SSA relative to seawater composition was observed more than 65 years ago and attributed to surface reactions (acidification) by acids produced from gaseous SO_2 and its oxidation onto sea salt (in analogy react methane sulphonic acid CH_3HSO_4 and nitric acid HNO_3):

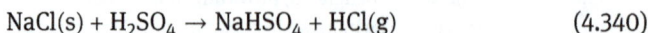

$$NaCl(s) + H_2SO_4 \rightarrow NaHSO_4 + HCl(g) \qquad (4.340)$$

with the overall reaction equation (4.341), where only strong acids with $H_{eff} >$ 10^{-3} mol L^{-1} · atm^{-1} let to HCl degassing:

$$NaCl(p) + H^+(p) \rightarrow Na^+(p) + HCl(g). \tag{4.341}$$

Soluble aerosols exist as either dry particles or solution droplets whose concentration is controlled by the ambient relative humidity. Very high solute concentrations may occur in the droplets, and modelling acid gas exchange in aerosol systems is therefore complicated by the nonideal behaviour of the solutions. Acids that are much less soluble than HCI cause it to degas because chloride concentrations are always very high in a marine aerosol, about 80 % of the total ionic strength. Hydrochloric acid is about three orders of magnitude more volatile than HI and HBr. Because of this and the fact that CI$^-$ is present in seawater at a higher concentration than both Br$^-$ and I$^-$, the loss of H$^+$ to the gas phase by degassing from the sea salt aerosol will be as HC1. The weak acid HF is almost five orders of magnitude less soluble than HCI, and calculations suggest HF will be degassed from the aerosol preferentially to HC1. Displacement of HC1 is likely to be small only for acids with Henry's law constants less than about 10^3 mol^2 kg^{-2} atm^{-1} (Brimblecombe and Clegg 1988).

In the marine remote boundary layer, the Cl depletion from individual particles is highly variable, in the range from sub-µm to above-µm particles. Polluted air masses may result in a complete loss of Cl. It has also been observed HCl degassing in continental PM, mainly believed to be by gaseous HNO$_3$ sticking onto the particles. This acid displacement has been studied in the laboratory to be diffusion-limited in the gas phase. In other words, after uptake of HNO$_3$, chloride is readily displaced as HCl into the gas phase. Because of recrystallisation, nitrate can replace all chloride even in the deeper layers of sea salt crystals.

Consequently, the particulate matter is relatively enriched in sodium (or, in other words, depleted in chloride) quantified by the ratio R_{meas} = [Na]/[Cl], which is in seawater $R_{seawater}$ = 0.86 (molar ratio; mass ratio amounts 0.56). Thus, deviations to higher values indicate Cl loss. However, over continents and in polluted air masses, so-called excess chloride may occur, mainly caused by human activity (coal combustion, waste incineration, salt industries). This excess chloride can be calculated according to equation (4.342); however, there are two preconditions: (a) that there are no sodium sources other than sea salt, and (b) that the local reference value of $R_{seasalt}$ is known. In older literature, instead of $R_{seasalt}$, authors used the seawater bulk value $R_{seawater}$ = 0.86.

$$[Cl^-]_{ex} = [Cl^-]_{seasalt}\left(\frac{R_{seasalt}}{R_{sample}} - 1\right) = [Cl^-]_{sample} - \frac{[Na^+]_{sample}}{R_{seasalt}}. \tag{4.342}$$

The Cl loss x (in %) in a sample can be calculated according to:

$$x = 100\left(1 - \frac{0.86}{R_{seasalt}}\right), \tag{4.343}$$

where $R_{seasalt}$ is the reference value at a given site. Based on experimental data for sea salt entering the Northern European continent, $R_{seasalt} = 1.16$ has been estimated, which corresponds to $x = 26\%$ mean Cl loss. Taking into account that sea salt is already depleted in Cl largely (50–75 %) when entering the continents (and will be further depleted by acid reaction during transport over the continents), the HCl flux from acid sea salt degassing could be globally 200–400 $\mathrm{Tg\,yr}^{-1}$.

> In coal, chlorine is found mostly as a mineral (NaCl) and usually comprises as much as 70–80 % of the total chlorine; further, sorbed chloride makes up 5–10 % of the total chlorine. Organic chlorine, comprising 0.5 % to 25 % of the total chlorine, may be covalent-bonded Cl in coal organic macromolecules but is mainly represented by "semi-organic" Cl, as anion Cl^-, sorbed on the organic coal surface in pores and being surrounded by pore moisture. These are HCl-complexes bonded with bases, such as quaternary nitrogen. The world-average chlorine content of hard and brown coal amounts 340 ppm and 120 ppm, and of their ashes 770 ppm and 2100 ppm, respectively (but having a large range of concentrations). Salty coal (which was often used in England and East Germany) contains 10-times more chloride. In some coals, organic chlorine makes up the main part of the total chlorine.

Thus, coal combustion leads to HCl emission directly and indirectly due to contact of flue ash with sulphuric acid gained from SO_2, according to equation (4.340). While evaporating solutions of $MgCl_2$ (potash industry and marine saline management), HCl is released:

$$MgCl_2 + H_2O \rightleftharpoons Mg(OH)Cl + HCl. \tag{4.344}$$

Students well know from laboratory praxis that condensed fine matter forms when opened near bottles of aqueous ammonia solution and hydrochloric acid, a process also well observed in atmospheric air.

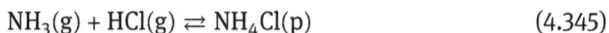

$$NH_3(g) + HCl(g) \rightleftharpoons NH_4Cl(p) \tag{4.345}$$

The majority of the current studies on gaseous hydrochloric (HCl) and hydrobromic acid (HBr) were conducted in marine or coastal areas. Recent measurements in urban Beijing confirmed some of our early findings (see below) that the HCl and HBr concentrations are enhanced along with the increase of atmospheric temperature, UVB, and levels of gaseous HNO_3, and that the gas-aerosol partitioning may also play a dominant role in the elevated daytime HCl and HBr. In summer 2006, we studied the Cl partitioning (Möller and Acker 2007) during a campaign at the research station Melpitz near Leipzig (Germany). The results demonstrate the important role of continental Cl degassing by an acid replacement process very likely only driven by gaseous HNO_3.

For the first time, HNO_3 and HCl in the gas phase, chloride and sodium in the particle phase were measured with high time-resolution and simultaneously together with a number of other atmospheric components (in gas and particulate phase) as well meteorological parameters. On most of the 19 measurement days, the HCl concentration showed a broad maximum around noon/afternoon (on average $0.1\,g\,m^{-3}$) and much lower concentrations during the night ($0.01\,\mu g\,m^{-3}$) with high correlation to HNO_3. The data support that: (a) HNO_3 is responsible for Cl depletion, (b) there is an increase in the Na/Cl ratio due to faster HCl removal during continental air mass transport, and (c) on average, 50 % of total Cl exists as gas-phase HCl. The time series for HCl and HNO_3 are shown in Figure 4.23, exhibiting a pronounced diurnal variation with usually low nocturnal values and high values at daytime reaching a maximum around noon. Remarkably, even the fine structure in temporal variation is identical for HCl and HNO_3, resulting in a high correlation: $[HCl] = 0.001 + 0.05[HNO_3]$; $r_2 = 0.79$ $(n = 800)$. On average, 83 % Cl depletion, calculated by $Cl_{depl} = 1 - R_{sea}/R_{sample}$, have been observed in PM with only small variation (Figure 4.23), mainly given by diurnal cycles.

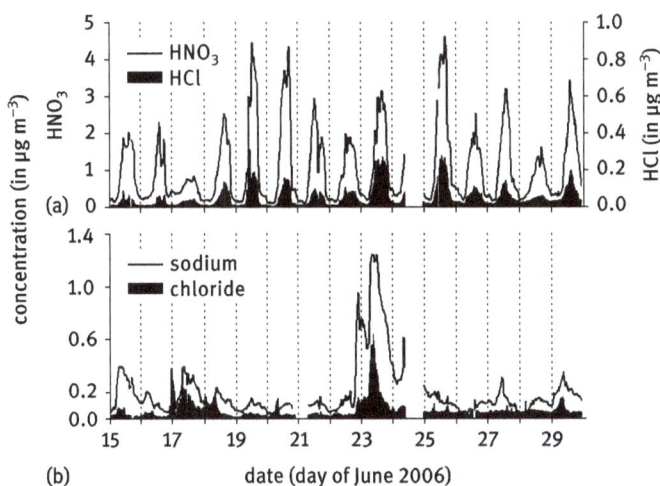

Figure 4.23: Time series of HNO, HCl and particulate sodium (Na^+) and chloride (Cl^-) in Melpitz, June 2006.

Less than 5 % of chlorine in coal could release in the form of gaseous Cl_2 during combustion. Chlorine dioxide ClO_2 is commonly used as a bleaching agent in kraft pulp mills; other sources that are hypochlorous acid $HOCl$ and chlorinated hydrocarbons, originating from volatilization of disinfectants during water and wastewater treatment in cooling towers, tap water and swimming pools. Chlorine radicals can be produced from photodissociation and oxidation of many of the most common chlorinated organic species, but these reactions are generally not fast enough to contribute significantly to the concentrations of chlorine radicals.

The main source of chlorine atoms in the atmosphere is the photolysis of nitryl chloride ($ClNO_2$) and nitroxyl chloride ($ClNO$), which are formed at night by reactions of N_2O_5 and N_2O_3 on particles containing chloride (see equations (4.114)–(4.117)). High $ClNO_2$ concentrations have been observed in coastal regions and ship plumes places where chloride from sea salt meets pollution. On the next day, sunlight breaks $ClNO_2$ into chlorine radicals (Cl) and nitrogen dioxide (NO_2).

$$ClNO_2 + hv \rightarrow Cl + NO_2, \tag{4.346}$$

$$Cl_2 + hv \rightarrow 2\,Cl, \tag{4.347}$$

while photolysis of $ClNO_2$ and Cl_2 occurs in the morning, the reaction of HCl with OH occurs in the afternoon, but it is relatively slow (only HI reacts quickly):

$$HCl + OH \rightarrow Cl + H_2O. \tag{4.348}$$

Chlorine atoms (Cl), even if present in the troposphere in small concentrations, can have a significant effect on tropospheric oxidation and can impact the production of ozone in urban environments. Atomic chlorine is extremely reactive towards volatile organic compounds (VOCs), with rate coefficients that are, with a few exceptions, at least an order of magnitude larger than those of hydroxyl radicals (OH). In addition, Cl reacts rapidly with some compounds, such as alkanes, with which OH is relatively unreactive. Even in the presence of strong photochemical sources, predicted ambient Cl concentrations are small ($<10^6$ atoms cm^{-3}). The spatially and temporally averaged (i. e. hemispheric to global scale over months to years) concentrations in the marine boundary layer were estimated to be on the order of 10^3 atoms cm^{-3}. The main sink of Cl atoms is via the reaction with the greenhouse gas methane:

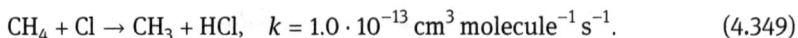

$$CH_4 + Cl \rightarrow CH_3 + HCl, \quad k = 1.0 \cdot 10^{-13}\ cm^3\,molecule^{-1}\,s^{-1}. \tag{4.349}$$

The photolysis of organic halogenated compounds in the troposphere (i. e. at wavelength $> 300\,nm$) is insignificant (stratospheric chlorine chemistry is described in Chapter 5.3.2). The only possible decomposition pathway goes via OH attack, but at relatively slow rates leading to lifetimes of about two years for CH_3Cl. Nevertheless, less than 10 % of the methyl chloride emitted reaches the stratosphere. The OH oxidation pathway ($k = 3.6 \cdot 10^{-14}\ cm^3\,molecule^{-1}\,s^{-1}$) produces Cl atoms:

$$CH_3Cl \xrightarrow[-H_2O]{OH} CH_2Cl \xrightarrow{O_2} O_2CH_2Cl \xrightarrow[-NO_2]{NO} HCHO + Cl. \tag{4.350}$$

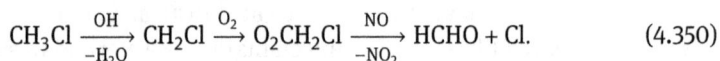

Principally, Cl reacts similarly to OH with all hydrocarbons under H abstraction in the aqueous phase, used for disinfection (see below). Further, the following reactions in the troposphere are not very high atmospheric importance. Whereas the role of halogens in the depletion of stratospheric ozone was studied for several decades, the role of halogens in tropospheric O_3 reactions in marine and polar environments has only been studied in the past 20 years (see Chapter 5.3.3.1). The competitive fate of halogen atoms (F, Cl, Br and I) is their reaction with ozone

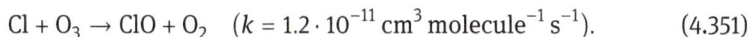

$$Cl + O_3 \rightarrow ClO + O_2 \quad (k = 1.2 \cdot 10^{-11}\ cm^3\ molecule^{-1}\ s^{-1}). \tag{4.351}$$

The radical chlorine monoxide undergoes many reactions, as also BrO and IO. As bi-radical reactions (interhalogen collisions, reactions with OH and HO$_2$ etc.) are only important near the radical formation, further reaction with ozone is the main route. Thus, higher chlorine oxides and finally the stable perchloric acid gained (see also Table 4.29 for oxides and oxoacids of chlorine):

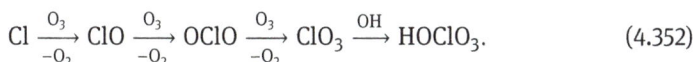

$$Cl \xrightarrow[-O_2]{O_3} ClO \xrightarrow[-O_2]{O_3} OClO \xrightarrow[-O_2]{O_3} ClO_3 \xrightarrow{OH} HOClO_3. \tag{4.352}$$

Perchloric acid exists in the gaseous phase in air and is a stable end product of atmospheric chemistry in the condensed phase due to its resistance to photolysis. Perchlorate is found at some arid places on Earth, dry deposited and accumulated over geological times (Atacama Desert); more likely are heterogeneous pathways via SSA and cloud droplets. The chlorine chemistry is too slow for ozone depletion in the troposphere but bromine and, in particular, iodine – combining with heterogeneous chemistry – lead to ozone depletion in marine environments. In a polluted environment,

Table 4.29: Oxides and oxoacids of chlorine. Note that the acids and their corresponding anhydrides are written inline (2 HClO$_n$ $\rightleftharpoons$ Cl$_2$O$_{n-1}$ + H$_2$O). Cl$_2$O$_2$ (2 ClO), Cl$_2$O$_4$ (2 ClO$_2$) and Cl$_2$O$_6$ (2 ClO$_3$) are "mixed" anhydrides (comparable with N$_2$O$_3$ and N$_2$O$_4$); Cl$_2$O$_5$ is unknown. ClO and ClO$_4$ are short-lived radicals, and ClO$_2$ is also analogue to NO$_2$ radical. From Cl$_2$O, the intermediate radical Cl–O–O is known from photolysis of Cl$_2$O$_2$; the latter is gained via dimerization of ClO. ClO (from Cl + O$_3$), ClO$_3$ (from Cl$_2$O$_4$ photolysis), ClO$_4$ (from Cl$_2$O$_7$ photolysis) and Cl$_2$O$_2$ (from ClO + ClO) are only known in gas phase. Of technical importance are only ClO$_2$ and Cl$_2$O. The (slow) formation of peroxo-hypochlorous acid HOO–Cl have been proposed in the atmosphere and aqueous phase.

oxides		oxoacids	
formula	name	formula	name
Cl$_2$O [Cl–O–Cl]	dichlorine monoxide	HO–Cl	hypochlorous acid
ClO	clorine monoxide		
ClO$_2$ [O–Cl–O]	chlorine dioxide		
ClO$_3$ [Cl(=O)$_3$]	chlorine trioxide		
ClO$_4$ [Cl(=O)$_4$]	chlorine tetroxide		
Cl$_2$O$_2$ [Cl–O–O–Cl]	dichlorine dioxide		
Cl$_2$O$_3$ [Cl–O–Cl(=O)$_2$]	dichlorine trioxide	HClO$_2$ [HO–Cl=O]	chlorous acid
Cl$_2$O$_4$ [Cl–O–Cl(=O)$_3$]	dichlorine tetroxide		
Cl$_2$O$_5$ [ClO$_2$ + ClO$_3$]	(unknown)	HClO$_3$ [HO–Cl(=O)$_2$]	chloric acid
Cl$_2$O$_6$ [(O=)$_2$Cl–O–Cl(=O$_3$)]	dichlorine hexoxide		
Cl$_2$O$_7$ [[(O=)$_3$Cl–O–Cl(=O$_3$)]	dichlorine heptoxide	HClO$_4$ [HOO–Cl(=O)$_2$]	perchloric acid

reactions with NO_2 are important:

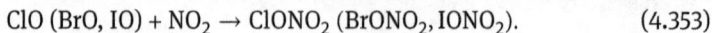

$$ClO \ (BrO, IO) + NO_2 \rightarrow ClONO_2 \ (BrONO_2, IONO_2). \tag{4.353}$$

The oxidation of HCl (or chloride, respectively) in the aqueous phase by H_2O_2 is described in older literature:

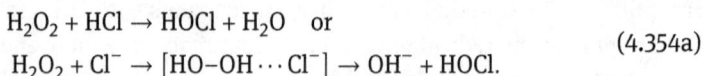

$$H_2O_2 + HCl \rightarrow HOCl + H_2O \quad \text{or}$$
$$H_2O_2 + Cl^- \rightarrow [HO\text{-}OH \cdots Cl^-] \rightarrow OH^- + HOCl. \tag{4.354a}$$

Assuming photocatalytic H_2O_2 formation in liquid films (see equation (4.35)), the above scheme would represent another radical source via the aqueous phase without the consumption of gas phase-produced oxidants and recycling between HCl and atomic Cl. This scheme would also support the findings of increasing chlorination together with (we know: TiO_2 containing) desert dust deposition onto oceans Superoxide O_2^- as a precursor of H_2O_2 has been measured directly in the ocean; thus, the following reaction scheme also would be possible (it corresponds to the net reaction $HO_2 + Cl^-$):

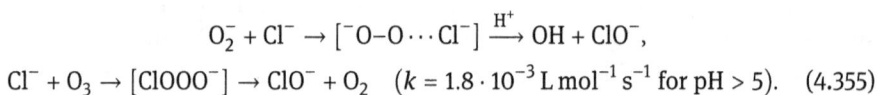

$$O_2^- + Cl^- \rightarrow [^-O\text{-}O \cdots Cl^-] \xrightarrow{H^+} OH + ClO^-,$$
$$Cl^- + O_3 \rightarrow [ClOOO^-] \rightarrow ClO^- + O_2 \quad (k = 1.8 \cdot 10^{-3} \, L \, mol^{-1} \, s^{-1} \text{ for pH} > 5). \tag{4.355}$$

Hypochlorite (ClO^-) was for the first time gained in 1789 by Claude Louis Berthollet (1748–1822) by passing of chlorine Cl_2 through a solution of sodium carbonate and named "Eau de Javel" (from his laboratory at Quai Javel in Paris):

$$Cl_2 + H_2O \rightleftharpoons HCl \ (\rightarrow H^+ + Cl^-) + HOCl \ (\rightleftharpoons H^+ + ClO^-).$$

With respect to environmental chemistry (water treatment and atmospheric chemistry), the reaction with ozone (see equation (4.69)) is most important, here shown for bromide (the first step is slow):

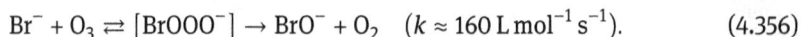

$$Br^- + O_3 \rightleftharpoons [BrOOO^-] \rightarrow BrO^- + O_2 \quad (k \approx 160 \, L \, mol^{-1} \, s^{-1}). \tag{4.356}$$

Hypobromite further oxidizes by O_3 finally to bromate BrO_3^-:

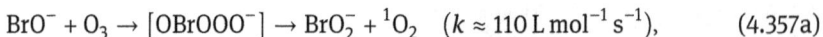

$$BrO^- + O_3 \rightarrow [OBrOOO^-] \rightarrow BrO_2^- + {}^1O_2 \quad (k \approx 110 \, L \, mol^{-1} \, s^{-1}), \tag{4.357a}$$
$$BrO^- + O_3 \rightarrow [OBrOOO^-] \rightarrow Br^- + 2O_2 \quad (k \approx 330 \, L \, mol^{-1} \, s^{-1}). \tag{4.357b}$$

The subsequent oxidation of bromite BrO_2^- to bromate BrO_3^- is fast:

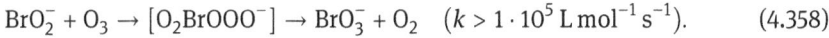

$$BrO_2^- + O_3 \rightarrow [O_2BrOOO^-] \rightarrow BrO_3^- + O_2 \quad (k > 1 \cdot 10^5 \, L\,mol^{-1}\,s^{-1}). \tag{4.358}$$

Bromate[88] has been identified as a carcinogenic compound with additional ecotoxicological impacts. In an environment with halogenide excess such as seawater and sea salt, hypobromous acid HOBr is in equilibrium (in an acid medium) with bromine:

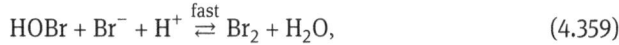

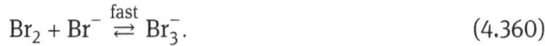

$$HOBr + Br^- + H^+ \overset{fast}{\rightleftarrows} Br_2 + H_2O, \tag{4.359}$$

$$Br_2 + Br^- \overset{fast}{\rightleftarrows} Br_3^-. \tag{4.360}$$

Similarly, the reaction of iodide with ozone occurs (also finally to iodate), but the first step is much faster ($k \approx 1 \cdot 10^9 \, L\,mol^{-1}\,s^{-1}$). Hypobromous acid HOBr and hypoiodous acid HOI are also gained by hydrolysis of Br_2 and I_2, respectively. The acid and its salts are unstable; HOCl is a very weak acid ($pK_{HOCl} = 7.5$; $pK_{HOBr} = 8.8$ and $pK_{HOI} = 10.6$), thereby is almost non protolysed in solution. Because dichlorine monoxide is regarded as the anhydride, the following equilibrium exists:

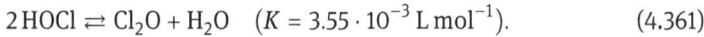

$$2\,HOCl \rightleftarrows Cl_2O + H_2O \quad (K = 3.55 \cdot 10^{-3} \, L\,mol^{-1}). \tag{4.361}$$

While heating the aqueous solution, disproportioning occurs into chloride and chlorate:

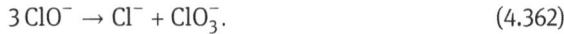

$$3\,ClO^- \rightarrow Cl^- + ClO_3^-. \tag{4.362}$$

HOCl is a strong oxidising agent ($E_{HOCl/Cl^-} = 1.49\,V$ at pH $= 0$) in acid but not in alkaline solutions. Hence, oxidation goes only from HOCl or a protonated form $HOClH^+$ with the direct transfer of Cl (chlorination), similar to the reaction of HOBr and HOI. It disproportionates into chloride (Cl^-) and chlorite (ClO_2^-) and afterwards chlorate (ClO_3^-) is gained. Finally, singlet dioxygen is then partly formed (see equation (4.30)), which can oxidise organic compounds, explaining the biocide application of hypochlorite.

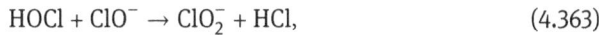

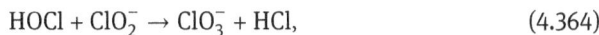

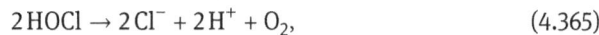

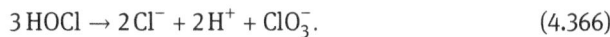

$$HOCl + ClO^- \rightarrow ClO_2^- + HCl, \tag{4.363}$$

$$HOCl + ClO_2^- \rightarrow ClO_3^- + HCl, \tag{4.364}$$

$$2\,HOCl \rightarrow 2\,Cl^- + 2\,H^+ + O_2, \tag{4.365}$$

$$3\,HOCl \rightarrow 2\,Cl^- + 2\,H^+ + ClO_3^-. \tag{4.366}$$

---

88 There are bromate-reducing aquatic bacteria where nitrate is a preferred electron acceptor, also applied for biological drinking water treatment process.

Chlorate (ClO_3^-) is relatively stable;[89] it decays slowly (especially when heating) into chloride and perchlorate (ClO_4^-):

$$4\,ClO_3^- \rightarrow Cl^- + 3\,ClO_4^-. \tag{4.367}$$

The formation reaction equation (4.355) goes in reverse in acid solutions:

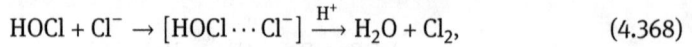

$$HOCl + Cl^- \rightarrow [HOCl \cdots Cl^-] \xrightarrow{H^+} H_2O + Cl_2, \tag{4.368}$$

Furthermore, in acid solution, HOCl decays according to (and explaining the chlorination function):

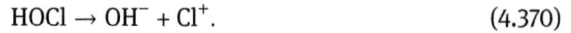

$$HOCl \rightarrow OH + Cl, \tag{4.369}$$
$$HOCl \rightarrow OH^- + Cl^+. \tag{4.370}$$

The chlorine cation (Cl^+) is a short-lived radical, combining with Cl^- in the following way:

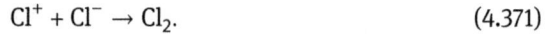

$$Cl^+ + Cl^- \rightarrow Cl_2. \tag{4.371}$$

It has been reported that Cl^+ can replace hydrogen in amino groups (which provides important biomolecules); it is clear that such process is life damaging:

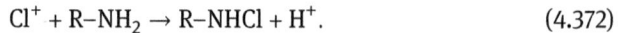

$$Cl^+ + R-NH_2 \rightarrow R-NHCl + H^+. \tag{4.372}$$

Chlorine radicals are also produced in solution from chloride by other radicals (X), such as nitrate (NO_3), sulphate (SO_4^-) and hydroxyl (OH) with $k \approx 10^7 \ldots 10^8 \, cm^3$ molecule$^{-1}\,s^{-1}$:

$$Cl^- + X \rightarrow Cl + X^-. \tag{4.373}$$

The process including OH is also described as equilibrium:

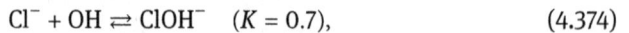

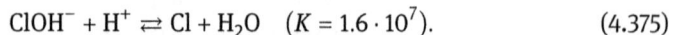

$$Cl^- + OH \rightleftharpoons ClOH^- \quad (K = 0.7), \tag{4.374}$$
$$ClOH^- + H^+ \rightleftharpoons Cl + H_2O \quad (K = 1.6 \cdot 10^7). \tag{4.375}$$

---

89 Aerobic microorganisms capable of using chlorate as an alternative electron acceptor can be found in fresh water, sludge, and soil. The biochemical pathway of these microorganisms appears to involve a chlorate reductase that catalyses the reduction of chlorate to chlorite. Chlorite is subsequently disproportionated into chloride and oxygen by chlorite dismutase. Also, perchlorate is transformed into chloride by perchlorate-reducing bacteria.

Chloride ion and chlorine atom form as adduct the dichlorine radical in an equilibrium. Therefore, all reactions go from Cl_2^-.

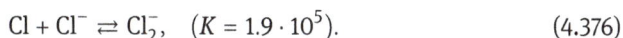

$$Cl + Cl^- \rightleftharpoons Cl_2^-, \quad (K = 1.9 \cdot 10^5). \tag{4.376}$$

Due to its strong oxidising properties, chlorine has long been used as a cleansing and disinfecting agent. Chlorination is an ancient process used to treat drinking water as well as wastewater.

Sodium hypochlorite is a cheap and powerful oxidising agent widely used in household bleaches. Typical household bleaches may contain between 1–5 % NaOCl under alkaline conditions to maintain hypochlorite stability. Hypochlorite exhibits useful actions such as decolourisation of soil/stains, breaking of soil matrix and killing of microorganisms. While it is clear that these actions are due to oxidative properties of hypochlorite, precise mechanisms are still subject to debate. Much of the confusion arises because of the relatively complex aqueous chemistry of chlorine. The precise speciation of hypochlorite solution depends on pH. It may contain ClO^-, $HOCl$, $Cl_2(aq)$, Cl_2O as well as low levels of chlorites and chlorates. All these species are strong oxidants, so establishing the nature of active oxidising species and their mode of attack under any given set of conditions is difficult. $HOCl$, ClO^- and Cl_2O are considered as the main oxidising species. $HOCl$ predominates in solution between pH 4 and 7.5, below which Cl_2 gas evolution becomes dominant. Hypochlorite ion is the main solution species above pH 7.5. Chlorite and chlorate are less effective oxidising agents, and hence their levels should be low in hypochlorite solutions. This is done by ensuring that hypochlorite solutions are maintained at high pH as the rate of their formation is low at high alkalinities. The ClO^- ion oxidises chromophores in coloured materials and is itself reduced to chloride and hydroxide ions.

The presence of peroxides leads to radical termination:

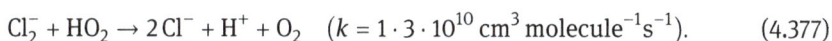

$$Cl_2^- + HO_2 \rightarrow 2\,Cl^- + H^+ + O_2 \quad (k = 1 \cdot 3 \cdot 10^{10}\ \text{cm}^3\,\text{molecule}^{-1}\text{s}^{-1}). \tag{4.377}$$

The dichlorine radical Cl_2^- decays in water and gives the hydroxyl radical, explaining also the oxidative capacity of water chlorination for removal of organic substances:

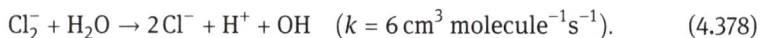

$$Cl_2^- + H_2O \rightarrow 2\,Cl^- + H^+ + OH \quad (k = 6\ \text{cm}^3\,\text{molecule}^{-1}\text{s}^{-1}). \tag{4.378}$$

However, it should be mentioned that chlorination also leads to harmful chlorinated organic substances. Atomic chlorine (Cl) is important in the oxidation of hydrocarbons (equation (4.374)), but can also add to R radicals forming unwanted organic halides (equation (4.375)):

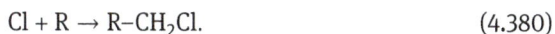

$$Cl + RH \rightarrow HCl + R, \tag{4.379}$$

$$Cl + R \rightarrow R\text{--}CH_2Cl. \tag{4.380}$$

The reactions equation (4.372) and equation (4.380) show the principal mechanism of unwanted by-product formation in water treatment by chlorination. Hence, in the case of pollution of waters with organic compounds (e. g., in swimming pools, wastewaters), it is likely that halogenated organic compounds are gained. Moreover, toxic chloramines are produced. The following species have been found in swimming pools: chloroform ($CHCl_3$), carbon tetrachloride (CCl_4), dichloromethane (CH_2Cl_2), trichloroethylene (CCl_2CHCl), bromoform ($CHBr_3$), and tetrachloroethylene (CCl_2CCl_2).

Hypochlorous acid directly reacts with organic nucleophilic compounds (X = O, N, S) or via the intermediate formation of the chlorine cation Cl^+ ($HOCl \rightleftharpoons OH^- + Cl^+$). Such reactions might also explain the formation of halogenated organic compounds in soils from OM, such as humic material where the volatile fraction (for example, chloroform) emits into the air. All olefinic hydrocarbons and aromatic compounds add Cl and OH from HOCl (see, for example, equation (4.383)).

$$HOCl + RX^- \rightarrow OH^- + RXCl, \tag{4.381}$$

$$Cl^+ + RXH \rightarrow RXCl + H^+, \tag{4.382}$$

$$HOCl + CH_3CH=CHCH_3 \rightarrow CH_3CH(Cl)-CH(OH)CH_3. \tag{4.383}$$

Recently, a reaction between hypochlorite and hydrogen peroxide was proposed in alkaline solution as an electron transfer where the ClO radical can couple together with OH to form peroxohypochlorous acid ClOOH, which decomposes with a yield of singlet dioxygen:

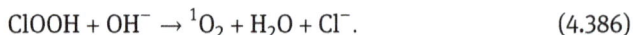

$$ClO^- + H_2O_2 \rightarrow ClO + OH + OH^-, \tag{4.384}$$

$$ClO + OH \rightarrow ClOOH, \tag{4.385}$$

$$ClOOH + OH^- \rightarrow {}^1O_2 + H_2O + Cl^-. \tag{4.386}$$

However, this pathway of chemical singlet dioxygen formation from MO_2 molecules (such as in the reaction between hypochlorite and hydrogen peroxide) where O_2 eliminates with the conservation of its total spin has long been known; oxygen formation in this reaction has been known since 1847.

4.8 Phosphorus

In line with oxygen, nitrogen and sulphur, phosphorus is another important life-essential element (e. g., RNA, DNA, phospholipids, and ATP/ADP) and therefore plays important roles in replication, information transfer, and metabolism. In contrast to the first three listed elements, it does not naturally occur elementally in nature. Moreover, P only (according to textbook knowledge) occurs in derivatives of phosphorous acid in nature. Phosphorus (white) was first produced as an element (but not recognised as an element) by the German alchemist Hennig Brand (c. 1630–1692) in Hamburg in 1669 from the heating of urine distillate remaining with sand. Bernhardt Siegfried Albinus (1697–1770) isolated phosphorus from the charcoal of mustard plants and cress, confirmed by Andreas Sigismund Marggraf (1709–1782) in 1743, and Scheele likely found it from the bone in 1769. But only Lavoisier recognised P as an element.

Terrestrial abiotic phosphorus is essentially ubiquitous on Earth in the oxidised form of phosphate (PO_4^{3-}); it is bound in minerals such as apatite, which are effectively insoluble in water, which severely limits their bio-availability. Phosphate salts that are released from rocks through weathering usually dissolve in soil water and will be absorbed by plants. A phosphate ion enters into its organic combination largely unaltered. From phosphorous acid H_3PO_4, many esters are formed simply by the exchange of H through organic residuals.

Because the quantities of phosphorus in soil are generally small, it is often the limiting factor for plant growth. That is why humans often apply phosphate fertilizers on farmland. Phosphates are also limiting factors for plant growth in marine ecosystems because they are not very water-soluble. Animals absorb phosphates by eating plants or plant-eating animals. Phosphorus cycles through plants and animals much faster than it does through rocks and sediments. When animals and plants die, phosphates return to the soils or oceans again during decay. After that, phosphorus will end up in sediments or rock formations again, remaining there for millions of years. Eventually, phosphorus is released again through weathering, and the cycle starts over.

It was thought that phosphorus could cycle in the atmosphere only as phosphate bound to aerosol particles such as pollen, soil dust and sea spray. Similarly to nitrogen, chemical forms of P have been found in an oxidation state −3 to +5. It is generally accepted that in contrast to O, S, N and C, the phosphorous cycle does not contain redox processes (i. e., remains in the state of phosphate) or volatile compounds (i. e. the atmosphere is excluded from the P cycle).

In continental rainwater, phosphate has been found in such concentrations that it could not be explained only by the scavenging of particulate phosphate, and it was concluded that a terrestrial source of volatile P compounds exists. Atmospheric deposition can supply an important amount of P to ecosystems.

Since the detection of phosphane in the wastewater treatment plants in 1988, more and more investigations have revealed that phosphane is closely related to ecological activities on a global scale.

Actually, as one of the critical elements, phosphorus, is cycling in a specific form, i. e. gaseous phosphane (PH_3), and getting involved actively in ecological interactions. Now phosphorus might be one of the important elements participating in global climate change together with carbon and nitrogen.

Phosphane PH_3 (formerly phosphine, also termed hydrogen phosphide)[90] has been detected in air in the range $pg\,m^{-3}$ to $ng\,m^{-3}$. Close to identified emission sources (paddy fields, water reservoirs and animal slurry), reported concentrations are significantly higher. Besides, PH_3 was found in remote air samples (low $ng\,m^{-3}$ range) in the high troposphere of the north Atlantic. In the lower troposphere, PH_3 is observed at night in the $1\,ng\,m^{-3}$ range, with peaks of $100\,ng\,m^{-3}$ in populated areas. During the day, the concentration is much lower (in the $pg\,m^{-3}$ range). Phosphane has a low water solubility, $H = 8.1 \cdot 10^{-3}\,L\,mol^{-1}\,atm^{-1}$, which is about three times less than Henry's law coefficient for NO_2. PH_3 cannot be photolysed in the troposphere but reacts quickly with OH; $k_{4.387} = 1.4 \cdot 10^{-11}\,cm^3\,molecule^{-1}\,s^{-1}$ (much faster than OH + NH_3):

$$PH_3 + OH \rightarrow PH_2 + H_2O. \tag{4.387}$$

The literature is devoid of studies on atmospheric PH_3 oxidation in detail. In $PH_3 + O_2$ explosions, the radicals PH_2 and PO have been detected, and PO has been found in interstellar clouds. The final product of PH_3 oxidation in the air is a phosphate ion, but nothing is known about the reaction steps. We can only further speculate that the oxidation proceeds in solution and/or interfacial, i. e. in the cloud and aerosol layer, and this might explain why PH_3 is found in the upper troposphere. The solubility does not control the 'washout' but the interfacial chemistry for low-soluble species as we have largely discussed.

PH$_3$ is a very weak base (p$K_b \approx 27$ compared with 4.5 for NH_3) but the phosphonium ion is known in solid salts, for example, with chloride ($PH_3 + HCl\ PH_4Cl$). Because the PH_3 mixing ratio is 10–100 times lower than that of NH_3, it is likely that during inorganic gas-to-particle conversion in the late morning ($SO_2\ H_2SO_4$) ammonium and phosphonium are taken up by the evolving particulate phase. The equilibrium is generally on the left-hand side (PH_3) but, because of acid excess and dynamic processes, it might be shifted to salt formation. Moreover, in the condensed phase, under strong oxidation conditions, OH can stepwise oxidise PH_3/PH_4^+ to phosphate. This process can better explain the observed daytime decrease of PH_3 concentrations and the relatively high PH_3 concentrations in the upper and remote atmosphere.

However, to gain the PO_4^{3-} ion, a very complex P chemistry has to be assumed, including all oxidants of interest (O_2^-, OH, O_3, and H_2O_2). We can speculate that alternate OH attacks abstract H from P and subsequent O_3 attacks add O onto P according to the known chain of known oxoacids:

---

90 P forms many compounds with hydrogen according the general formula P_nH_{n+m} (n in whole numbers, $m = 2, 0, -2, -4, \dots$) but few compounds have been isolated, of which the most important are PH_3 and P_2H_4, which are both volatile and self-igniting at high concentrations.

$$\left[\begin{array}{c}H\\|\\H-P-H\\|\\H\end{array}\right]^{+} \rightarrow \left[\begin{array}{c}H\\|\\O-P-H\\|\\H\end{array}\right]^{0} \rightarrow \left[\begin{array}{c}O\\|\\O-P-H\\|\\H\end{array}\right]^{-} \rightarrow \left[\begin{array}{c}O\\|\\O-P-O\\|\\H\end{array}\right]^{2-} \rightarrow \left[\begin{array}{c}O\\|\\O-P-O\\|\\O\end{array}\right]^{3-}$$

  phosphonium   phosphaneoxide   phosphinate   phosphonate   phosphate

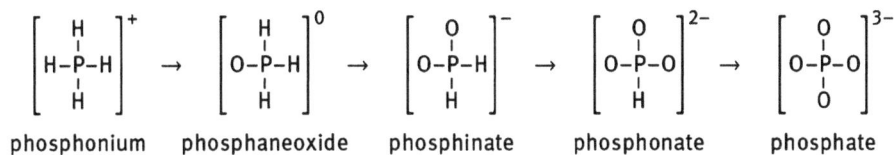

The fine particulate matter might be transported, and PH_3 can be released, whereas particulate acidity decreases and thereby explains why PH_3 seems to be ubiquitous in the air. Moreover, PH_2 probably returns to PH_3 via H abstraction from hydrocarbons in the gas phase:

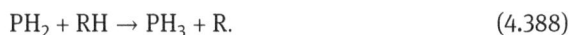

$$PH_2 + RH \rightarrow PH_3 + R. \tag{4.388}$$

In biomass, phosphorus is as important as ATP and ADP for providing energy transfers and in DNA and RNA where phosphate interlinks the nucleotides. In bones and teeth, it is found as calcium phosphates (approximately 85 % of phosphorus in the body). In contrast to nitrogen, plants do not reduce phosphate. However, PH_3 (and likely P_2H_4) exist in air and are emitted from decomposing biomass under anaerobic conditions. Even exotic formation via lightning, shown by simulated lightning in the presence of organic matter, which provides a reducing medium, was suggested, but this is hard to accept under atmospheric conditions. Natural rock and mineral samples release trace amounts of phosphane during dissolution in mineral acid. Strong circumstantial evidence has been gathered on the reduction of phosphate in the rock via mechanochemical or 'tribochemical' weathering at quartz and calcite/marble inclusions. However, this can also be because of traces of phosphides[91] in rocks, which produce PH_3 in acidic hydrolysis. Although phosphorus could be expected to occur naturally as a phosphide, the only phosphide in the earth's crust is found in iron meteorites as the mineral schreibersite $(Fe, Ni)_3P$, in which cobalt and copper might also be found.

When wetted, schreibersite forms hydrous, activated phosphate capable of forming key basic organic molecules, such as glycerol phosphate, nucleosides and phosphocholine, and intermediate phosphorus species, such as hypophosphite $(H_2PO_2^-)$ and phosphite (HPO_4^{2-}). While such intermediate phosphorous species would hinder organic reactions, they may still play an important role in the origin of life by efficiently reacting with solar ultraviolet (UV) radiation and dissolved HS^- to form orthophosphate (PO_4^{3-}). Thus, the schreibersite is one commonly accepted source of phosphate for the terrestrial prebiotic synthesis of essential organic phosphate molecules. A very recent study has shown that cloud-ground lightning strikes reduce phosphate in meteorites and estimated that lightning strikes on early Earth might have formed 10–1,000 kg of phosphide and 100–10,000 kg of phosphite and hypophosphite annually.

---

91 A phosphide is a compound containing the P^{3-} ion or its equivalent.

Many papers have suggested that PH_3 forms by microbial processes in soils and sediments, but no bacteria responsible or any chemical mechanism is known. The question arose 40 years ago when earlier evidence for the evolution of phosphane through the microbial reduction of phosphate in waterlogged soils could not be confirmed. It also shows that phosphane is sorbed by soil constituents and might not escape to the atmosphere if produced in soils. Recently, as reported above, PH_3 has been detected at surprisingly high concentrations in the marine atmosphere. The measurement technique to detect such low PH_3 concentrations became available around 1993; therefore, the absence of evidence of PH_3 before is no longer evidence of PH_3 absence.

> In previous studies, phosphorus usually is thought not to undergo redox reactions, and the removal of phosphorus in water occurs only from adsorption, complexation, and precipitation. However, the reduction of phosphate to phosphane occurs commonly in wetlands and paddy fields. It is well-established that the formation of phosphane in nature follows thermodynamics in terms of the redox potential levels. Anaerobic microorganisms use a sequence of terminal electron acceptors instead of oxygen during their respiration under the anaerobic condition. With decreasing redox potential, they reduce nitrates to nitrogen or N_2O, sulphates to sulphides, and carbonate to methane. Similarly, the production of PH_3 occurs under reducing conditions while it requires more energy and higher reducibility. When redox potential falls below $-300\,mV$, the phosphate may act as an electron receptor and finally get reduced to phosphane.

It is estimated that there are about $4 \cdot 10^6\,kg\,yr^{-1}$ of phosphane released to the atmosphere. At certain temperatures and under lightning conditions, most of the gaseous phosphane will be oxidised to phosphorus oxides, which might be deposited in clouds or rainwater, becoming a major source of phosphorus for the ecosystems that are poor in phosphorus.

> Eutrophication is a serious problem, causing the degradation of aquatic environments, destruction of aquatic ecosystems and restrictions on drinking water supplies. Phosphorus is viewed as the most critical factor for eutrophication in most lake ecosystems attracting numerous studies. Sewage is one of the major sources of phosphorus release into natural waters, and biological processes in wastewater treatment plants contribute to phosphorus removal.

It seems that only yet unknown microbial processes where phosphate is used as an oxygen source under anaerobic conditions (similar to the sulphate reduction) might explain PH_3 formation. This is still an open field of research and needs more specific microbial studies. Organophosphines PR_3, where R = alkyl or aryl (for example, triphenylphosphine oxide $OPPH_3$ and trimethylphosphine oxide $OP(CH_3)_3$) are known to form metal complexes and have been found in fungi. Organophosphines are easily oxidised to the corresponding phosphine oxide OPR_3, which are considered to be the most stable organophosphorous compounds. The identification that organophosphines are responsible for the 'typical smell' when touching metals might give evidence for such compounds in the human body. For centuries *ignis fatuus* was the name given to a phosphorescent light seen over marshy ground and around graveyards at night (ghostly lights). It is now known to be caused by the spontaneous combustion of gases emitted by decomposing organic matter where phosphanes would act as a 'chemical match'.

4.9 Metals and half-metals

So far, we have discussed the chemistry of almost all volatile compounds from non-metals such as hydrogen, oxygen, nitrogen, sulphur, carbon, phosphorus and halogens (noble gases as permanent constituents of the atmosphere play no role in environmental chemistry). Non-metals[92] form a large group of anions, such as hydroxides, nitrates, sulphates, carbonates, phosphates, and halides. Metal hydrides and oxides convert in an aqueous solution into hydroxides. Some non-metal hydrides (H_2O, NH_3, H_2S, HCl CH_4, PH_3) form acids, and in solution, the cation H^+, but NH_3 forms the environmentally important cation NH_4^+, whereas only CH_4 is insoluble. From the half-metals (in the past called metalloids having properties between metals and non-metals) are boron (B) and silicon (Si) of crucial natural importance, as borate (BO_3^-) in seawater (buffer capacity) and silicate (SiO_4^{4-}) in rocks (see also Chapter 5.1.5). Metals provide the cations to form with anions salts and minerals.

4.9.1 General remarks

In seawater, *all* elements are found (see Table 2.5 for some elements). The most abundant metals and half-metals are (concentration in ppm): sodium (10.8), boron (1–4), magnesium (1.3), calcium (0.44), potassium (0.40), lithium (0.17), strontium (0.008), and silicon (0.003). The chemical composition of the oceans is a result of rock dissolution (metals and half-metals) and scavenging of atmospheric constituents (non-metals) in early earth's evolution. In the crust remain significant concentrations of calcium, sodium, magnesium, and potassium (together 12%). Three main elements (besides oxygen) compose the crust: silicon (27.7%), aluminium (8.1%), and iron (5.0%), essential constituents of minerals. Going deeper into the Earth, in the mantle, we see a depletion of aluminium, sodium and potassium, silicon and iron remain, but magnesium increases. Finally, the earth's core consists of iron (89%), nickel (6%), and sulphur (4%). Volcanoes emit with the dust many metals: Cd, Hg, Ni, Pb, Zn, Ca, Sb,

---

92 Belong non-metals are H, C, N, O, P, S, Se, halogens and noble gases; C, P, and Se already have properties of half-metals. Typical half-metals are B, Si, As, Te, but Ge, Sb, Bi and Po are further called belong half-metals. It is further separated between semimetals (typically As, Sb, Bi, C, Sn) and half-metals, but more from physical views in terms of conductivity; note that the definitions are not sharply and transitions are possible. Half-metals are typically amphoteric, i. e. can be dissolved in acid and bases.

As and Cr; for Cd and Hg metals, volcanoes contribute 40–50 % of global emission and for the latter 20–40 %. Based on global emissions estimates, the emission of heavy metals amounts to 2–50 kt yr^{-1}.

> Volcanic eruptions are important sources of elements in the atmosphere, almost globally distributed and deposited. The soils are the interface between atmosphere and crust, therefore enriched in many elements. Consequently, soil dust is an important source of metals (and all elements) to the atmosphere to be distributed over larges areas. Hence, metals and half-metals are ubiquitous in the environment.

Vital elements (life-essential)[93] occurring in many but not all organisms are metals such as Li, Be, Na, Mg, K, Ca, V, Cr, Mn, Fe, Co, Ni, Cu, Zn, Mo, Sb and Sr, non-metals such as O, H, C, N, S, P, Se, halogens (F, Br, Cl, and I) and the half-metals B, Si and As. Those likely without any vital functions are (with increasing ordinal number) Al, Ti, Ga, Ge, Rb, Y, Zr, Nb, Ru, Os, Pd, Ag, Cd, In, Sb, Te, Cs, La, Hf, Ta, W, Re, Pt, Au, Hg, Tl, Bi and all radioactive elements. Some of them are non-toxic (Al, Zr, Ru, Pd, Ag,[94] Re, Pt, and Au), others are low toxic (e. g. Bi and Os) or high toxic (Hg and W). Some of the trace elements, being essential, such as arsenic, selenium, and chromium, are toxic and can even cause cancer. The toxicity of an element often depends on its chemical form (e. g., only Cr(V) is carcinogenic). *Every* element has three possible levels of dietary intake: *deficient, optimum,* and *toxic* in order of increasing dose. For humans, *all* elements, exceeding toxicologically relevant limits (which are, however, very hard to determine and controversially discussed among experts), can represent reproductive dangers and even suspected cancerogenic and teratogenic properties. In the human body, trace elements (Rb, Sr, Al, Ba, B, Li, V and others) are present in very small amounts, ranging from a few grams to a few milligrams, but are *not* required for growth or good health. Examples are rubidium (Rb) and strontium (Sr), whose chemistry is similar to that of the elements immediately above them in the periodic table (potassium and calcium, respectively, which are essential elements).

In the aqueous-phase chemistry of oxygen and sulphur, we have emphasised the importance of redox processes and oxidation states of transition metal ions such as iron, copper and manganese. We have mentioned the role of semiconductor metal oxides (Ti, Zn, Fe, Cu, Sn and others), which have been recognised in photosensitised electron transfers onto important gas molecules such as O_2, O_3, NO, NO_2, CO and CO_2. Elements with an impact on biogeochemical redox processes are (in line with increasing atom number) Se, Fe, Mn, Co, Cu, Cr, Hg, Tc, As, Sb, U and Pu; some of them are

---

93 There is no scientific consensus on which elements are life-essential or not; it is under permanent scientific progress. Some trace elements can replace known elements in their life-essential functions, e. g. tin (Sn). Organisms do not need all elements and some elements are essential only to plants, microorganisms or primitive animals.

94 Only to humans; very toxic for microorganisms.

only anthropogenically released (Hg, Tc, U, Pu). The basic elements in soil dust and rainwater are Na, K, Ca and Mg. Nevertheless, in minor and trace concentrations, all stable elements can be detected, such as (in decreasing concentrations) Fe, Zn, Mn, Cu, Ti, Pb, Ni, Sb, V, Cr, Sb, Hg and Th. They are called crystal elements and are thereby found in dust emissions from coal-fired power plants, soil dust and sea salt. There are 'typical' metals, which can be attributed to different sources (however, it can vary):

- tires and brakes): Fe, Cu, Sb, Zn,
- flue ash (coal- soil dust: Ti, Fe and Ni,
- combustion of fossil fuels (strong enrichment compared to geogenic sources): Sb, Pb, Zn, Cu, Hg, V,
- traffic (especially fired burners): Fe, Ti, Mn, Zn, V, Cu, Cr, Ni.

Tables 4.30 and 4.31 show some metal concentrations in rain and cloud water and atmospheric dust. Rainwater and cloud water concentrations (from different regions in Germany) are very similar. The derived residual from Mt. Brocken (Table 4.30) is not far from the dust concentration found at a Berlin tower (Table 4.31), indicating the large-scale and widely homogeneous distribution of metals. Interesting is the comparison of 'atmospheric' metals with those in seawater; in seawater, Fe, Mn, Cu and Pb are depleted, whereas V and Ni are enriched relative to the metals in the atmosphere. Fe, Mn, Cu and Pb are dominantly anthropogenically emitted by the combustion of fossil fuels.

Table 4.30: Trace metals in the rain (near Frankfurt am Main, Germany) and cloud water (Mt. Brocken, Germany); in g L^{-1} (long-term measurements). Residual is c(aq) · LWC, representing the CCN composition (in ng m^{-3}) at Mt. Brocken. Seawater composition in ppb (ng L^{-1}) for comparison.

metal	seawater (in ng L^{-1})	rainwater	cloud water[a] (in µg L^{-1})	residual (in ng m^{-3})
Fe	3.4	79.6	134	31
Zn	5	13.2	37.2	8.5
Pb	0.03	2.3	11.0	2.5
Mn	0.4	5.4	7.8	1.8
Cu	0.9	3.4	5.7	1.3
Ti	1	2.4	2.3	0.5
V	1.9	0.3	1.8	0.4
Ni	6.6	1.3	1.0	0.2
Sb	0.33	1.0	0.36	0.08
Cr	0.2	0.2	0.3	0.07
Co	0.4	0.1	0.08	0.02

[a]Additionally (not listed here), we found Sr and Ba in the range 2–3 mg L^{-1}, As, Mo, Cd, Sn, and Ba in the range 0.3–0.4 mg L^{-1}, Li, Be, Ga, Ge, Rb, Y, Zr, Nb, Cs, La, Ce, Pr, Nd, Sm, Eu, Gd, Tb, Dy, Ho, Er, Tm, Yb, Lu, Hf, Ti, Bi, Th and U almost <0.1 mg L^{-1}.

Table 4.31: Atmospheric trace metal concentration (in ng m^{-3}) in Berlin and surroundings (Germany); average from one-year (2001–2002) measurements in PM$_{10}$.

metal	street, heavy traffic	rural background	tower, 300 m
Fe	680 (±263)	150 (±130)	100 (±61)
Pb	16.3 (±9.6)	8.0 (±5.5)	8.0 (±4.7)
Ni	1.9 (±1.1)	1.3 (±0.5)	1.7 (±1.5)
Cd	0.35 (±0.20)	0.29 (±0.16)	0.27 (±0.15)
As	1.6 (±2.0)	1.2 (±1.2)	1.3 (±1.2)

Potentially toxic metals are now far below thresholds given by WHO for water, air and food. However, in the early time of industrialisation, without toxicological knowledge and no pollution control, serious damages to health because of pollution events (catastrophes) and long-term impact by high concentrations in the working environment have been reported. Doubtless, the air of settlements and towns was extremely polluted in the past. Heavy metals have been found in Greenland ice cores dating back to the Roman Empire, thus demonstrating that metallurgical operations of immense volume took place in that era.

Many metals form complexes, highly dependent on *pH*, soluble and insoluble minerals, and thereby vary in their availability for life (essential or not) and redox processes. The interested reader is referred to as special literature. Most scientific literature is found on iron, the fourth most abundant element on Earth. In the following chapter, only the main features of some environmentally important metals are presented.

4.9.2 Alkali and alkaline earth like metals: Na, K, Mg, and Ca

From the alkali metals (group 1 of the periodic table of the elements): Li, Na, K, Rb, Cs, and Fr, only sodium (Na) and potassium (K) are distributed in significant amounts in soils, the crust and waters. As mentioned, Li is found in high concentrations in seawater.

Alkali metals form monovalent cations and salts, which are all soluble. Their oxides and hydroxides provide the strongest bases in water. Ca, Mg, Na and K are the main metals in atmospheric samples (rain, clouds, dust), giving together almost 100 % of the metallic composition; therefore, we call all others *trace metals*.

Sodium is deposited from oceans in huge pools as rock salt (halite) and potassium as sylvine (which is an important fertiliser). The main mineral of rare Li is spudomen LiAl[Si$_2$O$_6$]. From the alkaline earth metals (Be, Mg, Ca, Sr, Ba, and Ra) are – together with Na and K – Ca and Mg the main metals (99 %) in atmospheric samples such as rain, clouds and dust (another important cation is ammonium NH$_4^+$). Sr and Ba are found in relatively high concentrations in seawater. The main mineral is calcite $CaCO_3$, forming worldwide large rocky mountains, and furthermore dolomite $MgCa(CO_3)_2$ and magnetite $MgCO_3$. Many silicates contain Mg and Ca (remember that Mg is enriched in the earth's mantle). Gypsum is a soft sulphate mineral composed of $CaSO_4 \, 2 H_2O$; it is also used as building material (it is the final and commercial product of flue-gas desulphurisation).

In contrast to the alkali metals, carbonates of all alkaline earth metals are difficultly soluble; fluorides of Ca, Ba and Sr are insoluble. The hydroxides of Ca and Mg (and other metals of group 2) are relatively difficultly soluble. Much more soluble is $Ca(HCO_3)_2$, which is important for washout processes in carbonate mountains and forming limestone caves:

$$CaCO_3 + H_2O + CO_2 \rightarrow Ca(HCO_3)_2. \tag{4.389}$$

The 'equilibrium' between better-soluble bicarbonate and shell-forming insoluble $CaCO_3$ is essential in seawater for many organisms (*biomineralisation*). Ca and Mg are extremely important life-essential elements for plants and animals (including humans).

4.9.3 Iron: Fe

Iron belongs to the first row of the transition metals together with Ti, V, Cr, Mn, Co and Ni. Iron is mostly bound together with silicates; the dark colours of primary rocks and the red-brown colour of soils is due to Fe compounds, such as haematite Fe_2O_3. Other important minerals are magnetite Fe_3O_4 and pyrite FeS_2. Other metals of this group (Ti, V, Cr, Ni) form minerals together with Fe (and explains why all these metals are found approximately according to their abundance together with Fe): ilmenite $FeTiO_3$, chromite $FeCr_2O_4$, pendlandite $(Ni, Fe)_9S_8$. Manganese is found almost as pyrolusite MnO_2 and manganese nodules on the sea floor.

All above-listed metals occur positive bi- and trivalent: M^{2+} and M^{3+}. The change of the oxidations state in redox processes is likely the most important chemical property of Fe and Mn (and Cu: Cu^+/Cu^{2+}) as presented in Chapter 4.3.3. Iron does not form well-defined oxides FeO, F_3O_4 ($FeOFe_2O_3$) and Fe_2O_3, furthermore hydroxides, the unstable white $Fe(OH)_2$, and the black $Fe(OH)_3$ and $FeO(OH)$, are all insoluble (see Table 3.3). We mentioned in Chapter 3.3.6.4 the rusting of iron and its redox properties, providing electron transfers via the biologically important Fenton reaction, see equation (4.45). The iron complexes undergo oxidative additions and reductive eliminations. Most important are $[Fe^{II}(CN)_6]^{4-}$, $[Fe^{III}(CN)_6]^{3-}$, $[Fe^{II}(C_2O_4)_3]^{4-}$, $[Fe^{II}(H_2O)_6]^{2+}$, and $[Fe^0(CO)_5]$. The iron oxalate complexes, often found in surface waters, photolyse gaining peroxo radicals (equation (4.34)).

> The ferric/ferrous pair (Fe^{3+}/Fe^{2+}) – because of the significant iron concentration compared to other transition metals – is the most important redox couple in natural waters.

Besides its important role in the living organisms' electron transfer process, iron holds and transfers oxygen in hemoglobin (the red pigment in blood) and myoglobin (a protein found in the muscle cells of animals). Other iron porphine proteins (e. g. cytochrome) and iron-sulphur proteins are responsible for electron transfers in complex biochemical processes such as nitrogen and carbon dioxide fixation, photosynthesis, and respiration.

The ion $[Fe^{III}(H_2O)_6]^{3+}$ is only for pH < 0 stable (not found under environmental conditions) and transfers between pH 0 and 2 into yellow-brown $[Fe(H_2O)_5(OH)]^{2+}$ and $[Fe(H_2O)_4(OH)_2]^+$:

$$[Fe^{III}(H_2O)_6]^{3+} \overset{\pm H^+}{\rightleftharpoons} [Fe(H_2O)_5(OH)]^{2+} \overset{\pm H^+}{\rightleftharpoons} [Fe(H_2O)_4(OH)_2]^+. \tag{4.390}$$

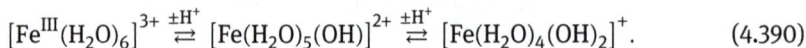

The solutions become colloidal. Shifting the aqueous solution more alkaline, amorphous red-brown $Fe_2O_3 \cdot xH_2O$ deposits. Such processes can be seen in lakes, created from former mining areas, which are very acid at the beginning.

In the early Earth, all metals were in their lower oxidation states, such as Fe^{2+}. With the evolution of oxygen (Chapter 5.1.4), FeO oxidises to Fe_2O_3 (depending on the modification, the colour is between red-brown and black). The relationship between oxidant availability and iron mobility during weathering is the primary tool for estimating oxygen levels. In most soils, H_2CO_3 is the most important weathering acid. If oxygen is supplied much more rapidly to a weathering horizon than carbon dioxide is consumed, then essentially Fe^{2+} in the weathering horizon will be oxidised to Fe^{3+} and retained as a component of ferric oxides and oxohydroxides.

4.9.4 Mercury: Hg

Mercury (Hg) is one of the few elements (Be and Cd are the others) that is *not essential* for organisms. It is the only liquid metal at room temperature.[95] In nature, it occurs mostly (it is found rarely elementally as small droplets) as sulphide, such as cinnabar HgS. Hg is very toxic, but the toxicity depends on the chemical form of bonding; methylmercury (and other organic Hg compounds) is the most toxic form of Hg. Methylmercury[96] is formed from inorganic mercury by the action of anaerobic organisms that live in aquatic systems, bioaccumulated and transferred via the food chain.

---

95 Gallium (Ga) and caesium (Cs) melt at 30 °C and 28.5 °C, respectively.

96 Methylmercury) is an ion, CH_3Hg^+, i. e., it exists as chloride (mostly), nitrate, and so on. (e. g., CH_3HgCl). It forms crystalline solids and is not volatile (in many publications methylmercury – often denoted as MeHg – is described wrongly as a molecular compound). Dimethylmercury (CH_3HgCH_3) is a strongly toxic liquid. Both compounds are enriched at seawater surfaces and easily photolysed

The only natural sources of Hg are volcanoes, fumaroles and hot springs. Due to the global geochemical distribution over the earth's age, Hg became a constituent of soils (soil dust emission) and as (main source nowadays but with low specific emission) oceanic emission of volatile organic Hg (Table 4.32). The surface seawater contains globally around $6\,ng\,L^{-1}$ Hg. Today, natural and anthropogenic sources are hard to separate because artificial activities have led to global Hg redistribution. There is no doubt that coal combustion is the largest anthropogenic source, and power plants (or burners) without gas treatments are the largest single emitters of Hg. The Hg content of coal amounts to 0.02–1.0 ppm, that of flue ash 0.62 ppm and the emission factor amounts to 0.04–$0.3\,g\,Hg\,t^{-1}$ coal (all global averages).[97] In coal, Hg is mineralogically bound as Hg(II), likely as HgS. During combustion, Hg is quantitatively released, first as a gaseous element. However, during the power plant process (including gas treatment), it is converted into particulate and bonded Hg(II).

Table 4.32: Global source fluxes of mercury (in Mt Hg yr^{-1}).

source	emission
natural sources:	
ocean	2.7
soils	1.0
volcanic activities	0.09
subtotal	3.8
anthropogenic sources:	
combustion processes	1.3
biomass burning	0.7
production of gold	0.4
non-iron metallurgy	0.3
others	0.3
subtotal	3.0

Mercury exists in four different forms in the environment:
- elementally (almost gaseous): Hg(g),
- as a particulate (soluble and insoluble): Hg(p),[98]

---

to CH_3Cl and Hg(0), where Hg escapes into air due to its large vapour pressure (aqueous-phase Hg concentration is larger than in air).

97 For West German power stations, fired with hard coal, the Hg content amounts 0.04 (0.18–0.48) ppm and in flue ash 0.23 (0.4–1.9) ppm.

98 All forms of bonding are possible but most likely is Hg(II).

- oxidised (mostly as chloride): $Hg(I)^{99}$ and Hg(II),
- organic (mostly as CH_3Hg^+): Hg(org).

In power plant chemistry, the following reactions are taken into account (g – gaseous, s – solid):

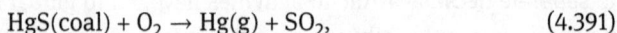

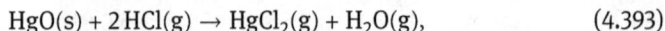

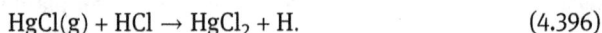

$$HgS(coal) + O_2 \rightarrow Hg(g) + SO_2, \tag{4.391}$$
$$2\,Hg(0) + O_2 \rightleftharpoons 2\,HgO(s), \tag{4.392}$$
$$HgO(s) + 2\,HCl(g) \rightarrow HgCl_2(g) + H_2O(g), \tag{4.393}$$
$$Hg(g) + Cl + M \rightleftharpoons HgCl(g) + M, \tag{4.394}$$
$$Cl_2 + M \rightleftharpoons 2\,Cl + M, \tag{4.395}$$
$$HgCl(g) + HCl \rightarrow HgCl_2 + H. \tag{4.396}$$

HgO is only decomposed into the elements >500 °C. Mercury chloride can exist at high temperatures. Hg(g) passes the plant and emits quantitatively; hence, the more gaseous $HgCl_2$/HgCl is gained the more efficiently it is removed by scrubbing. We see that the chloride content of the coal is important for Hg speciation. Particulate HgO forms after decreasing the flue gas temperature below 500 °C. Together with $HgCl_2$ it adsorbs on flue ash particles. Globally, we might assume that the Hg emissions split about in 50 % Hg(g) and in 50 % Hg(II).

Hg(g) is relatively rapidly (residence time of one month) oxidised in the atmosphere to Hg(II) by OH ($k = 9 \cdot 10^{14}$ cm^{-3} molecules^{-1} s^{-1}) and O_3 ($k = 3 \cdot 10^{20}$ cm^{-3} molecules^{-1} s^{-1}); the mechanism is not known, but HgO is assumed as a product:

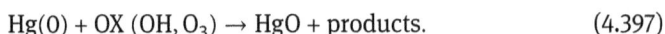

$$Hg(0) + OX\,(OH, O_3) \rightarrow HgO + products. \tag{4.397}$$

The rate of oxidation of elemental mercury is fundamental to atmospheric mercury chemistry because the oxidised mercury compounds (such as HgO and $HgCl_2$) produced are more soluble (and so are readily scavenged by clouds), less volatile (and therefore more rapidly scavenged by particulates), and have a higher deposition velocity. HgO is very soluble, would be scavenged by wetted aerosol particles, clouds and rain, and dissociates to Hg^{2+}. Under most atmospheric conditions, chloride concentrations in the aqueous phase are sufficiently high to drive recomplexation to $HgCl_2$. Oxidised mercury can also be reduced to elemental mercury in atmospheric droplets, thus limiting the overall rate of oxidation and deposition:

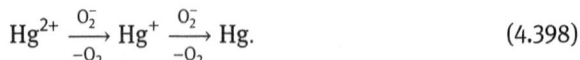

$$Hg^{2+} \xrightarrow[-O_2]{O_2^-} Hg^+ \xrightarrow[-O_2]{O_2^-} Hg. \tag{4.398}$$

---

99 Monovalent mercury exists always as two atoms: Hg_2X_2 (dimer Hg_2^{2+}). In contrast to $HgCl_2$ (sublimate), Hg_2Cl_2), (calomel) is difficultly soluble (by contrast, $HgNO_3$ is very soluble). Sublimate finds wide applications in medicine, a nice example that only the amount makes a substance poisonous.

However, the intermediate Hg^+ can also be quickly back-oxidised. Mercury dichloride has a large Henry's law coefficient ($H = 1.4 \cdot 10^6$ M atm^{-1}) and escapes from droplets. In air, it is quickly photolysed to Hg and chlorine. Hence, a dynamic equilibrium Hg $\rightleftarrows$ Hg(II) establishes; about 90 % of total Hg is as Hg(g) in the lower troposphere. Only in the upper troposphere (where there are no or only insignificant clouds) Hg is more in the oxidised form.

4.9.5 Cadmium: Cd

In the so-called zinc group (subgroup 2 or group 12 of the periodic table of the elements) are counted zinc (Zn), cadmium (Cd) and mercury (Hg). In contrast to Hg and Cd, Zn is essential for all living organisms – biologically most important after Fe. From Table 4.30, we see that Zn is after Fe most abundant in the atmosphere, and it is a common element in the earth's crust, soils, and waters. Global cycling is due to soil dust, volcanic emissions, and anthropogenic particulate emissions (mining, metallurgic operations, use of commercial products containing zinc[100]).

Cadmium exerts toxic effects on the kidney, the skeletal system and the respiratory system and is classified as a human carcinogen.

> Itai-itai disease was the name given to the mass cadmium poisoning of Toyama Prefecture, Japan, starting around 1912. Mining companies in the mountains released the cadmium into rivers. The causes of the poisoning were not well understood and, up to 1946, it was thought to be simply a regional disease or a type of bacterial infection. Medical tests started in the 1940s and 1950s, searching for the cause of the disease. In 1968, the Ministry of Health and Welfare issued a statement about the symptoms of *itai-itai* disease caused by cadmium poisoning.

Volcanic activity has been placed as high as 820 Mt per year. Cadmium emissions either may be considered as arising from point sources such as large manufacturing or production facilities or from diffuse sources such as may occur from the use and disposal of products by many consumers over large areas. Combustion of coal and oil are major sources of atmospheric cadmium pollution. Emissions from point sources have been stringently regulated since the 1970s, and cadmium emissions from point sources decreased dramatically during that period.

The vast majority of cadmium emissions, approximately 80 % to 90 %, are partitioned initially to soils. In soils, cadmium is largely bound to the non-exchangeable fraction, for example, on clays, manganese and iron oxides. For this reason, its mobility and transfer into the animal and human food chain are limited.

---

100 When collecting rainwater near a tower or metallic construction that is galvanised, you will always detect increased Zn concentration in solution!

4.9.6 Lead: Pb

Lead is one out of four metals (among Cd, Hg, and As) that have the most damaging effects on human health. Lead as an element was already known to ancient civilised peoples 3,000 B. C. and used in many everyday objects. Therefore, lead poisoning is as old as its use. The downfall of Romans is often attributed to their use of Pb as tableware and water pipes; they produced each year around 60,000 t. However, it is a chronic disease due to the small uptake of Pb by organisms when given high doses. Lead pipes were in use still after World War II.

Native lead is rare in nature. Currently, lead is usually found in ores with zinc, silver and copper, and it is extracted together with these metals; the main lead mineral is galena (PbS). Native lead is still used in daily life and industry. However, the environmental problems arose from tetraethyl lead ($Pb(C_2H_5)_4$) in gasoline to achieve higher octane ratings. While burning gasoline in the engine, Pb is released and combines rapidly with oxygen to form PbO and PbO_2. These particles provide surfaces for the terminated radical chain reaction. To avoid Pb deposits in the engine, additional gasoline organochalogens have been added. Therefore, volatile $PbCl_2$ and $PbBr_2$ are released into the air, which soon deposit on particles (in that time mostly sulphate); $PbSO_4$ is insoluble in water.

Due to the application of lead in gasoline (first used in 1923), an unnatural lead cycle has existed. The phase-out of Pb in gasoline ultimately became dependent on two factors: first, the ability of refiners to produce gasoline with a higher octane rating, and second, widespread use of catalytic converters, which are incompatible with the use of lead.

4.9.7 Arsenic: As

Half-metal arsenic is essential for humans and is found in all tissues up to 0.008 %. Biologically it acts as an inhibitor of free SH groups of certain enzymes. Elemental As and insoluble compounds (such as sulphides) are not toxic but As^{3+} insoluble substances such as As_2O_3 and AsH_3 are. As with the other volatile pollutants such as Cd, Pb and Hg, As is emitted from coal combustion.

Arsenic is most commonly found in groundwaters in many regions of the world; it is of natural origin and is released from the sediment into the groundwater. The WHO standard limit is 10 ppb, but studies suggest that significant increases in cancer mortality appear only at levels above 150 ppb. Generally, the main forms of As under oxic conditions are H_3AsO_4 (pH 2), $H_2AsO_4^-$ (pH 2–7), $HAsO_4^{2-}$ (pH 7–11), and AsO_4^{3-} (pH 11), respectively. Under reducing conditions, H_3AsO_4 is predominant at pH 2–9.

4.9.8 Silicon (Si) and aluminium (Al)

Both elements are the main constituents (besides oxygen) of all rocks (alumosilicates) in terms of SiO_2 and Al_2O_3. Silicon is life-essential, but aluminium is not. Diatoms (a major group of unicellular algae, which are among the most common types of phytoplankton) are enclosed within a cell wall made of silica (hydrated SiO_2). The cycle of Si weathering will be described in Chapter 5.1.5. Both elements are widely used by humans and are non-toxic. However, from long-term exposure (10 years or more) to relatively low concentrations of silica dust (SiO_2) and usually appearing 10–30 years after first exposure, chronic *silicosis*, an occupational lung disease (pneumoconiosis), occurs. Pulmonary complications of silicosis include chronic bronchitis and airflow limitation, the same symptoms as caused by smoking and asbestosis. This effect is not toxicological but 'simply' deposition of material in the lung, which cannot be removed by natural self-cleansing mechanisms. But it seems that freshly mined crystalline silica provides breaking of Si–O bonds and subsequent formation of superoxide and hydroxyl radicals:

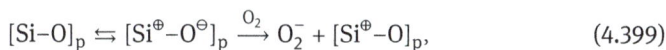

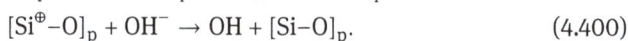

$$[Si-O]_p \leftrightarrows [Si^{\oplus}-O^{\ominus}]_p \xrightarrow{O_2} O_2^- + [Si^{\oplus}-O]_p, \tag{4.399}$$
$$[Si^{\oplus}-O]_p + OH^- \rightarrow OH + [Si-O]_p. \tag{4.400}$$

Long-time exposure in occupational environments to certain PM such as soot, coal, asbestos and minerals is well-known to cause serious lung diseases. Whereas silicosis was known since the nineteenth century, black lung disease was not well described until the 1950s. It is also known as coal workers' pneumoconiosis (CWP), which comes from inhaling coal mine dust that is usually less severe than silicosis. The other disease, silicosis, is caused by inhaling crystalline silica dust from crushed rocks. Black lung and silicosis often appear together because coal seams are found between rock layers that contain silica.

It is not quartz-sand such as in the Sahara (no Tuareg has been known to get silicosis even breathing huge amounts of sand corns over her life), almost spherically and polished particles. While mining, mineral dust particles with crystalline surfaces and broken, sharp edges are formed, which are fixed and accumulated on the lung tissue. Cigarette tar and asbestos fibres deposit in the same way, accumulating over decades before the lung function becomes limited. Moreover, inflammations are caused by additional adsorption of toxic substances, and finally, lung cancer occurs. In nature, free particles having such structures are not known. Therefore, the evolution created only very effective strategies for protection against natural soil dust because all animals and humans live on sandy surfaces.

When analysing atmospheric particulate matter (see Tables 2.13, 2.16 and 4.30), normally only the water-soluble fraction is determined, and in the case of metals, an acidic fusion is necessary (simply by adding nitric acid and getting pH around 1). The residual of the total sum of analysed compounds to the total weighted PM are insoluble silicates contributing to about 1/3 to total PM, that is 5–10 g m^{-3}. SiO_2 can only under fusion with hydrofluoric acid (in special devices under increased pressure and temperature) transfer into an aqueous solution.

Aluminium is not an essential element for either plants or animals but common in all soils and waters. As Al is mostly in insoluble compounds, available excess Al^{3+} can occur due to acidification. It can reduce the availability of phosphorus and sulphur for plants due to the formation of insoluble phosphates ($AlPO_4$) and sulphates ($Al_2(SO_4)_3$). The amount of soluble Al increases dramatically in nearly all soils as the soil pH drops below pH 5.0. In soils, Al occurs as hydroxides $Al(OH)_3$ and $AlO(OH)$, which are (in contrast to acidic silicates) amphoteric. Besides, aluminates are found, several hydroxyaluminates such as $[Al_n(OH)_m]$ and oxoaluminates $[AlO_m]$. From hydroxide, free Al^{3+} forms under acidic conditions (equation (4.401)) and under alkaline conditions aluminates (equation (4.402)):

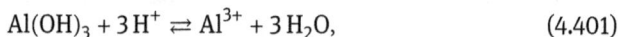

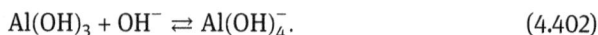

$$Al(OH)_3 + 3\,H^+ \rightleftharpoons Al^{3+} + 3\,H_2O, \tag{4.401}$$

$$Al(OH)_3 + OH^- \rightleftharpoons Al(OH)_4^-. \tag{4.402}$$

The amphoteric character of Al is seen by:

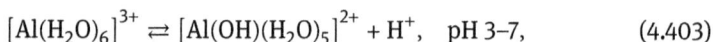

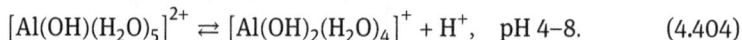

$$[Al(H_2O)_6]^{3+} \rightleftharpoons [Al(OH)(H_2O)_5]^{2+} + H^+, \quad pH\ 3\text{–}7, \tag{4.403}$$

$$[Al(OH)(H_2O)_5]^{2+} \rightleftharpoons [Al(OH)_2(H_2O)_4]^+ + H^+, \quad pH\ 4\text{–}8. \tag{4.404}$$

5 Chemical processes in the environment

At the beginning of this book, we stated that *environmental chemistry* is all chemistry outside the laboratory and industrial reactors. Needless to say, this chemistry was changing over the entire evolution of our Earth. Our currently *Changing World*, is often referred to as the Anthropocene Epoch, is a unit of geologic time used to describe the most recent period in earth's history when the human activity started to have a significant impact on the planet's climate and ecosystem. Officially, the current epoch is called the Holocene, which began 11,700 years ago after the last major ice age.

> To those scientists who do think the Anthropocene describes a new geological time period, the next question is, when did it begin, which also has been widely debated. A popular theory is that it began at the start of the Industrial Revolution of the 1800s when human activity had a great impact on carbon and methane in the earth's atmosphere. In 2016, the Anthropocene Working Group agreed that the Anthropocene is different from the Holocene, and it began in the year 1950 when the Great Acceleration, a dramatic increase in a human activity affecting the planet, took off.

There are two criteria that characterize the *Great Acceleration*, first the characteristic time of changing environmental parameters (such as temperature, glacier removal, ocean acidification, severe weather events and others) is much shorter than in previous geological epochs, and second, humans' impacts became either a comparable or even larger quantity as natural processes (such as global emissions) or on much smaller time-scale (such as land-use change). Fortunately, emissions of some pollutants have decreased and have been stagnant for some time. Conversely, some decreasing emissions result again in changing environmental chemistry. Furthermore, past increasing emissions led to changing natural emissions; for example, increasing marine iodine emissions due to increasing human-induced tropospheric ozone production. The complex (in other terms, non-linear), chemical, environmental system results in positive and negative feedbacks and loops.

> For example, tropospheric ozone shows large increases until the late 1980s in Europe. Since then, ozone values have declined and have now stabilized on a medium level. Recent years might indicate a slight increase. This evolution of tropospheric ozone is not understood in detail. On the other hand, climate change increases stomatal conductance thanks to the positive effects of higher air temperature and solar radiation on stomata opening. Even if the mean ozone concentrations decreased, higher phytotoxic ozone dose levels were observed over time, leading to higher ozone risk to European forests.

It is obvious that understanding *pollution chemistry* (often to what environmental chemistry is reduced) needs the knowledge of a short history of the earth's chemical evolution.

https://doi.org/10.1515/9783110735178-005

> i Nature comprises vast space within the atmosphere and hydrosphere and plenty of time for chemical evolution; hence low conversion rates over large volumes and times also provide huge global turnover (remember that the absolute chemical flux rate is defined as an amount per volume and time). Efficiency and high yield play no role in the biosphere, but speciation and the process of approaching a steady state within global cycling processes are based on a large diversity of species.

Changing these natural system properties (for example, human influences due to acidification and oxidative stress or natural catastrophic events) will shift or even interrupt naturally evolved biogeochemical cycles. In recent decades, humans have become a very important force in the earth's system, demonstrating that emissions and land-use change cause many of our environmental issues. These emissions are responsible for the major global reorganisations of biogeochemical cycles. Nevertheless, humans also do have all the facilities to turn the 'chemical revolution' into a sustainable chemical evolution. Let us define a *sustainable society* as one that balances the environment, other life forms and human interactions over an indefinite time.

> i The basic principle of global *sustainable chemistry*, however, is to transfer matter for energetic and material use only within global cycles. This provides no changing reservoir concentrations above a critical level, which is a quantitative estimate of an exposure to one or more pollutants below which significant harmful effects on specified sensitive elements of the environment do not occur according to present knowledge.

With humans as part of nature and the evolution of an artificial changes to the earth's systems, we also have to accept that we are unable to revert the present system back to a preindustrial or even prehuman state because this means disestablishing humans. The key question is which parameters of the environment allow the existence of humans under which specific conditions. The chemical composition of air is now contributed to by both natural and artificial sources. Nevertheless, major regional and global environmental issues, such as acid rain, stratospheric ozone depletion, pollution by POPs, and tropospheric ozone pollution, resulting in adverse effects on human health, plant growth and ecosystem diversity, have been identified and controlled to different extents by various measures in the last few decades. With respect to atmospheric pollution, the last unsolved issues (remaining pollutants) are greenhouse gases (GHG), namely CO_2, which contributes to about 70 % of anthropogenic global warming (other important gases such as CH_4 and N_2O are contributing about 25 % to warming; these gases are mainly associated with agricultural activities). Moreover, it seems that 'hot-spots' of tropospheric ozone are past, but the mean atmospheric concentrations

With the growth of 'megacities', local pollution will have a renaissance, and this will inevitably contribute to regional and subsequently global pollution by large plumes, such as 'brown clouds'. Thus, it is important to find answers to the following questions:

(a) What is the ratio of natural to artificial emissions?
(b) What are the concentration variations on different timescales?
(c) What are the true trends of species of artificial origin?
(d) What are the concentration thresholds for the effects we cannot tolerate?

The chemical composition of air has been changing since the settlement of humans. In addition to the scale problem (from local to global), we have to regard the timescale. Natural climate variations (e. g. due to ice ages) have a minimum timescale of 10,000 years. The artificial changes in our atmosphere over the last 2,000 years were relatively small before the 1850s. In the past 150 years (but almost all after 1950), however, the chemical composition has changed drastically. For many atmospheric compounds, anthropogenic emissions have grown to the same or even larger order of magnitude than natural ones. Because of the enormous population density, the need (or consumption) of materials and energy has drastically forced the earth's system.

The timescale of the adaptation and restoration of natural systems is much larger than the timescale of artificial stresses (or changes) to the environment. We should not forget that 'nature' could not assess its condition. In other words, the biosphere will accept all chemical and physical conditions, even worse (catastrophic) ones. Only humans possess the facility to evaluate the situation, accepting it or not, and concluding to make it sustainable.

Looking at the chemical evolution on Earth over geological periods, resulting in our present natural environment, will give us two signs. First, changing the natural environment by humans is much faster than natural adaptation and reorganisation. Second, the exploitation of natural resources is at such a level that it will no longer be possible to continue further business-as-usual-economy for the survival of mankind.

5.1 Chemical evolution

The term evolution[101] was used first in the field of biology at the end of the nineteenth century. In biology, evolution is simply the genetic change in populations of organisms over successive generations. Evolution is widely understood as a process that results in greater quality or complexity (a process in which something passes by degrees to a different stage, especially a more advanced or mature stage). However, depending on the situation, the complexity of organisms can increase, decrease, or stay the same, and all three of these trends have been observed in biological evolution. Nowadays, the word has a number of different meanings in different fields. *Geological evolution* is the scientific study of the Earth, including its composition, structure, physical prop-

---

101 From Greek ἐξελίγμός and ἐξελίσσω (Latin *evolutio* and *evolvere*), to evolve (develop, generate, process, originate, educe).

erties, and history; in other terms: the Earth change over time or the process of how the Earth has changed over time. The term *chemical evolution* is not well defined and is used in different senses.

Chemical evolution is not simply the change and transformation of chemical elements, molecules and compounds as is often asserted – that is the nature of chemistry itself. It is essentially the process by which increasingly complex elements, molecules and compounds develop from the simpler chemical elements that were created in the Big Bang. The chemical history of the universe began with the generation of simple chemicals in the Big Bang. Depending on the size and density of the star, the fusion reactions can end with the formation of carbon, or they can continue to form all the elements up to iron.

The origin of life is a necessary precursor for biological evolution, but understanding that evolution occurred once organisms appeared and investigating how this happens does not depend on understanding exactly how life began. The current scientific consensus is that the complex biochemistry that makes up life came from simpler chemical reactions, but it is unclear how this occurred. Not much is certain about the earliest developments in life, the structure of the first living things, or the identity and nature of any last universal common ancestor or ancestral gene pool. Consequently, there is no scientific consensus on how life began, but proposals include self-replicating molecules such as RNA and the assembly of simple cells. Astronomers have recently discovered the existence of complex organic molecules in space. Small organic molecules were found to have evolved into complex aromatic molecules over a period of several thousand years. Chemical evolution is an exciting topic of study because it yields insight into the processes that led to the generation of the chemical materials essential for the development of life. If the chemical evolution of organic molecules is a universal process, life is unlikely to be a uniquely terrestrial phenomenon and is instead likely to be found wherever the essential chemical ingredients occur.

5.1.1 Origin of elements and molecules

Our galaxy is probably 13.8 ± 0.06 Gyr old and was formed by the hot Big Bang, assuming that the whole mass of the galaxy was concentrated in a primordial core. Based on the principles of physics, it is assumed that density and temperature were about 10^{94} g cm^{-3} and 10^{32} K, respectively. The initial products of the Big Bang were neutrons which, when released from dense confinement (quarks), began to decay into protons and electrons: $n_0 = e^- + p^+$. As the half-life for this reaction is 12.8 minutes, we can assume that soon after the Big Bang, half of all the matter in the universe was protons and half electrons. Temperatures and pressures were still high and nuclear reactions possibly led to the production of helium via the interaction of neutrons and protons (remember that the proton already represents hydrogen); see Figure 5.1. Recall that it

^{4}He ^{4}He ^{4}He ^{4}He ^{4}He ^{4}He ^{4}He
^{56}Fe ← ^{52}Cr ← ^{48}Ti ← ^{44}Ca ← ^{40}Ca ← ^{36}Ar ← ^{32}S ← ^{28}Si

from 10^9 K   ↑ ^{4}He

^{12}C → ^{24}Mg

^{12}C → ^{23}Na

from $5 \cdot 10^8$ K ^{12}C → ^{20}Ne

^{1}H ^{1}H ^{3}He ^{4}He ^{4}He ^{1}H
^{1}H → ^{2}H(D) → ^{3}He → ^{4}He   → ^{8}Be → ^{12}C → ^{13}C
             $-2\,^1$H

from 10^7 K   ↓ ^{4}He   ↓ ^{1}H

^{16}O ^{14}N

↓ ^{1}H   ↓ ^{1}H

^{15}N ← ^{15}O

from $2 \cdot 10^8$ K

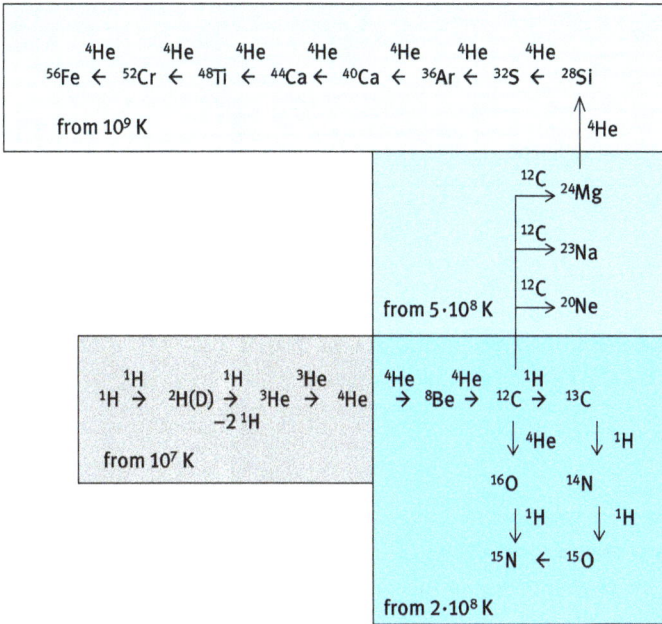

Figure 5.1: Scheme of thermonuclear formation of chemical elements (fusion reactions in stars).

is the number of protons in the nucleus that defines an element, not the number of protons plus neutrons (which determines its weight). Elements with different numbers of neutrons are termed isotopes, and different elements with the same number of neutrons plus protons (nucleons) are termed isobars.

Hydrogen and helium produced in the Big Bang served as the 'feed stock' from which all heavier elements were later created in stars. The fusion of protons to form helium is the major energy source in the Solar System.

This proceeds at a very slow and uniform rate, with the lifetime of the proton before it is fused to deuterium of about 10 Gyr (note that the proton lifetime concerning its decay is $>10^{30}$ yr). From He to Fe, the binding energy per nucleon increases with atomic number, and fusions are usually exothermic and provide an energy source. Beyond Fe, the binding energy decreases and exothermic reactions do not occur; elements are formed through scavenging of fast neutrons until ^{209}Bi. Heavier elements only are produced in shock waves of supernova explosions.

The most abundant elements (Figure 5.2) up to Fe are multiples of ^{4}He (^{12}C, ^{16}O, ^{24}Mg, ^{28}Si, ^{32}S etc.). During the red giant phase of stellar evolution, free neutrons are generated that can interact with all nuclei and build up all the heavy elements up to Bi; all nuclides with the atomic number > 83 are radioactive. It has been found that even ^{209}Bi decays, but extremely slowly ($\tau_{1/2} = 1.9 \cdot 10^{19}$ yr). The build-up of elements of ev-

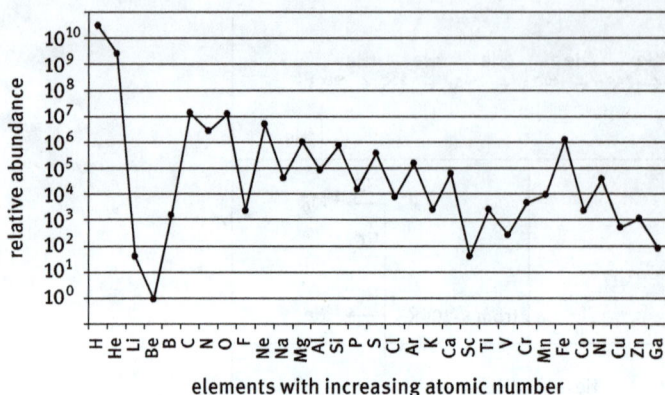

Figure 5.2: The abundance of chemical elements in space.

ery known stable isotope depends on different conditions of density and temperature. Thus, the production process required cycles of star formation, element formation in stellar cores, and ejection of matter to produce a gas enriched with heavy elements from which new generations of stars could form. The synthesis of material and subsequent mixing of dust and gas between stars produced the solar mix of elements in the proportions that are called 'cosmic abundance' (Figure 5.2 and Table 4.1). In addition to stable elements, radioactive elements are also produced in stars.

> The formation of molecules is impossible in stars because of the high temperature. However, in the interstellar medium chemical, reactions are possible that can create molecules.

Most of the molecular material in our galaxy and elsewhere occurs in *giant molecular clouds*. The heterogeneity of interstellar and circumstellar regions gives rise to a variety of chemistries. The interstellar medium, the region between the stars in a galaxy, has very low densities but is filled with gas, dust, magnetic fields, and charged particles. Approximately 99 % of the mass of the interstellar medium is in the form of gas (where denser regions are termed interstellar clouds), with the remainder primarily in the form of dust. The total mass of the gas and dust in the interstellar medium is about 15 % of the total mass of visible matter in the Milky Way. The exact nature and origin of interstellar dust grains are unknown, but they are presumably ejected from stars. One likely source is from red giant stars late in their lives. Interstellar dust grains are typically a fraction of a micron across, irregularly shaped, and composed of carbon and/or silicates. In these regions, temperatures are 10–20 K and the molecular processes, not being at thermodynamic equilibrium, require energy input to initiate.

However, high-energy cosmic rays penetrate and produce volume ionisation. The chemistry is initiated by the primary ionisation of H_2 and He, which constitute >99 % of the cosmic material in molecular clouds, providing primarily H_2^+ and He^+. H_2^+ is very rapidly converted to H_3^+ by reaction with H_2. Initially, the presence of nonpolar H_3^+

was surmised from observations of rotational transitions of the very abundant highly polar ion HCO^+, produced by proton transfer from H_3^+ to CO. Because the abundance of a collision complex will scale with the abundances of the collision partners, their collision frequency, and the binding energy of the complex, it appears that the species most likely to attract are an ion and H_2. The most abundant ion in dense molecular clouds is HCO^+. Thus, the species of interest initially is $H_2 - HCO^+$.

The role of interstellar dust in molecular growth is important because the dust particles provide a surface (heterogeneous chemistry) where reactions may occur under much higher density (collision probability). There are essential differences between laboratory and interstellar chemistry, namely the much larger timescale available in interstellar space. Radiation can break down the surface molecules and produce a wider variety of molecules. The study of interstellar chemistry began in the late 1930s with the observation of molecular absorption spectra in distant stars within the Galaxy, now called 'Large Molecule Heimat' (LMH). The species CH, CH^+ and CN have electronic spectra in an accessible wavelength region where the earth's atmosphere is still transparent. The character of these observable interstellar clouds is low density and essentially atomic with a small diatomic molecular component. Our discovery that the universe is highly molecular is quite recent. At present, almost 200 molecular species are listed. Among others, the following non-organic molecules and radicals, which we already have highlighted in Chapter 4, have been detected in the interstellar medium: CO, CO_2, HCO, HCN, OCS, H_2S, SO_2, N_2, NH_2, NO, HNO, N_2O, O_2, O_3, OH, HO_2, H_2O, H_2O_2. The molecular abundance does not follow the cosmic abundance of the elements. The wide variety of observed species includes ions and radicals. In particular, of the observed species with six or more atoms (presently 73 species), all contain carbon. Of the 4- and 5-atom species, only H_2O_2, HNO_2, NH_3, and SiH_4 are nonorganic. Thus, the chemistry of positively identified polyatomic species observed in the gas phase is carbon chemistry.

From the abundance of 'reactive' volatile elements in space in the order H–O–C–N, it appears that the simplest molecules derived (apart from H_2, O_2 and N_2) are bonds between the following elements (the bonding energy in kJ mol^{-1} is given in parenthesis); H–C (416), H–O (464), H–N (391), C–O (360) and N–O (181). Because of the excess of hydrogen in space, the hydrides (OH_2, CH_4, NH_3, SH_2) should have the highest molecular abundance among the compounds derived from such elements; furthermore, other simple gaseous molecules are CO, CO_2 and HCN. Because of the hydrogen excess, highly oxidised compounds (e. g. nitrate, sulphate, phosphate) are unlikely. Moreover, gaseous NO_x molecules are much more unstable compared with the other listed compounds; most of the oxygen is bonded in H_2O, CO_x, FeO and SiO_2.

Figure 5.3 shows schematically the possible reactions established by modelling as well as kinetic and thermodynamic considerations. All these reactions are sufficient to produce and destroy polyatomic species such as H_2O, HCN, NH_3 and HCHO. Overall, the original nebula is likely to have been composed of about 98 % gases (H, He, and noble gases), 1.5 % ice (H_2O, NH_3, and CH_4), and 0.5 % solid materials. Space consists

$$H + H \xrightarrow[-h\nu]{} H_2 \xrightarrow[-e]{h\nu} H_2^+ \xrightarrow[-H]{H_2} H_3^+ \xrightarrow[-H]{e} H_2$$

$$O \xrightarrow[-H_2]{H_3^+} OH^+ \xrightarrow[-H]{H_2} H_2O^+ \xrightarrow[-H]{H_2} H_3O^+ \xrightarrow[-H]{e} H_2O \xrightarrow{h\nu} OH \xrightarrow[-H]{O} O_2$$

$$C \xrightarrow[-e]{h\nu} C^+ \xrightarrow[-H]{OH} CO^+ \xrightarrow{e} CO \xrightarrow[-H]{OH} CO_2 \xrightarrow{h\nu} CO + O$$

$$C \xrightarrow{H_3^+} CH_2^+ \xrightarrow{H_2} CH_4^+ \xrightarrow{e} CH_4$$

$$C \xrightarrow{H_3^+} CH_3^+ \xrightarrow[-H]{H_2} CH_4^+ \xrightarrow{e} CH_4 \xrightarrow[-2H]{h\nu} CH_2 \xrightarrow[-H_2]{CH_2} C_2H_2$$

$$C \xrightarrow[-e]{h\nu} C^+ \xrightarrow[-H]{H_2O} HCO^+ \xrightarrow{\text{multistep}} HCHO$$

$$N \xrightarrow{H_3^+} NH_3^+ \xrightarrow{e} NH_3$$

$$N \xrightarrow[-H]{CH_2} HCN$$

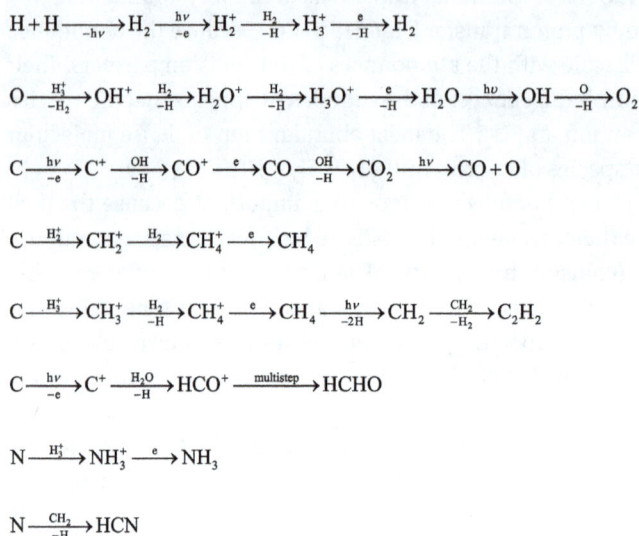

Figure 5.3: Interstellar formation of molecules.

of 98 % hydrogen (3/4) and helium (1/4); of the remaining 2 %, three-quarters is composed of just two elements, namely oxygen (2/3) and carbon (1/3). Based on the molar ratios, a formula for the 'space molecule' would be about $H_{2600}C_2O_3$.

5.1.2 Formation of the Earth

According to conventional astrophysical theory, our Solar System (the Sun and its planetary system) was formed from a cloud of gas and dust that coalesced under the force of gravitational attraction approximately 5 Gyr ago. This matter was formed from a collapsed supernova core, a neutron star with radiant energy and protons in the solar wind. High temperatures and violent conditions accompanied the formation of planetesimals and planets in many cases, and most interstellar dust particles were destroyed. However, the class of meteorites known as carbonaceous chondrites contain small particles with unusual isotopic ratios that indicate that they did not form in the solar nebula but rather must have been formed in a region with an anomalous composition (e. g. as the outflow from an evolved star) long before the formation of the Solar System. Therefore, these particles must have been part of the interstellar grain population prior to the formation of the solar nebula (see the last Chapter on the formation of molecules). Other debris from the supernova remains as gases and particulate matter termed *solar nebula*. This system cooled, particles rose by condensation growth, and the Sun grew by gravitational settlement about 4.6 Gyr ago. Cooling and subsequent

Table 5.1: Temperature-dependent condensation of compounds and formation of minerals.

T (in K)	elements, compounds, reactions	mineral
1600	CaO, Al$_2$O$_3$, REE oxides[a]	oxides (e. g., perovskite)
1300	Fe, Ni alloy metals	Fe–Ni
1200	MgO + SiO$_2$ → MgSiO$_3$	enstatite (pyroxene)
1000	alkali oxides + Al$_2$O$_3$ + SiO$_2$	feldspar
490–1200	Fe + O → FeO; FeO + MgSiO$_3$	olivine
680	H$_2$S + Fe → FeS	troilite
550	Ca minerals + H$_2$O	tremolite
425	olivine + H$_2$O	serpentine
175	ice-H$_2$O crystallise	water-ice
150	gaseous NH$_3$ + ice-H$_2$O → [NH$_3$ · H$_2$O]	ammonia-hydrate
120	gaseous CH$_4$ + ice-H$_2$O → [CH$_4$ · 7 H$_2$O]	methane-hydrate
65	CH$_4$, Ar crystallise	methane and argon ice

[a]REE = rare earth element

condensation occurred with distance from the protosun, resulting in an enlargement of heavier elements (e. g. Fe) at the inner circle, corresponding to the condensation temperature of matter (Table 5.1). Mercury formed closest to the Sun, mostly from iron and other materials in the solar nebula that condense at high temperatures (above 1,400 K).

Mercury's mean density is the second-highest in the Solar System, which is estimated to be 5.427 g cm^{-3} – only slightly less than earth's density of 5.515 g cm^{-3}. However, if the effects of gravitational compression – in which the effects of gravity reduce the size of an object and increase its density – then Mercury is, in fact, denser than Earth, with an uncompressed density of 5.3 g cm^{-3} compared to Earth's 4.4 g cm^{-3}. Venus has a mean density of 5.243 g cm^{-3}; also, like Earth, the interior is thought to be composed of iron-rich minerals, while silicate minerals make up the mantle and crust. As a terrestrial planet, Mars is also divided into layers that are differentiated based on their chemical and physical properties – a dense metallic core, a silicate mantle and a crust. The planet's overall density is lower than that of Earth's, estimated at 3.933 g cm^{-3}. As a gas giant Jupiter has a lower mean density (1.326 g cm^{-3}) than any of the terrestrial planets. Its core is believed to be composed of rock and surrounded by a layer of metallic hydrogen, and the outermost layer is are made up of elemental hydrogen and helium. At 0.687 g cm^{-3}, Saturn is the least dense of the gas giants.

The Earth, like the other solid planetary bodies, is formed by the accretion of large solid objects in a short time between 10 and 100 Myr (Figure 5.4); the postaccrecationary period dates from ~4.5 Gyr ago. Earlier theories suggest that the Earth was formed largely in the form of small grains but interspersed with occasional major pieces. The largest particles (protoplanets) developed a gravitational field and attracted further material to add to its growth. We assume that all this primary material was cold (10 K) at first. The energy of the collisions between the larger microplanets, as well as interior radioactive and gravitational heating, generated a huge amount of heat, and the

temperature (in K)

| | 20–50 | 300 ±20 | ≥ 1000 | ≫ 1000 |

~ 13.7 — formation of molecules and dust (in interstellar clouds) ← formation of elements (fusion in stars)

≥ 4.7 — formation of planetary bodies → formation of primitive life (complex organic molecules) ← ---- evolution of stars

4.7–3.8 — formation of the Earth → degassing: formation of a primitive atmosphere

3.8–2.2 — formation of a biosphere

2.2–0.1 — biosphere-atmosphere interaction

0 — modern Earth system ↔ human induced climate change

time (in Gy ago)

? (future)   ? (future)   ? (future)

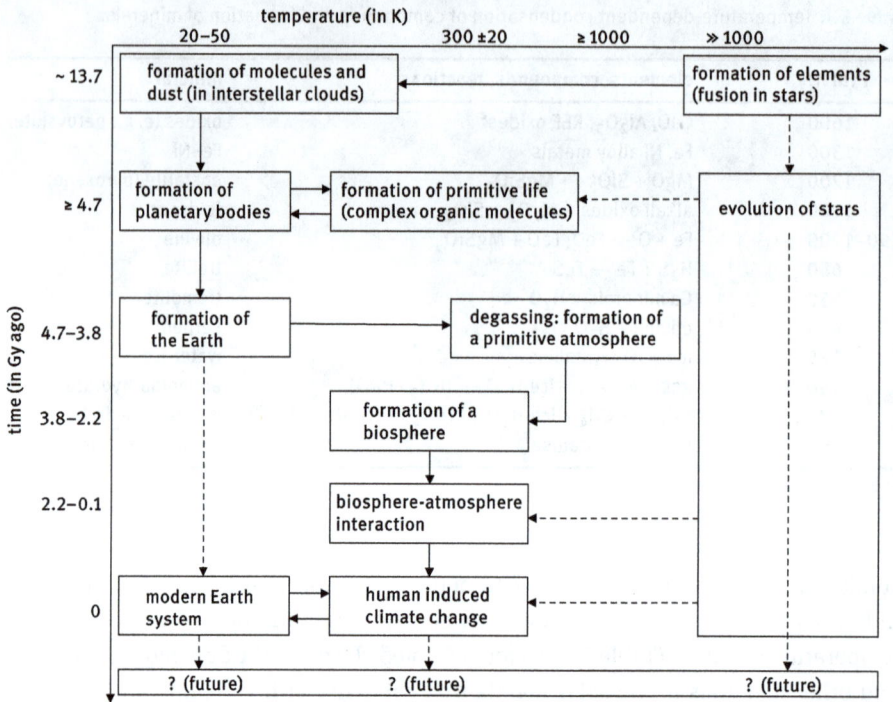

Figure 5.4: Evolution of the Solar System.

Earth and other planets would have been initially molten. The molten materials were also inhomogeneously distributed over the protoplanet. The Moon formed rather late in this process, about 45 Myr after the inner planets began to form. The current theory is that a Mars-sized planetoid, sometimes named Theia, collided with the Earth at this time. As astronomical collisions go, this was a mere cosmic fender-bender. The bodies, both molten, merged fairly smoothly, adding about 10 % to the Earth's mass. The Moon formed from the minimal orbiting debris (about 0.01 Earth's masses) resulting from this low-speed crash. During the formation of the Earth by the accumulation of cold solids, very little gaseous material was incorporated. Evidence of this comes from the extremely low level of the non-radiogenic noble gases in the atmosphere of the Earth. Among those, only helium could have escaped into space, and only xenon could have been significantly removed by absorption into rocks. Neon, argon and krypton would have been maintained as atmospheric components. Most of the helium found on Earth is ^{4}He, the result of the radioactive decay of uranium and thorium (see Chapter 5.3.4); primordial helium is ^{3}He.

The heavier molten iron sank to become the core, while materials of lower density (particularly the silicates) made their way to the surface. The lightest of all became the crust as a sort of 'scum' on the surface. This crust melted and reformed numerous times because it was continuously broken up by gigantic magma currents that erupted from the depths of the planet and tore the thin crust. With the dissipation of heat into space, the cooling of our planet began. In the magma ocean, blocks began to appear, formed from high-melting-point minerals sinking again into the heart of the Earth. Approximately 500 million years after the birth of the Earth, this incandescent landscape began to cool down. When the temperature fell below 1,000 °C, the regions of lower temperatures consolidated, became more stable and initiated the assembly of the future crust. Only with the further cooling of the planet did those fragments become numerous and large enough to form a first, thin, solid cover, a true primitive crust. This primordial crust might have developed as a warm expanse of rocks (some hundreds of degrees Celsius), interrupted by numerous large breaks, from which enormous quantities of magma continued to erupt. The composition of the crust began to change by a sort of distillation. Disrupted by highly energetic convective movements, the thin lithospheric covering would have been fragmented into numerous small plates in continuous mutual movements, separated and deformed by bands of intense volcanism. During this continuous remelting of the 'protocrust' heavier rock gradually sank deeper into the mantle, leaving behind a lighter magma richer in silicates. Thus, around the basalts appeared andesites: fine granular volcanic rocks, whose name derives from the Andes, where several volcanoes are known to form rocks of this type. Gradually, a granitic crust emerged.

Just seven elements (Si, Al, Fe, Ca, Na, Mg, K) in oxidised form comprise 97 % of the earth's crust (Table 4.1); it is notable that silica contributes 53 % of the total. With the exception of oxygen (which amounts to 46 % of all crust elements), none of such elements is in a volatile form in nature. In space, carbon, nitrogen and sulphur amount to 33 % of the total abundance of material, but in the earth's crust, they only constitute 0.057 % (0.02 %, 0.002 %, and 0.035 %, respectively).

This fact of the depletion of C, N and S by about two orders of magnitude in the earth's crust shows that these elements are partitioning among different composites. Today, the more external part of the crust or lithosphere constitutes the superficial covering of the Earth. Two kinds of crust are easily distinguished by composition, thickness and consistency; continental crust and oceanic crust. Continental crust has a thickness that, in mountain chains, may reach 40 kilometres. It is composed mainly of metamorphic rock and igneous blocks enriched with potassium, uranium, thorium and silicon. This forms the diffuse granitic bedrock of 45 % of the land surface of the Earth. The oceanic crust has a more modest thickness, in the order of 5–6 kilometres, and is made up of basaltic blocks composed of silicates enriched with aluminium, iron and manganese. It is continuously renewed along mid-ocean ridges (see Table 5.2).

At this early point in the history of the Solar System, there was a relatively short period (50 Myr or so) of intense meteoric bombardment (termed the *late heavy bombardment* LHB) which would have continually opened new holes in the crust, immediately filled by magma. The scars left by this intense meteoric bombardment have been

Table 5.2: Geographic quantities of the atmosphere, ocean and continents.

mass of the Earth	$6.0 \cdot 10^{27}$ g (density 5.52 g cm^{-3})
mass of the atmosphere	$5.2 \cdot 10^{21}$ g
mass of the troposphere (up to 11 km)	$4.0 \cdot 10^{21}$ g
volume of the Earth	$1.08 \cdot 10^{21}$ m^3
volume of the troposphere (up to 11 km)	$5.75 \cdot 10^{18}$ m^3
volume of the world's ocean	$1.37 \cdot 10^{18}$ m^3 (density 1.036 g cm^{-3})
area of northern hemispheric ocean	$1.54 \cdot 10^{14}$ m^2
area of southern hemispheric ocean	$2.10 \cdot 10^{14}$ m^2 [b]
area of continents northern hemisphere	$1.03 \cdot 10^{14}$ m^2
area of continents southern hemisphere	$0.46 \cdot 10^{14}$ m^2
depth of the crust[c]	35 km (locally varies between 5 and 70 km)
mass of the crust[c]	$4.9 \cdot 10^{25}$ g[d]
depth of the upper mantle[c]	60 km (locally varies between 5 and 200 km)
mass of the upper mantle[c]	$4.3 \cdot 10^{25}$ g[e]
depth of the mantle	2890 km
mass of the lower mantle[f]	$3.4 \cdot 10^{27}$ g[f]
thickness of the earth's atmosphere[a]	1000 km

[a] it is not a definite number – there is no set boundary where the atmosphere ends.
[b] total ocean area $3.62 \cdot 10^{14}$ m^2 after Schlesinger (1997)
[c] the lithosphere comprises the crust and the upper mantle
[d] assuming 35 km depth and 2.7 g cm^3 density
[e] assuming 60 km depth and 3.3 g cm^3 density
[f] between 60 and 2890 km; density about 6.0 g cm^{-3}

almost totally erased on the Earth by subsequent reworking of the crust. The evidence for the LHB is quite strong, however. It comes mostly from lunar astronomy (big craters formed significantly later than the large lunar maria, which are dark, basaltic plains on the Moon, formed by ancient volcanic eruptions), and the lunar rocks recovered from space exploration. The implication is that the post-Hadean granitic crust was not the product of gradual distillation but of catastrophic reworking after the protocrust was destroyed by the LHB. Water was likely carried by icy ammonia hydrate bodies to the Earth not only at the very beginning of the earth's formation around 4.6 Gy ago but also during the LHB.

> **i** Meteorites or their parent asteroids, as well as comets, ferry water, carbon (including organic compounds) and nitrogen to Earth; reactions under high pressure and temperature provide volatile substances.

It is a scientific consensus that most of the LHB was due to carbonaceous meteorites, which are today very rare, and, if fallen down, are soon oxidised. It is supposed that between 100 and 300 km depth below the earth's surface, we have a patchwork in which the carbonaceous chondrite material comprises 20 % on average (Gold 1999). In the last 25 years, however, several carbonaceous meteorites have been freshly found and analysed. At least 80 organic compounds are known to occur in them. Both aliphatic and aromatic compounds were detected; in addition, carbonyl groups (>C=O) appeared to be present as well as unsaturated groups of the vinyl or allylic type. Amino acids and sugars were encountered in all the meteorites studied. Seventeen amino acids were detected; serine, glycine, alanine and leucines. Glutamic acid, aspartic acid and threonine were found to be the most abundant. The absence of rotation, the type and distribution pattern of amino compounds in chondrules and matrix, the lack of pigments, fatty acids and presumably nucleic acids, in addition to other biochemical criteria, suggest that the organic material has been synthesised by chemical rather than known biochemical processes. The analysis of organic compounds in carbonaceous meteorites provides information about chemical evolution in an extraterrestrial environment and the possible compounds that could have been present on the Earth before and during the origin of life.

> The dominant fraction of carbon on Earth is termed kerogen; a mixture of organic chemical compounds that make up a portion of the organic matter in sedimentary rocks. Kerogen materials have also been detected in interstellar clouds and dust around stars. Terrestrial kerogen is almost all collected in sedimentary layers, which lie near the earth's surface, showing $H/C \sim 0.5$ (different kerogen types are distinguished according to H/C from <0.5 to >1.25). It is insoluble in normal organic solvents because of its huge molecular weight (upwards of 1,000 Daltons). The soluble portion is known as bitumen (petroleum belongs chemically to bitumen, as the liquid form). When heated to the certain temperatures in the earth's crust, some types of kerogen release crude oil or natural gas, collectively known as hydrocarbons (fossil fuels). When such kerogens are present in high concentrations in rocks such as shale and have not been heated to a sufficient temperature to release their hydrocarbons, they may form oil-shale deposits.

5.1.3 Degassing the Earth and formation of the atmosphere

As discussed in the previous Chapter 5.1.2, we are forced to conclude that the acquisition of gases or substances that would be gaseous at the pressures and temperatures that prevailed in the region of the formation of the Earth was limited to the small value implied by the low noble gas values. Assuming that gaseous material, except noble gases, was absent or of less importance in the mass budget of the initial Earth, all gases believed to have been present in the primordial or primitive atmosphere must be a result of the volatilisation of materials from the inner part of the Earth. Assuming that very little gaseous material (e. g., CO_2, CH_4, NH_3, H_2S, HCl) was incorporated into the primary earth's aggregate, it is likely that corresponding solid substances from which, under the current conditions (heat and pressure), gases could evolve were available in the crust or the inner Earth. Most scientists assume that the earth's atmosphere, about 4.5 billion years ago, consisted mainly of CO_2 under high pressure ($\sim$250 bars) and temperature (>300 °C), with N_2 and H_2O (and a little HCl) being important minor species. Those volatile elements and compounds were degassed from the inner Earth.

An earlier hypothesis suggested that the primitive (or better termed, primary) atmosphere consisted of NH_3 and CH_4. This idea was supported by the finding of both species in some meteorites and the belief that the solar nebula also contains a small amount of ammonia and methane. The existence of a $NH_3 - CH_4$ atmosphere was believed to be a precondition for the origin of life. The well-known Miller–Urey experiment in 1953 showed that under UV radiation, organic molecules could be formed in such an atmosphere. However, the intensive UV radiation at the earth's beginning would have destroyed NH_3 and CH_4 soon after, and no processes are known to chemically form both species in air. Furthermore, it became evident that it is difficult to synthesise prebiotic compounds in a non-reducing atmosphere. Whether the mixture of gases used in the Miller–Urey experiment truly reflects the atmospheric content of the early Earth is a controversial topic. Other less reducing gases produce a lower yield and variety. It was once thought that appreciable amounts of molecular oxygen were present in the prebiotic atmosphere, which would have essentially prevented the formation of organic molecules; however, the current scientific consensus is that such was not the case.

With the carbonaceous chondrite type of material as the prime source of the surface carbon, the question arises as to the fate of this material under heat and pressure, and in these conditions, it would encounter buoyancy forces that drove some of it toward the surface. The detailed mix of molecules will depend on pressure and temperature and on the carbon-hydrogen ratio present.

What would be the fate of such a mix? Would it all be oxidised with oxygen from the rocks, as some chemical equilibrium calculations have suggested? Evidently not, for we have clear evidence that unoxidised carbon exists at depths between 150 km and 300 km in the form of diamonds. We know diamonds come from there because it is only in this depth range that the pressures would be adequate for their formation. Diamonds are known to have high-pressure inclusions that contain CH_4 and heavier hydrocarbons, as well as CO_2 and nitrogen. The presence of at least centimetre-sized pieces of very pure carbon implies that carbon-bearing fluids exist there and that they must be able to move through pore spaces at that depth so that a dissociation process may deposit the pure carbon selectively; a process akin to mineralisation processes as we know them at shallower levels. The fluid responsible cannot be CO_2, since this has a higher dissociation temperature than the hydrocarbons that co-exist in the diamonds; it must therefore have been a hydrocarbon that laid down the diamonds: CH_4.

The destruction of hydrocarbons under pressure and higher temperatures produces CH_4 as well as elemental C (in oxygen-poor conditions) and CO_2 as well as H_2O (in oxygen-rich conditions) as a continuous process over geological epochs. Assuming for the carbonaceous matter, the formula $C_nH_mN_xO_y$, the following products could be produced under thermal dissociation (equations (5.1)–(5.4)). Under oxygen-free conditions, the products from thermal dissociation (see equation (5.1)) are $C + CO_2 + H_2$. Hydrogen can also be produced via reactions equation (5.3) and transform deep carbon into CH_4 and H_2O (equation (5.4)). Reaction equation (5.2) can invert under the conditions deep within the Earth (see equation (5.5)). The decomposition may be oxidative (equation (5.2)) or reductive (equation (5.4) or neutral equation (5.3)) where oxygen and hydrogen are produced even from the carbon substrate (equation (5.3)):

$$C_nH_mN_xO_y \xrightarrow{T,p} CH_4, C, H_2, CO, CO_2, N_2, H_2O, O_2, \tag{5.1}$$

$$C_nH_mN_xO_y \xrightarrow{O_2,T,p} CO_2 + H_2O, \tag{5.2}$$

$$C_nH_mN_xO_y \xrightarrow{H_2O,T,p} CH_4 + O_2 + H_2, \tag{5.3}$$

$$C_nH_mN_xO_y \xrightarrow{H_2,T,p} CH_4 + H_2O. \tag{5.4}$$

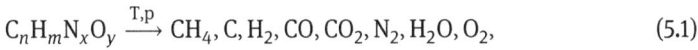

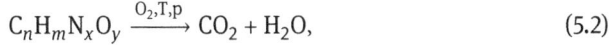

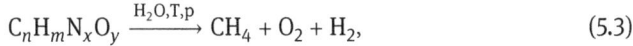

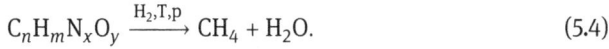

In other words, the process, shown below in equation (5.5), represents an inorganic formation of hydrocarbons ('fossil fuels'). Although the biogenic theory for petroleum was first proposed by Georg Agricola (1494–1555) in the sixteenth century, various abiogenic hypotheses were proposed in the nineteenth century, most notably by Alexander von Humboldt (1769–1859), Dmitri Mendeleev (1834–1878) and Berthelot, and renewed in the 1950s.

$$CO_2 (+H_2O) \underset{O_2}{\overset{H_2}{\rightleftarrows}} CO (+H_2O) \underset{O_2}{\overset{H_2}{\rightleftarrows}} C_nH_m (+H_2O). \tag{5.5}$$

Other atoms that may also be present, such as oxygen and nitrogen, will form a variety of complex molecules with carbon and hydrogen. Thus, it is easy to understand that reduced carbon in the form of CH_4, as well as in oxidised form (CO_2) and H_2O, will be produced. At sufficient depth, methane will behave chemically as a liquid, and it will dissolve the heavier hydrocarbons that may be present and, therefore, greatly reduce the viscosity of the entire fluid. The continuing upward stream would acquire more and more of such unchangeable molecules, and the final product that may be caught in the reservoirs we tap for oil and gas, is the end product of this process.

Under pressure and high temperature ($>900\,°C$), equilibriums are established between CO, CO_2, H_2O, H_2 and CH_4:

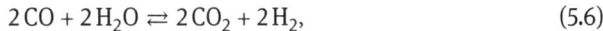

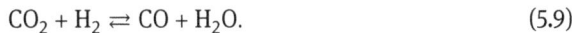

$$2\,CO + 2\,H_2O \rightleftarrows 2\,CO_2 + 2\,H_2, \tag{5.6}$$

$$4\,CO + 2\,H_2 \rightleftarrows 2\,H_2O + CO_2 + 3\,C, \tag{5.7}$$

$$2\,CO + 2\,H_2 \rightleftarrows CO_2 + CH_4, \tag{5.8}$$

$$CO_2 + H_2 \rightleftarrows CO + H_2O. \tag{5.9}$$

Now we see that CO_2, CH_4 (and H_2O) are available as conversion products from chondritic material and NH_3 and H_2O from icy meteorites in the time of LHB. Nitrogen (N_2), the principal constituent of the earth's atmosphere today, is believed to be produced from ammonia photolysis in the pre-biological atmosphere:

$$2\,NH_3 + h\nu \rightarrow N_2 + 3\,H_2 \uparrow. \tag{5.10}$$

Other constituents of the primitive atmosphere (besides CO_2, CH_4, N_2, NH_3 and H_2) were HCl, H_2S and SO_2. The thermal hydrolysis of chlorides, which may be primordial, can also explain degassing of HCl ($FeCl_2$ has been detected in meteorites):

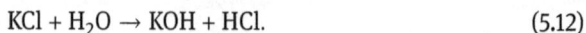

$$FeCl_2 + H_2O \rightarrow FeO + 2\,HCl, \tag{5.11}$$
$$KCl + H_2O \rightarrow KOH + HCl. \tag{5.12}$$

Today chlorides only exist dissolved in the oceans and marine sediments. The reactions equation (5.11) and equation (5.12) are remarkable through a reservoir separation of alkalinity into the crust (oxides and hydroxides) and acidity (HCl, potentially later SO_2) into the atmosphere. Further outgassing of H_2, H_2S and SO_2 is gained through iron(II) sulphide and oxide:

$$FeS + 3\,H_2O \rightarrow FeO + SO_2 + 3\,H_2, \tag{5.13}$$
$$FeS + H_2O \rightarrow FeO + H_2S, \tag{5.14}$$
$$3\,FeO + H_2O \rightarrow Fe_3O_4 + H_2 \tag{5.15}$$

These reactions rapidly changed the mantle redox state to a more oxidised level. Another initial production of free oxygen in the earth's mantle (beside equation (5.3)) can also be explained by the thermal decomposition of metal oxides, transported to hotter regions; for example, FeO, giving oxygen and metallic iron; the heavy iron moving toward the earth's core, leaving the oxygen to escape. The free oxygen, however, could have oxidised the reduced carbon existing in heavy hydrocarbons into carbon dioxide and water:

$$C_6H_2O + 6\,O_2 \rightarrow 6\,CO_2 + H_2O \quad \text{or generally} \quad (CH_2O)_n + n\,O_2 \rightarrow n\,CO_2 + n\,H_2O.$$

Rocks give off on average 5–10 times their own volume of gases (excluding H_2O vapour) when heated above 1,000 °C. Before heating, the rocks were dried in the experiments to remove hygroscopic moisture. The produced steam (H_2O) however was dominant and exceeded by a factor of 4–5 all other gases from the rocks. The evolved gases from granite and basalt are (% of volume in parenthesis):

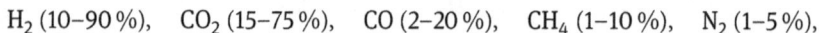

H_2 (10–90 %), CO_2 (15–75 %),   CO (2–20 %), CH_4 (1–10 %), N_2 (1–5 %),

and in traces, ($\ll 1\%$) HCl, SiF_4 and H_2S have been detected. Taking mean values, about 0.2 % of the rocks' mass is volatile (corresponding to 200 ppm hydrogen, 600 ppm carbon and 200 ppm nitrogen). Water, however, is the dominant volatile compound in rocks alongside traces of sulphur, nitrogen and halogens. The mean amount of water liberated from heated rocks corresponds to about 2 %, much more than is nowadays found in rocks as 'free' water (see below), suggesting that most of the water vapour produced while heating the rocks originates from OH-bonded water due to silicate condensation, as discussed below.

Water is also stored in silicates and liberated according to the conditions of temperature and free water. The most important ion in rocks is the silicate SiO_4^{4-} (similarly to sulphate SO_4^{2-}) from the weak orthosilicic acid $H_4SiO_4 = Si(OH)_4$. It is in equilibrium with silicon dioxide (SiO_2) by condensation (and liberation of H_2O) via metasilicic acid (H_2SiO_3):

$$H_4SiO_4\,[Si(OH)_4] \rightleftharpoons H_2SiO_3\,[O{=}Si(OH)_2] + H_2O \rightleftharpoons SiO_2 + 2\,H_2O. \tag{5.16}$$

SiO_2 (silica) finally is the anhydride of the acids; its solubility in water is about $0.12\,g\,L^{-1}$ (and increases strongly with temperature). Orthosilicic acid condenses to amorphic and/or polymeric SiO_2.

$$n\,H_4SiO_4 \rightarrow (SiO_2)_n + 2n\,H_2O. \tag{5.17}$$

Thus, water can be 'stored' in silicates and liberated by heating hydrated silicates. This group of metamorphic rocks includes serpentine $(Mg, Fe)_3Si_2O_5(OH)_4$ and tremolite $Ca_2Mg_5Si_8O_{22}(OH)_2$. Serpentine, a basic orthosilicate, is a very common secondary mineral, resulting from a hot water alteration of magnesium silicates (mostly peridotite), present in magma, a process termed *serpentinisation*:

$$3\,MgCaSi_2O_6 + 3\,CO_2 + 2\,H_2O \rightarrow Mg_3Si_2O_5(OH)_4 + 3\,CaCO_3 + 4\,SiO_2. \tag{5.18}$$

The water content in serpentines lies between 5 and 20 %. The decomposition of serpentine may be written as follows:

$$Mg_3Si_2O_5(OH)_4 \rightarrow Mg_2SiO_4 + MgSiO_3 + 2\,H_2O. \tag{5.19}$$

We now can explain the main components of the atmosphere before 4 Gyr ago (Table 5.3): CO_2, N_2, and H_2O as well as in traces CH_4, H_2S and SO_2 (note that the concentrations of all these 'trace' gases were several times higher than today because of the high atmospheric pressure). Very small traces of O_2 (see below) must be assumed through the photolysis of H_2O in the atmosphere.

In that time (no ozone and oxygen to prevent radiation <300 nm), solar radiation was able to dissociate all atmospheric compounds. This happens with photolysis of water (eq. (5.20)) and hydrides such as CH_4, NH_3 and H_2S (eq. (5.21)), X = C, N, and S:

$$H_2O \xrightarrow{h\nu} O + H_2(\uparrow), \tag{5.20}$$

$$XH_n \xrightarrow{h\nu} X + \frac{n}{2}H_2(-) \xrightarrow{m\,O} XO_m \text{ (oxide and anhydride)} \xrightarrow{H_2O} H_2XO_{m+1} \text{ (oxoacid).} \tag{5.21}$$

Table 5.3: Evolution of the earth's atmosphere (main composition); time in Gyr.

atmosphere	time ago	composition	origin	fate
primordial[a]	4.5–4.6	H_2, He	solar nebula	erosion to space
primitive (first)	~4.5	NH_3(?), CH_4(?), CO_2, H_2O[b]	degassing	photolysis
secondary	4.5–4.0	N_2, CO_2, H_2O[b]	degassing	washout
intermediate (third)	4.0–2.3	N_2, CO_2[c]	secondary phase	remaining
present (fourth)	2.3–0.5	N_2, O_2[d]	photosynthesis	biosphere–atmosphere equilibrium

[a]speculative
[b]in traces: H_2S, SO_2, HCl, and O_2
[c]in traces: H_2S, S_x; increasing O_2 and O_3
[d]in traces: O_3

The loss of hydrogen to space (and later its deep burial in hydrocarbons) is the reason for the changing redox state from low oxygen to a more oxidising environment. With an increasing state of oxidation, a rise of acidity also occurs, and the two combine until an equilibrium state is achieved in geochemical evolution.

> Oxidation/reduction and the acidity potential are interlinked where organisms create a biogeochemical evolution by separating oxidative and reductive processes among different living species.

Table 5.4 summarises the most important chemical relationships between such components. It is remarkable that only C, N and S compounds are gaseous and/or dissolved in water in all redox states, making these compounds globally distributable and exchangeable among different reservoirs to provide global cycles (see Chapter 5.2). The other minor elements listed in Table 5.4 provide important compounds for life and the geogenic (abiotic) environment but are much less volatile or almost immobile (Si, P). Some oxygenated halogens are unstable. Chemical evolution alone can change the distribution of the elements among different molecules and reservoirs, creating a heterogeneous world (see Figure 5.4).

Hence, all gases compiling and cycling through the atmosphere (compounds of nitrogen, carbon and sulphur, and water) originally volatilised from the crust in the degassing period to create a first "primitive" atmosphere (H_2O, NH_3, CH_4, CO_2, HCl, H_2S, SO_2). This was characterised by intensive atmospheric photochemical processes forming a secondary atmosphere (Tables 5.3 and 5.5) that was slightly oxidised ($NH_3 \rightarrow N_2$ and $CH_4 \rightarrow CO_2$). It is assumed that water soon condensed, creating the first hot oceans. Simultaneously with the first rain, however, all soluble gases (CO_2, NH_3, HCl, Cl_2, H_2S, and SO_2) were washed out to a different extent through significant differences in their solubility.

The parts of the earth's crust becoming the ocean bottom were likely to be highly alkaline because of NaO and MgO and, of much less importance, CaO (according to

Table 5.4: Substances in the reduced state (hydrides) and in oxidised form (oxides) as well as the corresponding oxoacids. If not mentioned, the species are gaseous under standard conditions; aq – exists only dissolved in water.

hydrogen excess (reduced state)	oxygen excess (oxidised state)	oxoacids
H_2	OH, HO_2	H_2O (liquid)
OH_2 (H_2O, liquid)	O_2, O_3	H_2O_2
CH_4	CO, CO_2	H_2CO_3 (aq)
NH_3	NO, NO_2, NO_3, N_2O_5	HNO_2, HNO_3
SH_2 (H_2S)	SO_2, SO_3	H_2SO_3 (aq), H_2SO_4 (aq)
SiH_4	SiO, SiO_2 (solid, insoluble)	$Si(OH)_2$, H_4SiO_4 (solid, insoluble)
PH_3	P_2O_5 (solid)	H_3PO_4 (solid)
AsH_3	As_2O_3 (solid)	$As(OH)_3$ (aq)
HBr	BrO, Br_2O, Br_2O_3 (solid)	HOBr, $HBrO_3$ (aq)
HCl	ClO, Cl_2O, ClO_2, Cl_2O_6 (liquid)	HOCl (aq), $HClO_2$ (aq), $HClO_3$ (aq)
HJ	JO, J_2O_5 (solid)	HJO_3 (solid)

Table 5.5: Composition of the prebiotic earth's atmosphere.

substance	concentration (in ppm)	change with altitude
N_2	800000	constant
CO_2	200000	constant
H_2O	8000	decrease
H_2	1000	constant
CO	70	increase
CH_4	0.5	decrease
O_2	0.000001	increase

the abundance of the cations in seawater. Large amounts of soluble oxides led to dissolved Na^+, Mg^{2+} and Ca^{2+} and OH^-, which converts bicarbonate into less soluble carbonates (equation (4.251)) as well as ammonium back to NH_3 (equation (3.100)) with subsequent degassing from the ocean. This is simply the explanation for the chemical composition of the seawater (Table 2.5).

Contrary to the air depletion by scavenging, the air was enriched relatively with insoluble N_2 and less soluble compounds such as CO_2 and H_2S. As described later (Chapter 5.2.3), there is a continuous flux of CO_2 through the oceans to the sediments converted as carbonate. Due to the low oxygen level, H_2S remains in the atmosphere for the first half of the earth's history. Small amounts of SO_2 from volcanic exhalations may have been in the air and seawater after wet deposition. It is likely that reduced matter (e. g. S-IV, Fe^{2+}) still existed in seawater because of the continued absence of oxygenic photosynthesis by cyanobacteria (see below). Because O_2 was not yet produced by water dissociation via photosynthesis, the earth's surface was a strong oxygen sink through oxidation of reduced metals (e. g. Fe and U), Table 5.5.

5.1.4 Evolution of life and atmospheric oxygen

The dominant scientific view is that the early atmosphere had 0.1 % oxygen or less. Assuming an O_2 level of 10^{-8} of the present level or less before 4 Gyr due to photochemical steady-states, with the evolution of biological life, it is believed that there was a concentration increase to 10^{-5}. The oxygen levels in the Archean probably remain low: less than 10^{-5} the present atmospheric level in the upper atmosphere and 10^{-12} near the surface. Much later (~2.2–2.4 Gyr ago), significant levels of oxygen arose in the atmosphere establishing the present (fourth) atmosphere. Our present oxygen is the result of *life*, specifically photosynthesis (see Chapter 5.2.2).

Today, there is no doubt that bacterial life is created, exists and survives in space. However, what is life? Where did we come from? These two fundamental questions remain (still) unanswered in science. The existence of humans and all animals depends on free oxygen in the atmosphere, and this compound is almost completely produced by oceanic cyanobacteria. Hence, the origin of life lies in the darkness of the evolution of molecules in structured systems (a chemical plant we call a *cell*) to provide work-sharing synthesis via non-equilibrium electron transfer processes (in other terms, redox processes; see Chapter 3.3.5.1). Cells represent a dissipative structure whose organisation and stability are provided by irreversible processes running far from equilibrium. One of the fundamental requirements for life, as we know, is the presence of liquid water on (or below) a planet's surface. Life began very early in earth's history, perhaps before 4 Gyr ago, and achieved remarkable levels of metabolic sophistication before the end of the Archean, around 2.5 Gyr. Wherever life developed, the conditions can be characterised as follows:
–   liquid water at about 40 °C,
–   dissolved nutrients (ammonium, carbonate, sulphide),
–   hydrogen and basic organic molecules,
–   protection against hard radiation,
–   inorganic substrate for fixing.

However, a homogeneous mixture such as aqueous solutions (ocean) or gases such as the atmosphere (Miller–Urey experiment) providing all necessary educts can only synthesise molecules that are much less complex than those found in organisms; a heterogeneous and very likely interfacial surrounding is essential. There were two fundamental problems: first, to explain how the giant polymers essential to life, especially proteins and nucleic acids, were synthesised under natural conditions from

their sub-units and, second, to understand the origin of cells. *Cell Theory* is one of the foundations of modern biology. Its major tenets are:
– all living things are composed of one or more cells;
– the chemical reactions of living cells take place within cells;
– all cells originate from pre-existing cells; and
– cells contain hereditary information, which is passed from one generation to another.

The debate is ongoing about how cell membranes and hereditary material (DNA and RNA) first evolved. Membranes are essential to separate the inner parts of the cell from the outer environment and are a selectively permeable barrier for certain chemicals. Both DNA and RNA are needed for a cell to be able to replicate and/or reproduce. Most organisms use DNA (deoxyribonucleic acid). DNA is a stable macromolecule consisting (usually) of two strands running in opposite directions. These strands twist around one another in the form of a double helix and are built up from components known as nucleotides. Biologists believe that RNA (ribonucleic acid) existed on Earth before modern cells arose.

> According to this hypothesis, RNA stored both genetic information and catalysed the chemical reactions in primitive cells. Only later in evolutionary time did DNA take over as the genetic material and proteins become the major catalyst and structural component of cells. RNA still catalyses several fundamental reactions in modern-day cells, which can be viewed as molecular fossils of an earlier world. Although RNA seems well suited to form the basis for a self-replicating set of biochemical catalysts, it is unlikely that RNA was the first kind of molecule to do so. From a purely chemical standpoint, it is difficult to imagine how long RNA molecules could be formed initially by purely nonenzymatic means. Given these problems, it has been suggested that the first molecules to possess both catalytic activity and information storage capabilities may have been polymers that resemble RNA but are chemically simpler. Presumably, pre-RNA polymers also catalysed the formation of ribonucleotide precursors from simpler molecules. Once the first RNA molecules had been produced, they could have diversified gradually to take over the functions originally carried out by the pre-RNA polymers, leading eventually to the postulated RNA world. An increasingly strident view is that protein either preceded RNA in evolution or, at the very least, that RNA and protein coevolved, in what is known as the 'proteins (or peptides) first' hypothesis.

DNA could maintain its structure in a vacuum, perhaps almost indefinitely, in the very low temperatures of space. Freeze drying in a vacuum (as exists in space) would ensure that free water in the cell diffuses out. The ability of bacteria to remain viable after exposure to high vacuum and extreme cold suggests the nuclei of comets are ideal sites to search for potentially viable microbes. Comets are formed from interstellar gases and grains containing interstellar bacteria and organic molecules. Radiogenic heating by nuclides such as ^{26}Al maintains a warm liquid interior for nearly one million years, and this is enough for bacterial replication. It seems that primordial cells were delivered to Earth for further evolution. Organic compounds were synthesised from the elements, in space *and* on Earth. Conditions for the development of self-organising organic matter (what we call life) were manifold and may not be specific to the Earth alone.

> i   The hypothesis called "panspermia" proposes an interplanetary transfer of life. Recent results from experiments at the International Space Station (ISS) indicated the importance of the aggregated form of cells for surviving in the harsh space environment.

With the assumption of primordial complex organic molecules, life could also arise deep in the Earth – protected against collisions and atmospheric phenomena. The 'soup' needed for the formation of life or development from simpler extraterrestrial bacteria within the carbonaceous chondrites was available: H_2O, NH_3 and organic compounds.

> i   Meteorites were carriers of prebiotic organic molecules to the early Earth; thus, the detection of extraterrestrial sugars in meteorites implies the possibility that extraterrestrial sugars may have contributed to forming functional biopolymers like RNA.

For the further evolution of the earth's atmosphere, the final answer to the question of where life originated is not so important. Today's atmosphere results from the evolution of the earth's biosphere and is developed under special physical and chemical conditions that have changed over time. It is now believed that life appeared very early on Earth, 3.8 Gyr ago or earlier. In biological evolution, the first primitive organisms must have based their development and growth on already existent organic compounds by re-synthesis. We know that the first forms of life must have existed under anaerobic conditions at the sea bottom. *Fermentation* is the process of deriving energy from the oxidation of organic compounds, a very inefficient process, where bacteria produce ethanol and carbon dioxide from fructose (and other organic material), but many other products (e. g. acids) and carbon dioxide may have been produced:

$$C_6H_{12}O_6 \rightarrow 2\,CH_3CH_2OH + 2\,CO_2. \tag{5.22}$$

An important success was achieved by the first autotrophic forms of life (methanogens and acetogens), which transfer carbon from its oxidised (inorganic) form (CO_2) to the reduced (organic) forms that results in bacterial growth (in contrast, heterotrophic organisms can use carbon only from living or dead biomass: higher plants, animals, mushrooms, most bacteria). This process is termed *anoxygenic photosynthesis*:

$$2\,CO_2 + 4\,H_2 \rightarrow CH_3COOH + 2\,H_2O, \tag{5.23}$$

$$CO_2 + 4\,H_2 \rightarrow CH_4 + 2\,H_2O. \tag{5.24}$$

Serpentinisation, arc volcanism and ridge-axis volcanism have provided hydrogen, where the geochemical processes may involve primordial hydrocarbon and water destruction.

The next step in biological evolution, *oxygenic photosynthesis*, sharply increased the productivity of the biosphere. Today the first photosynthetic prokaryotes range from cyanobacterial and algal plankton to large kelp. Such organisms have used H_2O as electron donator:

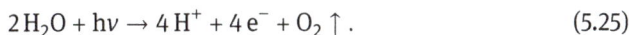

$$2\,H_2O + h\nu \rightarrow 4\,H^+ + 4\,e^- + O_2 \uparrow . \tag{5.25}$$

Generally, the process of photosynthesis is written as

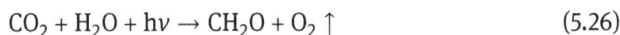

$$CO_2 + H_2O + h\nu \rightarrow CH_2O + O_2 \uparrow \tag{5.26}$$

where CH_2O is a synonym for organic matter (a building block of sugar $C_6H_{12}O_6$).

The creation of a photosynthetic apparatus capable of splitting water into O_2, protons and electrons were the pivotal innovation in the evolution of life on Earth. For the first time, photosynthesis had an unlimited source of electrons and protons by using water as the reductant.

By freeing photosynthesis from the availability of volatile-reduced chemical substances (such as H_2S, CH_4 and H_2), the global production of organic carbon could be enormously increased, and new environments opened for photosynthesis to occur. The CO_2 concentration (dissolved CO_2 and bicarbonate in seawater, respectively) began to decrease with the accumulation of biomass produced via photosynthesis because the organic carbon is buried in marine sediments,[102] leaving excess oxygen behind in the atmosphere – this excess oxygen would otherwise be used up as the organism decayed. Thus, for every carbon atom laid down as biological debris, approximately two oxygen atoms (as O_2) would be liberated.

The substantial deposition rates of ferric iron in massive banded iron sediment formations before 2.5 Gyr are clearly consistent with an abundant biological source of free oxygen. Indeed, vast sedimentary deposits of organic carbon, reduced sulphide, ferric iron, and sulphate on continental platforms and along coastal margins are among the most prominent and enduring legacies of billions of years of oxygenic photosynthetic activity. It is seen from the chemistry of photosynthesis that the process is a net source of oxygen. Dead organic material sank down to the sea bottom. The excess O_2 oxidised reduced compounds dissolved in seawater (Fe^{2+} to Fe^{3+} and other reduced metals, SO_3^{2-} to sulphate, NH_4^+ to nitrate) and not before reaching redox equilibrium, O_2 escaped from oceans into the atmosphere. Oxygen escaped into the

---

102 Note that respiration, which does return all the carbon and hydrogen contained in plant debris to the atmosphere as CO_2 and H_2O (the form in which it was taken up by the plants) was not yet available.

atmosphere was consumed for the production of CO_2 from CH_4. As mentioned above, another reduced gas was accumulated in the atmosphere from the very beginning, namely H_2S. In the ancient atmosphere, H_2S would be photolysed to H (which is escaping) and sulphur, which form S_{2n} molecules ($n = 1...4$) surviving and accumulating in the air. With the rise of atmospheric oxygen, therefore, the reduced sulphur pool must be oxidised first quantitatively before the biogenic oxygen production leads to rise atmospheric levels. Between 2.2 and 2.4 Gyr ago, a huge and rapid rise in atmospheric oxygen levels from less than 0.0001 % to at least 0.03 % is assumed, now often called the 'Great Oxidation Event'.

With increasing oxygen levels in the atmosphere, the ozone concentration rose – and as we have learned from photochemical modelling – faster than that of O_2. O_3 and O_2 are linked within a photo-stationary equilibrium (see Chapter 4.3.2.1). With increasing oxygen (and subsequent O_3), the absorption of UV(B) became complete. Before oxygen levels in the atmosphere were significant, a water column of about 10 m was sufficient to protect the layers below against UV. Only with reduced UV were aquatic organisms able to live near the surface, and finally, they were able to enter dry land and cover the continents. Thus it is necessary to state that neither missing nor present O_2 prevents colonisation of the land but the presence of hard UV radiation.

The accumulation of O_2 in the atmosphere led to the biological innovation of aerobic respiration, which harnesses a more powerful metabolic energy source. The toxic O_2 and the oxygen-containing radicals also caused different biological problems, now termed *oxidative stress*. The organisms answered this stress by developing mechanisms to protect themselves against oxidants (*antioxidants*). The organisms in existence at around 2 Gyr ago had two ways: firstly, to go back to anaerobic regions and live without oxygen, or secondly, to live in tolerance of oxygen. Choosing the second, evolution created with the respiration by heterotrophic organisms (biotic back reaction of equation (5.26)) a unique, biogenic-controlled equilibrium between atmosphere and biosphere, between reducing and oxidising regions of the Earth. The 'cycle' is closed by respiration, the process of liberation of chemical energy in the oxidation of organic compounds:

$$CH_2O + O_2 \rightarrow CO_2 \uparrow + H_2O. \tag{5.27}$$

It is remarkable that in this way, a stoichiometric ratio of 1 : 1 between fixed carbon and released oxygen is established. Therefore, net oxygen production is only possible when the rate of reaction equation (5.27) is lower than that of reaction equation (5.26), or in other words, the organic matter produced must be buried and protected against oxidation. This was the first closed biogeochemical cycle (Figure 5.5).

Oxygen probably continuously increased to about 2 % with the beginning of the Cambrian (600 Myr ago). This O_2 level would absorb 100 % of solar light with wavelength < 250 nm and 89 % of the wavelength < 302 nm (today, 97 % of the irradiation with the wavelength < 302 nm is absorbed). The water column necessary

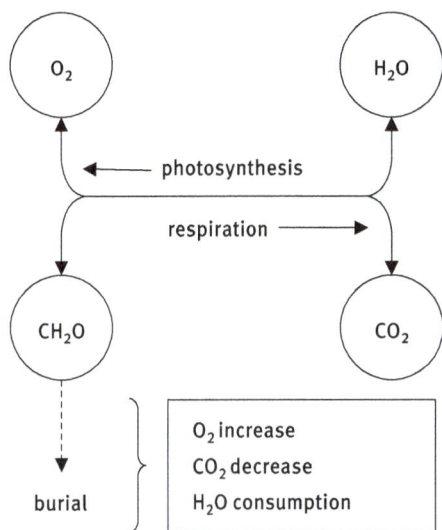

Figure 5.5: Schematic $CO_2 - O_2$ linkage: photosynthesis, respiration and organic carbon burial.

for protection reduced at this time to about one metre, and it is assumed that just after this time (0.5 Gyr ago) an erratic biological development on land began. At that time, at the end of the Ordovician and the beginning of the Silurian, the land was desolate and empty. It cannot be excluded that bacteria, lichen and algae already covered some parts of the land. The evolution from algae to land plants must have been a lengthy process. Then, after a short time, land plant photosynthesis increased O_2 in the atmosphere, and we assume that with the beginning of the Silurian (400 Myr ago), the ozone layer was sufficient to protect all life.

A second step in the rise of O_2 up to the present, 21 % was in connection with the colonisation of the land. From the Early Devonian (380 Myr ago), the evolution of flora gained momentum. The appearance of the first true trees is dated to about 370 Myr ago. According to the scientific consensus, the first verified land animal was a one-centimetre myriapod which appeared 428 million years ago. The earliest land animals probably lived in oxygen-poor shallow pools. The land at that time would have been much more nutrient-rich than the water, as plants colonised the land before animals and left their decaying plant matter everywhere. About 180 million years ago, the first mammals began to develop on land along with primitive birds. It took about 20 million years for animals to develop the art of breathing air and so to live on land. During the early Jurassic (warm tropical greenhouse conditions worldwide), then, evolution seems to have polarised: on the one hand, there were the ruling land animals, the great dinosaurs (for the next 135 million years), which filled the ecological roles now taken up by medium-sized and large mammals; on the other hand, the first mammals had appeared.

On land, carbon can be buried from litter and stepwise accumulation in soils but only at very low rates because of the presence of oxygen, and thus mineralisation was favoured. Hence only biomass under more or less anaerobic conditions (in lakes, marshes and the sea) can be deposited on the bottom and form sediments. Microorganisms, however, may facilitate the oxidation of sedimentary organic matter to inorganic carbon when sedimentary rocks are exposed by erosion. Thus, microorganisms may play a more active role in the biogeochemical carbon cycle than previously recognised, with profound implications for control on the abundance of oxygen and carbon dioxide in the earth's atmosphere over geological time.

On land, most photosynthesis is carried out by higher plants, not by microorganisms; but terrestrial photosynthesis has now little effect on atmospheric O_2 because it is nearly balanced by the reverse processes of respiration and decay.

> **i** Accepting that biological life (it remains a hypothesis) is causatively related to the changing air composition, we also have to take into account the feedback mechanisms. It is not important for our understanding to believe that the continental life was a result of the protecting ozone layer or whether the beginning land plants first created the protecting ozone layer via O_2 production (chicken-and-egg problem). What remains important is the idea that there exists a close relationship between biota and air. The evolution of one reservoir is the history of the evolution of the other one.

Holland (2006) divided the last 3.85 Gyr of earth's history into five stages:
1. During stage 1 (3.85–2.45 Gyr ago), the atmosphere was largely or entirely anoxic, as were the oceans, with the possible exception of oxygen oases in the shallow oceans.
2. During stage 2 (2.45–1.85 Gyr), atmospheric oxygen levels rose to values estimated to have been between 0.02 and 0.04 atm. The shallow oceans became mildly oxygenated, while the deep oceans continued to be anoxic.
3. During stage 3 (1.85–0.85 Gyr), atmospheric oxygen levels did not change significantly. Most of the surface oceans were mildly oxygenated, as were the deep oceans.
4. Stage 4 (0.85–0.54 Gyr) saw a rise in atmospheric oxygen to values not much less than 0.2 atm. The shallow oceans followed suit, but the deep oceans were anoxic, at least during the intense Neoproterozoic ice ages.
5. Atmospheric oxygen levels during stage 5 (0.54 Gyr-present) probably rose to a maximum value of ~0.3 atm during the Carboniferous before returning to the present value (0.21 atm). The shallow oceans were oxygenated, while the oxygenation of the deep oceans fluctuated considerably, perhaps on rather geologically short timescales.

5.1.5 Volcanism and weathering: inorganic CO_2 cycling

There are about 500 active volcanoes on the Earth. Of these, about 3 % erupt each year and of that number about 10 % have sufficient explosive power to transport gases and particles to the stratosphere. Magmatic gases released from volcanoes today contain water vapour and carbon dioxide as the main components, with smaller contributions of SO_2/H_2S, HCl, HF, CO, H_2, and N_2 but also traces of organic compounds, volatile metal chlorides and SiF_4. Modern volcanic gases are believed to be more oxidised than those at an early time in the earth's formation. Care is also required concerning the composition of volcanic exhalations at the earth's beginning, i. e. the present composition of volcanic exhalations may not absolutely represent the former one due to the recycling of rocky materials through volcanoes.

> Volcanism from the crust to the earth's surface is the driving force in recycling rocky material today. Subduction is the process in which one tectonic plate is pushed downward beneath another plate into the underlying mantle when plates move towards each other. The plate that is denser will slide under the thicker, less dense plate. Faulting (the process in which rocks break and move or are displaced along with the fractures) occurs in the process. The subducted plate usually moves in jerks, resulting in earthquakes. The area where the subduction occurs is the subduction zone. Magma is produced by the melting plate. It rises through fractures in the crust and reaches the surface to form volcanoes. It has been suggested that oceanic crust recycled into the mantle during subduction could be the source of plume volcanism. The oceanic crust sinks into the deeper mantle and accumulates at some level of density compensation, possibly at the core-mantle boundary. The accumulated layer locally reaches thicknesses exceeding 100 km. This model has proved to be very successful and is now widely accepted by the scientific community. The oceanic material recycled into the mantle is a combination of oceanic basalts from mid-ocean ridges, seamounts and ocean islands as well as sedimentary material deposited on the ocean floor. Moreover, a large amount of seawater (including dissolved matter) flows into the magma. In this way, atmospheric gases (oxygen, nitrogen and noble gases) dissolved in seawater can also go through subduction zones into the mantle.

The composition of volcanic exhalations differs from volcano to volcano. One has to draw the conclusion that modern volcanism provides a mixture of recycled atmospheric and surface material with primordial rocky gas evolution. It is clear that volcanic activity during the early history of the Earth (degassing period) was orders of magnitude higher than it is today (see Chapter 5.1.3). From 1975 to 1985, an average of 56 volcanoes erupted yearly. While some showed continuous activity, others erupted less frequently or only once, so that 158 volcanoes actually erupted over this time period. This number increases to 380 volcanoes with known eruptions in the twentieth century, 534 volcanoes with eruptions in historical times, and more than 1,500 volcanoes with documented eruptions in the last 10,000 years. Hence emission estimates found in the literature are always averages over different long timescales. Therefore, the true annual emission for a given year is almost unknown and varies considerably from year to year. In years with large volcanic eruptions, the emission could be many times higher than the average.

Volcanic gases are globally imbalanced on timescales compared to biogenic processes. Closure of the volcanic cycle via oceanic subduction and magma transformation is very slow; thus, we talk in geological timescales.

Volcanoes regulate the climate through CO_2 emissions.[103] Carbon dioxide emissions from volcanoes are given between 75 Tg yr^{-1} and 500 Tg yr^{-1}. The Mt. Etna CO_2 plume emission and diffuse emission combined amounts to 25 Mt yr^{-1}. Volcanoes are the only net source (remember: biospheric actions provide closed cycles due to emission = uptake globally) of reduced substances (mainly sulphur species) that influence the oxidation capacity of the atmosphere. There are a large number of global estimates of volcanic SO_2, with a variation between 0.75 and 30 Tg S yr^{-1}; the value with the most agreement seems to be 10 ± 5 Tg S yr^{-1}.

The volcanic CO_2 emission is very small compared to the CO_2 emission by fossil-fuel burning today. Burning of fossil fuels amounts now to $\sim 8 \cdot 10^{15}$ g C yr^{-1}, which is a mere 10 % of the terrestrial carbon uptake by photosynthesis; however, it interrupts the carbon cycle due to the large residence time of CO_2 in the atmosphere. The oceans mitigate this increase by acting as a sink for atmospheric CO_2. It is estimated that the oceans remove about $2 \cdot 10^{15}$ g C yr^{-1} from the atmosphere. This carbon is eventually stored on the ocean floor. Although these estimates of sources and sinks are uncertain, the net global CO_2 concentration is increasing. Direct measurements show that currently, each year, the atmospheric carbon content is increasing by about $3 \cdot 10^{15}$ g. Over the past two hundred years, CO_2 in the atmosphere has increased from about 280 parts per million (ppm) to its current level of 412 ppm (2020).

The CO_2 cycle has one major problem in the atmosphere – there is no direct chemical sink. In nature, CO_2 can only be assimilated by plants (biological sink) through conversion into hydrocarbons and stored in calcareous organisms, partly buried in sediments but almost completely turned back into CO_2 by respiration; hence, CO_2 partitions between the biosphere and atmosphere. The only definitive carbon sink is the transport of DIC to the deep ocean – when the ocean-atmosphere system is not in equilibrium, i. e. in case of increasing atmospheric CO_2 levels (due to anthropogenic and volcanic activities).

The only driving forces behind abiogenic removal of CO_2 from the atmosphere are dry deposition (absorption by the ocean, rivers, lakes, soils and rocky environment) and wet deposition (CO_2 scavenging by clouds and precipitation). River run-off is about $0.46 \cdot 10^{15}$ g yr^{-1} carbon and is much larger than the total wet deposited carbonate ($0.13 \cdot 10^{15}$ g yr^{-1} carbon), but is accounted for by volcanic emissions (and likely a small part of man-made CO_2).

---

103 On a short timescale, volcanic emissions play no role compared with other sources of emissions – with the exception of supervolcanic events. However due to emission into the upper troposphere, and occasionally direct into the lower stratosphere, volcanoes play an important role in providing trace species in layers of the atmosphere where the residence times increase significantly.

In the carbon cycle we have to consider long-term cycling, including rock weathering and volcanism. Over geological timescales, large (but very gradual) changes in atmospheric CO_2 result from changes in this balance between rock weathering and volcanism. CO_2 in the atmosphere is consumed in the weathering of rocks:

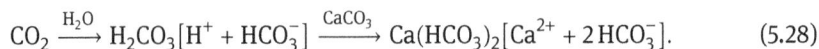

$$CO_2 \xrightarrow{\text{H}_2\text{O}} H_2CO_3[H^+ + HCO_3^-] \xrightarrow{\text{CaCO}_3} Ca(HCO_3)_2[Ca^{2+} + 2\,HCO_3^-]. \tag{5.28}$$

This comes about by the first global reaction, transforming silicates into carbonates:

$$CO_2 + (Ca, Mg)SiO_3 \rightarrow (Ca, Mg)CO_3 + SiO_2. \tag{5.29}$$

Carbonic acid is strong enough to dissolve silicate rocks – in small quantities, of course, and over long timescales. To illustrate this, we take an orthosilicate which is dissolved into orthosilicic acid (where SiO_2 is the anhydride) and bicarbonate:

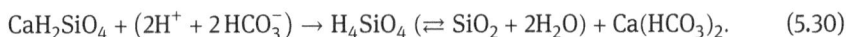

$$CaH_2SiO_4 + (2H^+ + 2\,HCO_3^-) \rightarrow H_4SiO_4 \,(\rightleftarrows SiO_2 + 2H_2O) + Ca(HCO_3)_2. \tag{5.30}$$

SiO_2 is moderately soluble in water (5–75 mg L^{-1} in river water and 4–14 mg L^{-1} in seawater, depending on pH and crystallite form). The products are then transported by river water to the oceans. There organisms such as foraminifera use calcium carbonate to make shells. Other organisms such as diatoms make their shells from silica. When these organisms die, they fall into the deepest oceans. Most of the shells redissolve but a fraction of them is buried in sediments on the sea floor. The overlying sediments are carried down to the depths by subduction. Temperature and pressure transform the shells back to silicate minerals, in the process releasing CO_2 back to the surface of the Earth through volcanoes and into the atmosphere to begin the cycle again, over a geological timescale. This inorganic (no photosynthesis) but biotic (mineral production) carbon cycle is not linked with the oxygen cycle but with water (H_2O) and acidity (H^+). Simply said, insoluble rock carbonate is transformed into more soluble bicarbonate where atmospheric CO_2 is fixed as dissolved bicarbonate. The volcanic carbon dioxide released is roughly equal to the amount removed by silicate weathering; so, the two processes, which are the chemical reverse of each other, sum to roughly zero, and do not affect the level of atmospheric carbon dioxide on timescales of less than about 10^6 years. As a planet's surface becomes colder, however, atmospheric CO_2 levels should tend to rise. The reason is that removal of CO_2 by silicate weathering followed by carbonate deposition should slow down as the climate cools, and would cease almost entirely if the planet were to glaciate globally. On planets like Earth that have abundant carbon (in carbonate rocks) and some mechanism, like plate tectonics, for recycling this carbon, volcanism should provide a more-or-less continuous input of CO_2 into the atmosphere.

5.2 Biogeochemistry

Biogeochemistry is the study of how chemical elements move through living systems and their physical environments, which also investigates the factors that influence cycles of key elements such as carbon, nitrogen, sulphur, phosphorus and others.

As the term "biogeochemistry" says, it is almost biochemistry and geochemistry of the lithosphere, pedosphere and hydrosphere in interaction with living organisms – mostly microorganisms, to a lesser importance plants and nearly negligible on a global scale, animals. However, humans became able to disturb and interrupt *natural* biogeochemical cycles, and thus artificial emissions also became a part of biogeochemical cycling. The atmosphere is a part of biogeochemical cycles; in short, *biogeochemistry* meets *atmospheric chemistry* at the borderline between the atmosphere and the earth's surface, studying the mass balance (or budget) including fluxes, turnovers and metabolism. The common interest concerns the sea-atmosphere, soil-atmosphere, and vegetation-atmosphere exchange and the impacts like acidification, "*Waldsterben*", eutrophication, global and climate change.

The chemical composition of the environment is determined by the emission of compounds from marine and terrestrial biospheres, anthropogenic sources and their chemistry and deposition processes (Figure 5.6). Biogenic emissions depend on physiological processes and the climate, and the atmospheric chemistry is governed by the atmospheric composition and climate involving feedback. Anthropogenic emissions depend on technical processes and the willingness of humans to control the climate and to keep the environment clean. Biogeochemical cycling, where the emissions from plants and microorganisms (in soils and waters) in terms of fluxes and substances specify the chemical regime, drives environmental chemistry. Behind the chemistry of the environment stands the chemistry in the lower atmosphere with interfaces *to* soils, waters and plants. Discussing the fate, transport and transformation of chemicals *in* soils, waters and plants is the task of biogeochemistry.

The Russian geochemist Vladimir Ivanovich Vernadsky [Владимир Иванович Вернадский] (1863–1945) is the founder of the new science *biogeochemistry*. Between 1909 and 1910, he established geochemistry as an independent branch of science, independently from Victor Moritz Goldschmidt (1888–1947), who pointed out in 1937 the importance of the work of Vernadsky. Whereas Frank Wigglesworth Clarke (1847–1931) collected an extensive geochemical data collection, Vernadsky treated the ways of dynamic geochemistry, already seeing biogeochemical cycling (yet without using this term), writing that understanding the chemical composition of the earth's crust needs the investigation of the natural processes and the role of chemical elements in the earth's history. It was not until the 1980s that western scientists were aware of Vernadsky's work. Still, Bert Bolin (1925–2007), who likely used first in the West the term "biogeochemistry" in the sense of science (1983), did not cite Vernadsky (further reading Möller (2022)).

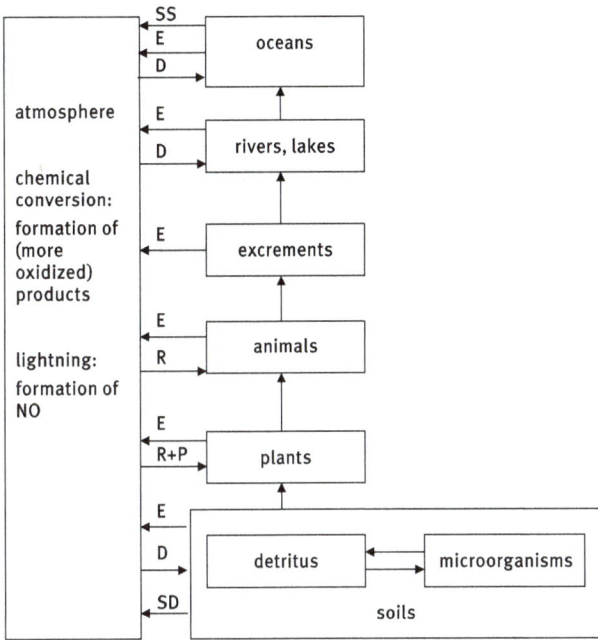

Figure 5.6: Biosphere–atmosphere interaction: fluxes of emission and deposition. E – emission of gaseous substances, SD – soils dust emission, SS – sea spray emission, D – deposition (dry, wet and sedimentation), R – respiration, P – photosynthesis.

The idea to establish a special project on global biogeochemical cycles within SCOPE (Scientific Committee of Problems of the Environment, founded 1969) arose in 1973 on a session in Paris, where Bert Bolin played a leading role in its project on biogeochemical cycles. However, Vernadsky was the first to propose the idea of biogeochemical cycling, having asked: what is the impact of life on the geology and chemistry of the Earth? Vernadsky understood by *noosphere*, called the *anthroposphere* by Paul Crutzen (1933–2021), a new dimension of the biosphere, developing under the evolutionary influence of humans on natural processes; consequently, Crutzen and Stoermer (2000) proposed to name the present epoch *Anthropocene*.

5.2.1 Biogeochemical cycling

The *biosphere*[104] is considered to represent the earth's crust, atmosphere, oceans, and ice caps, and the living organisms that survive within this habitat. Hence, the bio-

---

104 The term "biosphere" was coined in 1875 by the famous Austrian geologist Eduard Suess (1831–1914). The term "ecosystem" was coined in 1930 by Arthur Roy Clapham (1904–1990) to denote the combined physical and biological components of an environment.

sphere is more than a sphere in which life exists. It is the totality of living organisms with their environment, i. e. those layers of the Earth and the earth's atmosphere in which living organisms are located. Another common definition, such as 'the global sum of all ecosystems, however, calls for a definition of what is meant by an 'ecosystem'. An *ecosystem* is a natural unit consisting of all plants, animals and microorganisms (biotic factors) in an area functioning together with all of the non-living physical (abiotic) factors of the environment. Living phytomass becomes, via photosynthesis, the driving geological (in truth, biological) force moving material around the system naturally.

Some life scientists and earth scientists use the *biosphere* more limitedly. For example, geochemists define the biosphere as the total sum of living organisms (the 'biomass' or 'biota' referred to by biologists and ecologists). Thus, the three major biospheric pools are *live phytomass* (the mass of animals is negligible in a global context), *consumers* (animals) and *litter* (dead biomass). In this sense, the *biosphere* is one of four separate components of the geochemical model, the other three being the *lithosphere*, *hydrosphere* and *atmosphere*. The narrow meaning used by geochemists is one of the consequences of specialisation in modern science. Some might prefer the word *ecosphere*, coined in the 1960s, as all-encompassing of both biological and physical components of the planet.

Fundamentally, life on Earth is composed of six major elements, namely H, C, N, O, S and P. These elements are the building blocks of all the major biological macromolecules, including proteins, nucleic acids, lipids and carbohydrates. The production of macromolecules requires energy, which is almost exclusively derived from the Sun. The character of biological energy transduction is non-equilibrium redox chemistry. Besides energy, the production of macromolecules requires an input of a chemical substrate as the raw material, which has transportable (volatile, solvable) and transformable (oxidable, reducible, dissociable) geochemical conditions. Biological evolution gave rise to specified cells and organisms responsible for driving and maintaining global cycles. Because redox reactions always occur in a pair (oxidation-reduction), the resulting network is a linked chemical system of the elemental cycles. For example, reducing carbonate, sulphate and nitrate requires hydrogen and the oxidation of organic compounds (including carbon, sulphur and nitrogen) requires oxygen.

According to the biological cycle, which is that part of the biogeochemical cycle where biomass is produced and decomposed (Figure 5.7), all volatile compounds occurring in the biochemical cycles can be released to their physical surroundings. These are mainly:

- Carbon: CH_4, NMVOC (non-methane hydrocarbons), CO, CO_2
- Nitrogen: NH_3, N_2O, N_2, NO, RNH_2 (organic amines)
- Sulphur: H_2S, COS, $(CH_3)_2S$ (DMS), RSH (organic sulphides).

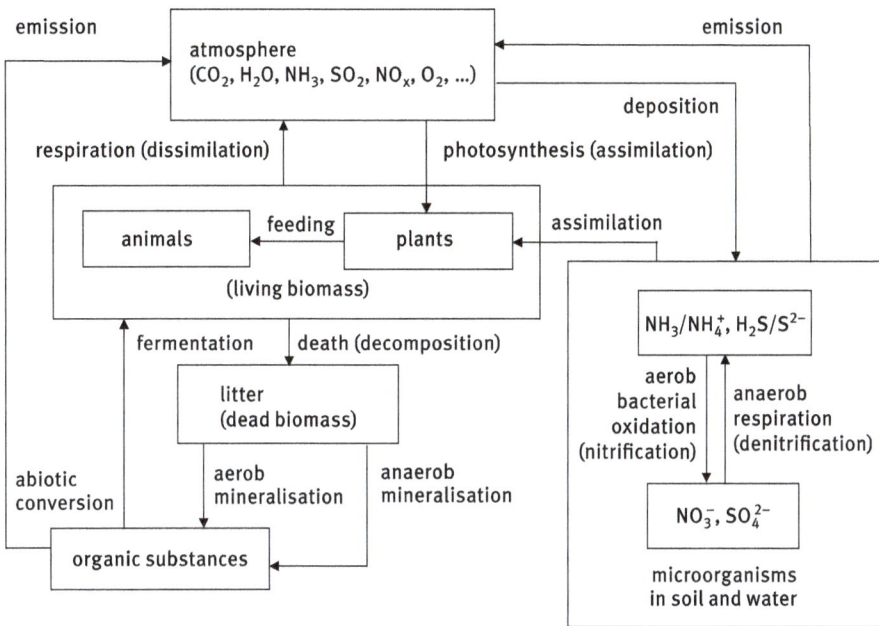

Figure 5.7: Scheme of global cycling and biosphere–atmosphere interaction.

The sulphur and nitrogen cycle is inherently linked with the carbon cycle. As discussed before (Chapter 5.1.4), oxygen is closely coupled with the carbon cycle and is chemically combined with compounds of nitrogen and sulphur (and others). Other cycles, such as those of phosphorus and of trace metals, are important for the biosphere but play a minor role in the atmosphere. With a few exceptions, most of the compounds of these cycles found in the atmosphere are condensed within particulate matter, only in recently appeared the understanding of how halogen cycling has begun.

Since the appearance of humankind, a new driving force has developed; the human matter turnover as a consequence of artificial (or anthropogenic) emissions. Therefore, today's biogeochemical cycles are very clearly no longer natural but anthropogenically-modified cycles. Moreover, human-induced *climate change* already causes permanent changes in natural fluxes; for example, through weathering, shifting redox and phase equilibriums.

A biogeochemical cycle comprises the sum of all transport and conversion processes that an element and its compounds can undergo in nature. The substances undergoing the biogeochemical cycle pass through several reservoirs (atmosphere, hydrosphere, pedosphere, lithosphere and biosphere), where certain concentrations accumulate because of flux rates determined by transport and reaction. The atmospheric reservoir plays a major role because of its high dynamic in transport and reaction processes and the global linkage between biosphere and atmosphere (Figure 5.7). A global cycle may be derived from the compositional balance of the individual reservoirs, where a (quasi) steady state is considered to exist. It shows variations on different timescales and may be disturbed by catastrophic events (e. g. volcanism, collision with other celestial bodies). The largest perturbation, however, is the one that can be observed over the past hundred years, caused by humans. Today, global observations from space result in almost three-dimensional concentration patterns, and by modelling, biogeochemical cycles can be simulated from the past until now under the influence of human activities, and furthermore, future scenarios of the changing world have been calculated.

Biological systems constantly synthesise, change and degrade organic and inorganic chemical species. The biological cycle includes the synthesis of living organic matter (plants) from inorganic compounds, which are mainly in the upper oxidised level (carbon dioxide, water, sulphate, and nitrate). During biogenic reduction to more reduced sulphur and nitrogen species, a variety of volatile S and N compounds are formed, which can enter the atmosphere; in other words, they undergo a phase transfer (emission).

> ℹ Biological processes can be considered as the driving force of atmospheric cycles. Emitted biogenic compounds were oxidised in the atmosphere and returned to ecosystems (mostly) in the oxidised form to be reduced and create a global redox couple.

From a physicochemical point of view, the driving forces in transport and transformation are *gradients* in pressure, temperature and concentration. Substances, once released from anthropogenic sources into the air, become an inherent part of biogeochemical cycles. Uptake of gases by *assimilation* occurs in plants because of photosynthesis and respiration and in animals due to respiration. Uptake of gases by sorption onto solid and aquatic surfaces of the Earth is a physicochemical process, termed *dry deposition*; however, assimilative uptake is also said to belong to dry deposition. Other processes of deposition (wet deposition and sedimentation) do not depend on surface properties but have to be taken into account as input fluxes to the biosphere. Biogenic emission is basically a loss of matter from the ecosystem. By turning into other ecosystems via atmospheric transport and transformation, however, it could have several functions (information by pheromones, self-protection, climatic regulation and nutrient spreading).

5.2.2 Principles of photosynthesis

> ℹ Photosynthesis is a series of processes in which electromagnetic energy is converted to chemical energy used for the biosynthesis of organic cell materials.

Somewhere around 3 billion years ago, an enzyme emerged that would dramatically change the chemical composition of our planet and set in motion an unprecedented explosion in biological activity. This enzyme used solar energy to power the thermodynamically and chemically demanding reaction of water splitting. In so doing, this enzyme provided biology with an unlimited supply of reducing equivalents (high-energy electrons) needed to convert carbon dioxide (CO_2) into the organic molecules of life. The by-product of the water-splitting reaction is molecular oxygen. The release of this gas also had dramatic consequences for the development of life since it created an oxygenic atmosphere and at the same time allowed the ozone layer to form. With oxygen available, the efficiency of cellular metabolism increased dramatically since, for a given amount of substrate, aerobic respiration provides in the region of 20 times more energy than anaerobic respiration. The overall process is described by the simple net equation

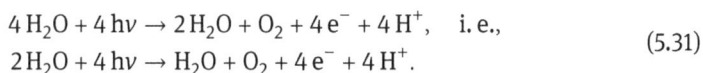

$$4\,H_2O + 4\,h\nu \rightarrow 2\,H_2O + O_2 + 4\,e^- + 4\,H^+, \quad \text{i.e.,}$$
$$2\,H_2O + 4\,h\nu \rightarrow H_2O + O_2 + 4\,e^- + 4\,H^+. \tag{5.31}$$

The splitting of two water molecules into oxygen using sunlight is the first step of photosynthesis, a process performed by plants, cyanobacteria, and algae. The biological water oxidation in nature is catalysed by a $CaMn_4O_5(H_2O)_4$ cluster housed in a protein environment that controls reaction coordinates, proton movement and water access. The cluster is the only biological catalyst that could oxidize water to molecular oxygen, and it appears that it has remained basically unchanged during 2 billion years of evolution. The enzyme which facilitates this reaction and therefore underpins virtually all life on our planet is known as Photosystem II (PSII). It is a pigment-binding, multisubunit protein complex embedded in the lipid environment of the thylakoid membranes of plants, algae and cyanobacteria. Today we have a detailed understanding of the structure and functioning of this key and unique enzyme. Many research groups investigate artificial photosynthesis as an alternative to electrochemical water splitting,[105] a key process in hydrogen production.

Here we will present the basic principles of photosynthesis and discuss the chemical evolution of the *assimilation* process. Let us understand assimilation is generally the conversion of nutrients into the fluid or solid substance of the body of an organism by the processes of digestion and absorption. It is not the aim here to discuss biological chemistry (*biochemistry*), but the *pathway* of inorganic molecules (CO_2, H_2O and O_2), which are the 'fundaments' of our climate and, therefore, environment through the organism. It is often said that our biosphere is far from redox equilibrium, or in other words, without photosynthesis, atmospheric oxygen would soon disappear. Establishing redox equilibrium requires that all redox couples (oxidants and reductants)

---

105 At the cathode, molecular hydrogen is produced by recombination of the protons and electrons. At the anode, water is split and molecular oxygen evolves together with protons and electrons. The anode reaction is found to be associated with slow kinetics, and an effective catalyst for the oxygen evolution reaction is needed.

in a natural system (such as waters, the atmosphere or within an organism) must be in equilibrium or, in other words, the rates of oxidation are equal to the rates of reduction – the net flux of electrons is zero. This is not the case in real systems because of different timescales between chemical kinetics (single reaction rates), transport rates and microbial catalysis. Moreover, many reactions are irreversible in a subsystem (for example, sulphate production in the atmosphere), and the products must transfer into another system (in that example, in soil having microbial anaerobic properties) for closing a cycle in the sense of dynamic but not thermodynamic equilibrium.

Without life on Earth, probably most of the geochemical redox potentials would reach equilibrium due to the tectonic mixing of all redox couples over the entire earth's history. Consequently, the role of green plants is the unique 'transfer' of photons from solar radiation into electrons and their transfer onto carrier molecules, creating electrochemical gradients and promoting synthesis and degradation. In Chapter 5.1.4, we shortly characterised the evolution of life from first organisms, which converted organic compounds by fermentation into biomolecules and the next steps via anoxygenic photosynthesis (using H_2 as a reducing agent) to oxygenic photosynthesis (water splitting process). The reactions equation (5.22) to equation (5.26), however, do not represent *elementary* chemical reactions but gross turnover mechanisms (termed in biology *metabolisation*). The (bio-)chemical processes consist of many steps (reactions chains) and include organic catalysts (*enzymes*), complex biomolecules being carriers of reducing (H) and oxidising (O) properties, as well as structured reactors with specific functions (hierarchic cell organs), transport channels and organic membranes (being the separating plates between different 'reactors').

Basically, the biological water-splitting process, where colouring matter (such as chlorophyll) is able to absorb photons and transfer them (similar to a photovoltaic cell) into electrons, works very similarly to an electrolytic cell (Figure 5.8) with a cathodic (electron donor) and an anodic site (electron acceptor). Chlorophyll (like other chromophores) consists of several conjugated π-electron systems containing electrons easily excitable by light absorption. For most compounds that absorb light, the excited electrons simply return to the ground energy level while transforming the energy into heat. However, if a suitable electron acceptor is nearby, the excited electron can move from the initial molecule to the acceptor. This process results in the formation of a positive charge on the initial molecule (due to the loss of an electron) and a negative charge on the acceptor and is, hence, referred to as *photoinduced charge separation*; The site where the separation change occurs is called the *reaction centre*:

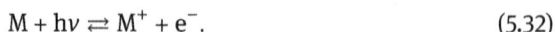

$$M + h\nu \rightleftarrows M^+ + e^-. \tag{5.32}$$

Thus, reaction (5.31) represents a chain of consecutive steps within a heterogeneous biological system and should not be confused with the simple inorganic aqueous-phase process such as described in Chapters 3.3.5.2 and 4.3.3. Formally, reduction of water

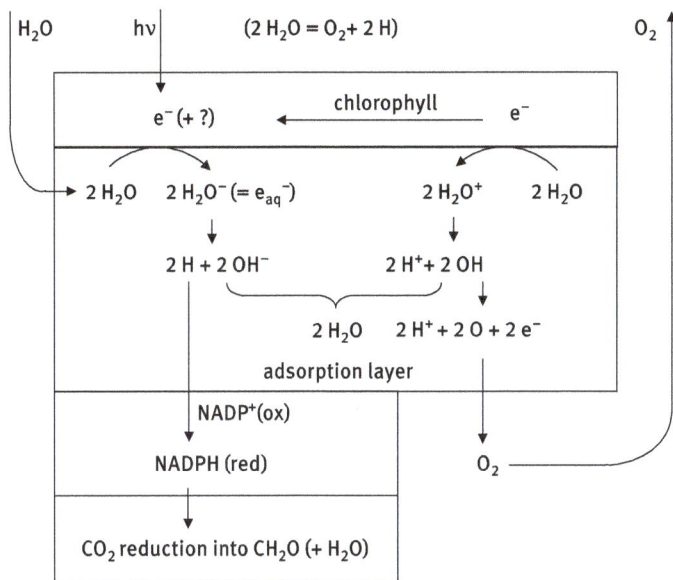

Figure 5.8: The basic scheme of the water-splitting process.

represents a redox couple: positive-charged hydrogen (H^+) is reduced to 'neutral' H and negative-charged oxygen (OH^-) is oxidised to 'neutral' O:

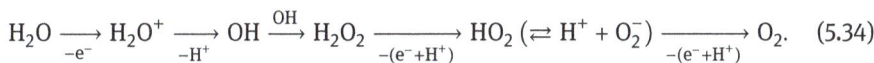

$$H_2O \xrightarrow{e^-} H_2O^- \rightleftharpoons H^+ + e^- + OH^-, \tag{5.33}$$

$$H_2O \xrightarrow[-e^-]{} H_2O^+ \xrightarrow[-H^+]{OH} OH \xrightarrow{} H_2O_2 \xrightarrow[-(e^-+H^+)]{} HO_2 \, (\rightleftharpoons H^+ + O_2^-) \xrightarrow[-(e^-+H^+)]{} O_2. \tag{5.34}$$

To avoid the production of high-energy intermediates, the natural system uses proton-coupled electron transfer mechanisms. Thus, either an electron transfer is followed by a proton transfer or a proton transfer is followed by an electron transfer. The electron continues moving along the membrane. Eventually, the electron reaches a molecule called $NADP^+$ (nicotinamide adenine dinucleotide phosphate: $C_{21}H_{29}N_7O_{17}P_3$),[106] and they combine to make NADPH (which is the reduced form of $NADP^+$). Next, the hydrogen ions move back into the fluid of the chloroplast. As the hydrogen ions move through a channel, a chemical reaction occurs, which changes the compound ADP and phosphates into ATP. It is in the light-independent reactions that NADPH and ATP are used to make glucose. This process is complex and called the Calvin cycle.

---

106 The IUPAC name is: [(2R,3R,4S,5S)-2-(6-aminopurin-9-yl)-5-[[[[(2S,3S,4R,5R)-5-(5-carbamoylpyridin-1-yl)-3,4-dihydroxy-oxolan-2-yl]methoxy-oxido-phosphoryl]oxy-hydroxy-phosphoryl]oxymethyl][-4-hydroxy-oxolan-3-yl]oxyphosphonic acid.

In plant, metabolism, reductive activation of molecular oxygen produces an array of reactive oxygen species (ROS), such as singlet oxygen ($^1O_2$), superoxide anion (O_2^-), hydrogen peroxide (H_2O_2) and hydroxyl radicals (OH). H_2O_2, O_2^-, and OH can convert to one another (see Chapter 4.3.3.2). H_2O_2 is produced as a by-product predominantly in plant cells during photosynthesis and photorespiration. It is the most stable of the so-called reactive oxygen species (ROS), and therefore plays a crucial role as a signalling molecule in various physiological processes. Growing evidence suggests that H_2O_2 plays a versatile role in plant defence and physiological reactions. To some extent, excess H_2O_2 accumulation can lead to oxidative stress in plants, which then triggers cell death. The evolution of all aerobic organisms is dependent on the development of efficient H_2O_2-scavenging mechanisms. Even under normal conditions, higher plants produce ROS during metabolic processes. Excess concentrations of ROS result in oxidative damage to or the apoptotic death of cells. The development of an antioxidant defence system in plants protects them against oxidative stress damage.

In summary, H_2O_2 plays a key role in regulating plant growth, development, resistance responses and signal transduction.

The energy provided by the donator (or, in other terms, the electrode potential) must be equivalent to the reaction enthalpy of equation (5.31). Consequently, the total system 'proton (wavelength) – chromophore (photon-electron-transfer) – electron (excited state or potential)' is a result of a coupled (quantum-) chemical and biological evolution. Subsequent to reaction equation (5.33) is the evolution of molecular oxygen and its release to the air. Biological evolution created the reaction system in such a way as to avoid oxygen diffusing to reducing (electronegative) sites.

Figure 5.9 shows simplified biochemistry, the organic synthesis during CO_2 reduction through photosynthesis. In a strong reducing medium, hydrogen (H) and carbon monoxide (CO) are the building blocks of organic compounds (similar to the Fischer–Tropsch synthesis). As seen, simple compounds in the reaction chain produced are also present in the main emitted substances: formaldehyde and acetaldehyde (HCHO and CH_3CHO), methanol (CH_3OH), methane (CH_4), acetone (CH_3OCH_3). The formation of acids (formic acid and acetic acid: HCOOH and CH_3COOH) needs an oxidative environment; remember, plants oxidise during respiration (which occurs the whole day, whereas photosynthesis only at light). The formyl radical HCO, a building block of sugars, is an important intermediate:

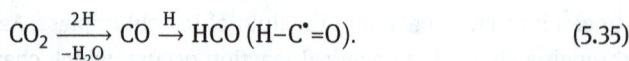

$$CO_2 \xrightarrow[-H_2O]{2H} CO \xrightarrow{H} HCO\ (H-C^\bullet{=}O). \tag{5.35}$$

The glucose made in photosynthesis travels around the plant as soluble sugars and gives energy to the plant's cells during respiration. The first stage of respiration is glycolysis, which splits the glucose molecule into two smaller molecules called pyruvate, and expels a small amount of ATP energy. This stage (anaerobic respiration) does not need oxygen. In the second stage, the pyruvate molecules are reorganized and fused over again in a cycle. While the molecules are being reorganized, carbon diox-

C_1 chemistry

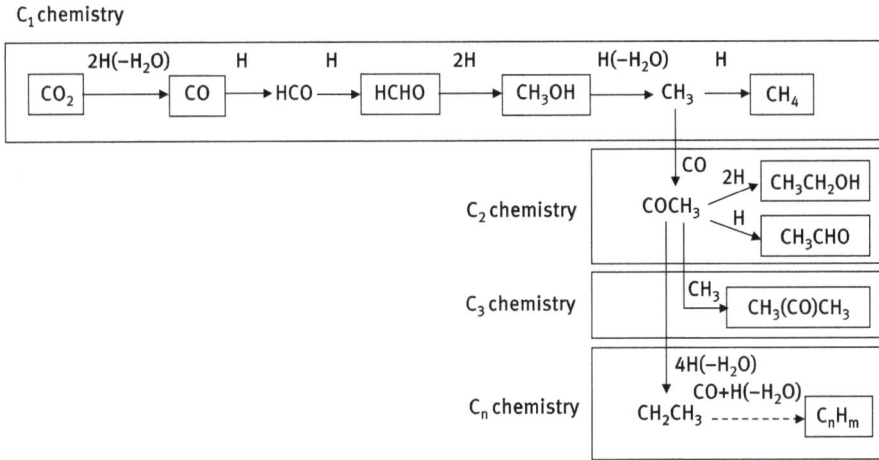

Figure 5.9: Simplified reaction scheme of photosynthetic formation of organic compounds. In boxes are stable substances, which are possibly emitted. The radical intermediates may undergo several competitive reactions, especially HCO (formyl) and CH_3 (methyl) depending on the ratio between reactants (H, CO, and O_2). In oxidative environments, acids are formed from aldehydes.

ide is formed, and electrons are removed and placed into an electron transport system, which (like in photosynthesis) produces a lot of ATP for the plant to use for growth and reproduction. This stage (aerobic respiration) does need oxygen. Plants respire at all times of the day and night because their cells need a constant energy source to stay alive. As well as being used by the plant to release energy via respiration, the glucose produced during photosynthesis is changed into starch, fats and oils for storage and used to make cellulose to grow and regenerate cell walls and proteins.

Although photosynthesis is one of the most vital chemical changes required for plant growth, respiration is equally important. Respiration is effectively photosynthesis backwards. In respiration, glucose molecules are broken down using ATP and enzymes to release energy. Plants are excellent at recycling. During the chemical changes in respiration, ATP is turned back into ADP plus a phosphate. This compound will be used again in photosynthesis.

5.2.3 Carbon cycle

In 1750, the atmospheric CO_2 level was 278 ± 5 ppm, which increased to 419 ppm in February 2022 (note that 1 ppmv of CO_2 = 2.13 Gt of carbon). Measurements and constructions of carbon balances, however, reveal that less than half of man-made emissions remain in the atmosphere. The anthropogenic CO_2 that did not accumulate in the atmosphere must have been taken up by the ocean, by the land biosphere or by a combination of both.

atmosphere (CO_2): 884 (+ 5 yr[1])

| 0.1 | 10.5 | 2.3 | 1.1 | 115 (P) (+3.4) | 60 (R_P) | 55 (R_S) | 79 (P) (+2.7) | 79 (R) |

volcanoes

fossil fuel and cement

biomass burning

land use change

land plants: 400–500

aboveground litter: 150–300
belowground litter: 120
soils, organic: 1600
soils, inorganic: 1100
permafrost: 1700
surface sediments: 1000

fossil fuels: 1000–2000
clathrates: 11000
organic C in sediments (kerogen): 15000
inorganic C in sediments (carbonates): 60000

river runoff: 0.8

DIC: 38000

DOC: 700
biota: 3

7.8 depth water

≤ 0.1 (burial)

sediments (seafloor): 150

~ 0.1   ≤ 0.1 (subduction)

mantle: ~500000 ($\tau \sim 3 \cdot 10^9$ yr)

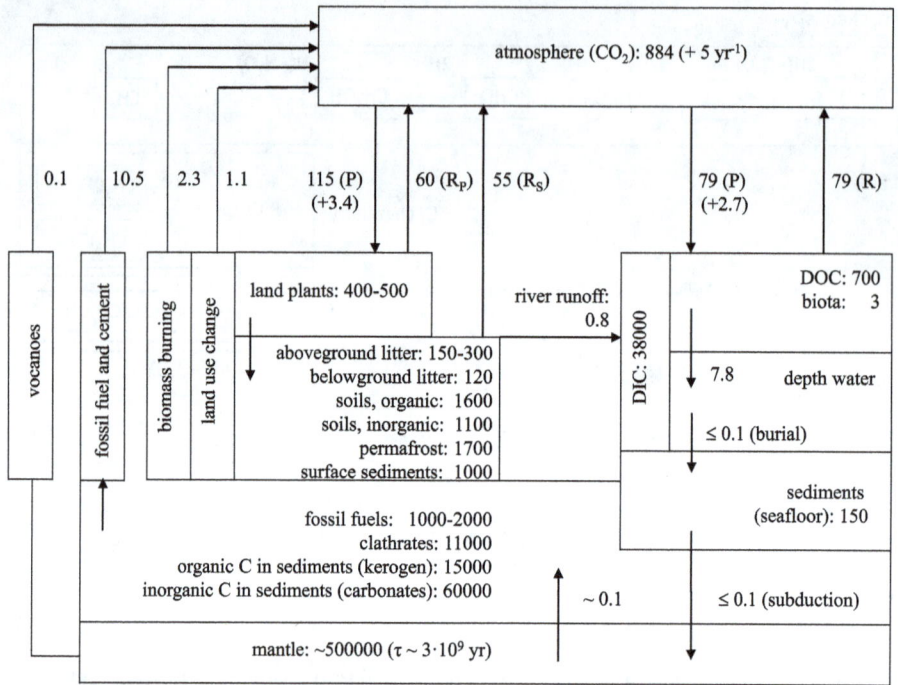

Figure 5.10: Scheme of the carbon cycle and reservoirs; fluxes in 10^{15} g yr^{-1} and pools in 10^{15} g. R_P – plant respiration, R_S – soil respiration, P – photosynthesis, DIC – dissolved inorganic carbon, DOC – dissolved organic carbon. River-runoff (0.8 Tg C yr^{-1}) consists of (all in Tg C yr^{-1}) 0.2 rock weathering, 0.4 soil weathering, and 0.2 hydrogen carbonate precipitation; The preindustrial ocean-atmosphere exchange amounts 70 Tg C yr^{-1} (20 anthropogenic additional), and the oceanic surface water contains 112 ± 17 Pg accumulated anthropogenic C. The atmospheric CO_2 mass corresponds to the actual CO_2 level of 415 ppm (November 2021). Data are compiled from different recent literature sources. Note that biomass CO_2 emission (also referred to as fire emission) is human-induced but within the natural carbon budget. The different uncertainties of fluxes do not allow a balanced global budget. Even the imbalance of human perturbance (difference between emissions and sinks) amounts −1 Tg C yr^{-1}. Despite very different values given in the literature for respiration and photosynthesis, the natural budget over a given longer time period is zero (i. e. balanced).

The amount of CO_2 removed from the atmosphere each year by oxygenic photosynthetic organisms is massive (Figure 5.10). It is estimated that photosynthetic organisms remove about $120 \cdot 10^{15}$ g C per year.[107] This is the *gross primary production* (GPP), the rate at which ecosystem producers capture and store a given amount of chemical energy as biomass in a given length of time. GPP is the primary driver of carbon cycling in vegetation and soils. Some fraction of this fixed energy is used by primary

---

107 This is equivalent to $4 \cdot 10^{18}$ kJ of free energy stored in reduced carbon, which is roughly 0.1 % of the visible radiant energy incident on the Earth per annum.

producers for cellular respiration and maintenance of existing tissues. Whereas not all cells contain chloroplasts for carrying the photosynthesis, all cells contain mitochondria for oxidising organic compounds, i. e. the yield of free energy (*respiration*). The remaining fixed energy is referred to as *net primary production* (NPP):

$$NPP = GPP - \text{plant respiration.}$$

A negative value of NPP means decomposition or respiration overpowered carbon absorption; more carbon was released to the atmosphere than the plants took in.

NPP is the primary driver of the coupled carbon and nutrient cycles and is the primary controller of the size of carbon and organic nitrogen stores in landscapes. The amount of carbon dioxide (CO_2) assimilated by photosynthesis is nearly equalled by the amount released back into the atmosphere by ecosystem respiration (defined as the sum of autotrophic plant respiration and heterotrophic microbial and animal respiration).

> Plant respiration consists of two parts: leaf respiration, which occurs in the light and darkness, and root respiration, which is counted to soil respiration. Root respiration accounts for approximately half of all soil respiration. Belowground autotrophic respiration (R_A), mainly originated from plant roots, mycorrhizae, and other microorganisms in the rhizosphere directly relying on the labile carbon component leaked from roots. Thus, respiration of the soil (R_S) can be partitioned into two processes: metabolic activity of plant roots (autotrophic respiration R_A) and the decomposition of dead organic material (heterotrophic respiration R_H). The global soil-to-atmosphere (or total soil respiration (R_S) carbon dioxide flux is increasing (which incorporates both R_H and belowground autotrophic respiration R_A). Recent findings suggest plant respiration is a larger source of carbon emissions than previously thought and warn that as the world warms, this may reduce the ability of the earth's land surface to absorb emissions due to fossil fuel burning.

Approximately half of the assimilated CO_2 is released by autotrophic (mostly dark) respiration (R_A), which varies with biotic and abiotic factors. Current values of R_A range between 44 and 64 Tg C yr^{-1}, which is approximately equal to the net primary production (NPP), and the decomposition of soil and litter is about 50 Tg C yr^{-1} (Fig. 5.8).[108] Global soils store at least twice as much carbon as the earth's atmosphere. Belowground autotrophic respiration is one of the largest but most highly uncertain carbon flux components in terrestrial ecosystems. Soil heterotrophic respiration (R_H) represents the carbon losses from the decomposition of litter detritus and soil organic matter by microorganisms.

---

108 Global carbon budgets usually ignored the respiration of human and animals due to a relatively small proportion compared with soil respiration on a global or regional scale.

The largest fraction of NPP is delivered to the soil as dead organic matter (litter), which is decomposed by microorganisms under the release of CO_2, H_2O, nutrients and a final resistant organic product, *humus*. Hence, a *net ecosystem production* (NEP) is defined:

$$NEP = NPP - \text{consumers respiration} (R_h).$$

Another part of NPP is lost by fires (biomass burning, by emission of volatile organic substances (VOC) and human use (food, fuel and shelter); loss of NPP (in 10^{15} g C yr^{-1}):
-   biomass burning: ~4.0
-   humans: 18.7
-   VOC emission: 1.2

NEP finally represents the burial carbon, a large flux at the beginning of the biospheric evolution but nowadays limited at about zero; the burial rate is very low, approximately between 60 Tg C yr^{-1} and 200 Tg C yr^{-1}. When carbon dioxide CO_2 is released into the atmosphere from the burning of fossil fuels and cement production, approximately 50 % remains in the atmosphere, while 25 % is absorbed by land plants and trees, and the other 25 % is absorbed into certain areas of the ocean.

It is due to the long residence time of that CO_2 in the atmosphere, which is not balanced within the biological carbon cycle (where the turnover time amounts only 2–3 years) that the carbon budget becomes out of balance, in other words resulting in an accumulation of carbon in the atmosphere, the ocean and terrestrial ecosystems.

Since the beginning of the Industrial Revolution, humans have emitted about $(365 \pm 30) \cdot 10^{15}$ g CO_2 – C from the combustion of fossil fuels and cement production, and about $(180 \pm 80) \cdot 10^{15}$ g CO_2–C from land-use change, mainly deforestation (period 1750–2011); Table 5.6. Since 2012, the yearly artificial CO_2 emissions were only very slightly increased, with a mean yearly value of $(9.7 \pm 0.8) \cdot 10^{15}$ g CO_2 – C yr^{-1}. Hence, now in 2021, humans have emitted since 1751 totally about 460 Tg C. More than half of all CO_2 emissions since 1751 have been emitted in the last 30 years. The atmospheric increase amounts until now to $292 \cdot 10^{15}$ g carbon; Table 5.6 shows two different global carbon budgets.

The Mauna Loa record (Figure 5.11), also known as the Keeling curve, is almost certainly the best-known icon illustrating the impact of humanity on the planet as a whole (Keeling et al. 1976). These measurements have been independently confirmed at many other sites around the world. By the early 1970s, this curve was gaining serious attention and played a key role in launching a research program into the effect of rising CO_2 on the climate. Since then, the rise has been relentless and shows a remarkably constant relationship with fossil fuel burning. Based on the simple premise that 57 % of fossil fuel emissions remain airborne, it can be well accounted for. Measurements of the changes in atmospheric molecular oxygen show that the O_2 content of air varies

Table 5.6: Global carbon budgets.

process	(Denman et al. 2007) in 10^{15} g C yr^{-1}		(Sabine et al. 2004) in 10^{15} g C		total in 10^{15} g C
	1990s	2000–2005	1800–1994	1980–1999	
industrial CO_2 emission[a]	6.4 ± 0.4	7.2 ± 0.3	244 ± 20	117 ± 5	329
land-use change	1.6[b]	1.6[b]	100–180	24 ± 12	156
sum	8.0	8.9	344–424	141	485
atmospheric increase	3.2 ± 0.1	4.1 ± 0.1	165 ± 4	65 ± 1	225[d]
difference (biospheric uptake)	4.8	4.8	179–259	76	260
ocean uptake	2.2 ± 0.4	2.2 ± 0.5	118 ± 19	37 ± 8	165[e]
terrestrial uptake	2.6[c]	2.6[c]	61–141	39 ± 18	95

[a] from fossil fuel use and cement production
[b] range 0.5–2.7
[c] range 0.9–4.3
[d] calculated from the difference 384 to 280 ppm CO_2
[e] to be assumed 50 % of cumulative industrial CO_2 emission

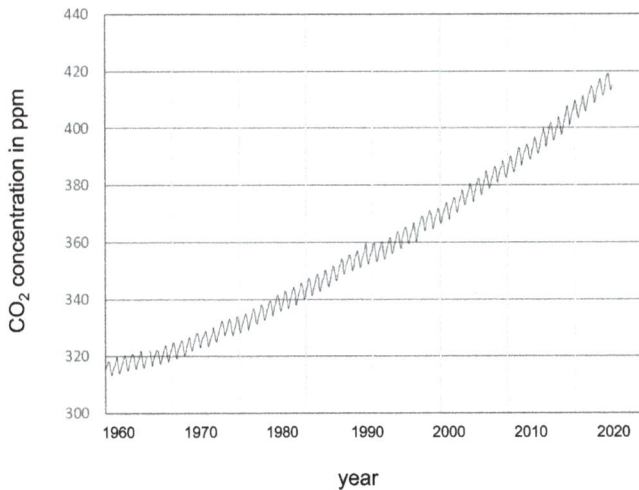

Figure 5.11: The Mauna Loa CO_2 record (Keeling curve) based on monthly averages (03/1958–11/2021). Data from: Dr Pieter Tans, NOAA/ESRL (www.esrl.noaa.gov/gmd/ccgg/trends/) and Dr Ralph Keeling, Scripps Institution of Oceanography (scrippsco2.ucsd.edu/).

(inversely with CO_2) seasonally in both the northern and southern hemispheres because of plant respiration and photosynthesis. The seasonal variations provide a new basis for estimating global rates of biological organic carbon production in the ocean, and the interannual decrease constrains estimates of the rate of anthropogenic CO_2 uptake by the oceans.

With constant increases expected, the atmospheric CO_2 mixing ratio is predicted to grow to 500 ppm in 2050 and 700 ppm in 2100. There is no doubt that CO_2 has been increasing since the Industrial Revolution and has reached a concentration unprecedented for over 400,000 years. However, by burning all the fossil fuels that are still available in a short time, the oxygen content of the atmosphere would be reduced by only 1 %. According to an estimate by Warneck (2000), the atmospheric CO_2 concentration would rise to about 800 ppm, twice the present level and about three times more than the preindustrial level. This is large compared with the CO_2 variation over the last few hundred thousand years but small concerning the timescale over epochs. The amount of combustible fossils fuels is only 40 times larger than the yearly biological turnover of carbon. The 'problem', however, consists in the human timescale of a few hundred years and the vulnerable infrastructural systems of mankind.

The most important reservoir is the back mixed surface layer of the ocean. The anthropogenic emissions are added to the atmosphere, continuously increasing the equilibrium carbon content of the surface layer as a result of the increased partial pressure of the carbon dioxide in the gas phase. The further transport of anthropogenic carbon from the surface to the ocean bulk (deep sea) is extremely slow (thousands of years), hence the limiting step. First quantification of the oceanic sink for anthropogenic CO_2 is based on a huge amount of measured data from two international ocean research programs (Table 5.6); the cumulative oceanic anthropogenic CO_2 sink for 1994 was estimated to be $(118 \pm 19) \cdot 10^{15}$ g CO_2 – C. The cumulative uptake for the 1750 to 2011 period is $\sim(155 \pm 30) \cdot 10^{15}$ g CO_2 – C from data-based studies, about 30 % of the total cumulative anthropogenic CO_2 emission.

The impact of global climate change on future carbon stocks is particularly complex. These changes might result in both positive and negative feedback on carbon stocks. For example, increases in atmospheric CO_2 are known to stimulate plant yields, either directly or via enhanced water use efficiency, thereby enhancing the amount of carbon added to soils. Higher CO_2 concentrations can also suppress the decomposition of stored carbon because C/N ratios in residues might increase and because more carbon might be allocated below ground. Predicting the long-term influence of elevated CO_2 concentrations on the carbon stocks of forest ecosystems remains a research challenge. The severity of damaging human-induced climate change depends not only on the magnitude of the change but also on the potential for irreversibility. Solomon et al. (2009) show that climate change that takes place because of increases in carbon dioxide concentration is largely irreversible for 1,000 years after emissions stop. There are strong arguments that the anthropogenic CO_2 increase is largely irreversible; hence, stopping emissions will not solve (though might smooth) climate change problems. The oceans have certainly been identified as the final sink of anthropogenic CO_2 but after thousands of years; moreover, the seawater uptake capacity will decrease, and oceanic acidification will result in serious ecological consequences.

There are strong arguments that the anthropogenic-caused CO_2 increase is largely irreversible; hence, stopping emissions will not solve (though might smooth) climate change problems. Consequently, CO_2 capture from the atmosphere remains a challenge for climate sustainability.

5.2.4 Nitrogen cycle

Even though the atmosphere is 78 % nitrogen (N_2), most biological systems are nitrogen-limited on a physiological timescale because most biota are unable to use molecular nitrogen (N_2). Two natural processes convert nonreactive N_2 to reactive N; *lightning* and *biological fixation*.

Atmospheric NO production via lightning is based on the same thermal equilibrium ($N_2 + O_2 \leftrightarrow 2\,NO$), which also takes place in all anthropogenic high-temperature processes (e. g. burning), see Chapter 4.4.2. This natural source, however, is too limited to supply the quantity of reactive nitrogen within the global biological nitrogen cycle (Figure 5.12). Biological nitrogen *fixation* by microorganisms in soils and oceans produces organic nitrogen (as reduced N^{3-} in the form of NH_2 bonding N) within the fixing organisms. This nitrogen is lost from the organisms after their death as ammonium (NH_4^+) via mineralisation. The fixed nitrogen budget can be viewed as a set of processes that transfer N between a large inert atmospheric N_2 pool and a much smaller pool of chemically diverse, biologically available, "fixed" nitrogen forms (NH_4^+/NH_3, R–NH_3, N2O, NO, NO_2^-/HNO_2, NO_2, NO_3^-/HNO_3) present in the biosphere, ocean, freshwaters, soils, and shallow sediments. The global environmental fixed nitrogen reservoir is primarily composed of dissolved inorganic and organic nitrogen in the ocean. Nitrate is by far the most abundant form of fixed nitrogen. Dissolved organic nitrogen (DON) in the ocean is the second-most abundant form of fixed nitrogen.

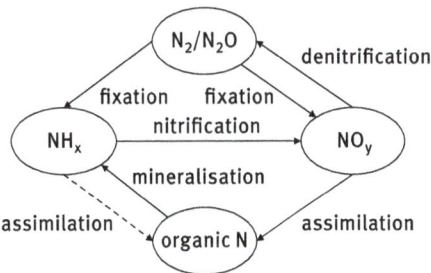

Figure 5.12: The biological nitrogen cycle.

Nitrification is the biological oxidation of ammonium (NH_4^+) to nitrate (NO_3^-), with nitrite (NO_2^-) as an intermediate under aerobic conditions: ($NH_4^+ \rightarrow NO_2^- \rightarrow NO_3^-$). During nitrification, because separate bacteria oxidise NH_4^+ into NO_2^- and NO_2^- into

NO_3^-, the process can lead to the temporary accumulation of NO_2^- in soil and water. Under oxygen-limited conditions, nitrifiers can also use NO_2^- or HNO_2 respectively as a terminal electron acceptor. These bacteria are generally chemoautotrophic, requiring only CO_2, H_2O and O_2. The nitrifying bacteria *Nitrosomas*, which converts NH_4^+ to NO_2^-, is also able to reduce NO_2^-. This nitrite can decompose abiotically, yielding NO or NO_2, substantially favoured in acidic soils (see Figures 4.7 and 4.12). The debate about whether NO is an intermediate or is produced by nitrifiers itself via NO_2^- reduction remains open. The reduction of NO_2^- to NO and N_2O by nitrifiers also prevents the accumulation of potentially toxic nitrite.

Denitrification is a microbial process for the growth of some bacteria in soil and water that reduces nitrate (NO_3^-) or nitrite (NO_2^-) to gaseous nitrogen oxides (almost all N_2O and NO) and molecular N_2 by essentially aerobic bacteria. The general requirements for denitrification to occur are: (a) the presence of bacteria possessing the metabolic capacity; (b) the availability of suitable reductants such as organic carbon; (c) the restriction of O_2 availability; and (d) the availability of N oxides. Current knowledge shows that the NO flux from the soil will depend both on physical transfer processes from the denitrification site to the atmosphere and on the relative rates of production and consumption of NO. Field measurements show that N_2 is the main product. NO_2 has been found as soil emission. However, it is not unlikely that the NO_2 detected was due to NO oxidation either in the soil air or in their chamber systems. Nevertheless, there are abiotic processes in soils that produce NO_2 from nitrite.

The accepted sequence for denitrification is demonstrated in Figure 4.7 – the inverse process corresponds to nitrification:

$$HNO_3 \text{ (nitric acid)} \rightarrow HNO_2 \text{ (nitrous acid)} \rightarrow HNO \text{ (nitroxyl)}.$$

Nitroxyl exists only as an intermediate and acts as a very weak acid ($pK = 11.4$) with NO^- being the protolytic anion, from which NO is formed via electron transfer (and vice versa). HNO is the transfer point to NO (as just described) and to N_2 and N_2O in parallel pathways (Figure 4.7). $H_2N_2O_2$ (hyponitrous acid), probably produced by enzymatic dimerisation of HNO, is formally the acid in the form of its anhydride N_2O (see for details Chapter 4.4.5.2).

Consequently, the biological nitrogen cycle is closed, beginning and ending with N_2: N_2 fixation $\rightarrow$ NH_4^+ *nitrification* $\rightarrow$ NO_3^- *denitrification* $\rightarrow$ N_2 (Figure 5.13). Each intermediate, which is produced within the biological cycle (NO, N_2O, NH_3, HNO_2, organic N species, see Figure 4.7), can escape the 'biosphere' (in terms of organisms, soil, water and plants) by physical exchange processes depending on many environmental factors. On the other hand, plants and organisms can take up the same species from their environment and, moreover, abiotically oxidised components (e. g. NO_2, HNO_3). The bacterial processes of denitrification and nitrification are the dominant sources of N_2O and NO in most systems. Only denitrification is recognised as a significant biological consumptive fate for N_2O and NO. The chemical decomposition of

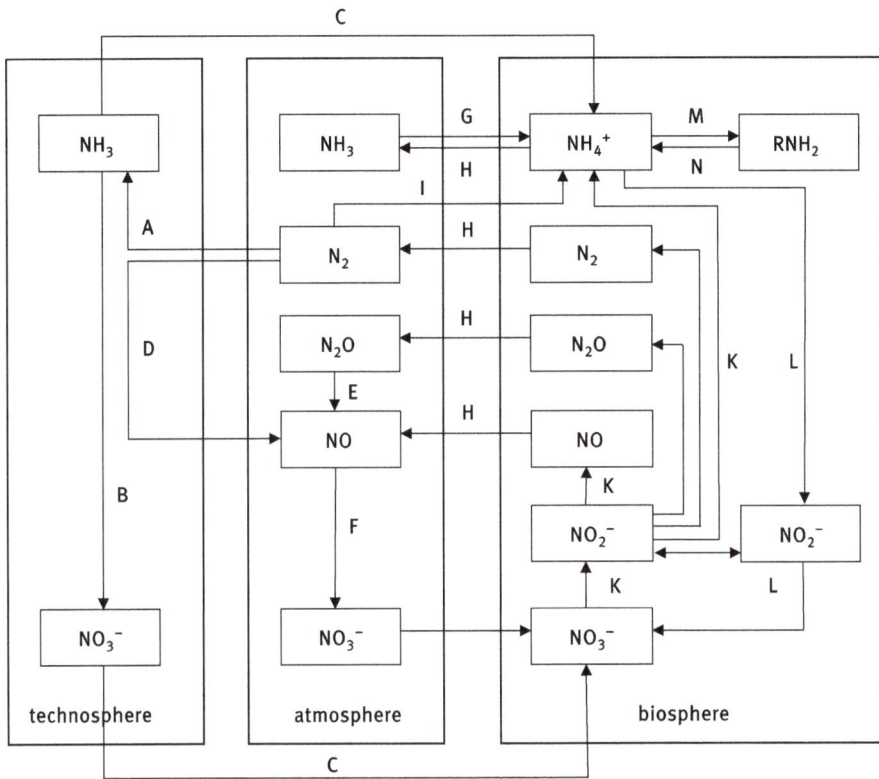

Figure 5.13: The biogeochemical nitrogen cycle. A – ammonia synthesis (artificial N fixation), B – oxidation of ammonia (industrial production of nitric acid), C – fertiliser application, D – formation of NO due to high-temperature processes, E – oxidation of N_2O within the stratosphere, F – oxidation of NO within the troposphere, G – ammonia deposition and transformation into ammonium, H – biogenic emission, I – biogenic N fixation, K – denitrification, L – nitrification, M – assimilation (biogenic formation of amino acids), N – mineralisation. RNH_2 organic bonded N (e. g., amines).

HNO_2 (or chemical denitrification), that is, the reduction of NO_2^- by chemical reductants under oxygen-limited conditions and at low pH, can also produce N_2, N_2O and NO. Chemical denitrification generally occurs when NO_2^- accumulates under oxygen-limited conditions, which may occur when nitrification rates are high, for example, after application of NH_4^+-based mineral fertilisers or animal manure. This process may account for 15–20 % of NO formation. Abiotic HNO_2 emission from soil NO_2^- also occurs because a simple chemical acid-base equilibrium occurs (see Chapter 3.2.2.3). HNO_2 is produced and emitted during nitrification occurring under low soil water content (less than 40 %) and may contribute up to 50 % of the reactive nitrogen released from soils. Emission and/or deposition (biosphere-atmosphere exchange) is, therefore, a complex function of biological, physical and chemical parameters describing the system.

In the process of *ammonification*, hydroxylamine (NH_2OH) is an important intermediate in two directions, denitrification direct to ammonium as well as the oxidation of ammonium to nitrate (nitrification). As long as nitrogen remains in its reduced form (NH_4^+), it remains in the local environment because of its affinity for soil absorption and its rapid uptake by biota. NH_4^+ is in equilibrium with NH_3, which can escape to the atmosphere, depending on pH, temperature, soil moisture, soil type and atmospheric NH_3 partial pressure. The equilibrium between emission and deposition (gas uptake) is called the *compensation point*; similar factors control the emission/dry deposition of NO. NH_3 is the major source of alkalinity in the atmosphere and acidity in soils. A small part of atmospheric NH_3 ($\leq 5\%$) is only oxidised by OH radicals, where the main product has been estimated to be N_2O, thus contributing around 5 % to estimated global N_2O production. Deposited NH_3/NH_4^+ will be *nitrified* in soils and water to NO_3^-, where two moles of H^+ are formed for each mole of NH_3/NH_4^+. Thus, any change in the rate of formation of reactive nitrogen (and N_2O), its global distribution or its accumulation rate can have a fundamental impact on many environmental processes.

In addition to being important to biological systems, reactive nitrogen also affects the chemistry of the atmosphere. At very low NO concentrations, ozone (O_3) is destroyed by reactions with radicals (especially HO_2), although at higher levels of NO (larger than 10 ppt), there is a net O_3 production (because HO_2 reacts with NO to form NO_2). The photolysis of NO_2 is the only precursor source of photochemically produced O_3 in the troposphere. Although N_2O is not viewed as a reactive form of nitrogen in the troposphere; it adsorbs IR radiation and acts as a greenhouse gas. N_2O, the product of microbial nitrification and denitrification, is a long-lived (114-year-lifetime) and potent greenhouse gas that, per molecule, is ~300 times stronger than CO_2. In the stratosphere, N_2O will be oxidised to NO_x and impacts the O_3 concentration.

The linkage between the biosphere and the atmosphere has to be assumed to have been in equilibrium prior to the current industrial age. Nitrogen species (NO_x, N_2O, N_2) emitted from plants, soils and water into the atmosphere, which are biogenically produced within the redox cycle between *nitrification* and *denitrification*, will be oxidised to nitrate and return to the biosphere. This cycle has been increasingly disturbed since the beginning of the Industrial Revolution more than 100 years ago.

> **i** Human activities related to food production and fossil fuel combustion have perturbed the global N cycle since the Industrial Revolution by introducing new N into terrestrial and marine fixed N inventories.

The most important aspect of the anthropogenically-modified N cycle is the worldwide increase of nitrogen fertiliser application. Without human activities, biotic fixation provides about 100 Tg N yr^{-1} on the continents, whereas human activities have resulted in the fixation of around an additional 200 Tg N yr^{-1} by fossil-fuel combustion (~40 Tg N yr^{-1}), fertiliser production (~120 Tg N yr^{-1}), and cultivation of crops (e. g.,

Table 5.7: Global turnover of nitrogen (in Tg yr^{-1}); data from several recent literature sources.

Natural nitrogen inputs	
terrestrial fixation	128
marine fixation	140–164
lightning	5
rock weathering	14–34
anthropogenic nitrogen inputs	
Haber–Bosch fertilizer production	120 (160 in 2100)
fossil fuel combustion	40
agricultural fixation	40
outputs	
terrestrial denitrification (land, rivers)	44
pyrodenitrifcation[a]	37
marine denitrification	200–300
marine sediment burial	20–58
terrestrial N_2O emission	13
marine N_2O emission	2–5
transport between reservoirs	
atmospheric emissions (NH_3, NO_x)	100
atmospheric deposition (land)	75
atmospheric deposition (ocean)	30–40
river runoff	30–40
to groundwater	18

[a]This is best defined as a way that nitrogen is lost or emitted by the use of heat or radiation, mainly, from the Sun, but also due to biomass burning. This can occur in the air or on the ground around vegetative regions. It is less likely to occur in arid regions, but it is possible

legumes, rice; about 40 Tg N yr^{-1}); Table 5.7. The oceans receive about 80 Tg N yr^{-1} (about 40 Tg N yr^{-1} by atmospheric deposition and about 40 Tg N yr^{-1} via rivers), which is incorporated into the oceanic nitrogen pool. The remaining about 220 Tg N yr^{-1} is either retained on continents, in water, soils and plants or denitrified to N_2. Thus, although anthropogenic nitrogen is clearly accumulating on continents, we do not know the rates of individual processes. It is predicted that the anthropogenic N fixation rate will increase by 60 % (based on 1990) by 2020, primarily due to increased fertiliser use and fossil-fuel combustion. About two-thirds of the increase will occur in Asia, which by 2020 will account for over half the global anthropogenic N fixation.

Excessive anthropogenic nutrients (N and P) from fertilizer runoff and atmospheric deposition have promoted widespread eutrophication of fresh and coastal water bodies.

In contrast to sulphur species, there are no differences in principle between natural and anthropogenic processes in the formation and release of reactive nitrogen species. Industrial nitrogen fixation (in separated steps: $N_2 \rightarrow NH_3$, $N_2 \rightarrow NO_x$, $NO_x \rightarrow NO_3$) proceeds via the same oxidation levels as biotic *fixation* and *nitrification*, either on purpose in chemical industries (ammonia synthesis, nitric acid production) or unintentionally in all high-temperature processes, namely combustion, as a by-product due to $N_2 + O_2 \rightarrow 2\,NO$. The input/output budget for fixed nitrogen is being dominated by biological processes: biological N_2 fixation as the nitrogen input and denitrification as the nitrogen loss. Moreover, fixed nitrogen has the interesting duality of being a critical factor for both biosynthesis and redox cycling. The net result appears to be a budget that is highly regulated by feedback. At this stage, the components of nitrogen cycling are relatively wellknown. However, their interactions and consequences remain mysterious, with conceptual arguments currently playing an outsized role.

5.2.5 Sulphur cycle

Like nitrogen, sulphur is an important element in biomolecules with specific functions. In contrast to nitrogen, where the largest pool is the atmosphere (molecular N_2), for sulphur, the largest pool is the ocean (as dissolved sulphate); both components are chemically stable. In the air, carbonyl sulphide (COS) represents the major sulphur component due to its long residence time. Similar to nitrogen and carbon, the sulphur content in the lithosphere is small because of degassing volatile sulphur compounds in the early history of the Earth. Primordial sulphides and elemental sulphur are almost all oxidised during atmospheric turnover. Moreover, humans have extracted them by mining from the lithosphere to such an extent that the remaining resources are negligible at present (Figure 5.14). The great 'role' of life again is the reduction of sulphate. Volcanism is an important source of sulphur dioxide (SO_2) and promotes the formation of a strong acid (H_2SO_4) which may play an important role in weathering. Consequently, the dominant anthropogenic SO_2 emission since the Industrial Revolution has resulted in significant acidification of many parts of the world (see also Chapter 5.3.1).

> Organisms involved in the biogeochemical cycling of sulphur compounds evolved early in the earth's history, and their activities have left an indelible mark in the geological record. This is in large part because the sulphur cycle interacts intimately with the cycling of many other elements of biogeochemical interest like oxygen, carbon, iron, and nitrogen. The sulphur cycle has also responded to changes in the ecology of the earth's surface environment as well as to changes brought about by fundamentally geological processes such as volcanic outgassing, hydrothermal interactions and subduction. Overall, the sulphur cycle has heavily influenced the evolving chemistry of the earth's surface. Indeed, one can argue that, to a great extent, the history of earth surface chemistry has been defined by the dynamics of the sulphur cycle.

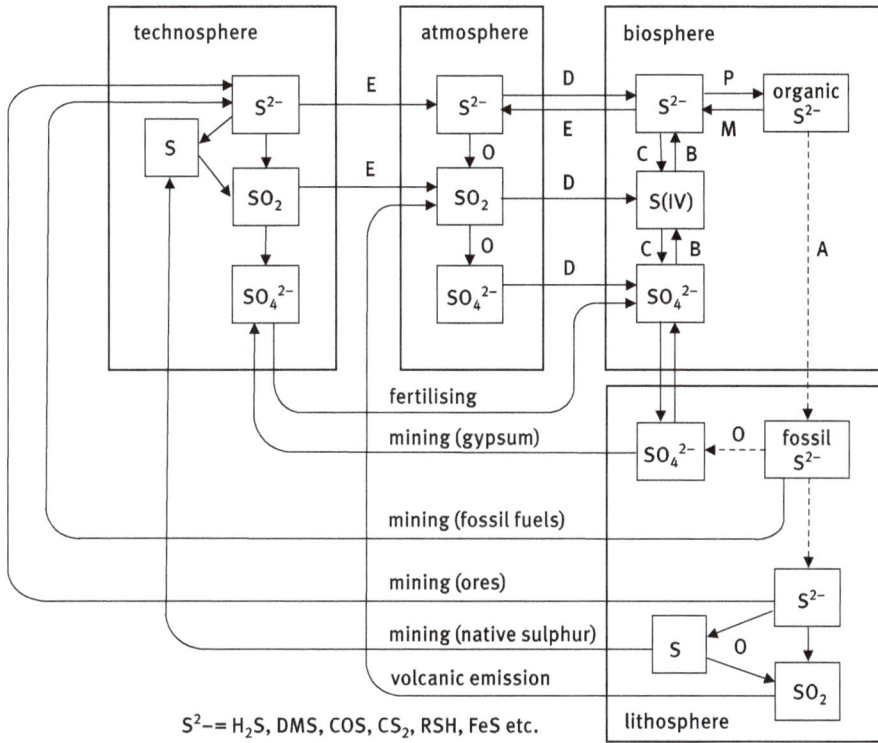

Figure 5.14: The biogeochemical sulphur cycle. A – burial (formation of sediments), B – assimilation (bacterial sulphate reduction), C – aerobic oxidation, D – deposition, E – emission, M – mineralisation, P – plant assimilation, O – oxidation.

In anaerobic environments (e. g. anoxic water basins, sediments of wetlands, lakes, coastal marine ecosystems), sulphur bacteria reduce sulphate to support respiratory metabolism, using sulphate as a terminal electron acceptor instead of molecular oxygen (*dissimilatory sulphate reduction*). This process is the major pathway for H_2S production globally. However, since the anaerobic environment is often not in direct contact with the atmosphere, the escape of H_2S is limited because of re-oxidation in an oxic layer. Reduced sulphur provides substrates for microbial oxidation to sulphate from which certain bacteria can obtain energy. Such microorganisms are present in high numbers at the oxic-anoxic interface and can completely oxidise H_2S and other reduced sulphur compounds in a layer smaller than a millimetre. Consequently, the large amount of H_2S that is produced in coastal areas cannot usually be transferred to the atmosphere. Within the biological sulphur cycle, immense amounts of sulphur are turned over: 550 Tg yr^{-1} in the oceanic and assimilative bacterial sulphate reduction (100–200 by land plants and 300–600 by sea algae) (see Table 5.8). COS is the most abundant tropospheric sulphur gas on Earth.

Table 5.8: Global turnover of sulphur. Data for sulphate reduction adapted from Andreae (1990).

process	Tg S yr^{-1}
bacterial dissimilatory sulphate reduction	
coastal zone	70
shelf sediments	190
depth sediments	290
assimilatory sulphate reduction	
land plants	100–200
ocean algae	300–600
anthropogenic SO_2 emission	60 ± 5
total natural sulphur emission (without sea salt)[a]	50 ± 25

[a]including 10 ± 5 from volcanism

Most biota, however, need sulphur to synthesise organosulphur compounds (cysteine and methionine as examples of the major sulphur amino acids). In contrast to animals, which depend on organosulphur compounds in their food to supply their sulphur requirement, other biota (bacteria, fungi, algae, plants) can obtain sulphur in the aerobic environment from sulphate reduction (*assimilatory sulphate reduction*). Most of the reduced sulphur is fixed by intracellular assimilation processes, and only a minor fraction is released as volatile gaseous compounds from living organisms. However, after the death of organisms, during microbial degradation, volatile sulphur compounds may escape to the atmosphere, mainly H_2S, but also organic sulphides like CH_3SH, CH_3SCH_3 (DMS), $CH_3S_2CH_3$ (DMDS) and CS_2, COS.

In open ocean waters, DMS is the predominant volatile sulphur compound formed by phytoplankton. The precursor of DMS is dimethylsulphoniopropionate (DMSP), which is produced within phytoplankton cells and is thought to have a number of important physiological functions. Unfortunately, the DMS emission from oceans is controversially discussed with the range of 19–58 Tg DMS-S yr^{-1}. The uncertainty factor in the DMS emission estimate of 2–3 is, therefore, an unresolved issue. This is a serious problem because of the dominant role of DMS in the natural sulphur budget (50–80 % of the total natural sulphur emission). The oxidation of DMS is described in Chapter 4.5.2 (equations (4.196)–(4.199)). DMS oxidation by OH is rapid, so a residence time equal to or less than one day could be assumed. DMSO and MSA also have some uncertain aspects. DMSO reacts rapidly with OH, probably resulting in the formation of SO_2 and MSA. The annual average wet deposition of MSA has been estimated to be 0.51 μeq m^{-2}d^{-1}, corresponding to a flux of 2 Tg S yr^{-1}, which is between 5 % and 10 % of the annual DMS emission. At the Cape Grim Baseline Air Pollution Station (Tasmania, Australia), a molar ratio of 6 % of MSA to excess sulphate in the aerosol phase has been found in agreement with the former value.

DMS is now believed to be the most probable natural sulphate precursor, and SO_2 from fossil-fuel combustion is the dominant artificial one. We already mentioned the importance of atmospheric sulphate acting as CCN. Could this natural DMS-derived 'sulphate function' result from the earth evolution, and are there feedback between the environment and the sulphur cycle (Lovelock and Margulis 1974, Charlson et al. 1987)? Has the natural functioning of the atmospheric sulphur cycle been perturbed by human activities? Temperature records supporting the hypothesis that anthropogenic sulphate aerosol influences clear-sky and cloud albedo, and thus climate, have been advanced by several investigators, who have suggested that any natural role of sulphur in climate has been subsumed by anthropogenic pollution. The direct climatic effect of sulphate aerosol is simply due to the reflection of sunlight back to space, while indirect climatic effects of sulphate result from aerosol influence on cloud albedo and/or extent. Sulphate aerosols contribute to cooling, either directly or indirectly, through their role in cloud formation. Also, changes in the chemical composition of aerosols may either increase or decrease the number of CCN. Changes in the concentration of the number of cloud droplets can affect not only the albedo but also the cloud lifetime and precipitation patterns. Precipitation is both an important aspect of climate and the ultimate sink for submicrometer particles and scavenged gaseous pollutants.

Sulphur, or more precisely sulphur dioxide (SO_2), is the oldest known pollutant. Without knowing the chemical species, its influence on the air quality was described several hundred years ago in European cities where it was prevalent because of coal burning. Nearly 200 years ago, SO_2 (and other gases such as HCl and NH_3) was identified as the cause of damages to plants in Great Britain and forests in the German Erzgebirge due to black-ash manufacturing and coal smoke. However, it became of huge environmental interest only in the middle of the twentieth century after the well-known London episode in 1952, where increased concentrations of SO_2 and particulate matter in the presence of fog led to an unusually high mortality rate for the particular time of year. Subsequently, the atmospheric chemistry of SO_2 and global sulphur distribution has been studied intensively. Since then, the anthropogenic sulphur emission (and, consequently, atmospheric SO_2 concentration) is continuously decreased in Europe (Table 5.9). Today, after the introduction of measures for the desulphurisation of flue gases from all power plants, SO_2 no longer plays a role as a pollutant in Europe.

Because of the fact that the only natural source of SO_2 (volcanism) shows large variations, and the mean annual estimate is around $10\,\mathrm{Tg\,yr^{-1}}$, anthropogenic SO_2 emissions currently still account for around 80 % of the total global flux of SO_2 and more than 90 % are injected into the northern hemisphere. Nevertheless, 10 % of the global anthropogenic sulphur emissions account for 50 % of the sulphur budget of the southern hemisphere. Thus, even in remote areas, we must assume that the sulphur budget is markedly disturbed by human activities.

Table 5.9: Global anthropogenic SO_2 emission (in Tg S yr^{-1}).

year	total	Europe and North America	Asia
1890	20	12	0
1910	32	30	1
1950	63	50	10
1960	97	62	17
1970	140	102	29
1980	152	102	37
1990	142	86	41
2000	105	37	43
2010	97	21	52

5.2.6 Acidity in the environment

Acidity is a chemical quantity that is essential for biological life. It results from the budget between *acids* and *bases* existing in the regarded reservoir, which finally is an equilibrium state because of the interaction of all biogeochemical cycles, including the water cycle. The term *acidification* is used to describe a process by which a given environment is made more acidic.

Expressing acidity in terms of pH (see Chapter 3.2.2.4) it ranges in the natural system between about 1 and 10:

- 8.2 in seawater,
- 5–9 of rivers and lakes (normally 7.4),
- 3.5–10 in soils (typically 5–7),
- 4–8.5 in atmospheric waters (typically 4–5.6),
- 1–3 as gastric acid in stomata of vertebrates.

Many biotas need a special range of pH for optimum growth as well as metabolic processes. Geochemical processes such as rock weathering proceed in a slightly acid medium. In soils and lakes, pH controls solubilities and thus bioavailability of nutrients but also toxic substances. Acid catalysis (proton transfer) is a vital process in organic chemistry (e. g. esterification and aldol reactions). Air pollution led to anthropogenic disruption of biogeochemical cycles and thus changing acidity. The large atmospheric concentrations (1–2 ppb) found in coniferous forests for formic and acetic acid and their relationship to plant physiological parameters (photosynthesis, transpiration and stomatal conductance) suggest an ecological controlling function. However, most acidity in ecosystem budgets is produced in soils from plant degradation to form humus via humic and fulvic acids.

The term *acidity* is often used to characterise the ability of a compound to release hydrogen ions (H^+) to water molecules as a measure, expressed by pH value, but sometimes denoted *acid capacity* or *acid strength*, not to be confused with the *acid constant*

K_a, although K_a is a measure for acid strength. By contrast, *alkalinity* (can also be called *basicity* or *basic capacity*) is used to characterise the ability of a compound to be a proton acceptor. Tables 3.4 and 3.5 list important acids and bases occurring in our natural environment. Remember that acidity is solely a property of aqueous solutions, i. e. natural waters in all their different forms such as rivers, lakes, ocean, atmospheric droplets, as well as surface, soil and ground water. Thus, the one source of acidity (and alkalinity, resp.) is the water molecule H_2O, making an imbalance between hydronium (symbolized as a proton or hydrogen cation H^+) and hydroxyl ions:

$$H_2O + H_2O \rightleftharpoons H_3O^+ + OH^-. \tag{3.102}$$

Remember that acids loose H^+ (proton donator), and bases gain H^+ (proton acceptor, simplified spoken). Generally, we separate between oxoacids $R-OH)^{109}$ and non-oxo acids (H–X), inorganic (or mineral) and organic acids. There are general rules:
-   the farther down an element X lies within the periodic system, the larger it's the acidity of its acid H–X (e. g. $H_2Se > H_2S > H_2O$),
-   the farther right X stays in a row of the periodic system, the larger it's the acidity of its acid H–X (e. g. $HCl > H_2SO_4 > H_3PO_4$),
-   positive charge within the anion increases the acidity; negative charges decrease it (e. g. $H_2O > HOOH$),
-   resonance stabilization within the anion increases the acidity (e. g. RCOOH > ROH).

Acids can be neutral compounds, anions and cations. Alcohols and phenols (–OH group) are weak acids; however, the most important organic acids are carboxylic acids, macromolecular humic and fulvic acids. Another class of important organic acids in nature are sulphonic acids, a combination between organic rest R and sulphuric acid H_2SO_4: $R-S(=O)_2OH$. The most important inorganic acids are hydrochloric acid HCl, nitric acid HNO_3 and sulphuric acid H_2SO_4. Besides acids, acid anhydrides, which form acid in reaction with H_2O (for example, CO_2, SO_2, SO_3, N_2O_3, N_2O_5, P_2O_5) are the most important in nature. Furthermore, so-called acid precursors, compounds that form an acid through oxidation (for example, NO, NO_2, hydrocarbons), must be considered. Some substances are amphoteric, i. e., they can react as acid (release a proton) or base (take up a proton). The best icon is water, according to equation (3.102), resulting in a pH of 7. Many metal aqua complexes are acids (but depending on the pH of the medium); see equations (4.402) and (4.307) for iron and alu-

---

109 Oxygen was called by Lavoisier from Greek ὀξύς [oxys = acid], intended to mean "acidifying (principle)," it was a Greeking of French *principe acidifiant*. oxygen was then considered essential in the formation of acids (it is now known not to be). In German (such as synonymously in many other languages), oxygen is name *Sauerstoff* [acid matter].

minium:

$$Me(H_2O)_n \rightleftharpoons Me(OH)(H_2O)_{n-1} + H^+. \tag{5.36}$$

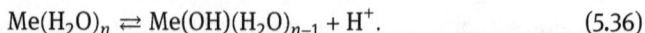

From a generic chemical point of view, water (H_2O) is the mother substance of acids, forming X–OH by replacing the H atom (formally); let us have a look at the formation of sulphurous acid from its anhydride SO_2 and H_2O (= H_2SO_3):

$$S(=O)_2 + OH^- \rightleftharpoons HO^{(-)}-S^{(+)}(=O)O^{(-)} \xrightarrow{\pm H^+} HO-S(=O)OH. \tag{5.37}$$

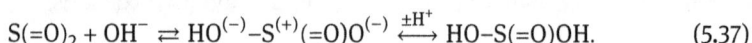

Gaining acidity (i. e. an excess of H^+ or in other terms, a depletion of OH^-) is closely linked with electron transfer processes (reduction-oxidation); remember that $H^+ + e^- = H$, the principle of reduction via linked proton and electron transfer: $H_2O + e^- = H_2O^- = H^+ + e^- + OH^-$. The hydroxyl ion (a base) is also formally an acid according to $OH^- \rightleftharpoons O^{2-} + H^+$, where the not free existing oxide anion (for example, in CaO) is the strongest base ($CaO + H_2O = Ca^{2+} + 2OH^-$). The interlinked proton-electron transfer and in the redox chemistry of water and oxygen the following schemes represent:

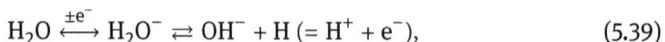

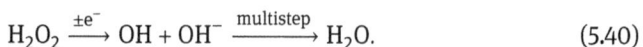

$$H_2O \xrightarrow{\pm H^+} OH^- \xrightarrow{\pm e^-} OH \xrightarrow{OH} H_2O_2 \xrightarrow{\pm H^+} HO_2^- \xrightarrow{\pm e^-} HO_2 \xrightarrow{\pm H^+} O_2^- \xrightarrow{\pm e^-} O_2, \tag{5.38}$$

$$H_2O \xrightarrow{\pm e^-} H_2O^- \rightleftharpoons OH^- + H (= H^+ + e^-), \tag{5.39}$$

$$H_2O_2 \xrightarrow{\pm e^-} OH + OH^- \xrightarrow{multistep} H_2O. \tag{5.40}$$

Also, non-oxo acids originated from water, e. g., via thermal hydrolysis (aquathermolysis)

$$NaCl + H_2O \rightarrow HCl + NaOH \tag{5.41}$$

and via an intramolecular redox process of halogens in water (valid also for Br and I):

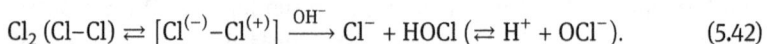

$$Cl_2 (Cl–Cl) \rightleftharpoons [Cl^{(-)}-Cl^{(+)}] \xrightarrow{OH^-} Cl^- + HOCl (\rightleftharpoons H^+ + OCl^-). \tag{5.42}$$

Organic acids in nature are produced in the photosynthetic process from $CO_2 + H_2O$. In the inorganic world, $CO_2 + H_2O$ forms the most important natural acidity (= $HCO_3^- + H^+$) in cloud, rain and groundwater, resulting in rock weathering. On the other hand, CO_2 produced in the early earth's history the most important natural base, the carbonate anion CO_3^{2-} (via CO_2 + CaO), to buffer acidity (for example, in seawater)

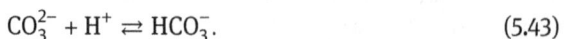

$$CO_3^{2-} + H^+ \rightleftharpoons HCO_3^-. \tag{5.43}$$

From a budget point of view, carbon dioxide provides global precipitation (assuming pH = 5.6 and taking into account the global rainfall from Figure 2.2) of 7–8 Teq yr^{-1} in the form of HCO_3^-. For global weathering, carbonic acid seems to be dominant, but other acids are not negligible. Moreover, only organic acids, sulphuric acid and nitric acid can produce acidic rain down to pH ~ 4 locally. By contrast, formic acid/formate and acetic acid/acetate are efficient buffers not only in the laboratory.

There are only three strong acids (HCl, HNO_3 and H_2SO_4) that are not directly emitted in nature (except probably a small amount of HCl from volcanoes). Gaseous HCl produced from sea salt particles is equivalent to the consumption of other strong acids (HNO_3 and H_2SO_4) because of heterogeneous reactions and HCl degassing. Hence, from NO and reduced sulphur emissions, globally strong acids will be produced via oxidation[110] (Chapters 4.4.5 and 4.5.3). Assuming a 50 % conversion of emitted NO and SO_2 into acids in the atmosphere (which surely represents a maximum), about 2–3 Teq yr^{-1} acidity was provided at the time with maximum SO_2 emission (1980) and further continuous decreasing.

Direct emissions are known for organic acids (HCOOH and CH_3COOH corresponding to 0.4–4.0 Teq yr^{-1}) and indirectly via HCHO oxidation; assuming that only 10 % of HCHO is converted into HCOOH via the cloud water phase (Chapter 4.6.3.2), an additional 3 Teq yr^{-1} is produced. The quantification of other organic acids (oxalic, propionic, butyric and many others) is impossible due to missing source information. Hence, including the acidity due to CO_2, the global produced acidity was always higher than anthropogenically gained, but *acid rain* (see next Chapter 5.3.1) was not a global but a regional problem. Atmospheric acidity is movable and thereby influences other ecosystems through weathering.

5.3 Atmospheric chemistry and air pollution

Between the middle of the eighteenth century and the end of the nineteenth century, the main constituents of air (N_2, O_2, CO_2 and noble gases) have been discovered and quantified. Some trace species (O_3, H_2O_2, HCl, H_2S, SO_2, HNO_2, HNO_3, NH_3, CH_4 and organic compounds, but not specified) have been identified but not quantified (only some ions in rainwater). Chemical processes and reactions in air could not yet be studied because of missing instrumental techniques. However, scientists have already recognised *cycling* matter. Clemens Winkler (1838–1904), who was the first to introduce (without success due to the complicated controlling procedure) a scrubbing technology to remove SO_2 from mining gases in Saxony, remained convinced that dilution

110 Note that dry deposited SO_2 produces acid in soils and lakes.

of exhaust gases was the best air pollution control and mentioned[111] very poetically (translated from German):

> The chemical plant out there, we call it nature, works restlessly for ages and ages day and night and remains yet spick and span. There is no ugly rubbish corner, no dirty rubble tip, and its streams remain clear, its atmosphere neutral, light and smokeless. There is a good reason behind and so presumptuous it might sound; it is no wonder because behind the Earth stands to Sun, being a powerful, inexhaustible power source. The Earth rotates restless off the Sun, presenting her side each moment. Here vapour rises, there it deposits, taking all foreign matter; there giant evaporation, here giant condensation – a unique, magnificent distillation process, surrounding the Earth! Hence, our Earth takes permanent a bath in its own distillate, washing out the atmosphere and the rocks, which flows without ceasing to the collecting tank of the ocean to disappear indistinguishably in her depth.

With the intense industrial development in the middle of the nineteenth century, air pollution as a new atmospheric aspect became the object of interest of researchers; more precisely, the impacts of air pollutants (forest decline, human health, corrosion) were the first focuses of research. Already in the late nineteenth century, some impacts could be related to individual air pollutants. Nevertheless, local air pollution, namely urban pollution in towns and cities, must be regarded at the end of the nineteenth century tremendously, likely by about two orders of magnitude higher in concentration than at present concerns the 'classical' pollutants (soot, dust, SO_2). Strong-smelling substances (likely organic sulphides), ammonia (NH_3) and organic amines must have been dominant in cities before the establishment of sewerage systems in the second half of the nineteenth century. *The Great Stink* or *The Big Stink* was a time in the summer of 1858 during which the smell of untreated sewage was very strong in central London.

However, in times when nothing was known about air composition (the air was regarded as a unique body), the inconvenience of air due to *foreign bodies* (the term pollution was not used) was well recognised, for example, by John Evelyn (1620–1706), who wrote on London: "*Traveller, at many Miles distance, sooner smells, than sees the City to which he repairs.*".[112] The terms 'smoake' and 'clouds' in Evelyn's booklet (only once he used the term 'fog') surely mean what we now call *smog*, an artificial expression coined by des Voeux in his paper *Fog and Smoke* for a meeting of the Public Health Congress in London in 1905. But in the second half of the nineteenth century, the pop-

---

111 *Über den Einfluss des Wasserdampfgehaltes saurer Gase auf deren Vegetationsschädlichkeit* (*On the influence of water vapour content of acidic gases on vegetation damages*). Lecture, held at the main meeting of the Society of German Chemists, Halle, May 31–June 3, 1896.
112 Evelyn, J. (1661) *Fumifugium: or, the inconvenience of the aer, and smoake of London dissipated together with some remedies humbly proposed.* Reprint (original by Godbid, London) by Swan Press, Haywards Heath, 1930, 50 pp. (p. 19).

ulation started to accept that the smoke plague was no longer the price of industrialisation and began to regard it as a problem.

> For centuries, until the end of the twentieth century, the air pollution problems associated with
> the combustion of fossil fuels, sulphur dioxide and soot (smoke) were the key air pollutants. Coal
> has been used in cities on a large scale since the beginning of the Middle Ages; this 'coal era' has
> not yet ended. Smoke and fog as contemporaneous phenomena were scientifically described by
> Julius Cohen (1859–1935), who had studied chemistry in Munich. He wrote: "Town fog is mist
> made white by Nature and painted any tint from yellow to black by her children; born of the air
> of particles of pure and transparent water, it is contaminated by a man with every imaginable
> abomination. That is town fog." Cohen conducted laboratory experiments and concluded: "The
> more dust particles there are, the thicker the fog". Carbonic acids (CO_2) and sulphurous acid (SO_2)
> were observed to increase rapidly during fog, and "although I have no determinations of soot to
> record, the fact that it increases also is sufficiently evident," he wrote. With these terms, the acid
> anhydrides CO_2 and SO_2 were mentioned in the literature of the nineteenth century and not the
> acids H_2CO_3 and H_2SO_3 (sometimes named gaseous carbonic acid). Fog water particles become
> coated with a film of sooty oil. Consequently, fog persists longer than under clean conditions.
> Francis Russell (1849–1914) used the expression 'smoky fog' and wrote: "town fogs contain an
> excess of chlorides and sulphates, and about double the normal, or more, of organic matter and
> ammonia salts".

The term 'atmospheric chemistry' appears soon after the discovery of the main chemical composition of our air and was first used ('Luftchemie' in German) by Johann
Friedrich Gmelin (1748–1804) in 1798 in the sense of a label for new science. It is remarkable that in an English version of this paper (1801), the phrase "Luftchemie" was
not translated. Augustus Allen Hayes (1806–1882), Assayer to the Commonwealth of
Massachusetts and a prominent but nowadays forgotten American chemist, write in
a Paper (1851) as first the phrase *chemistry of the atmosphere*: "The chemistry of our
atmosphere has, from the earliest time, been deemed of high interest, and its connection with the phenomena of animal and vegetable life has led to its study the
minds of the most eminent chemical investigators." In French, "chimique de l'atmo-
sphère" was first used by Jean-Baptiste Boussingault (1802–1887) in 1834, and further
by several scientists in the late 19th century (further reading Möller (2022)). About 50
years later, 'air chemistry' ('Luftchemie' in German) was introduced by Hans Cauer
(1899–1962) in 1949, who had no knowledge of its earlier use. It was soon used as
the label for a new discipline. The first monograph in the field of this new discipline
was written by Christian Junge (1912–1996), entitled *Air Chemistry and Radioactivity*
(New York and London 1963), soon after he had published a first long chapter entitled
Atmospheric Chemistry in a book in 1958. This clear term identifies a sub-discipline
of chemistry and not meteorology or physics. The 'discipline' was called 'chemical
meteorology' before that time. However, before the 1950s, chemical meteorology was
mainly looking for the relationship between condensation nuclei, their chemical composition and the formation of clouds and rain. Atmospheric chemistry as a scientific discipline using laboratory, field and (later) modelling studies was vigorously

developed after the identification of the Los Angeles smog at the beginning of the 1950s.

We will define air (or atmospheric) chemistry as the discipline dealing with the origin, distribution, transformation and deposition of gaseous, dissolved and solid substances in the air. This chain of matter provides the atmospheric part of the biogeochemical cycles. A more general definition, but one that is appealing as a wonderful phrase, is given by the German air chemist Christian Junge in his monograph (1963): "Air chemistry is defined as the branch of atmospheric science concerned with the constituents and chemical processes of the atmosphere". In other words, air chemistry is the science concerned with the origin and fate of the components in air. The origin of air constituents concerns not only all source and formation processes, the chemicals of air itself, but also emissions by natural and artificial processes into the atmosphere. The fate of air includes distribution (which is the main task of meteorology), chemical conversion, phase transfers and partitioning (reservoir distribution) and deposition of species. The deposition is going on via different mechanisms from gas, particulate and droplet phases to the earth's ground surface, including uptake by plants, animals and humans. Removal from the atmosphere is the input of matter to another sphere.

Atmospheric chemistry is again subdivided into sub-sub-disciplines such as tropospheric chemistry, stratospheric chemistry, cloud chemistry, precipitation chemistry, particle chemistry, polar chemistry, marine chemistry and so on. In the last two decades, much progress has been made in heterogenous or interfacial chemistry, such as air-water and air-particle exchange processes (for example, cloud droplets, ocean surface, aerosol particles), and the results suggest it controls to a large extent also the reservoir bulk chemistry.

In the main sections of Chapter 4, we described all important reactions occurring in the atmosphere not only in the gas phase but also in the aqueous phase of droplets. However, it remains difficult for the reader from that to understand complex processes in the atmosphere because *all* reactions with *all* compounds take place simultaneously in *all* phases, called *multiphase chemistry*. Moreover, besides chemical conversion, transportation occurs in the atmosphere, influencing concentrations and reactions.

5.3.1 Atmospheric acidification: "acid rain"

The term *acid rain* denotes one of the most serious environmental problems (about 1970–1990), i. e., the *acidification* of our environment. Fortunately, this problem no longer exists.

> Hippolyte Ducros (1805–1879), a pharmacist in Nimes (France), studied the composition of hail, which fell in a severe thunderstorm in April 1842 (Ducros 1845) and found free nitric acid; he entitled his paper in 1845 "*observations d'une pluie acide*" [observation of an acid rain]; this is the

first use of the term "acid rain" in literature. Robert Angus Smith (1817–1884, who is (wrongly) credited to coin the term „acid rain", used it only twice in his book "Air and Rain. The Beginnings of a Chemical Climatology" (1872) in connection with the effects of the atmosphere on stones and iron. The importance of atmospheric acidity, and especially *acid fog*, had already been identified at the end of the ninteenthth century as to be a cause of forest damage in Saxony by Hans Adolf Wislicenus (1867–1951). "Rediscovery" of acid rain is attributed to Eville Gorham (1925–2020), a Canadian-American biologist, also termed the "grandfather of acid rain" recognized in the late 1950s acidification of lakes and soils with subsequent ecological effects. The acid rain period – one of the most serious environmental issues of the twentieth century – finally was the trigger for the first large-scale measures in pollution control in the 1980s. The Convention on Long-range Transboundary Air pollution (CLRTAP) was signed in November 1979 (ratification in 1982), after considering the international problem that pollutants may originate in one country, with consequent effects felt in another country (but without binding effects). In the USA, the "Acid Precipitation Act of 1980" was signed. It is interesting to note that 15 years after air pollution abatement in central Europe, acid-sensitive Swedish lakes show slow recovery from historic acidification as shown from long-term (1987–2012) water quality monitoring. Overall, strong acid anion concentrations declined, primarily as a result of declines in sulphate. Chloride is now the dominant anion in many acid-sensitive lakes. Base cation concentrations have declined less rapidly than strong acid anion concentrations, leading to an increase in charge balance acid-neutralizing capacity.

The definition of acidity, given in the previous Chapter 5.2.6, is limited only to free H^+ and is not appropriate for the atmospheric multiphase system because of the gas-liquid equilibrium of acids and bases in the pH range of interest. Therefore, Waldman et al. (1982) defined *atmospheric acidity* to be the 'acidity' in the aqueous, gaseous and aerosol phases, representing the sum of individual compounds which are measured. They wrote: "The net acidity, however, is measured in solution, following eluation or extraction of the sample. Measurements of sample acidity are performed with a pH electrode".

Not all these definitions help clarify what we have to understand about *atmospheric* acidity. The term *acidifying capacity* is only of qualitative value and meets the same basic problems as for defining the oxidising capacity of the atmosphere. The problem lies in the different points of view between the analytical chemist and impact researcher. Svante Odén (1924–1986) wrote (Odén 1976): "The problem of air pollution or the impact of a specific pollutant can only be understood when reactions and interactions between all reservoirs are taken into account". The impact (acidification) is caused by acid deposition, which is a result of atmospheric acidity or, in other words, the *acidifying capacity* of the atmosphere. In atmospheric waters (cloud, fog and rain droplets), the following ten main ions must be taken into account: SO_4^{2-}, NO_3^-, Cl^-, HCO_3^-, NH_4^+, Ca^{2+}, Mg^{2+}, Na^+, K^+ and H^+. Of minor importance are HSO_4^-, HSO_3^-, SO_3^{2-}, NO_2^-, CO_3^{2-}, F^- and OH^-. Because of the *electroneutrality condition*, the following condition:

$$\sum_i [\text{cation}_i] = \sum_j [\text{anion}_j] \qquad (5.44)$$

must be valid.[113] In this and the following equations, only the equivalent concentrations will be used; otherwise, the stoichiometric coefficients must be considered, for example, $1\,eq\,SO_4^{2-} \equiv 0.5\,mol\,SO_4^{2-}$. The relationship equation (5.44) must not be confused with the definition of an acid $[H^+] = [A] - [B]$, where A represents acids and B bases. It follows from equation (5.44):

$$[H^+] = \sum_i [anion_i] - \sum_j [(cation\ without\ H^+)_j] = [A] - [B]. \qquad (5.45)$$

Werner Stumm (1924–1999) introduced the acidity as a *base neutralising capacity* (BNC), corresponding to the equivalent of all acids within the solution, titrated to a given reference point:

$$BNC \equiv [Acy] = [A] + [H^+] - [OH^-] \qquad (5.46)$$

and corresponding alkalinity as an *acid neutralising capacity* (ANC):

$$ANC = [Alk] = [B] + [OH^-] - [H^+]. \qquad (5.47)$$

Equation (5.46) and equation (5.47) are, therefore, not based on the electroneutrality equation (5.55) because they also include weak acids and bases which are not protolysed in solution at the given pH (that means in non-ionic form) but contribute to *acid* or *base titration*. The first attempt to determine airborne acidity was made by careful titration inserting microliter quantities of a NaOH solution using Gran's titration method (Brosset 1976). The reference points, however, are not objective criteria.[114] Zobrist (1987) extended this definition to the general equations:

$$[Acy]_{total} = [H^+] + [A] - [B] \approx [Acy]_{H^+} = [H^+] + [A]_{strong} - [B]_{strong}, \qquad (5.48)$$

where Acy_{total} denotes the *total acidity* (including weak acids) and Acy_{H^+} the *free acidity*.

These definitions might be useful for bulk water bodies such as rivers and lakes. Atmospheric water, however, as mentioned before, consists of single droplets that are mixed up while sampling. Moreover, individual samples collected in the field will furthermore mix, for example, when getting a time-averaged sample. All that means that

---

113 This equation can also be used as a quality control measure for the analytical procedure in atmospheric water: If the deviation of the quotient $\sum[anions]/\sum[cations]$ is $\geq 20\%$ from unit (1), the analysis must be repeated or the samples should be rejected.

114 The German DIN 38406 (1979) defined the acid capacity (in German *Säurekapazität*) to be equivalent to the hydrochloric acid consumption of pH = 4.3 and the sum of all carbonaceous bonded cations: $[Acy]_{H^+} = [HCO_3^-] + 2[CO_3^{2-}] + [OH^-] - [H^+]$.

individual droplets or samples with different acidity/alkalinity have been mixed, re-sulting in acid-base reactions and a shifting liquid-gas equilibrium with a new refer-ence point with an averaged 'final' acidity. Therefore, it is important to define acidity and alkalinity as conservative parameters, i. e. they are independent of pressure, tem-perature and ionic strength as well as CO_2 gas exchange. The acidity is then defined as:

$$[Acy] = [H^+] - [HCO_3^-] - [CO_3^{2-}] - [OH^-] = -[Alk] \qquad (5.49)$$

and, using equation (3.105):

$$[Acy] = [A^*] - [B], \qquad (5.50)$$

where A^* = anions without HCO_3^-, CO_3^{2-} and OH^-. It follows, using the equilibrium expressions for the ions (see Chapter 4.6.2 concerning carbonate protolysis chemistry, especially equations (4.239)–(4.242):

$$[Acy] = [H^+] - (K_{ap}H_{CO2}[CO_2(g)] + K_w)[H^+]^{-1} - K_{ap}K_2H_{CO2}[CO_2(g)][H^+]^{-2}. \qquad (5.51)$$

Adopting standard conditions (400 ppm CO_2, 298 K), it follows that:

$$[Acy] \approx [H^+] - 6.15 \cdot 10^{-12}[H^+]^{-1} - 2.6 \cdot 10^{-22}[H^+]^{-2}. \qquad (5.52)$$

Equation (5.53) is only valid without changing partial pressure (p_0), i. e. neglecting reservoir distribution between the gas and aqueous phases. The whole mass of the gas (expressed as p_0), however, is distributed between the gas phase (expressed as partial equilibrium pressure $p_{eq} = p_0 - p' = n(g)RT/V(g)$) and the aqueous phase, expressed as $n(aq) = V(aq)H_{eff}RT$, where $p' = p_{eq}H_{eff}RT$ and:

$$H(CO_2)_{eff} = \frac{[CO_2(aq)] + [H_2CO_3] + [HCO_3^-] + [CO_3^{2-}]}{[CO_2(g)]}. \qquad (5.53)$$

Figure 5.15 shows that under 'normal' atmospheric conditions with water pH between 4 and 5.7, almost all HNO_2 and NH_3 are dissolved into droplets, whereas SO_2 and CO_2 remain in the gas phase. Within the pH range 6–8, SO_2 is effectively scavenged but CO_2 is measurably transferred from the gas to the aqueous phase only above a pH of about 8. In Figure 5.16, the 'acidity' parameters $[H^+]$, $[OH^-]$, $[Acy]$ and $[Alk]$ are calcu-lated based on equation (5.49) and equation (5.51) and are then expressed as a loga-rithm with a dependency on pH. Figure 5.16 represents some characteristic points:

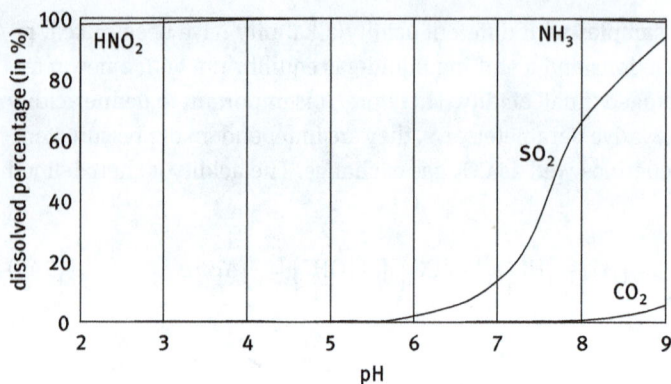

Figure 5.15: Reservoir distribution of gases between gas and liquid phases.

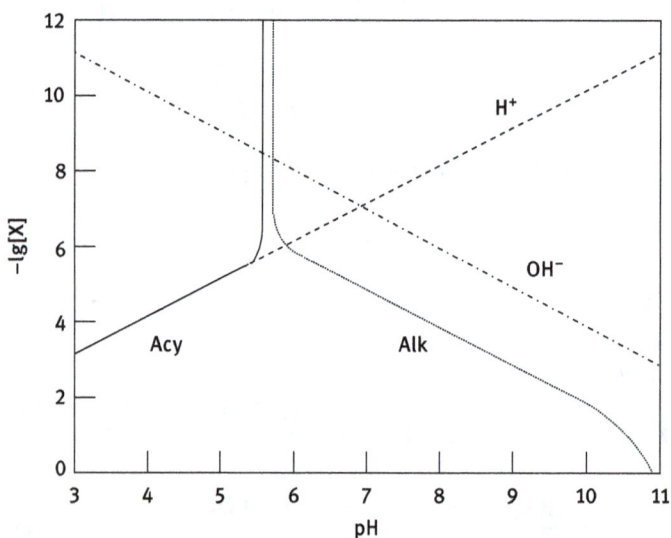

Figure 5.16: pH-dependent H_2O protolysis, acidity and alkalinity in a gas-aqueous atmospheric system.

pH	[Acy]	[Alk]
<5.68	$\approx[H^+]$	<0
=5.68	$\rightarrow 0$	$\rightarrow 0$
>5.68 (but <10.3)	<0	$\approx[HCO_3^-]$

From this table, we cannot conclude that pH = 5.68 ([Acy] $\rightarrow$ 0) represents the reference point for 'neutral' atmospheric water, i. e. water with a pH < 5.68 can be characterised as 'acid'. This reference point is valid only for the binary system CO_2/H_2O.

To quantify changes in the amounts of acids or bases that cause acidification, we also have to compare one state with another reference state. In natural systems, the most appropriate reference state is that which relates to the system itself. Other reference states have no meaning (physically or chemically) but might be helpful or even necessary from a computational point of view. Other atmospheric ions originating from the natural emission of gaseous species and PM lead to a natural acidity $[Acy]_{nat}$ or alkalinity, which varies in time and space. Therefore, the definition of *acid* rain or clouds gains by:

$$[Acy]_{measured} = [Acy]_{man\text{-}made} + [Acy]_{natural} \geq [Acy]_{natural}, \qquad (5.54)$$

where subscript 'measured' denotes the experimentally determined acidity in water samples. For pH > 10.3, CO_3^{2-} must be taken into the budget. However, this alkaline range is unlikely in natural waters; for pH $= 10.7$ it follows that $[Alk] = 1 \, mol \, L^{-1}$, i. e. in solution must be $1 \, mol \, L^{-1}$ cations (e. g. Ca^{2+}). Such highly concentrated solutions occur only in transition states (condensation/evaporation). Neglecting the CO_3^{2-} contribution to alkalinity, equation (5.51) transforms into (β constant):

$$[H^+] = 0.5[Acy] + \sqrt{0.25[Acy]^2 + \beta}. \qquad (5.55)$$

There is another complication. The samples of collected cloud and/or rainwater are most probably not in equilibrium with atmospheric CO_2. In this case, equation (5.49) is still valid for the acidity calculation; however, $[HCO_3^-]$ cannot substitute using Henry's law The only way to solve this problem is an analytical estimation of $[HCO_3^-]$, for example, using titration, electroanalytical methods or ion chromatography. Note that traditional titration methods result in the figure $([H_2CO_3/CO_2 - aq] + [HCO_3^-])$, where $[HCO_3^-]$ must be calculated using the equilibrium equation. Assuming a total carbonaceous concentration of $100 \, \mu eq \, L^{-1}$ at pH $= 4.5$, it follows that $[HCO_3^-] = 90 \, \mu eq \, L^{-1}$ and $[H_2CO_3/CO_2 - aq] = 10 \, \mu eq \, L^{-1}$.

CO_2 plays a special role in forming atmospheric acidity because of its high and constant concentration. Figure 4.15 shows that an important pathway in alkalinity (carbonate) formation goes via the CCN (nucleation and droplet formation) as well as aerosol scavenging. The last process, however, is of minor importance for clouds but contributes significantly to sub-cloud scavenging into falling raindrops. There is a variety of PM acting as ANC, including flue ash, soil dust and industrial dust. Carbonate particles could lead to an initially high alkaline aqueous phase that is an efficient absorber for gaseous acids and acidic precursors, especially SO_2. The solution becomes oversaturated with $CO_2(aq)$ and, consequently, CO_2 desorption occurs from the droplet. The neutralisation stoichiometry follows the overall reaction:

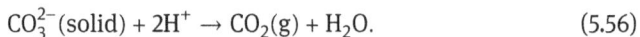

$$CO_3^{2-} (solid) + 2H^+ \rightarrow CO_2(g) + H_2O. \qquad (5.56)$$

As we have seen, acidity is a chemical quantity defined only in exclusively aqueous solutions in the environment (similar to the redox potential). However, it makes sense to introduce a *total acidity* for the whole volume of air, including the droplet, gaseous and particulate phases. This basic idea is applicable because it suggests a potential neutralisation of a reference reservoir (for example, soil or lake) and the *acidification* (or basification) of this reference reservoir. Let us define the atmospheric *acidifying capacity* as the sum of the *potential* acidity in gas $[Acy]_g$ and aerosols $[Acy]_a$ as well as the acidity in the aqueous phase $[Acy]_{aq}$ as follows:

$$[Acy]_{atm} = [Acy]_g + [Acy]_a + [Acy]_{aq}. \tag{5.57}$$

Whereas $[Acy]_{aq}$ is defined according to equation (5.56), the *potential aerosol acidity* can be defined similarly to equation (5.61):

$$[Acy]_a = [A^*] - [B] \approx [SO_4^{2-}] + [NO_3^-] + [Cl^-] - [NH_4^+] - [Ca^{2+}] - [Mg^{2+}] - [K^+]. \tag{5.58}$$

The ions listed in equation (5.69) represent the main constituents in cloud and rainwater and, thereby in the CCN. This list can be extended by minor species such as carbonate, bicarbonate, sulphite, nitrite, iron and others. Based on the electroneutrality condition, the soluble PM (ions listed in equation (5.58)) must be balanced with H^+, HCO_3^-, CO_3^{2-} and OH^-.

The definition of *potential gaseous acidity* is more complicated. Gas molecules form acidity only after dissolution in the cloud, fog and raindrops and subsequent protolysis reactions. The degree of dissolution – or in other words – phase partitioning depends on the initial droplet acidity (i. e. aerosol acidity) and from the gas phase composition of acidifying gases themselves. Therefore, it does not make sense to add only gaseous acids, according to Table 3.4. With a good approximation, we can assume that strong gaseous acids and bases are completely transferred to the aqueous phase and dissociated therein, whereas weak acids (e. g. carboxylic acids) contribute less to acidity. However, in the case of missing strong acids, they contribute significantly to acidity. Another problem originates from gaseous *anhydrides* that form acids after reaction with water (SO_2, SO_3 and N_2O_5)[115] only. Other gases, being not direct anhydrides (e. g. NO_2) can also produce acids in reaction with water. Taking into account only SO_2 besides strong gaseous acidic precursors and NH_3, the following equation represents the potential gaseous acidity, where ε (= $0 \dots 1$) represents the degree of acid formation by sulphite and sulphate:

$$[Acy]_g \approx [HNO_3] + [HNO_2] + [HCl] + \varepsilon[SO_2] - [NH_3]. \tag{5.59}$$

---

115 CO_2 is also an anhydride, however, because of its constant concentration the acidity contribution only depends on solution pH.

Generally, the term *acid deposition* could be better used to describe the acidity phase transfer (e. g. from the atmosphere to the biosphere).

5.3.2 Ozone

Ozone was discovered in 1839 by the German chemist Christian Friedrich Schönbein (1799–1868) in Basel while conducting electrolysis experiments with water. It was recognized to be a substance of unusual properties that has never been isolated in the pure state. The given names (before accepting *ozone*) show the relation to oxygen: electricized oxygen, allotropic oxygen, nascent oxygen, active oxygen, excited oxygen. The bleaching, antiseptic properties and its action as a deodorizer, disinfectant, and germicide were soon known and used in the nineteenth century. Before discovering ozone, other scientists observed that oxygen gas, when electrical sparks had been passed through it, acquired a peculiar smell. Dutch Martinus van Marum (1750–1837), doing experiments with his newly constructed electrifying machine in 1785, reported the odour of ozone, but he failed to identify it as a unique form of oxygen. As early as 1840, Schönbein proposed the name ozone associating with "sulphurous smell" [Θεείωι means divine and θήϊον – sulphurize, cleansing with sulphur but also divine, based on the Greek οξειν = smell] in association with a thunderbolt. Schönbein never elucidated its chemical constitution but was the first to get evidence for its presence in atmospheric air (1845). William Olding (1829–1921) first identified the formula O_3 in 1861. Gilbert Newton Lewis and Linus Pauling assumed in 1932 a linear structure O=O=O like that of SO_2, based on calculated formation enthalpy and stated that a triangle structure is not possible. Michael James Stuart Dewar (1918–1997) proposed in 1948 a π-complex structure for ozone which requires the molecule to be an acute-angled triangle in shape; the resonance behaviour does not show unpaired electrons and characteristics of a free radical (Dewar 1948). The bonding in ozone is discussed until the present, the ending twentieth century, however, the physical and chemical properties are well-known today for an understanding of atmospheric chemistry.

> The period 1850–1880 of uncounted ozone measurements using the Schönbein paper has often been criticized because test papers were unable to differentiate between oxidants. Qualitatively, however, Schönbein's measurements are fairly consistent in revealing that nineteenth-century seasonal ozone most often peaked in spring, followed by winter, with some spring peaks continuing into early summer. French physicist Albert Lévy (1844–1907), using the first reliable method[116] for determination of O_3 in air and observed the abundance of ozone almost continuously from 1877 to 1910 (where ozone averaged 11 ± 2 ppbv) at the municipal Observatory of Parc Montsouris in Paris. Cornelius Fox (1839–1922) used in 1873 the term "air-purifiers", which included ozone,

---

116 Bubbling the ozone containing air through a solution of potassium arsenite where it is transformed into arsenate, and the not converted arsenite titrated after *Mohr*'s method with potassium iodide solution.

peroxide of hydrogen, and nitrous acid and "purifying agents" German chemist Carl Oswald Viktor Engler (1842–1925) in Halle, coined the term "air cleanser" for oxidative substances like ozone
and hydrogen peroxide – hundred years before Crutzen in 1986 coined the phrase "detergent of
the atmosphere" to describe this important cleansing role of OH (O_3 and H_2O_2 are the OH precursors). The first direct evidence of ozone in the atmosphere was provided by Emil Schöne in 1884,
who measured the spectrum of the atmosphere in early winter 1883 near Moscow during intense
frost to avoid overlapping with the bands of water vapour (599–610 nm). The absorption spectrum of ozone (the principal band between 595 and 613 nm) was found to be well in accordance
with the description given by Joseph Chappuis (1854–1934) in 1880.

The formation mechanism of ozone in atmospheric air remained unsolved in the nineteenth century; from laboratory studies, it was known that it is gained through silent
electric discharges in air and while slow oxidation of oils. It was a long time confused
with other atmospheric oxidants such as hydrogen peroxide and nitrogen oxides because of its unsure detection with iodometric papers. Thus, many curious ideas on O_3
formation arose (e. g., from spraying water and plants); a belief, held for more than
100 years, on "ozone rich forests" was already proved to be false by Ernst Ebermayer
(1829–1908) in 1873 who found ozone above the canopy and close to the forest but not
inside the forest. The first idea that not electricity but solar radiation is the sole cause
of "activation of oxygen" comes from the German chemist Casimir Wurster (1854–1913)
in 1886. Only in 1900, it became evident that the oxygen photolysis needs wavelengths
that are not available in the troposphere.

Emil Gabriel Warburg (1846–1931) was the first who found in 1902 that during the silent discharge,
not only ozone is gained (*ozonized air*) but also decomposed (*deozonised air*). Sydney Chapman
(1888–1970), who was a British physicist in Manchester and London, presented the first complete
theory of stratospheric ozone (Chapman 1930); the main part of his communication was to explain the daily and seasonal variation of ozone as well the variation of the amount of ozone by
latitude. He stated that the production of ozone is solely by ultra-violet radiation and already assumed that the O atom formed by dissociation of O_2 is likely in a different electronic state, but
any knowledge about it was still missing. In the 1960s, it was assumed that the transport of ozone
from the stratosphere was thought of as the only major source of tropospheric ozone.

Before World War II, ozone formation was believed to only be occurring in the stratosphere.

5.3.2.1 Stratospheric ozone depletion: the "ozone whole"

The ozone layer discovery occurred at the faculty of sciences of Marseilles in 1913 by
the French physicists Charles Fabry (1867–1945) and Henri Buisson (1873–1944), who
determined the total amount of pure ozone to be 5 mm, and assumed that it exists in
the upper atmosphere. In 1924, Gordon Miller Bourne Dobson (1889–1975) and Douglas Neill Harrison (1901–1987) designed an instrument for the determination of the
ozone amount *from* photographic plates using a photometer, in 1931 further developed by Dobson, which made the observations much simpler, and which became the

standard instrument for measuring both total column ozone and the profiles.[117] The total column ozone (mass per square) is expressed in the Dobson unit (1 DU is defined as 0.01 mm thickness at standard conditions).[118]

> Alan West Brewer (1915–2007) was a British-Canadian physicist who joined Professor Dobson at Oxford in 1948. Dobson retired in 1950, and Brewer moved to the University of Toronto in 1962 and began to develop a new automated instrument for measuring ozone, introduced in 1986. The main problems with the Dobson Spectrophotometer are that it is manually operated, very large and heavy and can only measure ozone. In 1988 the Brewer became the World Meteorological Organisation (WMO) Global Atmosphere Watch (GAW) standard for Stratospheric Ozone Measurement. Ozone follows a distinct annual cycle, with substantial differences between altitudes. In the lower stratosphere, the photochemical lifetime of ozone is long – months to years. There, the annual cycle of lower stratospheric ozone (and total column ozone) is determined to a large degree by meteorological transports, especially by the large scale Brewer–Dobson circulation. This circulation is strongest in late winter and spring, when it transports ozone-rich air from low latitudes and high altitudes into the extratropical lower stratosphere, and results in a spring-time maximum of ozone in the lower stratosphere (e. g. at 22 km altitude). In the upper stratosphere, ozone is produced photo-chemically from oxygen and is destroyed in various cycles. These reactions are fast. The photo-chemical lifetime of ozone in the upper stratosphere is short (hours to weeks), and upper stratospheric ozone largely follows the annual cycle of solar irradiation. At 35 km, for example, the largest ozone densities occur in summer.

With the Dobson spectrophotometers, Joe Farman (1930–2013) and his team in the British Antarctic Survey discovered the ozone hole in 1984 (Farman et al. 1985). It is worth noting that the view of Farman on finding the ozone hole was based on long-term monitoring (BBC broadcast on July 6, 1999); Figure 5.17:

> The British Antarctic Survey set up stations in Antarctica. And so we had been monitoring many things in Antarctica for a long while. And suddenly, in 1985, it dawned on us that we were sitting on top of one of the biggest environmental discoveries of the decade, I suppose, or perhaps even of the century. We saw this little dip appearing, and then it just accelerated so rapidly that, within three or four years, we were talking about a 30 per cent drop in the thickness of the ozone above us, which was an enormous amount. We can be slightly proud of the fact. This was the first time anyone had shown that ozone levels had changed since the measurements began, way back in 1926 or thereabouts, when Dobson made his original pioneering measurements. The long-term monitoring of the environment is a very difficult subject. There are so many things you can monitor. And basically, it is quite expensive to do it. And, when nothing much was happen-

---

117 The vertical distribution of ozone is derived by the Umkehr method. Direct solar intensity is measured at two different wavelengths, one being more absorbed by ozone than the other. At sunrise and sunset, the intensities decrease at different rates. The ratio shows an inversion. This is called the Umkehreffekt and gives information about the vertical distribution of ozone in the atmosphere. An Umkehr measurement takes about three hours, and provides data up to an altitude of 48 km, with the most accurate information for altitudes above 30 km.

118 For example, 300 DU of ozone brought down to the surface of the Earth at 0 °C would occupy a layer only 3 mm thick. One DU is $2.69 \cdot 10^{16}$ ozone molecules cm^{-2}, or $2.69 \cdot 10^{20}$ m^{-2}. This is 0.4462 millimoles of ozone m^{-2}.

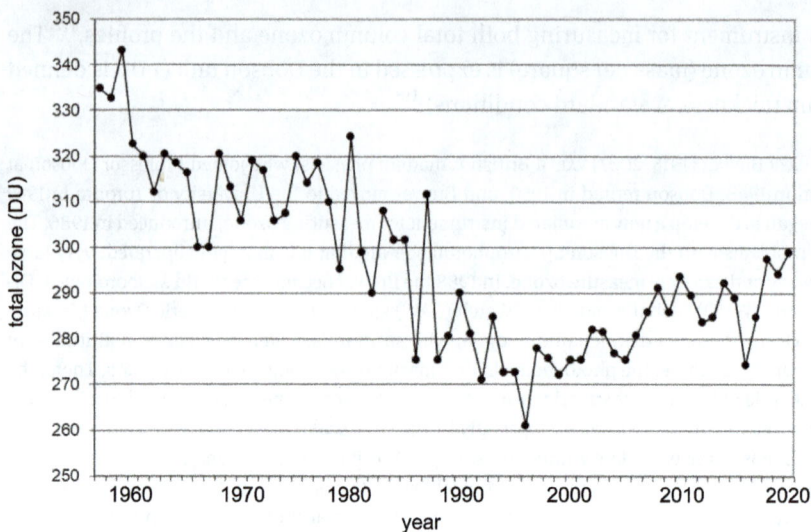

Figure 5.17: The historical record of total ozone at Halley Bay (Antarctica 76 °S); September average by the British Antarctic Survey from 1956–2020. Data after Zhang et al. (2021).

ing in the environmental field, all the politicians and funding agencies completely lost interest in it.

Chemistry in the stratosphere (which goes up to about 50 km in altitude) differs from that in the troposphere for three reasons:

– radiation below 300 nm is available for photodissociation that does not occur in the troposphere (Figure 5.18),
– not all trace gases (those with a short lifetime) are either available or found only in very small concentrations (especially water – the stratosphere is very dry),
– there is no precipitation and no liquid cloud (but special *polar stratospheric clouds*; PSCs).

The photolysis of O_3 and O_2 is crucial for the environment because they result in the continuous absorption of radiation below 300 nm (Figure 5.18). Whereas O_2 covers the wavelength range from less than 127 nm up to 242 nm (equation (4.7), equation (4.8) and equation (4.9)), O_3 that of less than 267 nm (equation (4.15) up to 310 nm (equation (4.14)). As products, a mixture of $O(^1D)$, $O(^3P)$ and excited O_2 is available. The Herzberg continuum (230–280 nm) consists of three bands. It is clear that O_3 formation (reaction equation (4.11)) as well as O_3 photolysis occurs. Another reaction is important in the stratosphere for O_3 removal (which plays no role in the troposphere):

$$O + O_3 \rightarrow O_2 + O_2. \tag{5.60}$$

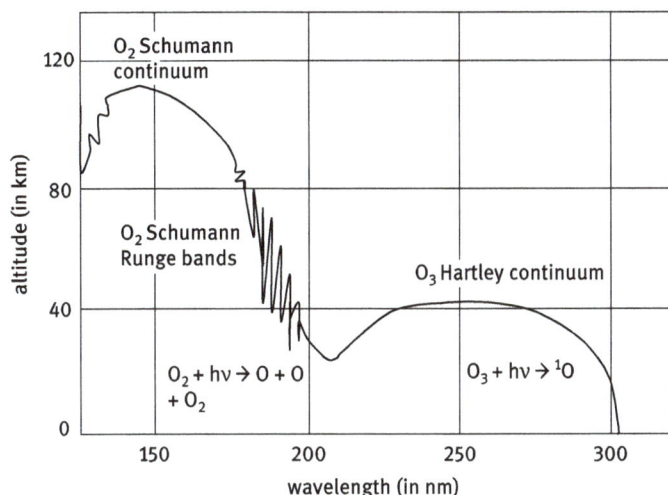

Figure 5.18: Penetration of solar UV radiation into the atmosphere as a function of wavelength and absorption by oxygen and ozone. The curve indicates the altitude at which incoming radiation is attenuated to about 1/10 of its initial intensity.

At 30 km (above the peak in the ozone layer), we observe substantial reductions in the amount of radiation received between 225 and 275 nm. The O_2 photodissociation in the Schumann–Runge continuum (125–175 nm) directly produces $O(^1D)$ and opens many radical reactions (but almost all molecules photodissociates at such hard radiation); this is in altitudes of 100–200 km. The solar Lyman-α line (121.6 nm) is, through O_2 photodissociation, an important source of $O(^1D)$ production throughout the mesosphere and lower thermosphere. The Lyman-α line also dissociates H_2O at altitudes between 80 and 85 km, where the supply of water vapour is maintained by methane oxidation:

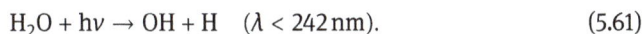

$$H_2O + h\nu \rightarrow OH + H \quad (\lambda < 242\,nm). \tag{5.61}$$

An increase in HO_x follows and thereby a sharp drop in the ozone concentration near the mesopause. The ozone concentration rapidly recovers above 85 km because of the rapid increase in O produced by the photodissociation of O_2 by the absorption in the Schumann–Runge bands and continuum. Above 90 km, there is a decrease in ozone because the three-body recombination of O_2 and O becomes slower with decreasing pressure.

There is another important difference between the stratosphere and troposphere. Whereas the troposphere is heated from the bottom (i. e. the earth's surface), the stratosphere heats from the top (i. e. by incoming solar radiation). This positive temperature gradient results in an extremely stable layering. Therefore, mixing and transport are much weaker than in the troposphere. At the bottom of the stratosphere, close to the tropopause, the lowest temperatures are found (200–220 K and at the poles down to ~180 K), whereas temperature rises up to 270 K can be found at 50 km altitude. At each point of the stratosphere, adiabatic radiation processes determine the temperature: O_3 absorbs UV radiation and heats the air and CO_2 absorbs IR radiation and cools the air. The chemical composition (mixing ratios) concerning the main constituents (N_2, O_2, CO_2 and CH_4) remains constant, but the pressure strongly decreases from about 200 hPa at 12 km to 1 hPa at 50 km altitude. The most important trace species is O_3; in the layer between 15 and 30 km altitude, about 90 % of atmospheric ozone is available. This is called the *ozone layer*.

Marcel Nicolet (1912–1996) and Sir David Robert Bates (1916–1994) proposed the first cycle of ozone destruction in the stratosphere (later named HO_x cycle). However, only through the implications of artificial influences on the stratospheric ozone cycle, Crutzen (1971), Johnston (1971), Molina and Rowland (1974), as well as Stolarski and Cicerone (1974), paid our attention to the stratospheric ozone layer drawn. Two of the above-listed reactions comprise the O_3 catalytic decomposition:

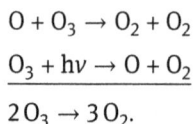

$$O + O_3 \rightarrow O_2 + O_2$$
$$O_3 + h\nu \rightarrow O + O_2$$
$$\overline{2\,O_3 \rightarrow 3\,O_2.}$$

By contrast, this decay is balanced with O_3 formation according to:

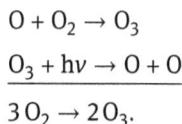

$$O + O_2 \rightarrow O_3$$
$$O_3 + h\nu \rightarrow O + O$$
$$\overline{3\,O_2 \rightarrow 2\,O_3.}$$

The O_3 photolysis to $O(^1D)$ (equation (4.14)) does not lead to net destruction of ozone. Instead, monooxygen is almost exclusively converted back to O_3 by reaction $O + O_2$ (equation (4.11). However, because O_2 dissociates to free oxygen atoms above about 30 km, below 30 km reaction equation (5.42) results in a net loss of odd oxygen (if the odd oxygen concentration is defined as the sum of the O_3 and O concentrations). The budget between the photodissociation of O_3 and its formation via equation (4.11) is zero (steady state). Because the rate of reaction equation (4.11) decreases with altitude, whereas that for reaction equation (4.15) increases, most of the odd oxygen below 60 km is in the form of O_3, whereas above 60 km, it is in the form of O. Odd oxygen is produced by reaction equation (4.7). It can be seen that reactions equation (4.11) and equation (4.14) do not affect the odd oxygen concentrations but merely define the ratio of O to O_3.

A significant fraction of the O_3 removal is caused by the presence of chemical radicals X, such as nitric oxide (NO), chlorine (Cl), bromine (Br), hydrogen (H) or hydroxyl

(OH), which serve to catalyse reaction equation (5.44), termed cycle 1:

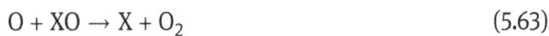

$$X + O_3 \rightarrow XO + O_2 \tag{5.62}$$
$$\underline{O + XO \rightarrow X + O_2} \tag{5.63}$$
$$O + O_3 \rightarrow 2O_2$$

When $k_{5.62} > k_{5.60}$ and substance X is not quickly lost by any termination reaction, the ozone decomposition flux according to cycle 1 becomes more prominent than it follows only from reaction equation (5.60). Reaction equation (5.62) is generally faster than equation (5.63) because of the stronger radical characteristics (with the exception of hydrogen atoms; Table 5.10). The species H and OH are natural constituents (in contrast to halogens and nitrogen species) of the stratosphere (but with lower concentrations than in the troposphere). Only the hydroxyl radical OH (no other XO species) can directly react with O_3 (reaction equation (4.20)), providing another effective O_3 decay cycle 2:

$$OH + O_3 \rightarrow HO_2 + O_2$$
$$\underline{HO_2 + O_3 \rightarrow OH + 2O_2}$$
$$2O_3 \rightarrow 3O_2$$

Table 5.10: Reaction rate constants for $X + O_3$ ($k_{5.62}$) and $O + XO$ ($k_{5.63}$) at 22 km altitude and 222 K (in 10^{-12} cm^3 molecule^{-1} s^{-1}).

X	O	H	OH	NO	F	Cl	Br	I
$k_{5.44}$	8	140	1.6	2	22	29	17	23
XO	–	OH	HO_2	NO_2	FO	ClO	BrO	IO
$k_{5.45}$	–	22	30	6.5	27	30	19	120

The reactive species, called *ozone-depleting substances* (ODS), come from different source gases such as H_2, H_2O, N_2O, CH_4 and halogenated organic compounds. It is noteworthy that influencing the stratosphere with water (from aircraft) and hydrogen (H_2 technology in discussion) results in ozone depletion. An important source of H_2O is methane, and a H_2O concentration maximum found between 50 and 70 km altitude results from CH_4 oxidation:

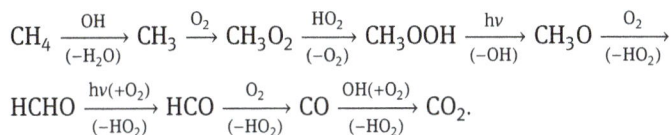

$$CH_4 \xrightarrow[(-H_2O)]{OH} CH_3 \xrightarrow{O_2} CH_3O_2 \xrightarrow[(-O_2)]{HO_2} CH_3OOH \xrightarrow[(-OH)]{h\nu} CH_3O \xrightarrow[(-HO_2)]{O_2}$$

$$HCHO \xrightarrow[(-HO_2)]{h\nu(+O_2)} HCO \xrightarrow[(-HO_2)]{O_2} CO \xrightarrow[(-HO_2)]{OH(+O_2)} CO_2.$$

An additional source of radicals (OH and HO_2) generates the photolysis of O_2 (equation (4.7) and equation (4.8)), O_3 (equation (4.15)) and H_2O (equation (5.43)) as well as CH_4, giving H (and subsequently HO_2):

$$CH_4 + h\nu \rightarrow CH_3 + H \quad (\lambda < 230\,nm). \tag{5.64}$$

Further, we have to consider the H_2O production reactions (which play no role in the troposphere:

$$OH + OH \rightarrow H_2O + O_2, \tag{5.65}$$
$$OH + HO_2 \rightarrow H_2O + O_2. \tag{5.66}$$

About 90 % of N_2O entering the stratosphere is photolysed to $N_2 + O$ (equation (4.95)); the remaining 10 % reacts with $O(^1D)$ to NO (equation (4.96a)), which can then enter the ozone decay cycle and consume O_3.

Halogens, however, are most important for ozone depletion. Below we will see that NO_y and halogens produce condensed reservoir species such as $ClONO_2$ and $BrONO_2$, which play a role in the 'ozone hole' (Figure 5.17). Precursors, such as halogenated organic compounds, are photolysed but also react with the oxygen atoms X = H, Br, F and Cl:

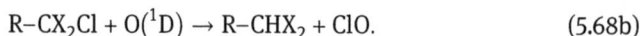

$$R-CX_2Cl + h\nu \rightarrow R-CHX + Cl, \tag{5.67}$$
$$R-CX_2Cl + O(^1D) \rightarrow R-CHX_2O + Cl, \tag{5.68a}$$
$$R-CX_2Cl + O(^1D) \rightarrow R-CHX_2 + ClO. \tag{5.68b}$$

The formation of ClO is dominant (about 60 %) because reactions with OH are too slow. Cycle 2 can run several thousand times before competing products (HOCl, HOBr, HOI and nitrates) are produced.

The ODS cycles, however, cannot explain the dramatic O_3 depressions observed in Antarctica every spring (from September to October) in the layer between 12 and 24 km because gas-phase chemistry (which stops in the arctic winter-like at night) is a continuous process. The simple explanation consists of an accumulation of radicals in a condensable matter, called *polar stratospheric cloud* (PSC), which can form in different types at T < −80 °C. In May and June, strong winds in the stratosphere begin to blow clockwise around the continent, called the Antarctic polar vortex. Isolated from warmer air outside the vortex, the air inside gets colder and colder in contrast to the Arctic, where only locally such low temperature is attained. Drifting around inside the polar vortex are the reservoir molecules. Type I consists of pure water, type II nitric acid trihydrate ($HNO_3 \cdot 3\,H_2O$) and type III a mixture ($HNO_3/H_2SO_4/H_2O$). These 'clouds' provide a surface for heterogeneous chemistry and for the absorption and storage of so-called reservoir gases HCl, HBr, HI, $ClONO_2$ and others. Easily photodissociable species are Cl_2, $ClNO_2$ and HOCl (similar to the other halogens). For example, the following reactions occur:

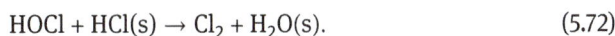

$$HCl(s) + ClONO_2 \rightarrow Cl_2 + HNO_3(s), \tag{5.69}$$

$$HCl(s) + N_2O_5 \rightarrow ClNO_2 + HNO_3(s), \tag{5.70}$$

$$ClONO_2 + H_2O(s) \rightarrow HOCl + HNO_3(s), \tag{5.71}$$

$$HOCl + HCl(s) \rightarrow Cl_2 + H_2O(s). \tag{5.72}$$

In the Antarctic spring (late August), PSCs evaporate and set free 'active' halogens ($Cl_2 + Cl + ClO + Cl_2O_2$; Figure 5.19) and remain only as liquid sulphuric acid particles. Most NO_x is stored as HNO_3 in solid PSCs and contributes only a little to O_3 depletion. By early November, the strong stratospheric winds circling Antarctica die down, and the polar vortex breaks up. As it does, ozone-rich air from outside the vortex flows in, and much of the ozone that was destroyed is replaced. In a sense, the hole in the ozone layer fills in. Usually, by the end of November, the amount of ozone in the stratosphere over Antarctica has almost returned to normal. Meanwhile, the 'normal' gas-phase cycles of ozone depletion control the steady-state concentration. The next winter, however, the cycle will begin again.

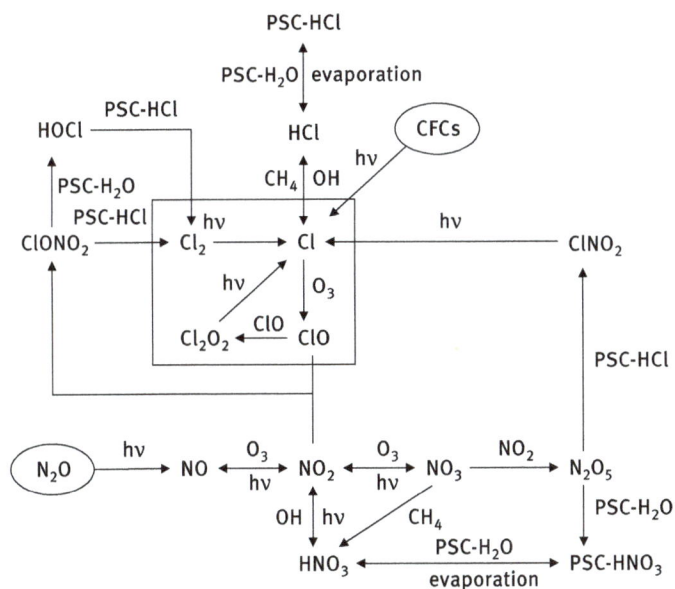

Figure 5.19: Scheme of stratospheric multiphase chemistry.

As seen from Figure 5.17, the Montreal Protocol banning ODSs shows the first result; a further decrease of ozone did stop around 1995, and after 2000 a slow recovery is seen. It is assessed that in 40–50 years, the ozone column will again be as it was before 1960.

5.3.2.2 Tropospheric ozone formation: "photochemical smog"

In 1944 damage to plants had been observed in the Los Angeles area, which for the first time were not related to 'classical' pollutants (such as SO_2 or fluorine compounds). Only a few years later, Arie Jan Haagen-Smit (1900–1977) and co-workers made automobile exhaust gases responsible for ozone formation, which then were considered as the impact species. Since that time, ozone has been regarded as the key species for the oxidation capacity of the atmosphere, and ozone episodes and summer smog have become highly relevant environmental issues.

Studies suggest that ozone in the troposphere has increased globally throughout much of the 20th century due largely to increases in anthropogenic emissions. Tropospheric ozone over the last four decades does indeed indicate increases that are global in nature yet highly regional due to the combined effects of regional pollution and transport. Based on measurements, it has been concluded that within the past 100 years (but mostly after 1950), ground-based ozone concentration has risen by about a factor of two. Models also show a doubling of the tropospheric ozone content because of human activities. Unfortunately, it remains unclear even today whether the surface-near concentration of ozone has already increased since the early twentieth century (such as proposed by Marenco et al. 1994); evident is an increase of tropospheric ozone worldwide with the beginning 1960s. However, like in Los Angeles in the late 1940s, it is very probable that at many industrial sites and towns, ozone was locally produced in the presence of nitrogen oxides and organic compounds much earlier – and likely soon later again consumed by air pollutants.

In East Germany (GDR) at the Meteorological Observatory Wahnsdorf near the city of Dresden at the beginning of 1950s, Friedrich Teichert (1905–1986), continued by Wolfgang Warmbt (1916–2008), expanded the surface ozone measurements by initiating a first worldwide network of sites. Together with results from other scientists, the measurements clearly show with today's knowledge that the source of ozone was exclusively from the stratosphere[119] (with a "normal" free tropospheric ozone value of 50–60 μg m^{-3}), that there was no photochemical ozone formation (which would have a maximum in July), and that lower ozone values were due to ozone degradation in the boundary layer. In other words, decades later, we were talking about the "ozone formation potential" in the time before air masses had a different "ozone destruction potential" – now resulting in a budget. Warmt reported in 1979 for the first time on the pronounced increase of the annual means of the ozone concentration from 30 μg m^{-3} (1956/57) to 48 μg m^{-3} (1976/77) at the station Arkona (Baltic Sea). The increase of ozone was contributed to the increasing pollution of the surface-near atmospheric layers from nitrogen oxides and reactive hydrocarbons, and Warmbt noted that the "observed increase of ozone seems to be a large-scale rather than a local phenomenon". Other observers (at that time based on shorter time-series) also observed a trend of increasing tropospheric ozone, and soon later measurements began at many sizes around the world.

---

119 An indication that ozone in the time before 1965 was originated from the upper atmosphere was also found by the positive correlation with artificial radioactivity due to nuclear tests.

The key finding and discussion in tropospheric ozone research in the second half of the twentieth century concerns the photochemical formation of tropospheric ozone, the increase of the ozone content in the troposphere, and strategies to control tropospheric ozone.

The longest quantitative ozone records are in Europe, which indicate that ozone doubled there between the 1950s and 2000. From 1950–1979 until 2000–2010, all available NH monitoring sites indicate increasing ozone, with 11 of 13 sites having statistically significant trends of 1–5 ppbv per decade, corresponding to >100 % ozone increases since the 1950s, and 9–55 % ozone increases since the 1970s. In the SH, only 6 sites are available, all indicating increasing ozone, with 3 having statistically significant trends of 2 ppbv per decade. Ozone monitoring in the free troposphere since the 1970s is even more limited than at the surface. Significant positive trends since 1971 have been observed using ozone sondes above Western Europe, Japan and coastal Antarctica (rates of increase range from 1–3 ppbv per decade), but not at all levels. In addition, aircraft have measured significant upper tropospheric trends in one or more seasons above the northeastern USA, the North Atlantic Ocean, Europe, the Middle East, northern India, southern China and Japan. Notably, no site or region has shown a significant negative ozone trend in the free troposphere since the 1970s. From 1990 until 2010, surface ozone trends have varied by region. Western Europe showed increasing ozone in the 1990s followed by a levelling off, or decrease since 2000. In the eastern US, surface ozone has decreased strongly in summer, is largely unchanged in spring, and has increased in winter, while ozone increases in the western US are strongest in spring. Surface ozone in East Asia is generally increasing. However, since 2013 decrease in ozone in the Chinese free troposphere have been observed and attributed to the reduction in NO_x anthropogenic emissions), contrary to the increase in ozone at the surface.

Ozone in the centre of the European continent typically experiences a summertime or broad spring/summer ozone peak, whereas sites located in less polluted regions of northern or western Europe experience a springtime peak. Ozone at rural sites across the eastern United States typically peaked in summer during the 1990s, while western US sites had a broader spring/summer peak. In contrast, the 2006–2010 time period shows that ozone in the eastern US has strongly decreased in summer while remaining constant in spring so that the summer maximum has been replaced by a broad spring/summer peak. This shift in the seasonal ozone cycle appears to be a response to emission reductions. Examination of the seasonal ozone cycle at four rural sites in Europe and one rural site in the western United States has shown that not only is springtime ozone greater in recent years (2005–2010) than in earlier decades (the 1970s through the early 1990s), but the seasonal maximum now occurs earlier in the year.

In terms of measurements, the World Meteorological Organization's Global Atmosphere Watch program coordinates an extensive worldwide observation network. Over 400 sites conduct long term measurements of physical, chemical, meteorological and radiation parameters and make their quality-tested data freely available to the scientific community. Approximately 100 GAW stations measure surface ozone, and the data are quality controlled to detect long-term trends and archived in the World Data Centre for Greenhouse Gases.

In Chapter 4.3.2.1, the elementary chemical reactions in the system $O - O_2 - O_3$ are described; in the following, our view is focused on the net formation of ozone in the troposphere and the understanding how the so-called ozone precursor control the ozone budget. In an atmosphere free of trace gases, ozone is photochemically self-destructed; the gross reaction is $2O_3 \rightarrow 3O_2$:

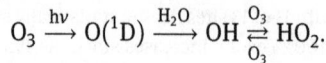

$$O_3 \xrightarrow{h\nu} O(^1D) \xrightarrow{H_2O} OH \underset{O_3}{\overset{O_3}{\rightleftarrows}} HO_2.$$

The other (main) photolysis pathway led to steady state O_3 concentration but not net production:

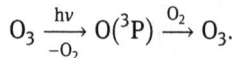

$$O_3 \xrightarrow[-O_2]{h\nu} O(^3P) \xrightarrow{O_2} O_3.$$

Net ozone production is only possible when $O(^3P)$ is formed from another reaction such as NO_2 photolysis:

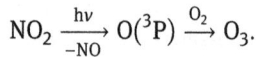

$$NO_2 \xrightarrow[-NO]{h\nu} O(^3P) \xrightarrow{O_2} O_3.$$

However, there appears a dilemma: First, NO can back oxidise to NO_2 by O_3 (an important reaction in polluted air) and second, the primary emission of NO_x is almost NO and not NO_2, i. e. to form ozone, NO must first be oxidised to NO_2, and when that happens by O_3, no net formation occurs:

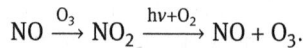

$$NO \xrightarrow{O_3} NO_2 \xrightarrow{h\nu+O_2} NO + O_3.$$

There are two key reactions of the radicals OH and HO_2, which are in competition with O_3, that turn permanent NO to NO_2 and reproduce HO_2:

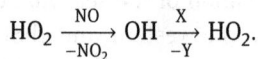

$$HO_2 \xrightarrow[-NO_2]{NO} OH \xrightarrow[-Y]{X} HO_2.$$

We already met the OH→HO_2 conversion by CO, CH_4 and other VOCs (equation (4.238) and equation (4.297):

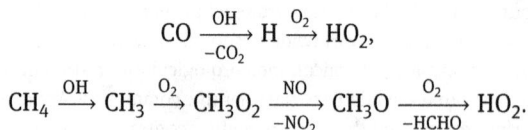

$$CO \xrightarrow[-CO_2]{OH} H \xrightarrow{O_2} HO_2,$$

$$CH_4 \xrightarrow{OH} CH_3 \xrightarrow{O_2} CH_3O_2 \xrightarrow[-NO_2]{NO} CH_3O \xrightarrow[-HCHO]{O_2} HO_2.$$

Now it is clearly seen that two interlinked cycles, $OH \rightleftarrows HO_2$ and $NO \rightleftarrows NO_2$, have been established (Figure 5.20). Moreover, it is seen that the NO_x cycle runs twice when

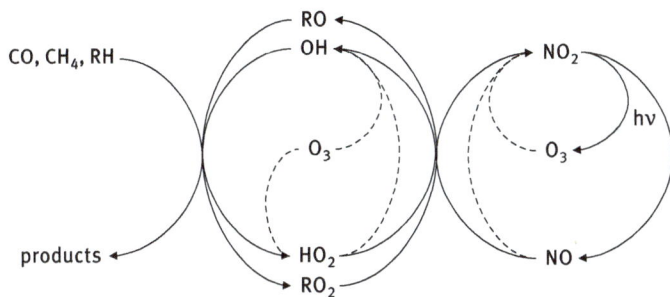

Figure 5.20: Scheme of gas-phase net O_3 production: HO_x, RO_x cycle and NO_x cycle; dotted lines denote competitive reaction $HO_2 + O_3$ and $NO + O_3$ (for large NO_x).

the OH reaction starts with CH_4 (or any other VOC, called NMVOC):

$$HO_2 + NO \rightarrow OH + NO_2,$$
$$RO_2 + NO \rightarrow RO + NO_2.$$

CO, CH_4 and NMVOC are the precursors of net ozone formation; NO_x plays the role of a catalyst. From CH_4 is gained first HCHO, which reacts much faster with OH (the same is valid for all hydrocarbons, producing aldehydes). That means, in a plume of pollutants with NO_x and VOC (typically for towns and industrial areas), the rate of ozone formation increases with the distance from the source; however, in "competition" with the dilution, i. e. the concentration of reactants decreases (k versus c in the reaction rate equation). Hence, the gross reactions are given by

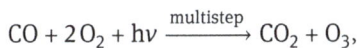

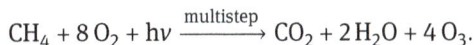

$$CO + 2O_2 + hv \xrightarrow{\text{multistep}} CO_2 + O_3,$$
$$CH_4 + 8O_2 + hv \xrightarrow{\text{multistep}} CO_2 + 2H_2O + 4O_3.$$

The last equation represents the maximum yield of ozone from CH_4 oxidation. All these reactions turn directly or in multiple steps OH $\rightarrow$ HO_2. When we now extend the budget by including the second NO_x cycle shown in Figure 5.21, we can establish the budget equation for net O_3 formation. The first oxidation step of CH_4 to HCHO results in:

$$CH_4 + 4O_2 + 2hv \xrightarrow{\text{multistep}} HCHO + H_2O + 2O_3.$$

Concerning formaldehyde, the budget amounts to:

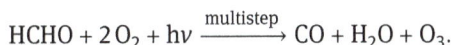

$$HCHO + 2O_2 + hv \xrightarrow{\text{multistep}} CO + H_2O + O_3.$$

The "perfect" ozone formation cycle (Fig. 5.20) is based on the odd oxygen cycling shown in Fig. 5.21. However, because of radical chains and competing reactions, OH

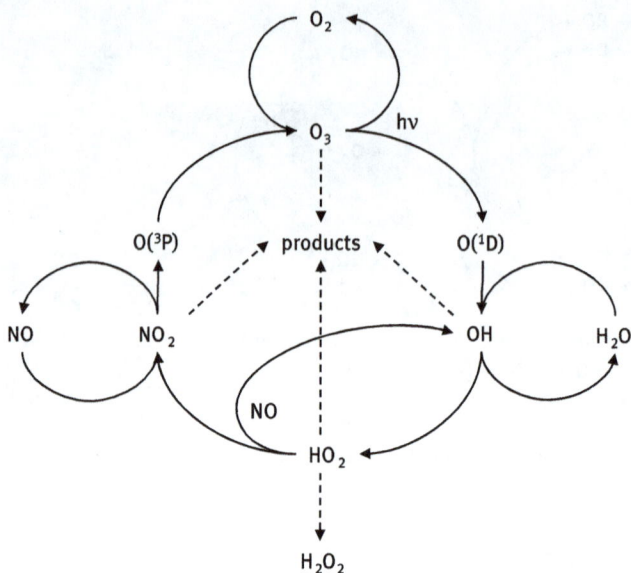

Figure 5.21: The cycle of odd-oxygen; dotted lines denote competitive reactions not shown.

is consumed through the formation of HNO_3 and HO_2 is consumed through the forma-tion of H_2O_2 and organic peroxides, interrupting the cycle shown in Figure 5.20 and reducing net ozone formation. Not shown in Figure 5.21 are the O_3 precursors (CO, CH_4 and RH), which maintain the OH $\rightleftharpoons$ HO_2 cycle without consuming O_3.

The role of NO_x in ozone formation is manifold. At very low NO concentrations, $NO + HO_2$ (eq. (4.97)) is in competition with $NO + O_3$ (eq. (4.98)), which results in a net decay of ozone. Both cycles are interrupted when $R_{4.98} \gg R_{4.97}$, which is given for a few tenths of ppb NO:

$$[NO] \ll \frac{k_{4.97}}{k_{4.98}} [O_3] \approx 10^{-2} [O_3].$$

This concentration is the threshold between the ozone-depleting chemical regimes (in the very remote air) and ozone-producing regimes (in NO_x air). Other important reactions of the HO_2 radical (with the exception of H_2O_2 formation via reaction eq. (4.23) are unknown. It also follows that H_2O_2 formation is favoured (and vice versa) in low NO_x and low O_3 environments, which are relatively restrictive conditions. At high NO_x concentrations (about >20 ppb), however, OH radicals are dominantly consumed by NO_x and NO_x is removed from the cycle through the formation of nitrous and nitric acid:

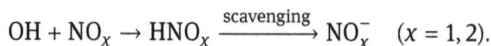

$$OH + NO_x \rightarrow HNO_x \xrightarrow{\text{scavenging}} NO_x^- \quad (x = 1, 2).$$

The ageing of air masses in terms of oxidation state is often described by ratios such as NO_y/NO_x. It can clearly be seen that in the daytime, the ratio NO_2/NO depends on radiation and O_3 concentration. Close to NO sources (e. g. traffic), O_3 can be totally depleted through (equation (4.98)). This is also called "ozone titration". NO_2 carries the oxygen and releases it via photodissociation. This was observed by the fact that O_3 concentration in suburban sites is often higher than in urban sites, thereby defining O_x, representing the sum of odd oxygen, which is therefore roughly constant for suburban and urban sites as:

$$O_x = O_3 + NO_2 \approx \text{constant.} \tag{5.73}$$

The net rate of ozone is given by three terms: the chemical production $P(O_3)$ in the gas phase, the sink term $S(O_3)$ by chemical conversion and deposition and the net transport term $T(O_3)$, given by influx and outflow:

$$\left(\frac{d[O_3]}{dt}\right) = P(O_3) - S(O_3) + T(O_3). \tag{5.74}$$

The chemical production term is given by the $OH–HO_2$ conversion reaction shown in Figure 5.20 of the whole ozone formation cycle as the rate-determining step:

$$P(O_3) = \left(k_{4.238}[CO] + k_{4.289}[CH_4] + \sum_i k_i[NMVOC] \right)[OH]. \tag{5.75}$$

Considering mean concentrations, it follows that O_3 formation rates of 0.2–2 ppb d^{-1} concerning CO and about 0.5 ppb d^{-1} concerning CH_4, respectively. This is close to the global mean net ozone formation rates found by modelling 1–5 ppb d^{-1}, suggesting that long-life carbon monoxide and methane are responsible for the global background ozone concentration. However, local and regional O_3 net production rates of about 15 ppb d^{-1} are observed under cloud-free conditions in summer. The *in-situ* formation rate can quickly reach up to 100 ppb h^{-1}. It is often neglected that the transport term in equation (5.40) can be dominant in the morning after the breakdown of the inversion layer resulting in vertical O_3 mixing, and in the case of changing air masses, vertical fluctuations can "produce" O_3 changes up to 10 ppb in minutes. During the night, dry deposition might reduce near-surface ozone to zero. Moreover, the reaction $NO + O_3$ "stores" odd oxygen in NO_2, which is not photolysed back to NO at night.

Generally, an increase of ozone is found with increasing height above sea level; remember that O_3 is not a primary emission but produced photochemically within the boundary layer and transported downwards from the stratosphere. In the first few 100 meters (within the mixing height) the strong diurnal variation can influence the mean value (daily average) by lowering. A "typical" diurnal ozone variation concurs with the intensity of solar radiation, whereas the maximum is shifted later to the afternoon. The diurnal variation, however, represents the budget (sources – sinks). Beginning at night (no photochemical production), the removal of O_3 proceeds by dry deposition and chemical reaction with NO_x. In the case of the nocturnal inversion layer, no ozone transport from the residual layer (above mixing height) occurs, and, consequently, O_3 concentration depletes (even to zero in urban areas). After sunrise, the inversion layer is broken down and, by vertical mixing, O_3 is transported down. Additionally, photochemical production increases. The daytime maximum O_3 concentration represents the well-mixed boundary layer atmosphere. In summer, the daily maximum net photochemical O_3 production is around 15 ppb; the ozone formation rate is proportional to the photolysis rate, assuming no variation in precursors such as NMVOCs. Thus, O_3 photochemical production shows a maximum at noon. After having the concentration maximum (which represents sources = sinks) in the late afternoon, the inversion layer builds up again (no more vertical O_3 mixing in), and finally, dry deposition and surface-based chemical removal reactions reduce the ozone. Thus, vertical transport is dominant in determining the diurnal variation, and, consequently, wind speed and temperature (again linked with radiation) are key meteorological parameters showing a correlation with O_3. Mean O_3 concentration is again well correlated with mean temperature. Similar to diurnal variation, seasonal variation is driven by photochemical ozone production with a maximum in summer. The winter minimum represents a reduced photochemical activity but might also show the chemical ozone depletion into cloud droplets.

In the presence of clouds, OH and HO_2 will be to a large extent scavenged by the droplets, not only due to its (not very large) solubility but especially because of aqueous-phase reactions that increase the radical flux into the aqueous phase. This reduces the gas-phase ozone formation, according to Figure 5.20 due to interruption of the OH $\rightleftharpoons$ HO_2 cycle. As for the gas phase, OH oxidises in aqueous phase organic compounds (RH), but HO_2 does not regenerate OH and turns into H_2O_2. The last compound is the main oxidiser of dissolved SO_2. Moreover, OH and O_3 also oxidise dissolved SO_2 (Figure 5.22). Therefore, the aqueous phase is an effective sink of atmospheric oxidants such as O_3.

The ozone loss in clouds was almost disregarded before the 1990s. Only at the beginning of the 1990s, modellers showed that clouds can effectively reduce gas-phase ozone. Based on long-term monitoring at Mt. Brocken (Harz, Germany), we generally found smaller O_3 concentrations (on average 30 %) under cloudy conditions than under comparable cloud-free conditions (Table 5.11).

At a given site, ozone can have two different sources: the local *in situ* photochemical formation and the *onsite* transportation (vertical and horizontal) of ozone. Local sinks of ozone are dry deposition (irreversible flux to ground) and chemical reactions in the gas phase, on aerosol particles and in droplets. Air masses have different potentials to form and destroy the ozone. Depending on air mass type, the past and fate of ozone (in relation to the site) can be characterized by production and/or destruction (positive or negative budget). With changing air masses, the ozone level can also drastically change. The atmospheric ozone residence time, depending only on

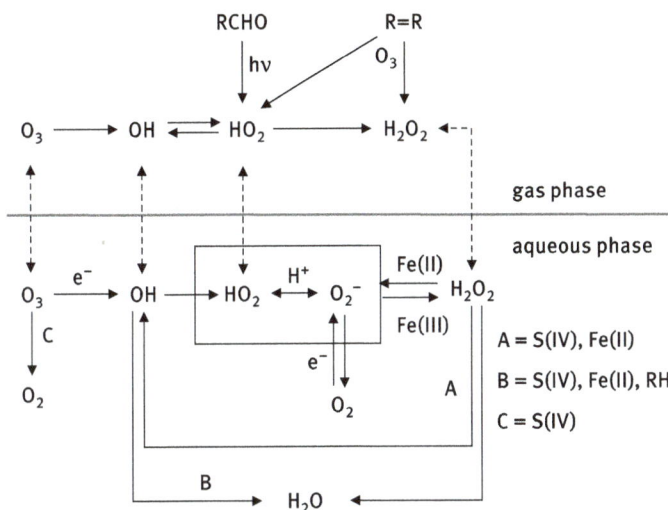

Figure 5.22: Scheme of O_3 multiphase chemistry.

Table 5.11: Statistical parameters for mean ozone concentrations in summer (mid-April until mid-October) and winter (mid-October until mid-April) and for cloud-free and cloudy condition ("station-in-cloud") at Mt. Brocken 1992–1997 (Harz Mt., Germany) 1142 m a. s. l., 51.80 °N and 10.67 °E, based on monthly means (in ppb).

parameter / situation	winter	summer	year
all events (ppb)	26.3 ± 4	44.0 ± 3	34.2 ± 3
cloud-free (ppb)	31.1 ± 5	47.1 ± 2	37.4 ± 4
station-in-cloud (ppb)	21.1 ± 4	33.5 ± 3	26.8 ± 4
LWC (in mg m^{-3})	272 ± 22	272 ± 27	263 ± 63
station-in-cloud (%)	59 ± 16	28 ± 10	45 ± 2
difference cloudy – cloud-free (ppb)	10.0	13.6	10.6
ratio cloudy / cloud-free	0.68	0.71	0.72

sink processes is extremely variable. It amounts to a few days near the ground and several months in the upper troposphere. Thus, ozone can be transported over long distances in the free troposphere. The advective transport to a receptor site is an important source of local ozone.

The mean (annual) background O_3 concentration for central Europe (the reference year 1995 ± 5) was 32 ± 3 ppb, where the following "sources" can contribute:

- 10 ± 2 ppb stratospheric ozone with small seasonal variation (spring peak);
- 6 ± 2 ppb natural biogenic ozone from natural VOCs with seasonal variation (0–12 ppb); and
- 16 ± 2 ppb artificial ozone from CH_4 and CO.

This "base" ozone of 32 ppb shows a seasonal variation between 26 ppb (winter) and 38 ppb (summer), where about 50 % is anthropogenic. Additional to the base ozone is 5 ppb of artificial "hot" ozone from fast-reacting NMVOCs with seasonal variation (0–15 ppb) and strong short-term variation (0–70 ppb).

In total, the "typical" European annual mean figure amounts to 37 ppb (winter 25 ppb and summer 47 ppb). It has been clearly shown that the "acute" air pollution problem is given from NMVOC precursors ("hot" ozone, see the third term in equation (5.75)), which contributed only 5 ppb (25 % of artificial contribution and 14 % of total ozone) to the long-term average in the 1990s. The "chronic" air pollution problem is caused by methane oxidation (and partly by CO), contributing to a likely continuously increasing base ozone (50 %). In Germany, and stepwise in all other European countries, the "hot" ozone problem has been solved. The problem of background ozone, connected with CH_4, cannot be solved. The reduction of "hot" ozone led to constant ozone levels or only minor ozone increases in the 1990s, but in the future, ozone could again rise according to the CH_4 emissions increase.

> Ozone as a secondary trace species exists in a non-linear relationship with its precursors. The ozone concentration is a result of sources and sinks. Besides the sources, the stratospheric O_3 input into the troposphere is much smaller (10–20 %) than believed some decades before, and the photochemical tropospheric production is the main source (80–90 %).

It is worth noting that the ozone trends vary broadly at various mountain sites. At the beginning of the 1990s, some stations in Europe showed no further O_3 increases. The number of days with high ozone concentrations (exceeding the thresholds of 180 µg m^{-3} and 240 µg m^{-3}) significantly decreased in Germany (and all across Europe). This coincides with the reduction of some precursor emissions (NMVOCs and NO). However, it seems that there are differences between seasons and site locations (background, rural, urban). In assessing ozone levels between different sites and different periods, it is essential to regard the "typical" variation of ozone with altitude, season and daytime.

5.3.3 Halogens

5.3.3.1 Tropospheric halogen chemistry and ozone removal

The recent focus of marine boundary layer halogen research has been on two atmospheric implications: a) participation of reactive halogen species in catalytic ozone destruction cycles including heterogeneous reaction in or on sea salt aerosol and b) the formation of new aerosol particles in the coastal boundary layer and their potential to act as cloud condensation nuclei (CCN).

In recent years, new insights have been found concerns reactions of ozone with halogenides dissolved in seawater, sea salt particles and seawater particles, having an important environmental role. Reactive iodine and bromine species, respectively, are known for altering atmospheric chemistry. Besides ozone depletion, oxidation of elemental Hg (following by HgO deposition) and dimethyl sulphide (DMS) has been considered.

We just have learned in Chapter 4.7.2 that gaseous hydrochloric acid (HCl) is permanently released from sea salt above the oceans as well as continents. The chlorine chemistry is too slow for ozone depletion in the troposphere; however, the chemistry of more weakly bonded bromine and iodine species dominates instead. The generation of sea salt aerosol at the ocean surface is the major tropospheric source producing about $6.2 \, \text{Tg yr}^{-1}$ of bromide. In particular, iodine – combining with heterogeneous chemistry – lead to ozone depletion in marine environments. In analogy to HCl acid displacement from sea salt, HBr (as shown by newest measurements) and HI (but at extremely low concentrations, so that gaseous iodine compounds originated from other processes, see below) must be expected in the atmosphere. All bromide and iodide have been found enriched in precipitations in comparison to seawater.

Measurements of the reactive iodine species (IO, OIO, CH_3I, and I_2) detected mixing ratios around the ppt level, reaching 100 ppt occasionally. The residence time of iodine species in the atmosphere ranges from a few days (CH_3I) to seconds (I_2), it is linked directly to the solar photolysis level, but it should be increased by the formation of particles. Satellite measurements showed high BrO concentrations over sea-ice with the appearance of frost flowers. Additionally, high BrO_x plumes in the Antarctic coasts were found to originate from sea-ice zones. Model studies presented the contribution of blowing snow as a BrO_x source in polar regions.

The catalytic destruction of O_3 (called *bromine explosion* some 20 years ago) by halogen atoms (Br, I) occurs via cycles I and II, here written only for bromine (note that interhalogens such as ICl, BrCl, IBr, IBr_3 also will be photolyzed into the atoms). In the troposphere, reactions of cycle I shows the main ozone destroying cycle of bromine (and iodine, resp.) in the troposphere. In polar regions, the reactive bromine (BrO_x) cycle is activated during the polar sunrise on the coasts.

Cycle I (net budget $2 \, O_3 \rightarrow 3 \, O_2$)

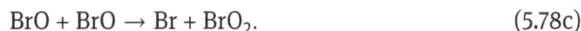

$$Br_2 + h\nu \rightarrow 2 \, Br, \tag{5.76}$$

$$Br + O_3 \rightarrow BrO + O_2, \tag{5.77}$$

$$BrO + BrO \rightarrow 2 \, Br + O_2, \tag{5.78a}$$

$$BrO + BrO \rightarrow Br_2 + O_2, \tag{5.78b}$$

$$BrO + BrO \rightarrow Br + BrO_2. \tag{5.78c}$$

The monoxide radicals XO can undergo photolysis to reform atoms X (Br, I) and O (which generates with O_2 ozone), thus reducing the O_3 decomposition capacity of this cycle:

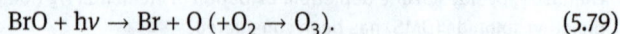

$$BrO + hv \rightarrow Br + O \ (+O_2 \rightarrow O_3).\tag{5.79}$$

The chemical pathway via reaction (5.78c) destroys only one O_3 molecule because of the tendency of BrO_2 to undergo rapid photolysis, resulting in both BrO and O_3 formation, following:

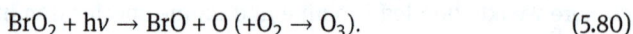

$$BrO_2 + hv \rightarrow BrO + O \ (+O_2 \rightarrow O_3).\tag{5.80}$$

Cycle II (net budget $HO_2 + O_3 \rightarrow OH + O_2$)

In competition with reaction (5.79), the halogen monoxide radical reacts further with a peroxo radical (HO_2) to form a hypohalous acid (HOX). The HOX rapidly breaks apart under light to regenerate the halogen radical, but not the ozone, thus destroying ozone.

$$BrO + HO_2 \rightarrow HOBr + O_2,\tag{5.81}$$

$$HOBr + hv \rightarrow OH + Br,\tag{5.82}$$

$$Br + O_3 \rightarrow BrO + O_2.\tag{5.77}$$

In polluted air mass, additional reactions with NO and NO_2 occur:

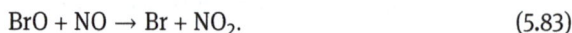

$$BrO + NO \rightarrow Br + NO_2.\tag{5.83}$$

During the night, the main reactions are driven by the nitrate radical (NO_3) instead of O_3. Formation of atoms due to missing radiation goes via

$$Br_2 + NO_3 \rightarrow Br + BrONO_2.\tag{5.84}$$

An important pathway goes via NO_2 (also for BrO_2)

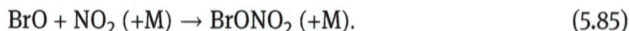

$$BrO + NO_2 \ (+M) \rightarrow BrONO_2 \ (+M).\tag{5.85}$$

In analogy to bromine, IO, HOI and $IONO_2$ are gained. Besides photolysis, the other important loss pathway of HOI and $IONO_2$ is its heterogeneous uptake onto halide-rich aerosol surfaces to produce interhalogens ICl and IBr in the marine atmosphere, which rapidly releases reactive halogen atoms which significantly affect atmospheric oxidizing capacity. Gas-phase formation of interhalogens is less favoured, for example, according to

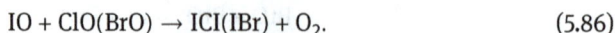

$$IO + ClO(BrO) \rightarrow ICl(IBr) + O_2.\tag{5.86}$$

Iodine oxides are promoted with a high release of I_2 during summer. This is due to a strong relationship between iodine oxides formation and ozone concentrations. Recent field measurements in various locations confirmed the presence of iodic acid (HIO_3), the anhydride of iodine pentoxide I_2O_5. I_2O_3 is considered as one of the gas-phase main precursors of higher iodine oxides species; its formation is obtained via

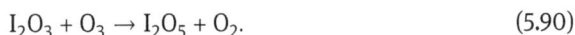

$$IO + OIO \rightarrow I_2O_3, \tag{5.87}$$
$$I_2O_2 + O_3 \rightarrow I_2O_3 + O_2, \tag{5.88}$$
$$IO + IO \rightarrow I_2O_2, \tag{5.89}$$
$$I_2O_3 + O_3 \rightarrow I_2O_5 + O_2. \tag{5.90}$$

Already 20 years ago, IO_2 (OIO) was recognized as one potential key species for new particle formation via inorganic polymeric structures as proposed for I_2O_4. Both undecomposed I_2O_4 ($[IO]^+[IO_3]^-$) and its thermal decomposition product (I_2O_5) are expected to result in the formation of iodate ions. Atmospheric removal of I_2O_5 or $HOIO_2$ goes as nucleation or condensation onto aerosols. At all, according to the newest research, gas-phase halogen chemistry in the troposphere is strongly interlinked with the heterogeneous process on water surfaces and particles.

The most important source of atmospheric bromine is due to the heterogeneous reaction between bromide and ozone. Results of earlier studies have shown that BrO_x sources are heterogeneous reactions occurring on the surfaces of blowing snow, sea salt aerosols (SSA), frost flowers, and surface snow with subsequent photolysis of Br_2 and BrCl. SSA is thought to be the largest source of tropospheric bromine. Volatilization of bromide from SSA can take place by heterogeneous reactions with HOBr, HOCl, N_2O_5, ozone, and $ClNO_3$ once alkalinity has been titrated and SSA is acidified:

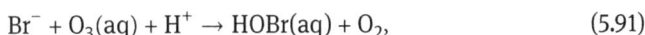

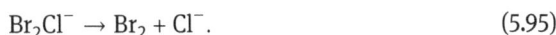

$$Br^- + O_3(aq) + H^+ \rightarrow HOBr(aq) + O_2, \tag{5.91}$$
$$HOBr + Br^- + H^+ \rightarrow Br_2 + H_2O, \tag{5.92}$$
$$HOBr + Cl^- + H^+ \rightarrow BrCl + H_2O, \tag{5.93}$$
$$BrCl + Br^- \rightarrow Br_2Cl^-, \tag{5.94}$$
$$Br_2Cl^- \rightarrow Br_2 + Cl^-. \tag{5.95}$$

Degassed bromine is fast photolysed into Br atoms that further react with O_3 to bromine monoxide BrO. Recent experimental results indicate that SSA enhancement by strong winds engenders activation of atmospheric BrO_x cycles via heterogeneous reactions on SSA. Earlier studies have indicated "blowing snow" as a source of atmospheric BrO_x. However, blowing snow had Br^- enrichment, in contrast to strong Br^- depletion in SSA. Although reactions described by equations (5.91)–(5.93) require acidity to release Br_2 and BrCl as a trigger of the atmospheric BrO_x cycle, blowing snow has higher pH. Furthermore, the lifetime of blowing snow is too short because of efficient dry deposition. In contrast to large amounts of acidic species supplied from anthropogenic processes in the Arctic, the source strength of anthropogenic acidic species is less in the Antarctic Circle. Therefore, acidity/alkalinity in the surface snow with high salinity is likely different in Antarctica and the Arctic.

Results show that enhancement of sea salt aerosols (SSA) and heterogeneous reactions on SSA are the main key processes for atmospheric BrO_x cycle activation. In-situ aerosol measurements and satellite BrO measurements demonstrated clearly that a BrO plume appeared simultaneously in SSA enhancement near the surface. Results show that surface O_3 depletion occurred in aerosol enhancement because of SSA dispersion during the polar sunrise.

> The key reaction for halogen induced ozone depletion is the heterogeneous reaction of dissolved ozone with halogenides to form hypohalite ions (XO^-), where the further fate of XO^- depends on pH of the aqueous solution.

The study of iodine on the earth system scale is extremely challenging because its biogeochemical cycles occur on a vast array of timescales – from seconds for some atmospheric processes to up to millennia in the ocean. These include the rates and controls of iodine cycling in the ocean, including in oxygen-depleted waters, and how iodide present at the very surface of the ocean is quantitatively transformed into iodine emissions to the atmosphere. Historically, the biogeochemical cycling of iodine has tended to be studied separately and by different scientific communities in its marine and atmospheric compartments. A consideration of iodine from such a perspective is necessary to understand the linkages and feedbacks between biogeochemical and physical processes in the ocean and ozone (and other oxidants) in the atmosphere, which have policy-relevant impacts arising from emissions of ozone precursors through to climate change, ocean acidification and stratospheric ozone (Carpenter et al. 2021).

Over about the last decade, evidence has emerged that oceanic emissions of iodinated organic compounds such as CH_3I, CH_2ICl and CH_2I_2 are likely not the primary source of atmospheric iodine, as was originally thought. The dominant fraction (80 %) is instead believed to arise from a heterogeneous reaction of iodide with gaseous O_3 at the sea surface. Note that O_3 is hardly soluble, but once in contact with the water surface, it reacts interfacial which results in an enhanced flux of ozone (dry deposition) to the surface:

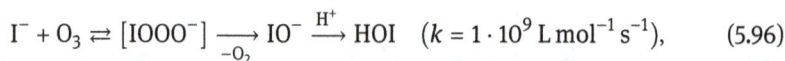

$$I^- + O_3 \rightleftharpoons [IOOO^-] \xrightarrow[-O_2]{} IO^- \xrightarrow{H^+} HOI \quad (k = 1 \cdot 10^9 \, L \, mol^{-1} \, s^{-1}), \qquad (5.96)$$

$$HOI + I^- + H^+ \rightleftharpoons I_2 + H_2O. \qquad (5.97)$$

The reaction of I^- is 10^7 and 10^{12} times faster than the corresponding reaction with Br^- and Cl^-, respectively. Thus, considering the concentrations of halides in seawater (in mol L^{-1}) 0.55 for Cl^-, $0.8 \cdot 10^{-3}$, and $5.0 \cdot 10^{-6}$ for I^-, the reactions rates concerning Br^- and I^- are much faster than for Cl^-, but the reaction between chloride and ozone remains fast enough to provide an effective sink for ozone due to the large chloride concentration (residence time about 20 s).

I_2 and HOI are in equilibrium between seawater and air. Thus, any subsequent gas-phase reaction of I_2 and HOI would result in a shift of the equilibrium towards the atmosphere, in other terms, in an increase in oceanic iodine emission. However, the equilibrium (or emissions, respectively) depends on many factors in the sea-surface microlayer, such as surfactants and organic matter, which is able to uptake iodine and thus reduce the emission. It is remarkable to note here that ozone in the air was first detected in the mid-nineteenth century by Schönbein's iodometric method, the colouring of starch-iodide paper (compare with footnote 87 on p. 269), however with several problems due to interference with humidity, and the dependency of exposure time and temperature (which can today be explained by the complex reaction mechanism).

The mechanism of the reactions in the determination of ozone via reaction with potassium iodide was not known before 1952 and was described in detail in the 1970s. Under neutral conditions, iodine is formed (which can evaporate under dry conditions), which can react with the base (OH^-) formed in this reaction gaining iodate ion. This reaction is reversed upon acidification. Iodate also can be formed in the reaction between iodide and ozone. Furthermore, under acidic conditions, the production of hydrogen peroxide has also been considered. The fast initial reaction is $I^- + O_3 \rightarrow I + O_3^-$ (ozonide decays into hydroxyl ion and the radical), but also direct oxidation to iodate occurs very fast: $I^- + O_3 \xrightarrow[(-O_2)]{} IO^- \xrightarrow[(-O_2)]{O_3} IO_3^-$. The oxidation of iodide by hydrogen peroxide is also fast: $I^- + H_2O_2 \rightarrow I + OH + OH^-$. The first investigation of the reaction mechanism between iodide and hydrogen peroxide was made by the Austrian-American chemist Hermann Alfred Liebhafsky (1905–1982) in the 1930s, supposing a series of observable and short-living intermediates.

Very recent results show that iodide (I^-) efficiently catalyses the oxidation of Br^- and Cl^- in aqueous nanodroplets exposed to ozone, the ever-present atmospheric oxidizer, under conditions resembling those encountered in marine aerosols. Br^- and Cl^-, which are rather unreactive toward O_3 and were previously deemed unlikely direct precursors of atmospheric halogens, are readily converted into IBr_2^- and ICl_2^- en route to $Br_2(g)$ and $Cl_2(g)$ in the presence of I^-. Fine sea salt aerosol particles, which are predictably and demonstrably enriched in I^- and Br^-, are thus expected to globally release photoactive halogen compounds into the atmosphere, even in the absence of sunlight. It has been strongly suggested that the heterogeneous reaction of HOI on aerosol (equation (5.98)) is the major source of ICl and IBr:

$$HOI + Cl^- \text{ (or } Br^-\text{)} \rightarrow ICl \text{ (or IBr)}. \tag{5.98}$$

Note that although ICl and IBr can also be formed via the gas-phase reactions of IO + ClO and IO + BrO, respectively (see eq. (5.86)), the calculated rate of gas-phase production for typical levels of halogen oxides measured at Mace Head is at least 40 times slower than the heterogeneous production of ICl and IBr via the reaction (5.98) of HOI on aerosol. The increase of ICl and IBr after sunset is likely due to the oxidation reaction of I_2 by the nitrate radical (NO_3), leading to the production of $IONO_2$ via reaction equation (5.99), which subsequently undergoes heterogeneous reaction with marine aerosol (equation (5.100)) to form ICl and IBr:

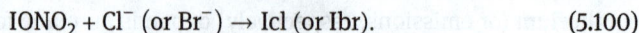

$$I_2 + NO_3 \rightarrow I + IONO_2, \tag{5.99}$$

$$IONO_2 + Cl^- \text{ (or } Br^-) \rightarrow ICl \text{ (or } IBr). \tag{5.100}$$

Photolysis of gaseous halogen compounds rapidly releases reactive halogen atoms, which significantly affect atmospheric oxidizing capacity. In summarizing the interlinked ocean-atmosphere halogen chemistry, we can identify the following main issues:

- The main emitted halogen compounds from the sea surface are HOBr and HOI.
- Permanent removal of atmospheric O_3 at the ocean surface due to oxidation of dissolved halogenides, mainly iodide and bromide (but also chloride on a longer time scale) leads to an enhanced O_3 dry deposition and increased iodine emission.
- Occasional removal of O_3 at the surface of sea salt particles in the lower marine boundary layer due to heterogeneous oxidation of halogenides and subsequent halogen loss, depending on sea salt events; this process very likely also occurs on halogen-containing continental particulate matter.
- Event-related O_3 depletion (tropospheric "ozone-whole") in the marine polar atmosphere after sun-rise due to accumulation of reactive halogen compounds in the polar night.

> **i** Increasing O_3 concentrations since the pre-industrial period (due primarily to increased anthropogenic emissions of nitrogen oxides) imply that atmospheric iodine should be substantially higher now than in the past.

Figure 5.23 summarises the aqueous phase chlorine chemistry (that of bromine is similar), and Figure 5.24 shows the multiphase chemistry of iodine (that of bromine is similar). In contrast to the other halogens, iodine forms polymeric oxides that can provide effective CCN in the marine environment.

5.3.3.2 Rethinking halogen loss from sea salt aerosol

In Chapter 4.7.2, we described the loss of chloride (that also occurs for bromide and iodide) in terms of HCl from sea-salt aerosol, which was attributed to more than 65 years of surface reactions (acidification) by acids produced from gaseous SO_2. Chlorine degassing also have been observed in a continental particular matter and almost attributed to the sticking of nitric acid HNO_3 onto the particles. In the previous Chapter, the release of reactive halogen compounds (not HCl) in the marine atmosphere was explained to be attributed to the surface reaction of O_3 with halogenides (X = Cl, Br, I).

$$X^- + O_3 \rightarrow XO^- + O_2. \tag{5.101}$$

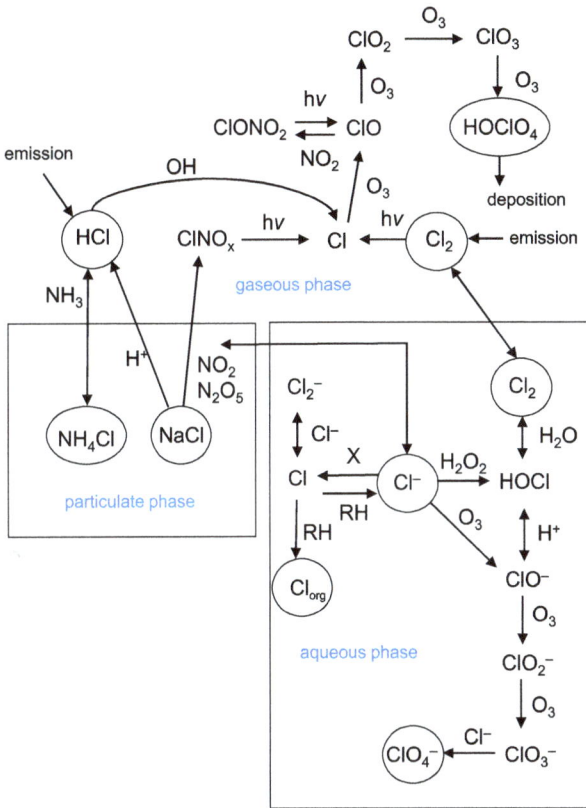

Figure 5.23: Simplified scheme of multiphase chlorine chemistry; $X = NO_3$; OH, $x = 1, 2$.

Whereas the first step does not depend on pH, the formation of HOCl (namely for HOCl degassing) needs slight acidification (remember $pKa = 7.7$); in contrast, HOBr and HOI are so weak acids that they are just formed in an aqueous solution ($YO^- + H_2O \rightarrow OHY + OH^-$). However, it was also notified that (note that sea salt is slightly alkaline) acidification is needed to form molecular halogens:

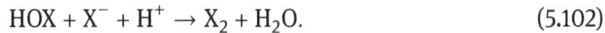

$$HOX + X^- + H^+ \rightarrow X_2 + H_2O. \tag{5.102}$$

Such conditions exist in polluted continental air, and we can set up the net reaction (p – particulate, g – gaseous)

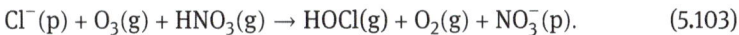

$$Cl^-(p) + O_3(g) + HNO_3(g) \rightarrow HOCl(g) + O_2(g) + NO_3^-(p). \tag{5.103}$$

Thus, there would not be HCl primarily degases, but hypochlorous acid which day-time is quickly photolyzed into OH + Cl, and Cl subsequent fast turns into HCl via Cl + RH (e. g. CH_4). This process also would increase the oxidation capacity through the

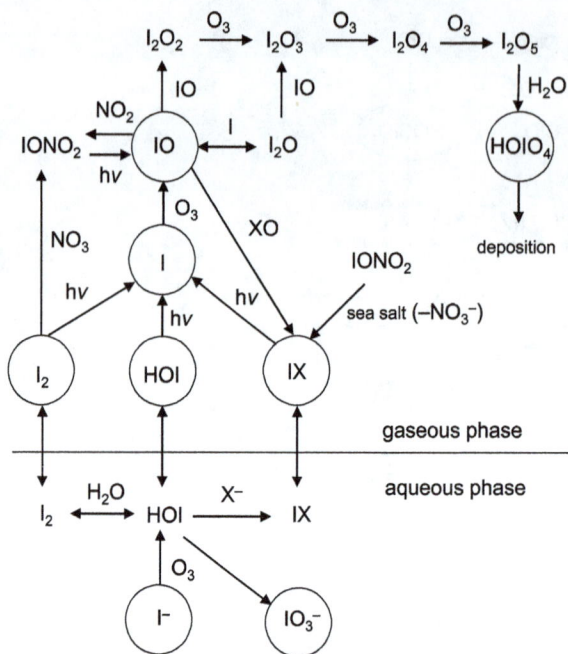

Figure 5.24: Simplified scheme of multiphase iodine chemistry; X = Cl, Br.

production of OH radicals which can turn NO_2 to HNO_3. Our experimental results (see Figure 4.23) show no correlation between HCl and particulate chloride (this is correlated with sodium showing its sea salt origin according to changing transport processes and marine source strength) but with HNO_3. This hypothesis does not exclude an acid displacement but suggests comparative pathways, depending on the environmental conditions such as humidity, concentrations of O_3, HNO_3 and SO_2. Note that SO_2 as an acid precursor also first needs oxidants (O_3 or H_2O_2 depending on the acidity of the reaction medium, see Chapter 4.5.3:

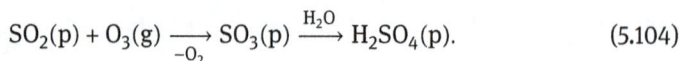

$$SO_2(p) + O_3(g) \xrightarrow[-O_2]{} SO_3(p) \xrightarrow{H_2O} H_2SO_4(p). \tag{5.104}$$

The reaction rate constant of (28) is about 10^{10} larger ($k \approx 10^7 \, L \, mol^{-1} \, s^{-1}$ at pH = 5) than reaction (27) so that first S(IV) oxidation after absorption of gaseous SO_2 occurs, but this results in a decreasing pH (more acidity) with a reduction of SO_2 absorption and – more important – a drastic decrease of the S(IV) oxidation by O_3. Thus, the second step could be the preferential oxidation of Cl^- by O_3 – and HCl degassing due to acidification of the particle.

5.3.4 Atmospheric removal: deposition processes

The deposition is the mass transfer from the atmosphere to the earth's surface; it is the opposite of emission (the escape of chemical species from the earth's surface into the air). It is a flux given in mass per unit of area and unit of time. According to the various forms and reservoirs of atmospheric chemical species (molecules in the gas, particulate and liquid phases), different physical and chemical processes are distinguished:

– *sedimentation* of matter because of the earth's gravitational force (this is valid only for particles of a certain size; larger about 5 µm);
– sorption of molecules and small particles at the earth's surface with subsequent vertical transport process called *dry deposition*;
– sorption of molecules and impaction of particles by falling hydrometeors called *wet deposition*; and
– *impaction* of particles from airflow at surfaces.

Gravitational settlement is only of interest for large particles in the upper range of the coarse mode. In addition, hydrometeors (raindrops, hail and snow particles) settle because of gravitation, but we do not consider water to be deposited (it is physically sedimented, but rather we call the process *precipitation*) but only the scavenged chemical trace species. Therefore, we consider three reservoirs of trace substances: gaseous (molecules), mixed in atmospheric particles and dissolved in hydrometeors. Despite the fact that sedimentation and impaction can be important removal processes under specific conditions, they play no significant role in the regional and global budget of fluxes in the environment. On local sites, however, such removal processes can be important, for example, fog droplet impaction by animals (e. g. *Stenocara* beetle) and montane cloud forests, and the sedimentation of dust after volcanic eruptions or soil dust after storm events. Particle removal is often parameterised as not sharply separating between dry deposition and sedimentation but including deposition processes such as turbulent transfer, Brownian diffusion, impaction, interception, gravitational settling and particle rebound. Unfortunately, in the scientific literature, the term 'dry' for all 'non-wet' deposition processes is widely distributed. Figure 5.25 shows the removal process of gases and particles from the atmosphere (impaction is not shown, but it is a simple collision impaction of solid particles and cloud droplets). Without further comments, it is clear that dry deposition also occurs during precipitation; for soluble gases, dry deposition velocity increases because of the reduced soil resistance (see below). Total deposition onto the earth's surface (normally in the sense of an artificial collector surface) is called *bulk deposition*:

bulk deposition = dry deposition + wet deposition + sedimentation.

From a measurement point of view, there are difficulties approaching the physically correct removal processes. Dry deposition strongly depends on the surface charac-

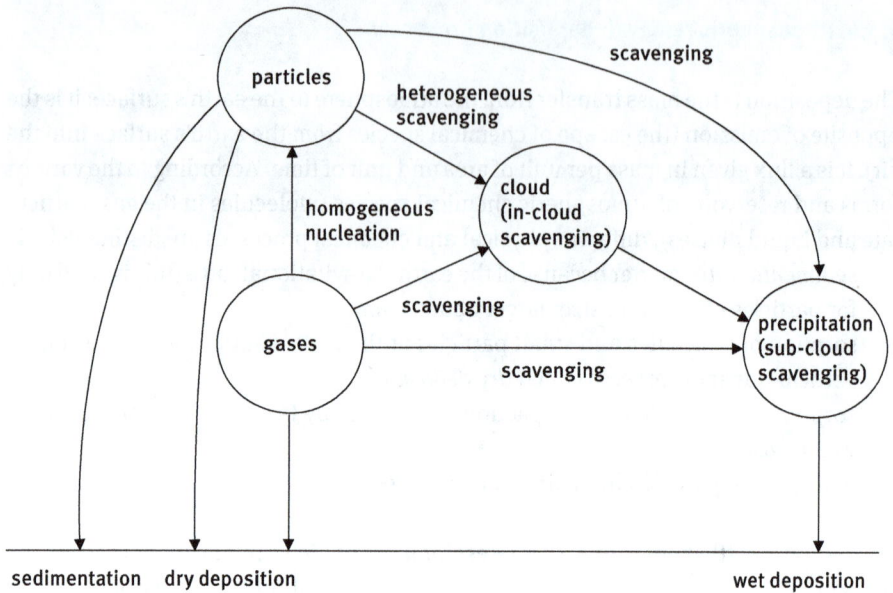

Figure 5.25: Scheme of deposition processes.

teristics; correct estimation is only possible by flux measurements (eddy correlation and gradient methods). Any so-called deposition sampler can be installed with a wet-only/dry-only cover plate to avoid bulk sampling, but it is never possible to avoid the collection of sedimentation dust in dry-only samplers or dry deposition or sedimentation by wet-only samplers. However, with enough accuracy, all other deposition processes contribute negligibly to the deposition flux under wet-only and dry-only conditions, respectively.

Dry deposition is a generic and very similar process to the mass transfer in the case of heterogeneous processes as shortly described in Chapter 3.3.7, where the transport axis is vertically downwards, and the absorbing object is the fixed earth's surface. The situation, however, is complicated by the possible upward flux of the substance due to plant and soil emissions. Only the net flux is measurable. The case that downward and upward fluxes (so-called bidirectional trace gas exchanges) are equal ($F_{dry} = F_Q$) is called the *compensation point*. It only depends on the concentration gradient ($c_h - c_0$). Additionally, the net flux can be influenced by fast gas-phase reactions within the diffusion layer. Let us now consider the surface like sink, independent of the specific process, which can be:

– surface adsorption;
– interfacial transfer (absorption including chemical reaction);
– stomatal uptake.

Table 5.12: Averaged dry deposition velocities (in cm s^{-1}) above land.

substance	SO_2	NO	NO_2	HNO_3	O_3	H_2O_2	CO	NH_3
v_d	0.8	<0.02	0.02	3	0.6	2	<0.02	1

The concept of dry deposition and the dry deposition flux is considered of first-order concerning atmospheric concentration:

$$F_{dry} = -v_d c, \tag{5.105}$$

where v_d is the dry deposition velocity (dimension: distance per time). The advantage of introducing the deposition velocity is to avoid a microphysical treatment of vertical diffusion and surface interfacial processes by using a single mass transfer coefficient, which is measurable (Table 5.12). The limiting application of equation (5.105) is because of the dependence of v_d on various parameters and states of the atmosphere and the earth's surface. Besides equation (4.312), the general diffusion equation is valid:

$$F_{dry} = K_z \left(\frac{dc}{dz} \right)_{z=h} = v_d c, \tag{5.106}$$

where K_z is the turbulent vertical diffusion coefficient. From this equation, it can be seen that v_d is experimentally quantifiable through $v_d = K_z (d \ln c/dz)$. The dry deposition process consists of three steps:

-   aerodynamic (turbulent) transport through the atmospheric surface layer to the molecular (diffusion) boundary layer close to the surface;
-   molecular diffusion transport onto the surface (quasi-laminar sub-layer);
-   uptake by the surface as the ultimate process of removal.

Each step contributes to v_d or the dry deposition resistance $r = 1/v_d$. According to the three layers, the total resistance r is given from the partial resistances (Figure 5.26), the aerodynamic r_a, the quasi-laminar r_b and the surface resistance r_c:

$$\frac{1}{v_d} = r = r_a + r_b + r_c. \tag{5.107}$$

This equation follows from the flux through different layers under steady-state conditions with the boundary condition $c_0 = 0$, rearranging after c_1, c_2 and c_3:

$$F = \frac{c_3 - c_2}{r_a} = \frac{c_2 - c_1}{r_b} = \frac{c_1 - c_0}{r_c} = \frac{c_3 - c_0}{r}. \tag{5.108}$$

The limitation of the resistance model lies in its application only for sufficient homogeneous surfaces such as forests, lakes and grasslands. Therefore, in dispersion models, dry deposition can be described using partial or weighted partial areas within the grid.

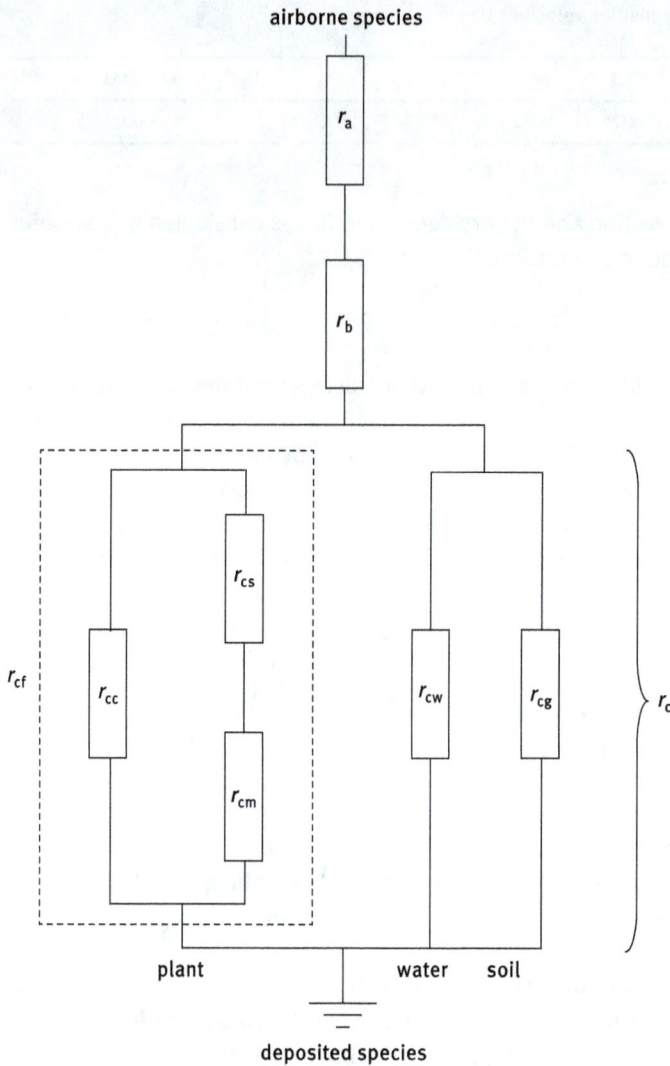

Figure 5.26: Resistance model of dry deposition. r_a – aerodynamic resistance, r_b – quasi-laminar resistance, r_c – soil resistance, r_{cw} – soil resistance (water), r_{cg} – soil resistance (other ground), r_{cf} – foliar resistance (weighted by leaf area index), r_{cs} – stomatal resistance, r_{cc} – cuticular resistance, r_{cm} – mesophylic resistance.

As seen in Figure 5.23, wet deposition is a complex process beginning with hetero-geneous scavenging during cloud droplet formation, cloud processing (in-cloud scav-enging) and precipitation (sub-cloud scavenging; sometimes called below-cloud scav-enging). Hence, several microphysical processes and chemical reactions must be con-sidered. It is remarkable that wet deposition is extremely difficult to model in contrast to dry deposition but is easier to measure by rain gauges (water sampling and analysing the dissolved substances), whereas dry deposition measurements (vertical flux quantifications) are laborious. The number of precipitation chemistry measurements is uncountable; many networks has existed since the 1950s. In contrast to dry deposition, wet deposition is occasional. In remote areas, the average fluxes concerning dry and wet deposition are similar, but as precipitation occurs for only 5–10 % of the year, the event-based wet flux is considerably larger than the dry flux within a comparable time. Despite the complexity of the specific steps in gaining wet deposited substances, we can parameterise the wet deposition flux F_{wet} similar to the approach concerning dry deposition:

$$F_{wet} = -\left(\frac{dc}{dt}\right)_{wet} = k_{wet}c.$$

(5.109)

By contrast, it is valid from the simple balancing of deposited rainwater that

$$F_{wet} = r \cdot c(aq),$$

(5.110)

where c_{aq} is the rainwater concentration of the substance (dimension: mass per litre), and r is the rainfall amount (dimension: litre per unit of time and unit of area).

5.3.5 Radioactivity

All nuclides that have an ordinal number larger than lead (Pb) are radioactive. The natural radioactive decay of atomic nuclei is a spontaneous process according to the following scheme and corresponds to a monomolecular reaction (first-order); however, we call radioactive decay not among *chemical* reactions:

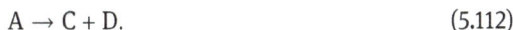

$$A \rightarrow B,$$

(5.111)

$$A \rightarrow C + D.$$

(5.112)

The decay rate is proportional to the amount N, and it follows that the half-life $\tau_{1/2}$ of the decay, i. e. the time when 50 % of the original amount is removed ($N = N_0/2$):

$$\frac{dN}{dt} = \lambda N,$$

(5.113)

$$\tau_{1/2} = \frac{\ln 2}{\lambda}.$$

(5.114)

The reciprocal value of the decay constant $1/\lambda$ corresponds to the mean lifetime $\bar{\tau}$ of the atoms, i. e. the arithmetic mean of the individual lifetimes of all atoms before they decay:

$$\bar{\tau} = \frac{1}{N_0} \int_{N_0}^{0} t \, dN. \tag{5.115}$$

The phenomenon of *natural* radioactive decay is normally characterised through eject-ing of either a positive helium core He^{2+} (alpha decay), which consists of two protons and two neutrons:

$$^{226}_{88}Ra \rightarrow\ ^{222}_{86}Rn +\ ^{4}_{2}He \quad (\tau_{1/2} = 1{,}600 \text{ years}) \tag{5.116}$$

or a negative electron e^- (beta decay):

$$^{227}_{89}Ac \rightarrow\ ^{227}_{90}Th + e^- \quad (\tau_{1/2} = 21{,}772 \text{ years}). \tag{5.117}$$

Some nuclei additionally emit (to balance the energy) gamma radiation. Other types occur but are much less probable. Radionuclides are also found in some elements with $z < 83$, such as K, Rb, Cd, In, Te, La, Nd, Sm, Gd, Lu, Hf, Ta, Re, Os, Pt and Pb. Well known is the nuclide ^{40}K, which forms through the rare electron capture Ar and ^{14}C that are used for age determination:

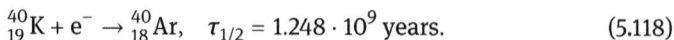

$$^{40}_{19}K + e^- \rightarrow\ ^{40}_{18}Ar, \quad \tau_{1/2} = 1.248 \cdot 10^9 \text{ years}. \tag{5.118}$$

Carbon-14 goes through radioactive beta decay by emitting an electron (beta particles) and an antineutrino:

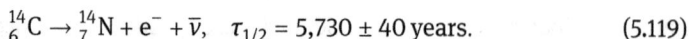

$$^{14}_{6}C \rightarrow\ ^{14}_{7}N + e^- + \bar{v}, \quad \tau_{1/2} = 5{,}730 \pm 40 \text{ years}. \tag{5.119}$$

The maximum distance travelled of the emitted beta particles is estimated to be 22 cm in air and 0.27 mm in body tissue. There are four natural radioactive decay series on Earth:

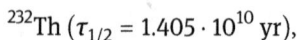

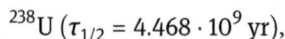

$$^{232}Th \ (\tau_{1/2} = 1.405 \cdot 10^{10} \text{ yr}),$$
$$^{238}U \ (\tau_{1/2} = 4.468 \cdot 10^9 \text{ yr}),$$
$$^{227}Ac \ (\tau_{1/2} = 21.77 \text{ yr}),$$
$$^{237}Np \ (\tau_{1/2} = 2.14 \cdot 10^6 \text{ yr}).$$

Np is extremely rare on Earth, however. The initial elements of the Th and Ac series are ^{244}Pu ($\tau_{1/2} = 1.4 \cdot 10^{11}$ yr) and ^{239}Pu ($\tau_{1/2} = 2.411 \cdot 10^4$ yr), respectively. Natural Pu is also extremely rare. Thus, the unstable but long-lived isotopes (^{232}Th, ^{235}U, ^{236}U) makes up the internal heat source that drives volcanic activity and processes related to internal convection in the terrestrial planets; these three nuclides (and ^{244}Pu) are used to determine the age of the Earth, meteorites and other celestial bodies. The four decay series naturally produce 16 radioactive elements with atomic numbers < 83 (Rb, K, Cd, In, Te, La, Nd, Sm, Gd, Lu, Hf, Ta, Re, Os, Pt, Pb), which all have extremely long lifetimes of from 10^{10} to 10^{24} years. The final product of all decay series is inactive lead (Pb), however of different atomic mass: ^{206}Pb (from uranium), ^{207}Pb (from actinium) and ^{208}Pb (from thorium). Note also that many unstable nuclides can be produced artificially.

> Natural radiation has always been part of the human environment. Its main components are cosmic and cosmogenic radiation, terrestrial gamma radiation from natural radionuclides in rocks and soil, and natural radioactive substances in our food and air.

Among the intermediates are of special interest the three gaseous nuclides actinon ($^{219}_{86}Rn$, $\tau_{1/2} = 3.96\,s$), thoron ($^{220}_{86}Rn$, $\tau_{1/2} = 55\,s$) and radon ($^{222}_{86}Rn$, $\tau_{1/2} = 3.8\,d$), all isotopes of radon, which are released from soils. They cause secondary radioactivity due to their ubiquitous occurrence through the formation of solid polonium (Po), which adsorbs on surfaces and again decays radioactively. The potential for radon entry from the ground depends mainly on the activity level of ^{226}Ra in the subsoil and its permeability with regard to airflow. Example of terrains with a high radon potential is alum shales, some granites and volcanic rocks. At some places in Europe and Northern America, indoor levels of Rn can cause a risk of lung cancer.

The Sun and stars send a constant stream of cosmic radiation to Earth. Of primary cosmic rays, which originate outside of the earth's atmosphere, about 99 % are the nuclei (stripped of their electron shells) of well-known atoms, and about 1 % are solitary electrons (similar to beta particles). Of the nuclei, about 90 % are simple protons, i. e. hydrogen nuclei; 9 % are alpha particles, and 1 % are the nuclei of heavier elements. Differences in elevation, atmospheric conditions, and the earth's magnetic field can change the amount (or dose) of cosmic radiation that we receive. Secondary cosmic rays (mostly low energy neutrons), formed by interactions in the earth's atmosphere, produce radionuclides (for example, ^{7}Be, ^{10}Be, and ^{14}C by interaction with atmospheric nitrogen and oxygen), and account for about 45 to 50 millirem of the 360 millirem of background radiation that an average individual receives in a year.

The third type of radioactivity is artificially produced by using radioactive nuclides in techniques and medicine, but in the early 1950s by nuclear bomb tests and nowadays by catastrophes such as the Chernobyl and Fukushima nuclear power plant disasters. The nuclear tests, as mentioned before, forced research in air chemistry to study long-range dispersion and deposition (called in that time *fallout* and *washout*). In all nuclear tests, ^{235}U, ^{238}U and ^{239}Pu produces most radioactive isotopes such as ^{137}Cs, ^{90}Sr, ^{144}Ce, ^{95}Zr, ^{89}Sr and ^{140}Ba. Between 1952 and 1962, more than 60 Mt of fission material was ejected into the stratosphere. It is estimated that the radiation and radioactive materials from atmospheric testing taken in by people up until the year

2000 would cause 430,000 cancer deaths. The study predicted that roughly 2.4 million people could eventually die from cancer due to atmospheric testing.

The Chernobyl disaster (1986) resulted in the release of large quantities of radioactive particles into the atmosphere, spreading over much of the western USSR and Europe. An estimate for total nuclear fuel material released into the environment corresponds to the atmospheric emission of 6 t of fragmented fuel. ^{131}I and ^{137}Cs are responsible for most of the radiation exposure received by the general population; there will be 50,000 excess cancer cases resulting in 25,000 excess cancer deaths. The Fukushima Daiichi nuclear disaster (2011) led to radioactive contamination by ^{134}Cs and ^{137}Cs of the Eastern Pacific. Both nuclear disasters were 'man made' but avoidable accidents in Chernobyl due to improper management and the Fukushima power plant was unprotected from potential tsunamis.

Krypton ^{85}Kr ($\tau_{1/2}$ = 10.76 years) has both natural (e. g. cosmic ray) and artificial sources (mostly nuclear reprocessing) and increases permanently in the atmosphere. Measurements carried out in Prague show an increase from 0.75 Bc m^{-3} (1982) to 1.1 Bc m^{-3} (1992) and at Mt. Zugspitze in 2001, an average of 1.45 Bc m^{-3} was found. However, the radiation dose is rather low and contributes only 1 % of the maximum permissible dose. Being an inert noble gas, it circulates globally, dissolves in rain and oceans, is absorbed by soils but is also released from soils, and appears as a tracer for radioactivity.

6 Green chemistry

Humans – by decoupling their life cycle from natural conditions – have altered "natural" biogeochemical cycles. In recent decades, humans have become a very important force in the earth system, demonstrating that emissions and land-use change are the causes of many of our environmental issues. These emissions are responsible for the major global reorganisation of biogeochemical cycles. With humans as part of nature and the evolution of an artificial changed earth system, we also have to accept that we are unable to remove the present system into a pre-industrial or even prehuman state because this means disestablishing humans. The key question is, which parameters of the climate system allow the existence of humans under which specific conditions. We can state that humans became a global force in the chemical evolution with respect to climate change by interrupting naturally evolved biogeochemical cycles. But humans also do have all the facilities to turn the "chemical revolution" into a sustainable chemical evolution. At the beginning of Chapter 5, we defined a sustainable society as being able to balance the environment, other life forms and human interactions over an indefinite period.

Global sustainable chemistry first needs a paradigm change, namely the awareness that growth drives each system towards a catastrophe. However, even adopting the future scenario of a constant world population and fixed per capita consumption (in other terms, no further growth), there remain two principal problems:

1. Environmental pollution: can be solved by new technologies and social/individual willingness.
2. Depleting non-renewable resources: is inevitable and only a question of time.

Our non-renewable resources include (note, minerals are silicates and non-silicates; the latter include carbonates, oxides, halides, sulphides, sulphates, phosphates and others of minor importance)
– fossil fuels (coal, oil, gas) for energetic use and chemical products,
– mineral ores for the production of metals and half-metals,
– minerals for the production of salts, cement, bricks, ceramics, glass, natural stones etc.

Renewable resources include
– biomass: for energetic use, building and texture materials as well products of organic chemistry,
– atmospheric air: nitrogen for ammonia synthesis and nitrate production, liquid water (from water vapour and droplet harvesting), carbon dioxide (near-future technology),
– freshwater: hydrogen (and oxygen) production,
– seawater: salt production, extraction of metals (forward-looking technology),
– solar radiation: for energetic use (direct and indirect),

https://doi.org/10.1515/9783110735178-006

–   waste: recycling and reuse.

Sustainable chemistry, also known as green chemistry, is a chemical philosophy encouraging the design of products and processes that reduce or eliminate the use and generation of hazardous substances. Anastas and Warner (1998) developed 12 principles of green chemistry:

1. *Prevention*: it is better to prevent waste than to treat or clean up waste after it has been created.
2. *Atom Economy*: synthetic methods should be designed to maximise the incorporation of all materials used in the process into the final product.
3. *Less Hazardous Chemical Syntheses*: wherever practicable, synthetic methods should be designed to use and generate substances that possess little or no toxicity to human health and the environment.
4. *Designing Safer Chemicals*: chemical products should be designed to affect their desired function while minimising their toxicity.
5. *Safer Solvents and Auxiliaries*: the use of auxiliary substances (e. g. solvents, separation agents etc.) should be made unnecessary wherever possible and innocuous when used.
6. *Design for Energy Efficiency*: energy requirements of chemical processes should be recognised for their environmental and economic impacts and should be minimised. If possible, synthetic methods should be conducted at ambient temperature and pressure.
7. *Use of Renewable Feedstocks*: raw material or feedstock should be renewable rather than depleting whenever technically and economically practicable.
8. *Reduce Derivatives*: unnecessary derivatisation (use of blocking groups, protection/ deprotection, temporary modification of physical/chemical processes) should be minimised or avoided if possible because these steps require additional reagents and can generate waste.
9. *Catalysis*: catalytic reagents (as selective as possible) are superior to stoichiometric reagents.
10. *Design for Degradation*: chemical products should be designed so that at the end of their function, they break down into innocuous degradation products and do not persist in the environment.
11. *Real-time Analysis for Pollution Prevention*: analytical methodologies need to be further developed to allow for real-time, in-process monitoring and control before the formation of hazardous substances.
12. *Inherently Safer Chemistry for Accident Prevention*: substances and the form of a substance used in a chemical process should be chosen to minimise the potential for chemical accidents, including releases, explosions and fires.

Today, unfortunately, there is much misuse with the label 'green', and, from a scientific point of view, chemistry is chemistry and whether green nor sustainable. Such as, in living nature, there is no waste because of carbon cycling; chemists of the nineteenth century already used coal (and later oil) completely to obtain a wide range of products, simply by the economic principle to maximise the yield and to minimise the energy expenditure. With the exception of principle 7 (use of renewable feedstocks), almost all other principles have been practically considered by the chemical industry.

> The current main problem is the energetic use of fossil fuels for electricity generation and chemical processing, turning them completely into carbon dioxide and other oxidised contents.

Thus, what we should understand behind *green* chemistry is that
a) the primary materials for chemical processing are no fossil fuels
b) but largely renewable and/or cycling feedstocks;
c) energy for chemical processing is not based on fossil fuels but almost on solar energy.

The basic principle of global sustainable chemistry, however, is to transfer matter for energetic and material use only within global cycles without changing reservoir concentrations above a critical level, which is "a quantitative estimate of an exposure to one or more pollutants below which significant harmful effects on specified sensitive elements of the environment do not occur according to present knowledge" (Nilsson and Grennfelt 1988). Chemistry solely, however, cannot solve all problems resulting from the 'Great Acceleration' such as excess production, inadequate products, marine plastic pollution, climate change, social inequalities and others. And nobody should believe that digitalisation and artificial intelligence are a panacea; what we need is a 'Great Deceleration' in terms of 'back to nature' but not in view of the Romantics but rather in a new modern approach, defined by coming generations and societies.

> The first Industrial Revolution (nineteenth century) is characterized as the period of mechanisation using water and steam power, followed by the second industrial revolution (first half of the twentieth century) using electricity for mass production via assembly lines which transferred to the end of the twentieth century to the third industrial revolution using computers and automatisation (digitalisation). This phase surely is going on with cyber-physical systems (also called 4th industrial revolution). However, since the second industrial revolution, electricity has been based up to the present on fossil fuel burning. In my opinion, we expect a fifth industrial revolution by providing electricity solely based on solar radiation (second half of twenty-first century). Together with that, a sixth industrial revolution is inevitable, the use of carbon as material and energy carrier from the atmosphere, the CO_2 cycling. This could be likely the last industrial revolution reaching global steady states and gaining sustainable societies.

It is trivial to state that humanity cannot exist infinite because of the definite limit of radiation transfer from the Sun. Life limits by catastrophic events (mass extinction) likely occur in future, and slow geotectonic processes will change the land cover of our planet Earth.

6.1 The carbon problem

We have stated that human evolution was inherently linked with hydrocarbons, used as fuel, food and material, gained from renewable sources (trees, plants and animals) such as doing by every creature, but also from (what the human problem is) geological stocks (coal, oil and gas). The current problem consists in the replacement of fossil fuels as an energy source, and the forward-looking problem comprises an alternative source of hydrocarbons.

> From the analysis of the history of anthropogenic trace compounds in the atmosphere, its impacts on the climate system – so far, we understand the processes of climate change – and recognising the introduced abatement, few atmospheric environmental problems remain that are connected with the compounds N_2O, CH_4 and CO_2. The carbon dioxide problem is by far the most serious. There is no or only very little hope that global CO_2 emissions will significantly decrease in the next two or three decades. In contrast to N_2O and CH_4, the atmospheric residence time of "anthropogenic" CO_2 is orders of magnitudes larger; in other words, even when there would be a zero CO_2 emission world soon, the subsequent "greenhouse" effect will still last several hundreds of years. That part of N_2O and CH_4 linked with fossil fuels will "automatically" be solved together with the CO_2 problem. Hence, because the remaining sources of N_2O and CH_4 are linked with agriculture and food production, they are likely to become the dominant residual problem in 100 years, but with a far lower impact factor than today.

Mining and combustion of fossils fuels now result in geological reservoir redistribution of carbon close to (or even passing?) the "tipping point". In the last two decades, we observed acceleration of CO_2 release due to economic growth, which presently seems to go further on a constant level. The large CO_2 residence times in air and seawater avoid reaching a steady-state (global cycle in-time) and a recovery (climate restoration) also after a full stop of fossil fuel use. Climate scientists have warned that to have a 50-to-50 chance of limiting global warming to not more than 2 °C above the average global temperature of pre-industrial times throughout the twenty-first-century cumulative carbon emissions between 2011 and 2050 need to be limited 300 Gt C. Recent calculations suggest that this necessitates one-third of oil reserves, half of the gas reserves and over four-fifths of coal reserves to remain untapped from 2010 to 2050. With business-as-usual global warming leads to unacceptable degrees of peak global warming, around 5 °C. This highlights the urgency and scale of the climate policy challenge. To keep the 2 °C climate warming goal, the cumulative CO_2 emission must be limited to about 1000 Gt carbon (at present, we are approaching 650 Gt). Hence the atmospheric CO_2 increase must be stopped in the next few decades, and the "energy turnaround" (renewable energy revolution) must be completed by the end of this century.

Therefore, much more forced by climate change and its uncertain but very likely catastrophic impacts after reaching the "tipping points" than by fossil resource limits, we need the transfer into the "solar era" as soon as possible. Without any doubt, electricity is the unique form of energy in future, and its direct application (also for mobility and heating) will increase – and can replace a large percentage of traditional fuel-based on fossil resources. However, there are some open questions that have to be answered and transferred into technical solutions to establish the solar era.

– Electricity will be produced not constant over time and not correlated with the demand of energy hence it must be stored, likely best by transfer into "chemical energy", to manage energy supply.
– Due to safety reasons, excess energy must be stored (for example, in water reservoirs, but this way is limited); again best way seems to transfer electricity into "chemical energy".
– There are transport applications (for example, air traffic, long-distance street traffic, ship traffic) where electricity cannot be taken directly from nets or storage units and will be neither ecological nor economic.
– There are chemical processes (for example, metallurgy, cement production) that cannot operate by electricity but need "chemical energy".
– Humans always need synthetic organic materials (polymers, drugs, chemicals etc.). These can be produced from remaining fossil resources, but also from biomass – and from CO_2.

The private electro vehicle booster presently pushed by policy and governments is likely not sustainable and should be re-analysed. Other transition scenarios must evaluate and compare, such as green mobility – cycling, public transit, walking, and compact, car-free cities – but also traditional internal-combustion engines using carbon-neutral fuels.

> A vast new infrastructure is needed, including utility rights-of-way, transformers, substations, circuits, conduits, wires, meters, electric panels, junction boxes, and hardware and software for payment, wireless connectivity, and metering. Larger charging stations at activity centres such as office parks or shopping centres might require bigger transformers and substations and more infrastructure. There will also be a need for consistent, standardised regulation and compatibility between chargers and vehicles. Virtually none of this infrastructure exists presently.

> There is an epochal decision about mobility in front of us, and based on recent IPCC findings, the next decade suggests that little room for error can be tolerated.

6.2 The carbon economy

There are several ways as a consequence of the nearly irreversible accumulation of anthropogenic CO_2 in air and the cognition that carbon from fossil fuels is limited but, on the other hand, the knowledge that carbon compounds are optimal carriers for energy conversion and materials:

– to reduce fossil fuel combustion (replacement by solar energy);
– to capture CO_2 from exhaust gases and storage/sequestration (CCS technology);
– to capture CO_2 from exhaust gases and utilization (CCU technology);

- to capture CO_2 from ambient air (DAC) air;
- to recycle CO_2 into utilisable carbon compounds (or sequestration achieving a negative flux).

6.2.1 Carbon capture and storage (CCS)

CCS^{120} is an essential element of any low carbon energy future and industrial future, but the policy is the main issue, not technology. CO_2 capture technology has been used since the 1920s for separating CO_2 sometimes found in natural gas reservoirs, from the saleable methane gas. Capturing CO_2 from power plants in order to produce electricity, however, is purely done for emissions reduction reasons. The basic idea of CCS – capturing CO_2 from flue-gases and preventing it from being released into the atmosphere was first suggested in 1977, using existing technology in new ways to achieve political targets of CO_2 limitation and even 'zero-CO_2 emission'.

> The technology consists of two parts, first the absorption tower (column) where CO_2 from the waste gas is stripped by a washing solution, and the desorber where CO_2 is freed almost by heating of the solution between 120 and 140 °C. As absorption solution, different amines have been investigated, and mainly monoethanolamine (MEA) is now used. The process required a significant amount of energy (reduction in energy efficiency by about 25 %). In addition, other environmental impacts associated with the toxicity and environmental fate of the solvent have to be considered. The current research on CO_2 absorption using MEA is mainly focused on the minimisation of energy consumption during solvent regeneration. Another field of recent research is the use of the absorption–desorption process but utilising a different solvent. Adsorption, CO_2 conversion, calcium looping, and membrane technology are the focus of alternatives to the conventional process for CO_2 capture in a post-combustion scenario.

As of the end of 2012, there were 5 large scale CCS projects in operation around the world, in 2015 already 22 large-scale CCS projects and currently 27 in operation or under construction, with the capacity to capture up to 40 million tons of CO_2 per year (Mtpa). From 75 Mtpa at the end of 2020, the capacity of projects in development grew to 111 Mtpa in September 2021. This must increase at least 40-fold by 2050 to meet the scenarios laid out by the IPCC. Historically, CCS projects tended to be vertically integrated, with a capture plant having its own dedicated downstream transport system. This favoured large-scale projects, where economies of scale made downstream costs reasonable. Recently, there has been a trend toward projects sharing CO_2 transport and storage infrastructure: pipelines, shipping, port facilities, and storage wells. These 'CCS networks' mean smaller projects can also benefit from economies of scale.

---

120 The letter 'S' stands mostly for 'storage' but also for 'sequestration'. The world 'sequestration' was used before 'CCS' in law and biology (uptake and separation of toxins by plant tissues). Because 'capture' already means 'separation' of CO_2 from exhaust gas, 'sequestration' is synonym with 'storage' in geological stocks (permanent or for reuse).

From the long-term perspective, the transmission to renewable energies will solve the problem of the depletion of fossil fuels, but it is surely too late to minimise climate change. Moreover, although future oil consumption might stagnate or even decrease, coal consumption remains important for a long time. Carbon capture and storage (CCS) technology is considered the only practical solution for early CO_2 control. CO_2 storage (and sequestration) is the ultimate technology for "neutralising" the CO_2 budget through the combustion of fossil fuels but also the only way to make the biosphere-atmosphere CO_2 budget "negative", i.e. reducing atmospheric CO_2 loading. CCS technology is now widely accepted as a solution for meeting the climate protection targets (CO_2 emission reduction) worldwide and a precondition for further use of coal-fired power plants beyond 2020 in Europe. CCS is an essential decarbonisation option for the world's industrial businesses. Key emission-intensive sectors such as chemicals, iron and steel, and cement are sometimes referred to as 'hard to abate'. These sectors cannot make their products without producing CO_2. Switching to renewable energy or focusing on energy efficiency is unable to solve a substantial fraction of their emissions.

Regardless of the chosen technology, it is important to keep in mind that the final objective of the development of any technological option to the conventional process should aim at negative carbon intensity (energy-related CO_2 emissions to the atmosphere in grams of carbon in relation to energy in Jouler). This means that more CO_2 has to be captured than emitted per unit of energy. Two large groups of potential carbon-negative renewable energy technologies (negative CO_2 emissions) have been identified: a) carbon-negative biofuels i.e. bioenergy with CCS (BECCS) and b) carbon-negative products derived from CO_2 and renewable energy (CCU), including both direct air capture (DACCS). Thus, the development of novel technology must always consider the overall impact of the entire process and the scale in which that technology can be applied to give solutions on a local or a global scale.

There is an increasing need for CO_2 transport and storage infrastructure. An emerging trend is the development of CCS networks that aggregate CO_2 from multiple sources. Networks offer economies of scale for individual CCS sites by sharing larger infrastructure for CO_2 liquefaction and port facilities, CO_2 compression and pipelines. After capture, CO_2 can be transported as a gas or in the dense phase. Where CO_2 shipping is preferable (like in coastal locations with long offshore distances to storage), economies of scale enable lower costs for each tonne of shared CO_2 liquefaction infrastructure and for port facilities loading and unloading liquid CO_2.

CO_2 storage is accompanied by several problems that have not yet been fully solved. The main objection, however, lies in the one-way function: the one-time use of fossil fuels by transforming them into carbon dioxide, which is at best deposited for a million or more years in geological stocks. Mineral carbonation is a geological process in which CO_2 reacts with rocks to form stable mineral products known as carbonates. Basalts are prevalent globally and have favourable morphology and mineralogy for mineral carbonation storage. These and many other CO_2 reactive rocks are conveniently located in regions where conventional CO_2 storage (for example, depleted oil and gas fields) is generally absent. The "safety" of CO_2 storage is realised by a small diffusive loss rate compared with the capture rate (catastrophic events of CO_2 eruptions from geological stocks are postulated not to occur). Hence, a CO_2 residence time of only a few hundred years in geological stocks is sufficient to compensate effectively for climate impacts from CO_2. Moreover, CO_2 storage is a resource for future carbon use. But even if CCS technology is deployed at all large industrial facilities, more than half of global CO_2 emissions would remain.

Instead of storing CO_2, it is suggested that it should be recycled by transformation into different reduced carbon species (CH_4, CO, C and many hydrocarbons), now termed *solar fuels* (see Chapter 6.2.4). The ultimate solution to the CO_2 problem in terms of atmospheric recovery and providing a carbon feedstock is direct air capture (DAC); see next Chapter 6.2.2.

6.2.2 Direct air capture (DAC)

The idea of CO_2 capture from ambient air using an alkaline solution is not new (Tepe and Dodge 1943) and was used as a pre-treatment before cryogenic air separation. In general, air capture includes all processes of CO_2 fixing and sequestration. In the past, it focused on, but it remains an option today and for the future too. Bio-energy with carbon storage (BECCS) is the term referring to a number of biofuel technologies, which are followed by carbon sequestration and yielding "negative emission energy". However, the key factor in CO_2 removal from the atmosphere is the specific carbon flux per time and square. Plant assimilation needs time and a large area, whereas bringing biomass (almost always wood) to biofuel power plants also needs energy. However, it is important to study all practical measures to avoid abrupt climate change (ACC) and ensure the safety of risky geoengineering.

Direct air capture projects are in an earlier stage of development. They involve direct removal of CO_2 from the atmosphere without photosynthesis (via biomass). Atmospheric CO_2 is very dilute and much harder to capture than industrial CO_2. Comparatively large volumes of air must be handled for each tonne captured. Larger capture equipment is needed, so projects cost more than industrial CCS applications with the same capacity. The thermodynamics of gas separation means that more dilute CO_2 also requires more energy to capture it. Carbon dioxide capture can be applied both in closed technical plant systems as well as in open-field technology (geo-engineering).

The idea of direct air capture (CO_2 extraction from the air) as a climate control strategy is now accepted and seriously considered in global ecological and economic models. Keith (2009) writes: "Air capture is an industrial process for capturing CO_2 from ambient air; it is one of an emerging set of technologies for CO_2 removal that includes geological storage of biotic carbon and the acceleration of geochemical weathering. Although air capture will cost more than capture from power plants when both are operated under the same economic conditions, air capture allows one to apply industrial economies of scale to small and mobile emission sources and enables a partial decoupling of carbon capture from the energy infrastructure, advantages that may compensate for the intrinsic difficulty of capturing carbon from the air."

The large-scale scrubbing of CO_2 from ambient air was first suggested by Lackner et al. (1999). Zeman and Lackner (2004) write: "It is not economically possible to perform a significant amount of work in air, which means one cannot heat or cool it, compress it or expand it. It would be possible to move the air mechanically but only at speeds that are also easily achieved by natural flows. Thus, one is virtually forced into considering physical or chemical adsorption from natural airflow passing over some recyclable sorbent."

> A complete air capture system requires both a contactor and a system for regenerating the absorbing solution. Almost all investigators suggested techniques based on sodium hydroxide, whereas sodium carbonate is converted back into NaOH by "causticisation," one of the oldest processes in the chemical industry. Different absorbers were proposed, such as large convective towers, packed scrubbing towers and a fine spray of the absorbing solution in open towers. CaO–CaCO₃ cycles have also been proposed using solar reactors. Moreover, it has been proposed to adopt technology used in large-scale cooling towers and waste treatment facilities, which are designed to efficiently bring very large quantities of ambient air into contact with fluids. The design they present assumes that absorber fluid is an aqueous solution that absorbs CO_2 from ambient air (typically of a 1–2 M NaOH solution) with flux across the surface of the liquid film of order 1 mg $m^{-2}s^{-1}$, and that, under typical operating conditions, each kilogram of solution absorbs about 20 g of CO_2 before it is returned for regeneration. A new generation of polymeric resins containing specific primary amine functionalised based sorbents have been developed. They are completely regenerated at temperatures in the order of 100 °C and show low H_2O adsorption of 1.5 mol kg^{-1}.

Generally, it is a huge challenge to believe that direct CO_2 extraction from the air can be achieved in quantities approaching an order of several Gt C yr^{-1}. The processed air volume is large (10^7 km^3 because of about 40 t CO_2-C km^{-3}) but corresponds to the air volume passing through about 100 cooling towers of large power plants. Remember that about 50 % (about 4 Gt C yr^{-1}) of technically emitted CO_2 comes from small and mobile units, a percentage likely to increase further in the future. Additionally, about 1–2 Gt C yr^{-1} comes from land-use change and wood fuel use, which are categories that should diminish in the future. Some 4 Gt C yr^{-1} is absorbed by the biosphere (ocean and forest) but with an anticipated decreasing capacity. This "uptake capacity" is not constant but at a certain percentage (likely non-linearly) of the total CO_2 released into the air. Hence (assuming full CO_2 capture from large stationery sources), there is a requirement of at least 2 Gt C yr^{-1} air capture.

> Compensation of atmospheric CO_2 buildup through engineered chemical sinkage was proposed, and the CO_2 removal from the air was calculated by asking for the area needed if this was a perfect, flat sink with a dry deposition velocity of 1 cm s^{-1}. It is a hundred thousand square kilometers value, which constitutes an upper limit for absorbing the annual anthropogenic CO_2 input. Roughness elements and vertical fences could increase the transfer velocity (by reducing the at-

mospheric residence) and increase the specific absorbing area per horizontal air column surface. A total square reduced by a factor of 10 might be able to be reached. Assuming CO_2 solvents having a surface resistance being zero, the atmospheric (dry deposition) flux is determined only by the quasi-laminar and atmospheric resistance, and a value between 0.4 and 1.2 kg CO_2-C m^{-2} d^{-1} can be estimated. This corresponds to an uptake rate of about 2000 t C ha^{-1}yr^{-1}; at least 50 times more than most manipulated algal aquacultures will yield.[121] To "capture" 1–2 Gt CO_2-C yearly, a square (assuming 50 % scrubbing efficiency) of 10^4 km^2 is only needed. However, in contrast to CCS in this approach, it is not the aim to extract CO_2 in a short time from a given volume of gas (air) but to reach saturation of the CO_2 solvent for further desorption and solvent cycling.

The world's first commercial DAC plant for the supply and sale of CO_2 opened in 2017 near Zurich in Switzerland. The commercial-scale DAC plant uses technology patented by Swiss company Climeworks that filters and captures pure carbon dioxide. The filter material is made of porous granulates modified with amines, which bind the CO_2 in conjunction with the moisture in the air. This bond is dissolved at temperatures of 100 °C. DAC projects are under development around the world:

- DAC technology firm Carbon Engineering, in collaboration with Oxy Low Carbon Ventures, is developing a 1 Mtpa project in the Permian Basin in Texas, US. With the DAC advantage of flexible location, this project is positioned adjacent to existing CO_2 transport and storage infrastructure.
- Swiss company Climeworks, in collaboration with geological storage company Carbfix, is constructing its commercial-scale Orca DACCS project in Iceland. Iceland offers low-cost renewable energy to power the capture. The Carbfix storage approach is also low cost compared to other locations – water flowing from an existing geothermal power plant will dissolve CO_2 before it is injected into a basalt formation underground. Mineral carbonation will transform the CO_2 into a solid for permanent storage.
- Plans to build Europe's first large scale DAC facility were unveiled by Storegga and Carbon Engineering in mid-2021. Scotland-based Dreamcatcher will take advantage of abundant renewable energy and anticipated CCS infrastructure nearby, capturing between 500,000 and one million tonnes of atmospheric CO_2 each year.

---

121 Algal productivity rates between 5 and 10 g C m^{-2} d^{-1} have normally been cited (Drapcho and Brune 2000), but were reported to be up to 15 C m^{-2} d^{-1} in highly modern farming systems. Again, to achieve 0.1 Mt C d^{-1}, a farming area of about 7000 km^2 is needed or 210,000 km^2 globally, which corresponds to an area roughly 50 % of the size of Germany. Surely there is a research need to optimise (and maximise) CO_2 capture by industrial biofarming in sun-belt countries. For example, nutrients for biofarming could be taken from municipal wastewater of nearby "solar cities" and/or recycled from the biomass conversion process into CO_2 (note that fixed CO_2 is the aim rather than biofuel).

The extremely low concentration of CO_2 in the air (and/or other reservoirs such as seawater[122]) makes any technical and economical solution of DAC challenging. Design and synthesis of new CO_2 absorbing materials is the key for the application of DAC.

> Any technical solution in a sustainable conception is based on the paradigm change to establish a zero-carbon budget (not zero emission!) and to measure the effect no longer on the energy efficiency (solar energy is in "excess") but on the matter budget with respect to climate sustainability.

The real "price" of CO_2 emitted from fossil fuels (and hence fossil fuel costs) must include climate change effects; this would force the energy transition and also DAC technologies. It is self-evident that only solar energy is used for DAC. In the Chapter 6.2.3, we will see that DAC is a basic technology for the carbon economy similar to biogenic assimilation.

6.2.3 Carbon dioxide cycling (DACCU)

CO_2 is unique,[123]
- as a final oxidation product of all organic matter and carbonaceous materials,
- because of its global cycling and homogeneous distribution in the atmosphere (but keeping a level before "tipping points"),
- as a resource for organic materials concerns carriers of energy and functional materials,
- carbon is the only element forming complex molecules and substances and being within a global dynamic[124] cycle and gaseous compounds on lowest (CH_4) and highest oxidation state (CO_2),

---

122 The desorption of dissolved CO_2 from seawater (being 90 % HCO_3^- with a mean total DIC concentration of ~28 mg C L^{-1} – by a factor of 1000 larger than atmospheric carbon) could be another approach for closing the carbon cycle (we currently call it *seawater capture*). Hence, in the case of a technology with 60 % desorption efficiency, "only" 150 km^3 seawater must be globally processed daily to attain the production mentioned above (1 Gt yr^{-1}). In our laboratory, an ultrasonic-based CO_2 desorption technique has been developed as an alternative to the thermal stripping in the CCS process, and it is not difficult to believe that this technology could be applied to seawater decarbonization.

123 In a certain sense hydrogen (H_2) can also play this role when we adopt the natural water splitting process, which was already proposed as "hydrogen technology" in the early 1980s. But there are several problems, a) safety in storage and transport, b) leakage and atmospheric implications and c) missing material supply. Water electrolysis will play an important role in this conception for oxy-fuel combustion (O_2 supply) and CO_2 reduction (H_2 supply).

124 In (biogeochemical) cycles move all elements and its compounds – but often on geological time scale (beside carbon only sulphur and nitrogen are in similar dynamic cycles).

- the only environmental problem of CO_2 is its rise in the atmosphere (and seawater) with climatic implications; hence controlling its level on acceptable values will overcome the environmental problem.

Let us combine DAC with CO_2 utilisation to gain a CO_2 loop (to name it DACCU – direct air capture and carbon utilisation)[125] – and build an artificial carbon (CO_2) cycle in analogy to the natural assimilation-respiration carbon cycle (Fig. 6.1) according to the scheme

$$CO_2 \text{ (from air capture)} + H_2 \text{ (from water electrolysis)} \rightarrow \text{hydrocarbons} \xrightarrow{\text{use}} CO_2 + H_2O.$$

It is evident that through the realisation of these principles a CO_2 "zero-budget world" rather than a "CO_2 free world" can be reached because there is a closed anthropogenic carbon cycle. All CO_2 still emitted – and also cannot be captured in future from mobile and small equipment – will be captured from the air and cycled for reuse. With this

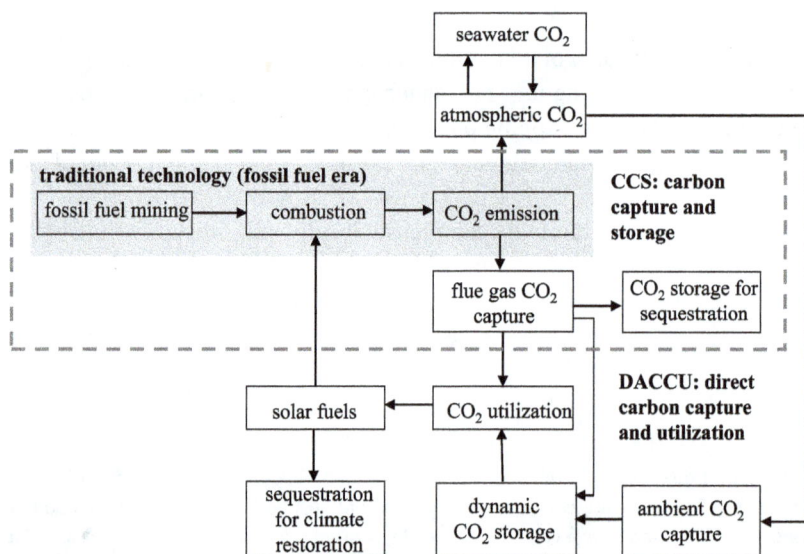

Figure 6.1: Scheme of energy transition from fossil to the solar era including a CO_2 economy. Three overlapping systems: fossil fuel burning without (grey box) and with carbon capture (dotted box) as well solar fuel production/use and global carbon cycling. The driving force is exclusively solar radiation; hence the CO_2 economy is interdependent with solar electricity conceptions. Elements of this concept can be introduced parallel with further use of fossil fuels aimed by its stepwise replacement.

---

125 This conception I have published under the name SONNE - SOlar-based maN-made carboN cyclE (Möller 2012); "Sonne" is the German word for sun.

in mind, CCS technology (carbon capture and storage/sequestration) makes (more) sense – despite the controversially discussed CO_2 storage problems – and provides considerable incentive because CO_2 storage is now only temporary (we call it dynamically) until it is recycled from waste to feedstock. The proposed DACCU technology allows a stepwise replacement of coal and other fossil fuels by solar fuels but keeps the carbon-based infrastructure such as pipelines, tankers, storage facilities, engines and allows the continued use of other available technical applications developed in the last hundred and more years, but within a CO_2 neutral closed loop.

Closure of the carbon cycle, however, is only possible when CO_2 will be extracted from natural reservoirs such as the atmosphere and seawater (DAC, see Chapter 6.2.2) because a complete "industrial" CO_2 capture will be impossible with regard to many small and mobile sources.

A global closed anthropogenic carbon cycle by using only solar energy for processing creates a global zero-carbon budget (but not a zero CO_2 world), solves the problem of electricity storage based on CO_2 utilization, provides carbon-based materials only from CO_2 utilisation, and further uses the infrastructure developed for the fossil fuel era.

The CO_2 cycling technology consists of several formally independent infrastructures but uses already existing networks:
a) green electricity generation and distribution;
b) DAC processing, CO_2 storage and transportation;
c) green hydrogen production and transportation;
d) CO_2 methanation, CH_4 storage and transportation;
e) CO_2 and CH_4 to hydrocarbon product conversion;
f) secondary green electricity generation in CH_4 hybrid power plants with CCS.

Despite the fact that some countries have electricity generation largely by water power, other countries by wind power and photovoltaic, great importance in future will get sites in sun-belt countries providing huge areas for almost permanent conversion of solar radiation in electricity (and heat for on-site chemical processing) by different technologies. Principally, it is possible to combine the solar site with other modules (DAC and methanation), but there are arguments to localise air capture close to CO_2 storage to avoid transportation to sites where climatic conditions support large CO_2 absorption and/or to sites of CO_2 processing.[126] Solar fuel (in contrast to CO_2) can be easily transported and stored using the existing traditional infrastructure for liquids and gases. Clean hydrogen can be produced in three ways:

---

[126] An interesting site would be Iceland, which might be able to provide power from geothermal heat and cold carbon-rich seawater for CO_2 extraction as well as air capture because of the large temperature differences in rich and lean CO_2 loading. Captured CO_2 could be reduced on site to solar fuels which are transported by tank ships to Central Europe.

- from fossil fuels with CCS (blue hydrogen);
- from biomass;
- from electrolysers powered by renewable electricity (green hydrogen) or nuclear power.

Keeping in mind sustainable approaches and that not energetic efficiency a zero-material budget is essential, further use of fossils fuels should obsolete and biomass is too valuable as resource for steam reforming, and only electrolysis remains despite its high electricity demands (55 kWh per kg H_2 compared to 2–4 kWh per kg H_2 based on fossil fuels with CCS). There is no need for long-range transportation of hydrogen (which is one feedstock for CO_2 reduction) because electricity can be transported advantageously in comparison to H_2. Hydrogen can also be produced at any site by water electrolysis (also using renewable sources other than direct solar electricity). There are clear advantages to CO_2 reduction close to the solar fuel consumers, including the avoidance of the expensive transportation of CO_2 back to the sun-belt countries and the possibility of mixing solar fuels directly with oxygen from the water electrolysis to create "oxyfuels". Oxyfuels can be burned in stationary power plants with the result that the flue gas is almost all pure CO_2. However, there is an interesting aspect for back transportation of CO_2 from industrial to sun-belt countries. The solar site depends on the delivery of feedstock CO_2. Thus, a geopolitical equilibrium can be reached in the sense of a win/win situation.

The specific approaches put together in this "CO_2 economy" are already known and/or proposed, but the creation of an artificial carbon cycle in such an integrative approach and with this rigorousness in linking energy with material economy adopting the principle of natural cycling but not copying natural processes (such as artificial photosynthesis), as suggested here, is worldwide unique and new (even much more complex than the "methanol economy").

> Olah (2005) proposed the "methanol economy", but in the DACCU concept, CH_3OH is only one possible product among C_1 chemicals; the Fischer–Tropsch synthesis (from $CO + H_2$) basically offers a wide range of organic compounds, including liquid fuels. Our "CO_2 economy" includes the "CH_3OH economy". Recently it was provided that the energetic efficiency of the overall energy conversion-storage system, including CH_3OH as a storage medium, is only half of that for CH_4 (about 30 %).
>
> However, taking into account ambient CO_2 capture, the overall energy efficiency will be drastically lower. Such as, in nature (the photosynthesis efficiency concerning solar light is only 2–3 %), we realise a closed carbon loop only with large energy input (in other words, low energetic efficiency), but based on incoming solar radiation, 1000 times higher than present global human energy demand. However, still unanswered is the question, what are the limits of solar use without getting other climate implications.

From methanation, the ultimate energy carrier CH_4 is gained from CO_2, captured from ambient air, but also recycled in so-called hybrid power stations to provide chemical storage of renewable electricity. Such power plant (like modern gas power stations) combusts CH_4 with CCS. A future scenario is oxyfuel combustion (O_2 is the byprod-uct in the water electrolysis) to gain high-concentrated CO_2 exhaust gas for recycling via methanation in periods of excess renewable electricity. Human's evolutionary re-sponsibility should also consider the retransfer of emitted CO_2 into geological stocks, for example, as elemental carbon for safe sequestration and stepwise but long-lasting climate recovery.

All CO_2 still emitted will be captured and cycled for reuse. Naturally, the energy needed for CO_2 reduction comes from renewable sources. The proposed DACCU tech-nology allows a stepwise replacement of coal and other fossil fuels by solar fuels. Fi-nally, there is a closed carbon cycle similar to natural photosynthesis, a respiration cycle (Figure 6.1). Carbon-based solar fuels (see next Chapter) solve the problem of energy storage and allow the continued use of available technical applications to pro-vide products for materials and are within a CO_2 neutral closed loop.

> Economic paradigm change: solar energy is "in excess" (comparing with the global human de-mand) and is naturally dissipated in the atmosphere; hence, large energy-consuming conversion processes and direct air capture can be carried out for resource generation and climate sustainabil-ity: a new economy-thinking based on sustainability (or closed carbon cycle) is needed. In other terms, not energy but material efficiency becomes the key factor.

6.2.4 Solar fuels: carbon as material and energy carrier

The idea of using CO_2 as a chemical raw material for the chemical and polymer indus-tries is not new and now widely discussed to contribute to a circular economy. How-ever, when using CO_2 from fossil-fuel gases, it is only climate-sustainable if the prod-ucts are "sequestrated", for example, by long-term use in carbon materials such as polyurethanes. CO_2 captured from fossil-fired power plants and "utilised" for storage of excess electricity (for example, from wind power) may help to improve the energy efficiency (because the excess electricity cannot be used on demand) but not solve the climate problem.

Thus, it is also suggested synthesising carbon-neutral hydrocarbons (CNHCs) from air-captured CO_2. This term (in fact CO_2 neutral) is inconsistent, and in its place, I will use "solar fuels" to express that the supply of process energy for the chemical reduction of captured CO_2 must be based on *solar* energy processing instead of fossil (geothermal heat is another option).

In last year, considerable progress was achieved in the catalytic hydrogenation of CO_2 (methanation). The possible synthesis of C_1-chemicals (CO, C, CH_4, CH_3OH, and HCHO) from CO_2 and directly further to C_3 in analogy to the assimilation process (see also Fig. 5.9) leads to a variety of important basic chemicals being available for either direct combustion or material use (industrial synthesis in organic chemistry); we now call them *solar fuels*. The possible synthesis of methanol and formic acid from CO_2

leads to important basic chemicals for industrial synthesis in organic chemistry. However, a global CO_2 economy must provide chemicals in order not only of a hundred million tons but also an amount of 1–2 orders of magnitude more gaseous and liquid fuels. By using high-temperature chemical processes (which were known for many years but due to the high energy consumption have hardly been mentioned before) based on solar thermal energy, it is also possible to remake "coal chemistry" (gasification and liquefaction) via CO_2 reduction. Namely, carbon monoxide (CO) and elemental carbon may be produced and inversely transformed. For example, elemental carbon could be stored better than carbon dioxide (for sequestration aiming climate abatement) but could also be reused directly. It is known that in high-temperature processes of conversion of carbon compounds to elemental carbon, the yield of polymeric carbon structure (fullerenes) are large and unforeseen changes in creating new carbon materials are made possible. It is self-evident that all energy for processing is solar-based. It is out of the focus of this Chapter to review the chemical processes of CO_2 utilisation – this Chapter only will present the basic ideas of chemical conversion in the sense of the chemical evolution sustained by humans – and the interested reader is referred to as special literature.

As mentioned, the world's infrastructure is based on fossil fuels and products derived from coal and petrochemistry. The following groups of substances are delivered from fossils fuels and can also be produced from CO_2 using solar-based processes:

- gaseous hydrocarbons (C_nH_{2n+2} with $n < 6$); methane as a key substance; all substances with $n > 2$ are easily liquefiable;
- liquid hydrocarbons (C_nH_{2n+2} with $n > 5$) such as alkanes but also oxygenated liquid compounds from C_1 upwards (for example, methanol);
- gaseous CO (the typical town gas in the past);
- elemental carbon (C in different modifications).

Energetic use is simply combustion in different burners and engines with, ideally, transformation back into CO_2. Reaction enthalpies are different and thereby provide possible usable energy.

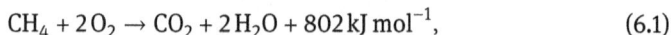

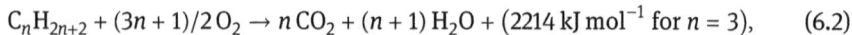

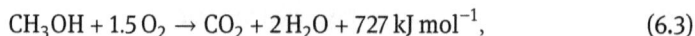

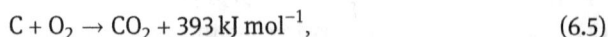

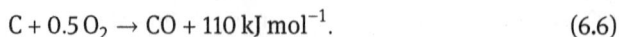

$$CH_4 + 2O_2 \rightarrow CO_2 + 2H_2O + 802\,kJ\,mol^{-1}, \tag{6.1}$$

$$C_nH_{2n+2} + (3n+1)/2O_2 \rightarrow nCO_2 + (n+1)H_2O + (2214\,kJ\,mol^{-1}\ \text{for}\ n = 3), \tag{6.2}$$

$$CH_3OH + 1.5O_2 \rightarrow CO_2 + 2H_2O + 727\,kJ\,mol^{-1}, \tag{6.3}$$

$$CO + 0.5O_2 \rightarrow CO_2 + 283\,kJ\,mol^{-1}, \tag{6.4}$$

$$C + O_2 \rightarrow CO_2 + 393\,kJ\,mol^{-1}, \tag{6.5}$$

$$C + 0.5O_2 \rightarrow CO + 110\,kJ\,mol^{-1}. \tag{6.6}$$

Despite the lower energy yield, it is evident that hydrogen-free carbon carriers such as CO and C have the large advantage of consuming much less oxygen and producing only

CO_2 when considering oxyfuel technology in the future. That could be an important feature in establishing the carbon cycle and turning CO_2 back into the feedstock for solar fuels. In the production of group one and group two compounds in the list above, some progress has already been made towards CO_2 hydrogenation (equations (5.5) and (5.9)). Equation (6.7) has already been discovered by Paul Sabatier (1854–1941) in the nineteenth century.

$$CO_2 + 4\,H_2 + 618\,\text{kJ mol}^{-1} \rightarrow CH_4 + 2\,H_2O, \tag{6.7}$$

$$CO_2 + 3\,H_2 + 523\,\text{kJ mol}^{-1} \rightarrow CH_3OH + H_2O, \tag{6.8a}$$

$$CO + 2\,H_2 + 308\,\text{kJ mol}^{-1} \rightarrow CH_3OH. \tag{6.8b}$$

What must be considered in carrying out reactions (6.7) and (6.8a), (6.8b) is the source of H_2. It is clear that "traditional" reactions (water-gas shift and CH_4 reforming) cannot be used and that H_2 must be generated via water electrolysis using renewable energy sources such as solar radiation. Thus, the argument that it is preferable to use H_2 directly as fuel is inconsistent with our aim of carbon cycling. Moreover, carbon carriers provide a range of products that are better suited to the available infrastructure than H_2. Overall, the photosynthesis-like reactions described by equations (6.9) and (6.10) are carried out:

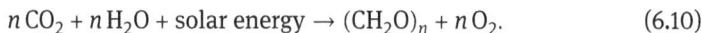

$$CO_2 + 2\,H_2O + \text{solar energy} \rightarrow CH_3OH + 1.5\,O_2, \tag{6.9}$$

$$n\,CO_2 + n\,H_2O + \text{solar energy} \rightarrow (CH_2O)_n + n\,O_2. \tag{6.10}$$

The formation of methanol from CO_2 hydrogenation is known as the CAMERE process. In the past, a thermochemical heat pipe application was proposed, which is based on equation (6.11a), where CO_2/CH_4 reforming (using solar energy) gives CO/H_2 gas, which can be converted back (6.11b) in an exothermic reactor with equivalent energy output. However, this technology needs entirely new infrastructure.

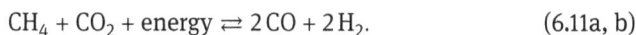

$$CH_4 + CO_2 + \text{energy} \rightleftarrows 2\,CO + 2\,H_2. \tag{6.11a, b}$$

The CO/H_2 gas (also termed water-gas) – depending on the $CO : H_2$ ratio – can also be converted (using the Fischer–Tropsch synthesis) to alkanes, alkenes and alcohols. Preferably, substances should be generated for application in known technical systems such as liquefied petroleum gas or low-pressure gas (LPG) and gasoline as well as natural gas. The production of solar substitutes for diesel and oils (C > 8), that is petrol products from the fractional distillation of crude oil between 200 °C and 350 °C, is also possible but offers no advantages in the solar fuel cycle and its stepwise replacement by gases and gasoline should be foreseen.

An interesting way of generating solar fuels can be seen in two "classical" inorganic carbon carriers: CO and carbon itself. The gasification of carbon Equation (6.12a) produces CO (generator gas) in the so-called Boudouard equilibrium, called after Octave Leopold Boudouard (1872–1923):

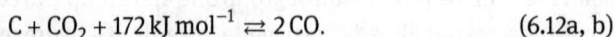

$$C + CO_2 + 172\,kJ\,mol^{-1} \rightleftharpoons 2\,CO. \qquad (6.12a, b)$$

The inverse reaction (6.12b) opens the way to finally convert CO_2 to elemental carbon via equations (6.7) and (6.11a). Alternatively, CO can be produced by the high-temperature electrolysis of CO_2:

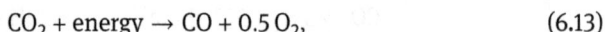

$$CO_2 + energy \rightarrow CO + 0.5\,O_2, \qquad (6.13)$$

where the overall reaction represents the inverse Equation (6.6). High-temperature electrolysis using solid oxide electrolytic cells offers absolute new synthesis pathways. In contrast to Equation (6.13), the electrolysis of CO_2/H_2O leads to CO and CH_4:

$$CO_2 + H_2O + energy \rightarrow CO + H_2 + 1.5\,O_2, \qquad (6.14)$$
$$CO_2 + 2\,H_2O + energy \rightarrow CH_4 + 2\,O_2. \qquad (6.15)$$

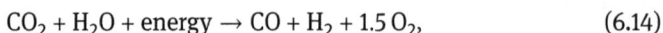

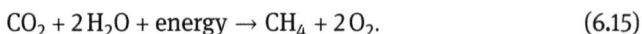

For long-term space missions, these reactions were considered to provide a closed cycle of production of oxygen and consumption of respiratory CO_2. A final pyrolysis Equation (6.16) recycles hydrogen, but more interesting for earth applications is the formation of elemental carbon according to the gross conversion process shown in Equation (6.17).

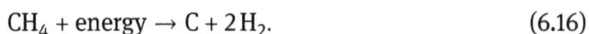

$$CH_4 + energy \rightarrow C + 2\,H_2. \qquad (6.16)$$

Equation (6.13) and (6.15) provide the following overall reaction:

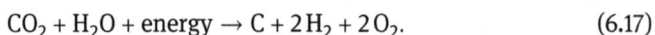

$$CO_2 + H_2O + energy \rightarrow C + 2\,H_2 + 2\,O_2. \qquad (6.17)$$

This set of reactions based on solid high-temperature electrolytic CO_2 reduction shows that more advantages are likely at desert solar sites than the conversion processes in detail decisions will be ever made. The general budget is described by

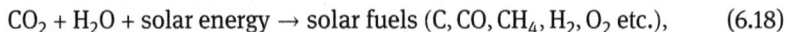

$$CO_2 + H_2O + solar\ energy \rightarrow solar\ fuels\ (C, CO, CH_4, H_2, O_2\ etc.), \qquad (6.18)$$

independent of the detailed conversion processes. In the inverse reaction described by Equation (6.18), solar fuels reconvert into CO_2 and H_2O via the combustion or electricity in fuel cells and set the energy free, primarily as heat, for energetic use.

A List of acronyms and abbreviations in environmental sciences found in literature[127]

AAS	atomic absorption spectrometry
ACC	abrupt climate change
AOT	accumulated dose over the threshold
ANC	acid neutralising capacity
AP	aerosol particle
BBOA	biomass burning organic aerosol
BC	black carbon (= soot)
BECCS	bioenergy with CCS
BL	boundary layer
BNC	base neutralising capacity
BOVOC	biogenic oxygenated volatile organic carbon
BP	before present (also B. P., e. g. kyr BP – thousand years before present)
Btu	British thermal unit (sometimes also BTU)
BVOC	biogenic volatile organic carbon
CCN	cloud condensation nuclei
CCS	carbon capture and storage (or: sequestration)
CCU	carbon capture and utilisation
CFCs	chlorofluorocarbons
CN	condensation nuclei
CNHC	carbon-neutral hydrocarbon
DAC	direct air capture
DACCU	direct air capture and carbon utilisation
DCA	dicarboxylic acid
DIC	dissolved inorganic carbon
DM	dry mass or dry matter
DMDS	dimethyl disulphide
DMS	dimethyl sulphide
DMSO	dimethyl sulphoxide
DNA	deoxyribonucleic acid
DOC	dissolved organic carbon
DON	dissolved organic nitrogen
DU	Dobson unit (total ozone column-concentration)
EC	elemental carbon (= soot)
EF	emission factor
EF	enrichment factor
EM	elemental matter (carbon-related, see OM)
EMEP	cooperative programme for monitoring and evaluation of the long-range transmission of air pollutants in Europe
EPA	U. S. environmental protection agency
EU	European Union
FAO	food and agriculture organisation of the United Nations
FGC	flue gas cleaning

---

127 Often the plural is formed by adding 's', for example: PSCs – polar stratospheric clouds.

https://doi.org/10.1515/9783110735178-007

FGD	flue-gas desulphurisation (power plant)
FT	free troposphere
GAW	global atmospheric watch (WMO monitoring program)
GC	gas chromatography
GCMS	gas chromatography, coupled with mass spectrometry
GCM	general circulation model
GEIA	global emission inventory activity (within IGAC)
GEP	gross ecosystem production
GEWEX	global energy and water cycle experiment
GHG	greenhouse gases
GVM	global vegetation monitoring
GPCP	global precipitation climatology project
GPP	gross primary production
GWP	global warming potential
HCFCs	hydrochlorofluorocarbons
HFCs	hydrofluorocarbons
HMS	hydroxymethanesulphonate
HMSA	hydroxymethanesulphonic acid
HOA	hydrocarbon-like organic aerosol
HULIS	humic-like substances
IEA	international energy agency
ICP	inductively coupled plasma
IGAC	international global atmospheric chemistry (project within IGBP)
IGBP	international geophere-biosphere program
IN	ice nuclei
IPCC	intergovernmental panel on climate change
IR	infrared
ITCZ	intertropical convergence zone
IUPAC	International Union of Pure and Applied Chemistry
IWC	cloud ice water content
LC	liquid chromatography
LHB	late heavy bombardment
LIA	little ice age
LNG	liquefied natural gas (CH_4)
LPG	liquefied petroleum/propane gas, also low-pressure gas (C3/C4 hydrocarbons)
LRTAP	Convention on long-range transboundary air pollution
LS	lower stratosphere
LWC	liquid water content
MEGAN	model of emissions of gases and aerosols from nature
MODIS	moderate resolution imaging spectroradiometer
MOPITT	measurement of pollution in the troposphere
Mtpa	million tons per year
MS	mass spectrometry
MSA	methanesulphonic acid
Mt	megaton (10^{12} g)
NASA	national aeronautics and space administration (USA)
NASN	national air surveillance network (by EPA in the USA)
NCAR	national centre for atmospheric research
NDACC	network for the detection of atmospheric composition change

NCEP	national centres for environmental prediction
NEP	net ecosystem production
NH	northern hemisphere
NMHC	nonmethane hydrocarbons
NMVOC	nonmethane volatile organic compound
NOAA	national oceanic and atmospheric administration (USA)
NOM	natural organic matter
NPP	net primary production
OA	organic aerosol (total, see also HOA, BBOA)
OC	organic carbon
ODP	ozone depletion potential
ODS	ozone-depleting substances
OECD	organisation for economic co-operation and development
OM	organic matter
OOA	oxygenated organic aerosol
OVOC	oxygenated volatile organic carbon
PAH	polycyclic aromatic hydrocarbons
PAN	peroxyacetyl nitrate
PAR	photosynthetic active radiation
PBL	planetary boundary layer
Pg	picogram (10^{15} g)
PFCs	perfluorocarbons
PM	particulate matter
PM_1	particulate matter with aerodynamic particle diameter $\leq 1\,\mu m$
PM_{10}	particulate matter with aerodynamic particle diameter $\leq 10\,\mu m$
$PM_{2.5}$	particulate matter with aerodynamic particle diameter $\leq 2.5\,\mu m$
POA	primary organic aerosol
POCP	photochemical ozone creation potential
POM	particulate organic matter (part of PM)
POPs	persistent organic pollutants
PSC	polar stratospheric cloud
RDS	rate-determining step
REE	rare earth element
RH	relative humidity
RCHO	aldehyde
RCOOH	organic acid
RNA	ribonucleic acid
RO	organic oxy radical (e. g. alkoxyl radical)
RO_2	organic peroxy radical
ROOH	organic peroxide
ROS	reactive oxygen species
SD	standard deviation
SH	southern hemisphere
SI	System Internationale
SIC	soil inorganic carbon
SOA	secondary organic aerosol
SOC	soil organic carbon
SOM	soil organic matter
SPM	suspended particulate matter (= dust, atmospheric aerosol)

SSA	sea salt aerosol
SST	sea surface temperature
STE	stratosphere-troposphere exchange
stp	standard temperature and pressure
SVOC	secondary volatile organic carbon
TC	total carbon (sum of organic and inorganic)
TCL	tropospheric chlorine loading
Tg	terragram (= Mt)
TIC	total ionic content
TMI	transition metal ion
TOC	total organic carbon
toe	tonne of oil equivalent
TOMS	total ozone mapping spectrometer
TPM	total particulate matter
TSP	total suspended matter (= PM)
TWC	three-way catalytic converters
UT	upper troposphere
UV	ultraviolet
UVOC	unsaturated volatile organic compounds
VOC	volatile organic compound
WHO	world health organisation of the United Nations
WMO	world meteorological organisation of the United Nations
WSOC	water-soluble organic compounds

B Quantities, units and some useful numerical values

The SI (abbreviated from the French *Le Système International d'Unités*), the modern metric system of measurement, was developed in 1960 from the old metre-kilogram-second (mks) system, rather than the centimetre-gram-second (cgs) system, which, in turn, had a few variants. Since the SI is not static, units are created, and definitions are modified through an international agreement among many nations as the measurement technology progresses, and the precision of measurements improves.

Long the language universally used in science, the SI has become the dominant international commerce and trade language. The system is nearly universally employed, and most countries do not even maintain official definitions of any other units. A notable exception is the United States, which continues to use customary units in addition to the SI. In the United Kingdom, conversion to metric units is government policy, but the transition is not quite complete. Those countries that still recognise non-SI units (e. g. the U. S. and U. K.) have redefined their traditional non-SI units in the SI units.

It is important to distinguish between the definition of a unit and its realisation. The definition of each base unit of the SI is carefully drawn up so that it is unique and provides a sound theoretical basis upon which the most accurate and reproducible measurements can be made. The realisation of the definition of a unit is the procedure by which the definition may be used to establish the value and associated uncertainty of a quantity of the same kind as the unit.

Some useful definitions

A *quantity in the general sense* is a property ascribed to phenomena, bodies or substances that can be quantified for or assigned to a particular phenomenon, body or substance. Examples are mass or electric charge.

A *quantity in the particular sense* is a quantifiable or assignable property ascribed to a particular phenomenon, body or substance. Examples are the mass of the Moon and the electric charge of the proton.

A *physical quantity* is a quantity that can be used in the mathematical equations used in science and technology.

A *unit* is a particular physical quantity, defined and adopted by the convention, against which other particular quantities of the same kind are compared to express their value.

The *value of a physical quantity* is the quantitative expression of a given physical quantity as the product of a number by a unit, and the number is its numerical value. Thus, the numerical value of a particular physical quantity depends on the unit in which it is expressed.

https://doi.org/10.1515/9783110735178-008

Table B.1: The seven basic units of SI.

quantity	name of unit	symbol
length	metre	m
mass	kilogram	kg
time	second	s
thermodynamic temperature	kelvin	K
amount of substance	mole	mol
electric current	ampere	A
luminous intensity	candela	cd

Other quantities, called *derived quantities*, are defined in terms of the seven base quantities using a system of quantitative equations. The 22 SI derived units for these derived quantities are obtained from these equations and the seven SI base units. Examples of such derived SI units are given in Table B.2, where it should be noted that the symbol 1 for quantities of dimension 1, such as mass fraction, is generally omitted.

Table B.2: Derived and other units (examples only; the list can be continued). There are 22 derived units accepted.

quantity	name of unit	symbol	special definition	definition from SI
force	newton	N	–	$kg\,m\,s^{-2}$
pressure	pascal	Pa	$N\,m^{-2}$	$kg\,m^{-1}\,s^{-2}$
mass density[b]	–	ρ^{c}	–	$kg\,m^{-3}$
chemical concentration[b]	molar	M	$mol\,L^{-1}$	$10^{-3}\,mol\,m^{-3}$
chemical concentration[b]	molal	m	$mol\,kg^{-1}$	$mol\,kg^{-1}$
energy, work, quantity of heat	joule	J	$N\,m$	$kg\,m^{2}\,s^{-2}$
power, radiant flux	watt	W	$J\,s^{-1}$	$kg\,m^{2}\,s^{-3}$
electric potential[a]	volt	V	$N\,m\,A^{-1}\,s^{-1}$	$kg\,m^{2}\,s^{-3}\,A^{-1}$
electric charge	coulomb	C	–	$A\,s$
electric conductance	siemens	S	$A\,V^{-1}$	$kg^{-1}\,m^{-2}\,s^{3}\,A^{2}$
electric resistance	ohm	Ω	$V\,A^{-1}$	$kg\,m^{2}\,s^{-3}\,A^{-2}$
electric capacitance	farad	F	$C\,V^{-1}$	$kg^{-1}\,m^{-2}\,s^{4}\,A^{2}$
dynamic viscosity[b]	poise	P	$Pa\,s$	$kg\,m^{-1}\,s^{-1}$
kinematic viscosity	stokes	St	–	$10^{-4}\,m^{2}\,s^{-1}$
moment of force[b]	–	–	$N\,m$	$kg\,m^{2}\,s^{-2}$
surface tension[b]	–	–	$N\,m^{-1}$	$kg\,s^{-2}$
heat flux density, irradiance[b]	–	–	$W\,m^{-2}$	$kg\,s^{-3}$
frequency	hertz	Hz	–	s^{-1}
wave number[b]	–	–	–	m^{-1}
area[b]	–	–	–	m^{2}
volume[b]	–	V^{c}	–	m^{3}
speed, velocity[b]	–	v^{c}	–	$m\,s^{-1}$
acceleration[b]	–	a^{c}	–	$m\,s^{-2}$

[a] also: potential difference (voltage), electromotive force
[b] not accepted as *derived* units
[c] generally accepted but could be different from author to author

Table B.3: Non-SI units approved for use with the SI.

quantity	symbol	value in SI units
minute (time)	min	$1\,min = 60\,s$
hour	h	$1\,h = 60\,min = 3600\,s$
day	d	$1\,d = 24\,h = 86{,}400\,s$
degree (angle)	°	$1° = (\pi/180)\,rad$
minute (angle	′	$1′ = (1/60)\,\check{r} = (\pi/10{,}800)\,rad$
second (angle)	″	$1″ = (1/60)′ = (\pi/648{,}000)\,rad$
litre	L[a]	$1\,L = 1\,dm^3 = 10^{-3}\,m^3$
metric ton[b]	t	$1\,t = 10^3\,kg$
electronvolt[c]	eV	$1\,eV = 1.602176565 \cdot 10^{-19}\,J$ (approximately)
nautical mile	–	1 nautical mile $= 1852\,m$ knot
knots	–	nautical mile per hour $= (1852/3600)\,ms^{-1}$
hectare	ha	$1\,ha = 1\,hm^2 = 10^4\,m^2$
bar	bar	$1\,bar = 0.1\,MPa = 100\,kPa = 1000\,hPa = 10^5\,Pa$
ångström	Å	$1\,Å = 0.1\,nm = 10^{-10}\,m$
curie	Ci	$1\,Ci = 3.7 \cdot 10^{10}\,Bq$
roentgen	R	$1\,R = 2.58 \cdot 10^{-4}\,C\,kg^{-1}$
rad	rad	$1\,rad = 1\,cGy = 10^{-2}\,Gy$
rem	rem	$1\,rem = 1cSv = 10^{-2}\,Sv$

[a]This unit and its symbol l were adopted by the CIPM in 1879. The alternative symbol for the litre, L, was adopted by the CGPM in 1979 in order to avoid the risk of confusion between the letter l and the number 1. Thus, although both l and L are internationally accepted symbols for the litre, to avoid this risk, the preferred symbol for use in the United States is L.
[b]In many countries, this unit is called 'tonne'.
[c]The electron volt is the kinetic energy acquired by an electron passing through a potential difference of 1 V in a vacuum. The value must be obtained by experiment and is therefore not known exactly.

Table B.4: Prefixes used to construct decimal multiples of units (SI).

Y	yotta	10^{24}
Z	zetta	10^{21}
E	eta	10^{18}
P	peta	10^{15}
T	tera	10^{12}
G	giga	10^{9}
M	mega	10^{6}
k	kilo	10^{3}
h	hecto	10^{2}
da	deca	10
d	deci	10^{-1}
c	centi	10^{-2}
m	milli	10^{-3}
μ	micro	10^{-6}
n	nano	10^{-9}
p	pico	10^{-12}
f	femto	10^{-15}
a	atto	10^{-18}
z	zepto	10^{-21}
y	yocto	10^{-24}

Table B.5: Some useful numerical values.

name	symbol	value	definition
Boltzmann constant	k	$1.3806488 \cdot 10^{-23}\,\mathrm{J\,K^{-1}}$	$k = R/N_A$
Avogadro constant	N_A	$6.02214129 \cdot 10^{23}\,\mathrm{mol^{-1}}$	[a]
Loschmidt's number	n_0	$2.6516462 \cdot 10^{25}\,\mathrm{m^{-3}}$	[b]
Gas constant	R	$8.3144621\,\mathrm{J\,K^{-1}\,mol^{-1}}$	$R = kN_A$
Planck constant	h	$6.62606957 \cdot 10^{-34}\,\mathrm{J\,s^{-1}}$	[c]
Stefan–Boltzmann constant	σ	$5.670373 \cdot 10^{-8}\,\mathrm{J\,m^{-2}\,K^{-4}\,s^{-1}}$	$\sigma = 2\pi^5 k^4/15h^3c^2$
Faraday constant	F	$96{,}485.3365\,\mathrm{C\,mol^{-1}}$	$F = eN_A$
elementary charge	e	$1.60217657 \cdot 10^{-19}\,\mathrm{C}$	[d]
gravitational constant	G	$6.67384 \cdot 10^{-11}\,\mathrm{m^3\,kg^{-1}\,s^{-2}}$	$G = f \cdot r^2/m_1 m_2$ [e]
standard gravity	g	$9.80665\,\mathrm{m\,s^{-1}}$	$g = G \cdot m_{earth}/r_{earth}^2$ [f]
speed of light	c	$299{,}792{,}458\,\mathrm{m\,s^{-1}}$	

[a] number of atoms in $0.012\,\mathrm{kg}\,^{12}\mathrm{C}$
[b] number of atoms/molecules of an ideal gas in $1\,\mathrm{m^{-3}}$ at $0\,°\mathrm{C}$ and 1 atm
[c] Planck–Einstein relation: $E = h\nu$
[d] fundamental constant, equal to the charge of a proton and used as the atomic unit of charge
[e] empirical constant (also called Newton's constant) gravitational attraction (force f) between objects with mass m_1 and m_2 and distance r
[f] nominal acceleration due to gravity at the earth's surface at sea level

C List of the elements (alphabetically) with the exception of transactinides (z = 103–118); blue: radioactive elements

element	ordinal number	symbol	relative atomic mass[a]	abundance (%)
actinium	89	Ac	227.028	$6 \cdot 10^{-14}$
aluminium	13	Al	26.9815385	7.7
americium	95	Am	241.061375	–
antimony	51	Sb	121.760	$2 \cdot 10^{-5}$
argon	18	Ar	39.948	$3.6 \cdot 10^{-4}$
arsenic	33	As	74.921595	$1.7 \cdot 10^{-4}$
astatine	85	At	209.987	$3 \cdot 10^{-24}$
barium	56	Ba	137.327	0.04
berkelium	97	Bk	247	–
beryllium	4	Be	9.012182	$2.7 \cdot 10^{-4}$
bismuth	83	Bi	208.98040	$2 \cdot 10^{-5}$
boron	5	B	10.81	0.001
bromine	35	Br	79.904	$6 \cdot 10^{-4}$
cadmium	48	Cd	112.411	$2 \cdot 10^{-5}$
caesium	55	Cs	132.90545	$3 \cdot 10^{-4}$
calcium	20	Ca	40.078	3.4
californium	98	Cf	242	–
carbon	6	C	12.011	0.02
cerium	58	Ce	140.116	0.006
chlorine	17	Cl	35.453	0.11
chromium	24	Cr	51.9961	0.01
cobalt	27	Co	58.933195	$2.4 \cdot 10^{-3}$
copper	29	Cu	63.546	0.005
curium	96	Cm	244.0703	–
dysprosium	66	Dy	162.50	$3 \cdot 10^{-4}$
einsteinium	99	Es	252	–
erbium	68	Er	167.259	$2.7 \cdot 10^{-4}$
europium	63	Eu	151.964	$1.1 \cdot 10^{-4}$
fermium	100	Fm	257	–
fluorine	9	F	18.998403	0.06
francium	87	Fr	223	–
gadolinium	64	Gd	157.25	$5.2 \cdot 10^{-4}$
gallium	31	Ga	69.723	$1.6 \cdot 10^{-3}$
germanium	32	Ge	72.64	$1.4 \cdot 10^{-4}$
gold	79	Au	196.96657	$4 \cdot 10^{-7}$
hafnium	72	Hf	178.49	$3 \cdot 10^{-4}$
helium	2	He	4.002602	$4.2 \cdot 10^{-7}$
holmium	67	Ho	164.93032	$1.1 \cdot 10^{-4}$
hydrogen	1	H	1.00794	0.74

https://doi.org/10.1515/9783110735178-009

element	ordinal number	symbol	relative atomic mass[a]	abundance (%)
indium	49	In	114.818	$1 \cdot 10^{-5}$
iodine	53	I	126.90447	$5 \cdot 10^{-5}$
iridium	77	Ir	192.217	$1 \cdot 10^{-7}$
iron	26	Fe	55.845	4.7
krypton	36	Kr	83.798	$1.9 \cdot 10^{-8}$
lanthanum	57	La	138.90547	0.003
lawrencium	103	Lr	262	–
lead	82	Pb	207.2	$1.2 \cdot 10^{-5}$
lithium	3	Li	6.941	0.002
lutetium	71	Lu	174.967	$5 \cdot 10^{-3}$
magnesium	12	Mg	24.305	2.0
manganese	25	Mn	54.938045	0.091
mendelevium	101	Md	258	–
mercury	80	Hg	200.592	$8 \cdot 10^{-6}$
molybdenum	42	Mo	95.94	$1.4 \cdot 10^{-4}$
neodymium	60	Nd	144.242	$2.7 \cdot 10^{-3}$
neon	10	Ne	20.1797	$5 \cdot 10^{-7}$
neptunium	93	Np	237	–
nickel	28	Ni	58.6934	$7.2 \cdot 10^{-3}$
niobium	41	Nb	92.90638	0.002
nitrogen	7	N	14.0067	0.017
nobelium	102	No	259	–
osmium	76	Os	190.23	$5 \cdot 10^{-7}$
oxygen	8	O	15.999	48.9
palladium	46	Pd	106.42	$1 \cdot 10^{-6}$
phosphorus	15	P	30.973761998	0.1
platinum	78	Pt	195.084	$1 \cdot 10^{-6}$
plutonium	94	Pu	244	$2 \cdot 10^{-19}$
polonium	84	Po	209	$2 \cdot 10^{-14}$
potassium	19	K	39.0983	2.4
praseodymium	59	Pr	140.90765	$8 \cdot 10^{-4}$
promethium	61	Pm	145	$1 \cdot 10^{-19}$
protactinium	91	Pa	231.03588	$9 \cdot 10^{-11}$
radium	88	Ra	226	$1 \cdot 10^{-10}$
radon	86	Rn	222	$6 \cdot 10^{-16}$
rhenium	75	Re	186.207	$1 \cdot 10^{-7}$
rhodium	45	Rh	102.9055	$5 \cdot 10^{-7}$
rubidium	37	Rb	85.4678	0.009
ruthenium	44	Ru	101.07	$1 \cdot 10^{-6}$
samarium	62	Sm	150.36	$6 \cdot 10^{-4}$
scandium	21	Sc	44.955912	$2.1 \cdot 10^{-3}$
selenium	34	Se	78.96	$5 \cdot 10^{-6}$
silicon	14	Si	28.0855	26.3
silver	47	Ag	107.8682	$7 \cdot 10^{-6}$
sodium	11	Na	22.989769	2.7
strontium	38	Sr	87.62	0.036
sulphur	16	S	32.065	0.03

element	ordinal number	symbol	relative atomic mass[a]	abundance (%)
tantalum	73	Ta	180.94788	$2 \cdot 10^{-4}$
technetium	43	Tc	98	–
tellurium	52	Te	127.6	$1 \cdot 10^{-6}$
terbium	65	Tb	158.92535	$9 \cdot 10^{-5}$
thallium	81	Tl	204.3833	$5 \cdot 10^{-5}$
thorium	90	Th	232.03806	0.0011
thulium	69	Tm	168.93421	$5 \cdot 10^{-5}$
tin	50	Sn	118.71	$2 \cdot 10^{-4}$
titanium	22	Ti	47.867	0.42
tungsten	74	W	183.84	$1.5 \cdot 10^{-4}$
uranium	92	U	238.02891	$1.7 \cdot 10^{-4}$
vanadium	23	V	50.9415	0.013
xenon	54	Xe	131.293	$2.5 \cdot 10^{-4}$
ytterbium	70	Yb	173.04	$3.3 \cdot 10^{-4}$
yttrium	39	Y	88.90585	$3.2 \cdot 10^{-3}$
zinc	30	Zn	63.38	0.007
zirconium	40	Zr	91.224	0.016

[a] rounded after 3rd decimal place

Bibliography

Anastas, P. T. and J. C. Warner (1998) Green chemistry: theory and practice. Oxford University Press, New York, 30 pp.

Andreae, M. O. (1990) Ocean-atmosphere interactions in the global biogeochemical sulphur cycle. *Marine Chemistry* **30**, 1–29.

Ardon, M. (1965) Oxygen. Elementary forms and hydrogen peroxide. W. A. Benjamin, New York, 106 pp.

Barnes, I., G. Solignac, A. Mellouki and K.-H. Becker (2010) Aspects of the atmospheric chemistry of amides. *ChemPhysChem* **11**, 3844–3857.

Bernal, J. D. and R. H. Fowler (1933) A theory of water and ionic solution, with particular reference to hydrogen and hydroxyl ions. *Journal of Chemical Physics* **1**, 515–548.

Boyle, R. (1680) Chymista Scepticus uel Dubia et Paradoxa Chymico-Physica, circa Spagyricorum Principia, vulgo dicta Hypostatica, Prout proproni & propugnari solent a Turba Alchymistarum... 5th edn. (in Latin), Samuel de Tournes, Geneve, 148 pp.

Brasseur, G. P., R. G. Prinn and A. P. Pszenny, eds. (2003) Atmospheric chemistry in a changing world. An integration and synthesis of a decade of tropospheric chemistry research. Springer-Verlag, Berlin, 300 pp.

Brimblecombe, P. and S. L. Clegg (1988) The solubility and behaviour of acid gases in the marine aerosol. *Journal of Atmospheric Chemistry* **7**, 1–18.

Brosset, C. (1976) A method of measuring airborne acidity: its application for the determination of acid content on long-distance transported particles and in drainage water from spruces. *Water, Air, and Soil Pollution* **6**, 259–275.

Carpenter, L. J., R. J. Chance, T. Sherwen, S. M. Ball, M. J. Evans, H. Hepach, L. D. J. Hollis, T. D. Jickells, A. Mahajan, D. P. Stevens, L. Tinel and M. R. Wadley (2021) Marine iodine emissions in a changing world. *Proceedings of the Royal Society* **A477**. https://doi.org/10.1098/rspa.2020.0824.

Chapman, S. (1930) A theory of upper atmosphere ozone. *Memoirs of the Royal Meteorological Society* **3**, 103–109.

Charlson, R. J., J. E. Lovelock, M. O. Andreae and S. G. Warren (1987) Oceanic phytoplankton, atmospheric sulphur, cloud albedo and climate: a geophysiological feedback. *Nature* **326**, 655–661.

Clarke, F. W. (1920) The data of geochemistry. Government Printing Office, Washington D. C. (USA), 832 pp.

Cossa, A. (1867) Ueber die Ozonometrie. *Zeitschrift für Analytische Chemie* **6**, 24–28.

Crutzen, P. J. (1971) Ozone production rates in an oxygen-hydrogen-nitrogen oxide atmosphere. *Journal of Geophysical Research* **76**, 7311–7327.

Crutzen, P. J. and E. F. Stoermer (2000) The "Anthropocene". *Global Change Newsletters* **41**, 17–18.

Decesari, S., M. C. Facchini, E. Matta, M. Mircea, S. Fuzzi, A. R. Chughtai and D. M. Smith (2002) Water soluble organic compounds formed by oxidation of soot. *Atmospheric Environment* **36**, 1827–1832.

Denman, K. L., G. Brasseur, A. Chidthaisong, P. Ciais, P. M. Cox, R. E. Dickinson, D. Hauglustaine, C. Heinze, E. Holland, l. D. Jacob, U. Lohmann, S. Ramachandran, P. L. da Silva Dias, S. C. Wofsy and X. Zhang (2007) Couplings between changes in the climate system and biogeochemistry. In: Climate change 2007: the physical science basis. Contribution of working group I to the fourth assessment report of the intergovernmental panel on climate change (eds. S. Solomon, D. Qin, M. Manning et al.), Cambridge University Press, Cambridge, U. K., pp. 499–588.

Derwent, R., P. Simmonds, S. O'Doherty, A. Manning, W. Collins and D. Stevenson (2006) Global environmental impacts of the hydrogen economy. *International Journal of Nuclear Hydrogen Production and Application* **1**, 57–67.

https://doi.org/10.1515/9783110735178-010

Dewar, J. M. S. (1948) The structure of ozone. *Journal of Chemical Society* 299–1305.

Drapcho C. M. and D. E. Brune (2000) The partitioned aquaculture system: impact of design and environmental parameters on algal productivity and photosynthetic oxygen production. *Aquacultural Engineering* 21, 151–168.

Drever, J. I. (1997) The geochemistry of natural waters: surface and groundwater environments, 3rd edn., 436 pp.

Dunstan, W. R., H. A. D. Jowett and E. Goulding (1905) CLIII. – The rusting of iron. *Transaction of the Chemical Society* 87, 1548–1574.

Farman, J. C., B. G. Gardiner and J. D. Shanklin (1985) Large losses of total ozone in Antarctica reveal seasonal ClO_x/NO_x interaction. *Nature* 315, 207–210.

Finlayson-Pitts, B. J. and J. N. Pitts, Jr. (2000) Chemistry of the upper and lower atmosphere – theory, experiments and applications. Academic Press, San Diego (USA), 969 pp.

Franco, B., T. Blumenstock, C. Cho et al. (2021) Ubiquitous atmospheric production of organic acids mediated by cloud droplets. *Nature* 593, 233–237.

Fukuto, J. M., J. Y. Cho and S. H. Switzer (2000) The chemical properties of nitric oxide and related nitrogen oxides. In: Nitric oxide, biology and pathobiology (ed. L. J. Ignarro), Academic Press, San Diego, pp. 23–40.

Ghosh, R. S., S. D. Ebbs, J. T. Bushey, E. F. Neuhauser and G. M. Wong-Chong (2006) Cyanide cycle in nature. In: Cyanides in water and soil (eds. D. A. Dzombak, R. S. Ghosh and G. M. Wong-Chong), CRC Press, Boca Raton, FL (USA), pp. 226–236.

Gmelin, L. G. (1827) Handbuch der theoretischen Chemie. Erster Band, 3rd edn. by F. Varrentrapp, Frankfurt am Main, 1454 pp.

Gmelin, L. G. (1848) Handbuch der Chemie. Vierter Band. Handbuch der organischen Chemie – Erster Band. Heidelberg, 936 pp.

Gmelin (1936) Gmelins Handbuch der anorganischen Chemie, 8. Auflage, System-Nr. 4, Stickstoff, Lieferung 3, p. 718.

Gold, Th. (1999) The deep hot biosphere. Springer-Verlag, Berlin, 235 pp.

Gräfenberg, L. (1902) Das Potential des Ozons. *Zeitschrift für Elektrochemie* 8, 297–301.

Gräfenberg, L. (1903) Beiträge zur Kenntnis des Ozons. *Zeitschrift für anorganische Chemie* 36, 355–379.

Graham, Th. (1861) Liquid diffusion applied to analysis. *Philosophical Transactions of the Royal Society London* 151, 183–224.

Gupta, V., and K. S. Carroll (2014) Sulfenic acid chemistry, detection and cellular lifetime. *Biochimica et Biophysica Acta* 1840, 847-875.

Hall, C., P. Tharakan, J. Hallock, C. Cleveland and M. Jefferson (2003) Hydrocarbons and the evolution of human culture. *Nature* 426, 318–322.

Hart, E. J. and M. Anbar (1970) The hydrated electron. John Wiley & Sons, New York, 267 pp.

Heckner, H. N. (1977) The cathodic reduction of nitrous acid in the second reduction step in high acid solutions by the potential step method (formation of HNO). *Journal of Electroanalytical Chemistry* 83, 51–63.

Herrmann, H., A. Tilgner, P. Barzaghi, Z. Majdik, S. Gligorovski, L. Poulain and A. Monod (2005) Towards a more detailed description of tropospheric aqueous phase organic chemistry: CAPRAM 3.0. *Atmospheric Environment* 39, 4351–4363.

Hoffmann, M. R. (1977) Kinetics and mechanism of oxidation of hydrogen sulphide by hydrogen peroxide in acidic solution. *Environmental Science and Technology* 11, 61–66.

Hoffmann M. R. and J. O. Edwards (1975) Kinetics of the oxidation of sulphite by hydrogen peroxide in acidic solution. *Journal of Physical Chemistry* 79, 2096–2098.

Holland, H. D. (1984) The chemical evolution of the atmosphere and oceans. Princeton Univ. Press, New Jersey (USA), 582 pp.

Holland, H. D. (2006) The oxygenation of the atmosphere and oceans. *Philosophical Transaction of the Royal Society, Series B*, **361**, 903–915.

Hough, A. M. (1988) The calculation of photolysis rates for use in global modelling studies. Technical Report, UK Atomic Energy Authority. Harwell, Oxon. (UK), 347 pp.

Hughes, G. and C. R. Lobb (1976) Reactions of solvated electrons. In: Comprehensive chemical kinetics, Vol. 18 (eds. C. H. Bamford and C. F. H. Tipper), Elsevier Sci. Publ. Amsterdam, pp. 429–461.

IUPAC (1981) Manual of symbols and terminology for physicochemical quantities and units, appendix V, symbolism and terminology in chemical kinetics. *Pure and Applied Chemistry* **53**, 753.

IUPAC (2006) Compendium of chemical terminology, 2nd edn. (the "Gold Book"). Compiled by A. D. McNaught and A. Wilkinson. Blackwell Scientific Publications, Oxford (1997). XML online corrected version: http://goldbook.iupac.org (2006–) created by M. Nic, J. Jirat, B. Kosata; updates compiled by A. Jenkins. ISBN 0-9678550-9-8. doi:10.1351/goldbook.

Johnston, H. S. (1971) Reduction of stratospheric ozone by nitrogen catalysts from supersonic transport exhaust. *Science* **173**, 517–522.

Keeling, R. F., R. B. Bacastow, A. E. Bainbridge, C. A. Ekdahl, P. R. Guenther and L. S. Waterman (1976) Atmospheric carbon dioxide variations at Mauna Loa Observatory, Hawaii. *Tellus* **28**, 538–551.

Keith, D. W. (2009) Why capture CO_2 from the atmosphere? *Science* **325**, 1654–1655.

Koop, T., B. Luo, A. Tsias and T. Peter (2000) Water activity as the determinant for homogeneous ice nucleation in aqueous solution. *Nature* **406**, 611–614.

Kopp, H. (1931) Geschichte der Chemie. Neudruck der Originalausgabe in zwei Bänden (1843–1847), A. Lorentz Leipzig, Dritter Teil, 372 pp.

Koppenol, W. H. (2000) Names for inorganic radicals. *Pure and Applied Chemistry* **72**, 437–446.

Lackner, K. S., P. Grimes and H. J. Ziock (1999) Capturing carbon dioxide from air. In: 24th annual technical conference on coal utilization, Clearwater, Florida (USA).

Langmuir, D. (1997) Aqueous environmental geochemistry. Prentice-Hall, 600 pp.

Lavoisier, A.-L. (1789) Elements of chemistry in a new systematic order, containing all the modern discoveries. Transl. by R. Kerr, reprint 1965, Dover Publ. New York, 511 pp.

Lee, C.-L. and P. Brimblecombe (2016) Anthropogenic contributions to global carbonyl sulfide, carbon disulfide and organosulfides fluxes. *Earth Science Reviews* doi:10.1016/j.earscirev.2016.06.005.

Libes, S. (2009) Introduction to marine biogeochemistry. 2nd edn., Elsevier, 928 pp.

Livingstone, D. A. (1963) Chemical composition of rivers and lakes. In: U. S. geological survey professional paper, 6th edn. (ed. M. Fleischer), pp. 1–64.

Lovelock, J. E. and L. Margulis (1974) Atmospheric homeostasis by and for the biosphere: the Gaia hypothesis. *Tellus* **26**, 2–10.

Mader, P. M. (1958) Kinetics of the hydrogen peroxide-sulphite reaction in alkaline solution. *Journal of the American Chemical Society* **80**, 2634–2639.

Madronich, S. (1987) Photodissociation in the atmosphere. 1. Actinic flux and the effects of ground reflection and clouds. *Journal of Geophysical Research* **92**, 9740–9752.

Marenco, A., H. Gouget, P. Nédélec and J.-P. Pagés (1994) Evidence of a long-term increase in tropospheric ozone from Pic du Midi series: consequences: positive radiative forcing. *JGR* **99** (1994) 16,617–16,632. *See also*: Marenco, A., N. Philippe and G. Hérve (1994) Ozone measurements at Pic du Midi observatory. *EUROTRAC Annual report part 9: TOR*, EUROTRAC ISS, Garmisch-Partenkirchen, pp. 121–130.

Mason, B. J. and C. B. Moore (1982) Principles of geochemistry, 4th edn. John Wiley & Sons, New York, 344 pp.

Miranda, K. M., M. G. Espey, D. Jourd'heuit, M. B. Grisham, J. M. Fukuta, M. Feelisch and D. A. Wick (2000) The chemical biology of nitric oxide. In: Nitric oxide, biology and pathobiology (ed. L. J. Ignarro), Academic Press, San Diego, pp. 41–57.

Molina, M. J. and F. S. Rowland (1974) Stratospheric sink for chlorofluoromethanes. Chlorine atom catalyzed destruction of ozone. *Nature* **249**, 810–812.

Möller, D. (1980) Kinetic model of atmospheric SO_2 oxidation based on published data. *Atmospheric Environment* **14**, 1067–1076.

Möller, D. (1983) The global sulphur cycle. *Idojaras* **87**, 121–143.

Möller, D. (1989) The possible role of H_2O_2 in new-type forest decline. *Atmospheric Environment* **23**, 1187–1193.

Möller, D. (2003) Luft. DeGruyter, Berlin, 750 pp.

Möller, D. (2008) On the history of the scientific exploration of fog, dew, rain and other atmospheric water. *Die Erde* **139**, 11–44.

Möller, D. (2009) Atmospheric hydrogen peroxide: evidence for aqueous-phase formation from a historic perspective and a one-year measurement campaign. *Atmospheric Environment* **43**, 5923–5936.

Möller, D. (2012) SONNE: solar-based man-made carbon cycle, and the carbon dioxide economy. *Ambio* **41**, 413–419.

Möller, D. (2019) Chemistry of the climate system. Vol. 1: fundamentals and processes, 3rd edn. DeGruyter, Berlin, 616 pp.

Möller, D. (2022) Atmospheric chemistry. A critical voyage through the history. DeGruyter, Berlin, xxx pp.

Möller, D. and K. Acker (2007) Chlorine phase partitioning at Melpitz near Leipzig. In: Nucleation and atmospheric aerosols (eds. C. D. O'Dowd and P. Wagner), Proceedings of 17th Int. Conference Galway, Ireland 2007, Springer, pp. 654–658.

Nilsson, J. and P. Grennfelt, eds. (1988) Critical loads for sulphur and nitrogen. UNECE/Nordic Council workshop report, Skokloster, Sweden. March 1988. Nordic Council of Ministers: Copenhagen.

Odén, S. (1976) The acidity problem – an outline of concepts. *Water, Air and Soil Pollution* **6**, 137–166.

Olah, G. A. (2005) Beyond oil and gas: the methanol economy. *Zeitschrift für Angewandte Chemie International Edition* **44**, 2636–2639.

Ostwald, Wo. (1909) Grundriss der Kolloidchemie. Theodor Steinkopff, Dresden, 525 pp.

Penkett, S. A., B. M. R. Jones, K. A. Brice and A. E. J. Eggleton (1979) The importance of atmospheric O_3 and H_2O_2 in oxidizing SO_2 in cloud and rainwater. *Atmospheric Environment* **13**, 123–137.

Popovicheva, O., D. Baumgardner, K. Gierens, R. Miake-Lye, R. Niessner, M. Rossi, M. Petters, J. Suzanne and E. Villenave (2007) Atmospheric soot: environmental fate and impact. ASEFI 2006 Meeting Summary and the Atmospheric Soot Network (ASN) definition.

Prasad, M. N. V., K. S. Sajwan and R. Naidu, eds. (2005) Trace elements in the environment: biogeochemistry, biotechnology, and bioremediation. CRC Press, Boca Raton, FL (USA), 726 pp.

Reuder, J. (1999) Untersuchungen zur Variabiliät von Photolysefrequenzen. Doctoral thesis, Brandenburg University of Technology (BTU), Cottbus (Germany).

Rick, S. W. (2004) A reoptimization of the five-site water potential (TIP5P) for use with Ewald sums. *Journal of Chemical Physics* **120**, 6085–6093.

Sabine, C. L., R. A. Feely, N. Gruber, R. M. Key, K. Lee, J. L. Bullister, R. Wanninkhof, C. S. Wong, D. W. R. Wallace, B. Tilbrook, F. J. Millero, T.-H. Peng, A. Kozyr, T. Ono and A. F. Rios (2004) The oceanic sink for anthropogenic CO_2. *Science* **305**, 367–371.

Sander, R. (2015) Compilation of Henry's law constants, version 3.99. *Atmospheric Chemistry and Physics* **15**, 4399–4981.

Schade, G. W. and P. J. Crutzen (1995) Emission of aliphatic amines from animal husbandry and their reactions: potential source of N_2O and HCN. *Journal of Atmospheric Chemistry* **22**, 319–346.

Schlesinger, W. H. (1997) Biogeochemistry – an analysis of global change. Academic Press, San Diego, 588 pp.

Schlesinger, W. H. (2013) Biogeochemistry: an analysis of global change, 3rd edn., 688 pp.

Schönbein, C. F. (1844) Ueber die Erzeugung des Ozons auf chemischem Wege. Basel, pp. 62, 94, 110.

Schönbein, C. F. (1861) Ueber die Bildung des Wasserstoffsuperoxydes während der langsamen Oxydation der Metalle in feuchtem gewöhnlichem Sauerstoff oder atmosphärischer Luft. *Annalen der Physik und Chemie* **188**, 445–451.

Sehested, K., J. Holcman and E. J. Hart (1983) Rate constants and products of the reactions e_{aq}^-, O_2^- and H with ozone in aqueous solution. *Journal of Physical Chemistry* **87**, 1951–1954.

Seinfeld, J. H. and S. N. Pandis (1998) Atmospheric chemistry and physics – from air pollution to climate change. John Wiley & Sons, New York, 1326 pp.

Solomon, S., G.-K. Plattner, R. Knutti and P. Friedlingstein (2009) Irreversible climate change due to carbon dioxide emissions. *Proceedings of the National Academy of Science* **106**, 1704–1709.

Soret, J.-L. (1864) Ueber die volumetrischen Beziehungen des Ozons. *Annalen der Chemie und Pharmacie (Lieb. Ann.)* **130**, 95–101; Ueber das volumetrische Verhalten des Ozons. *Annalen der Physik und Chemie (Pogg. Ann.)* **197**, 168–283.

Sparks, D. (2003) Environmental soil chemistry, 2nd edn., Academic Press. 352 pp.

Sposito, G. (2016) The chemistry of soils, 3rd edn., Oxford Univ. Press, 272 pp.

Staehelin, J. and J. Hoigné (1982) Decomposition of ozone in water: rate of initiation by hydroxide ions and hydrogen peroxide. *Environmental Science and Technology* **16**, 676–681.

Staehelin, J., R. E. Bühler and J. Hoigné (1984) Ozone decomposition in water studied by pulse radiolysis. 2. OH and SO_4 as chain intermediates. *Journal of Physical Chemistry* **88**, 5999–6004.

Stolarski, R. S. and R. J. Cicerone (1974) Stratospheric chlorine: a possible sink for ozone. *Canadian Journal of Chemistry* **52**, 1610–1615.

Struve, H. (1871) Studien über Ozon, Wasserstoffhyperoxyd und salpetrigsaures Ammoniak. *Zeitschrift für analytische Chemie* **10**, 292–298.

Stumm, W., ed. (1987) Aquatic surface chemistry: chemical processes at the particle-water interface. John Wiley & Sons, New York, 520 pp.

Stumm, W. and J. J. Morgan (1996) Aquatic chemistry: chemical equilibria and rates in natural waters. Wiley, 1040 pp.

Tan, K. H. (2010) Principles of soil chemistry, 4th edn. Series: Books in soils, plants, and the environment. CRC Press, 390 pp.

Tepe J. B. and B. F. Dodge (1943) Absorption of carbon dioxide by sodium hydroxide solutions in a packed column. *Transactions of the American Institute of Chemical Engineers* **39**, 255–276.

Traube, M. (1882) Über Aktivierung des Sauerstoffs. *Berichte der Deutschen Chemischen Gesellschaft* **15**, pp. 222, 659, 2421, 2423.

Walden, P. (1941) Geschichte der organischen Chemie seit 1880. Zweiter Band zu C. Graebe: Geschichte der organischen Chemie. Springer-Verlag, Berlin, 946 pp.

Waldman, J. M., J. W. Munger, D. J. Jacob, R. C. Flagan, J. J. Morgan and M. R. Hoffmann (1982) Chemical composition of acid fog. *Science* **218**, 677–679.

Warneck, P. (2000) Chemistry of the natural atmosphere. Academic Press, New York, 969 pp.

Warneck, P. (2003) In-cloud chemistry opens pathway to the formation of oxalic acid in the marine atmosphere. *Atmospheric Environment* **37**, 2423–2427.

Wedepohl, K. H. (1995) The composition of the continental crust. *Geochimica et Cosmochimica Acta* **59**, 1217–1232.

Weiss, J. (1935) Investigations on the radical HO_2 in solution. *Transactions of the Faraday Society* **31**, 668–681.

Wiberg, N., A. Holleman, E. Wiberg, eds. (2001) Hollemann–Wiberg's inorganic chemistry. Academic Press, 1924 pp.

Wright, M. R. (2007) An introduction to aqueous electrolyte solutions. Wiley, 602 pp.

Zeman, F. S. and K. S. Lackner (2004) Capturing carbon dioxide directly from the atmosphere. *World Resource Review* **16**, 62–68.

Zhang, L. N., S. Solomon, K. A. Stone, J. D. Shanklin, J. D. Eveson, S. Colwell, J. P. Burrows, M. Weber, P. F. Levelt, N. A. Kramarova and D. P. Haffner (2021) On the use of satellite observations to fill gaps in the Halley station total ozone record. *Atmospheric Chemistry and Physics* **21**, 9829–9838.

Zobrist, J. (1987) Methoden zur Bestimmung der Azidität in Niederschlagsproben. VDI Berichte 608, Düsseldorf, pp. 401–420.

Author index

Agricola, Georg (1494–1555) 313
Albinus, Siegfried (1697–1770) 283
Arrhenius, Svante August (1859–1927) 86, 87, 111

Balard, Antoine-Jérôme (1802–1876) 268
Bates, David Robert (1916–1994) 370
Beer, August (1825–1863) 129
Berthelot, Marcellin (1827–1907) 183, 313
Berthollet, Claude Louis (1748–1822) 278
Berzelius, Jöns Jacob (1779–1848) 4, 116, 183
Black, Joseph (1728–1799) 21
Bolin, Bert (1925–2007) 328
Boltzmann, Ludwig (1844–1906) 67
Bosch, Carl (1874–1940) 178
Boudouard, Octave Leopold (1872–1923) 416
Boyle, Robert (1627–1691) 2, 23
Brand, Hennig (c. 1630–1692) 282
Brewer, Alan West (1915–2007) 367
Brønsted, Johannes Nicolaus (1879–1947) 87, 89, 91
Buisson, Henri (1873–1944) 366

Carlyle, Thomas (1795–1881) 1
Cauer, Hans (1899–1962) 357
Chapman, Sydney (1888–1970) 366
Chappuis, Joseph (1854–1934) 366
Charles, Jacques (1746–1823) 23
Chatin, Gaspard-Adolphe (1813–1901) 269
Clapham, Arthur Roy (1904–1990) 329
Clarke, Frank Wigglesworth (1847–1931) 328
Clausius, Rudolf Julius Emanuel (1822–1888) 60
Clément, Nicolas (1779–1841) 269
Cohen, Julius (1859–1935) 357
Courtois, Bernard (1777–1838) 269
Crutzen, Paul (1933–2021 329

D'Ans, Jean (1881–1969) 4
Davy, Humphrey (1778–1829) 207
Dobson, Gordon (1889–1975) 367
Dobson, Gordon Miller Bourne (1889–1975) 366
Ducros, Hippolyte (1805–1879) 358
Döbereiner, Johann Wolfgang (1780–1849) 116

Ebermayer, Ernst (1829–1908) 366
Emich, Friedrich (1860–1940) 171

Engler, Carl Oswald Viktor (1842–1925) 366
Evelyn, John (1620–1706) 356

Fabry, Charles (1867–1945) 366
Fenton, Henry (1854–1929) 170
Fox, Cornelius (1839–1922) 365
Fresenius, Carl Remigius (1818–1897) 4

Gautier, Armand (1837–1920) 150
Gay-Lussac, Joseph (1778–1850) 23, 207
Gmelin, Johann Friedrich (1748–1804) 357
Goldschmidt, Victor Moritz (1888–1947) 328
Gorham, Eville (1925–2020) 359
Graham, Thomas (1805–1869) 15

Haagen-Smit, Arie Jan (1900–1977) 374
Haber, Fritz (1868–1934) 178
Harrison, Douglas Neill (1901–1987) 366
Hayes, Augustus Allen (1806–1882) 357
Helmont, Jan Baptist van (1580–1644) 21
Henry, William (1772–1836) 77
Humboldt, Alexander von (1769–1859) 313

Junge, Christian (1912–1996) 211, 357, 358

Kekulé, Friedrich August (1829–1896) 104

Lambert, Johann Heinrich (1727–1777) 129
Lampadius, Wilhelm August (1772–1842) 264
Lavoisier, Antoine Laurent de (1743–1794) 3, 4, 207, 283
Lewis, Gilbert Newton (1875–1946) 88, 365
Liebhafsky, Hermann Alfred (1905–1982) 387
Liebig, Justus von (1803–1873) 4, 140
Lowry, Thomas Martin (1874–1936) 87
Lémery, Nicolas (1645–1715) 4
Löwig, Carl Jacob (1803–1890) 268

Marchand, Eugène (1816–1895) 268
Marum, Martinus van (1750–1837) 365
Meissner, Georg (1829–1905) 169
Mendeleev, Dmitri (1834–1878) 313

Nernst, Walther Herman (1864–1941) 67
Newton, Isaac (1643–1727) 3
Nicolet, Marcel (1912–1996) 370

Odén, Svante (1924–1986) 359

https://doi.org/10.1515/9783110735178-011

Olding, William (1829–1921) 365
Ostwald, Wilhelm (1853–1932) 3, 86

Pauling, Linus (1901–1994) 3, 40
Pfaff, Christoph Heinrich (1773–1852) 269
Planck, Max (1858–1942) 67
Pošepný, František (1836–1895) 264
Priestley, Joseph (1733–1804) 21
Prout, William (1785–1850) 169

Ramsay, William (1852–1916) 21
Raoult, François Marie (1830–1901) 45
Russell, Francis (1849–1914) 357
Rutherford, Daniel (1749–1819) 21

Sabatier, Paul (1854–1941) 415
Scheele, Carl Wilhelm (1742–1786) 21, 283
Schönbein, Christian Friedrich (1799–1868)
    169, 172, 365
Schöne, Emil (1838–1896) 169
Schöne, Emil (1838-1896) 366
Strutt, John William (1627–1795) 21

Struve, Heinrich (1822–1908) 169
Stumm, Werner (1924–1999) 169, 360
Stöckhardt, Julius Adolf (1808–1886) 4
Suess, Eduard (1831–1914) 329
Sørensen, Søren Peder Lauritz (1868–1939) 92

Teichert, Friedrich(1905–1986) 374
Thomson, William, Lord Kelvin (1824–1907) 77
Thénard, Louis Jacques (1777–1857) 207

Uexküll, Jacob von (1864–1944) 1

van't Hoff, Jacob (1852–1911) 85
Vernadsky, Vladimir Ivanovich (1863–1945) 328

Warburg, Emil Gabriel (1846–1931) 366
Warmbt, Wolfgang (1916–2008) 374
Winkler, Clemens (1838–1904) 355
Wislicenus, Hans Adolf (1867–1951) 359
Wurster, Casimir (1854–1913) 366
Wöhler, Friedrich (1800–1882) 4, 140

Subject index

absolute humidity 48
absorption
– radiation 129
– UV radiation 322
absorption cross section 130, 133, 134
abundance of elements 145
acceptor acid 91
accretion 307
acetaldehyde
– photolysis 249
– photolysis rate 135
acethyl peroxide 250
acetic acid
– aqueous chemistry 251
– degassing 231
– natural acidity 352
acetone 258
acetyl radical 249
acetylene 256
acid constant 89
acid deposition 365
acid fog 359
acid neutralising capacity 360
acid rain 224, 300, 358, 363
acid strength 89, 352
acid-base theory 88
acidification 358
acidifying capacity 359, 364
acidity
– atmospheric 246
– definition 360
– ecosystem 352
– total and free 360
acidity constant 87
acids 87, 91
actinic flux 133
actinic radiation 127
activation energy 111, 114
activation enthalpy 114
activator 116
activity 72
activity coefficient 72
acyl radical 239, 242
acylperoxo radical 242
addition 106
adiabatic 62, 67

adiabatic isolated system 60
adsorption 82
adsorption and capillary condensation 84
adsorption isotherm 83
aerosol
– acidity 201, 364
– carbonaceous 54
– definition 15
– organic 53, 252, 253
– photosensitiser 193
– soil dust 52
– sulphate 208
air
– composition 22
– definition 8
air chemistry
– first use of the word 357
air cleanser 366
air pollution
– change 210
– definition 10, 22
– history 356
– ozone 382
– problems 227, 357
– soot 54, 226
– sulphur dioxide 207
– urban 53, 227
airs 21
albedo 129
alchemy 2, 183
alcohols
– dehydration 255
aldehydes 242, 243, 260
algal aquacultures 408
aliphatic compounds 236
alkali metals 290
alkaline earth metals 290
alkalinity 353, 363
alkanes
– bond 104
– emission 237
– in air 239
– reaction with OH 240
– systematic 240
alkanone 258

https://doi.org/10.1515/9783110735178-012

alkenes
– emission 237
– in air 239
– reaction with O3 257
– reaction with OH 255
– SOA formation 257
– systematics 240
alkenyl radicals 239
alkoxyl radicals 55, 239, 241–243
alkyl nitrites 205
alkyl peroxides 242
alkyl peroxyl radicals 205, 241
alkyl radicals 239–241
alkynes 254
alpha decay 396
alumina 158
aluminates 298
aluminium 298
– hydroxides 298
– silicates 309
– standard electrode potential 136
alumosilicates 297
alums 42
amide 201
amide anion 183
amidogen 182
amine 201, 203
amino acid 200
ammonia
– as hydrogen storage 178
– dissociation 89
– dry deposition velocity 393
– early atmosphere 312
– gas-to-particle formation 183
– oxidation 183
– particle formation 55, 274
– photodissociation 182
– photodissociation early earth 313
– reaction with OH 182
– sources 179
ammonia hydrate 307
ammonification 177, 179, 346
ammonium
– biological oxidation 343
– hydrolysis 94
ammonium nitrite, aqueous-phase chemistry
    183
ammonium sulphate 183
amount of substance 15

amphoteric 89, 298
amylamine 201
Angeli's salt 193
anharmonic oscillator 131
anion 14, 86
Anthropocene 299, 329
anthroposphere 329
antioxidant 215
aquathermolysis 210, 354
aquatic surface chemistry 169
aqueous solutions 44
aragonite 232
Arctic ozone depletion 266
arenes 258
argon
– discovery 21
– ice 307
– in air 308
aromatic compounds 236, 258
– emission 237
– in air 239
– in crude oil 238
– reaction with OH 258
Arrhenius constant 111
Arrhenius equation 111
Arrhenius factor 114
Arrhenius plot 115
Arrhenius–Ostwald theory 87
arsenic 296
assimilatory sulphate reduction 350
astatine 261
Atacama saltpetre 266
atmosphere
– definition 8
– early composition 316
atmospheric aerosol 51
atmospheric chemistry, definition 358
atmospheric water 3, 11, 37
atmospheric window 130
atom 13
atomic number 13
ATP 181
autotrophic 255, 320
autoxidation 139
Avogadro's law 25
azanide 183

base neutralising capacity 360
bases 87

benzene
– reaction with OH 258
– structure 104
benzeneoxide 260
benzo[a]pyrene 259
Bergeron-Findeisen process 46
BET isotherm 83
bicarbonate
– aqueous-phase chemistry 92
bicarbonyl 248
Big Bang 302
bioaerosol 54
biocatalyst 116
bioelement 148
biofuel 238
biogeochemical cycle 8, 322, 330, 331
biogeochemical evolution 316
biogeochemistry 328
biological chemistry
– nitrogen oxides 196
– oxygen 143
– sulphides 207
biological cycle 330
biological evolution 302, 323
biological fixation 176, 343
biological particle 54
biomass
– biogeochemistry 147
– burning 23, 151, 201
– definition 330
biomass burning
– CH3I emission 264
– combustion process 238
– emission of chlorine compounds 264
– Hg emission 293
– NPP loss 340
– VOC emission 237
biosphere 329
biosphere–atmosphere interaction 329, 331
biotite 51
bitumen 311
black carbon 226
black lung disease 297
bleaching 139, 169
blue haze 53, 257
blue plume 211, 215
Bodenstein principle 112
Boltzmann constant 26, 67, 425
Boltzmann distribution 29

bond length 100
borate, as buffer 232, 287
borax 95
Boudouard equilibrium 416
Boyle–Mariotte law 25
Boyle's law 25
Brewer–Dobson circulation 367
bromate 279
bromide
– loss from sea salt 383
– reaction with O3 278, 385
bromine
– aqueous-phase chemistry 385
– discovery 268
– ozone depletion 270
– physiology 262
bromine explosion 266, 383
bromine monoxide 383, 385
bromine radical
– from BrO + BrO 383
– reaction with O3 383
bromite 279
bromocarbons 268
brown clouds 227
Brownian motion 32
Brønsted theory 87
buffer capacity 95
buffer ratio 94
buffer solution 94
Bunsen absorption coefficient 78

C1 aqueous-phase chemistry 248
C1 gas-phase chemistry 247
C2 aqueous-phase chemistry 251
C2 gas-phase chemistry 250
cadmium
– emission 295
– standard electrode potential 136
– toxicity 295
– volcanic emission 288
calcite 51, 232, 285, 291
calcium carbonate
– from weathering 327
– hydrolysis 93
– in ocean 233
– solubility 231
calcium oxide, hydrolysis 94
calcium phosphate 285
calomel 294

CAMERE process 415
capillaries 84
carbon
– black 54
– buried organic 321
– earth 224
– elemental 54, 225
– elemental, reaction with O3 227
– elemental, reaction with OH 227
– organic in PM 227
– reservoir 338
– river run-off 338
carbon burial 340
carbon capture and storage 403
carbon capture and utilization 410
carbon cycle 323, 327, 338
carbon dioxide
– air capture 406
– aqueous-phase chemistry 229
– as resource 409
– atmospheric increase 342
– cumulative emission 402
– cycling 402
– cycling technology 411
– history 21
– hydrogenation 415
– in air 337
– reduction to fuels 234
– residence time 326
– solubility 230
– volcanic emission 326
carbon dioxide economy 412
carbon disulphide
– emission 211
– gas-phase chemistry 211
– reaction with OH 212
carbon economy 224
carbon monoxide
– discovery 228
– dry deposition velocity 393
– in air 229
– reaction with OH 229
carbon neutral hydrocarbons 413
carbon neutrality 224
carbon-14 396
carbonaceous meteorite 306, 310
carbonate
– alkalinity 363
– aqueous-phase chemistry 229, 232

– buffer capacity 232
– global wet deposition 326
– in rocks 327
– in seawater 231
– natural acidity 361
carbonate mountains 291
carbonate radical anion 233
carbonic acid 231
– dissociation diagram 92
carbonyl sulphide
– emission 211
– gas-phase chemistry 211
– in air 210
carboxylic acids
– affecting CCN 258
– global acidity 353
– global emission 237
– names 254
– reaction scheme 243
– sources 246
catalase 170
catalysis 116
catalyst 116
cation 14, 86
causticization 407
CCS see carbon capture and storage
CCU see carbon capture and utilization
cell 125, 166, 198, 297, 302, 318, 319, 334, 339
cell theory 319
chain reaction 143
Chappuis band 160
charge transfer 109
Charles's law 25
chemical amount 15
chemical bonding 100
chemical compound 13
chemical equilibrium 74, 84
chemical evolution 302
chemical kinetics 107
chemical potential 70
chemical reaction 106
chemical standard potential 71
chemistry, definition 2
Chernobyl disaster 398
Chile salpetre 199
chlorate 279
chloride
– biological chemistry 268
– cycling 264

– reaction with NO3 190
– reaction with O3 278
– seasalt 265
chlorination 279, 281
chlorine
– aqueous-phase chemistry 280, 388
– atomic *see* chlorine radical
– disinfecting agent 281
– in coal 274
– in soils 267
– organic compounds 264, 266
– photolysis rate 135
– physiology 261
– radical 281
chlorine cation 280
chlorine compounds
– emission 264
– use 267
chlorine cycle 265, 267
chlorine nitrate, photolysis rate 135
chlorine radical
– reaction with hydrocarbons 281
– reaction with O3 277
chlorite 279
chlorofluorocarbons 267, 271
chlorophyll 136, 259, 334
chromophoric 135, 173, 334
Clapeyron's equation 76
clathrate hydrates 34, 43
Clausius–Clapeyron equation 76
Clausius–Clapeyron plot 76
clean air 23
climate 301
climate change 237, 342
closed system 60
cloud chemistry 56
cloud condensation nuclei
– chemical composition 81
– cloud processing 11, 35, 37
– from iodine 388
cloud drop
– acidity 361
– formation 37, 77
– oxalic acid source 254
– size 37
– sulphate source 215
– surfactant 204
cloud processing 35, 56, 394

cloud water
– chemical composition 37
– oxalic acid formation 253
– sampling 20
– trace metals 289
coal chemistry 414
coal gas 228
coal mine dust 297
cobalt
– in minerals 285
– standard electrode potential 136
collision
– cosmic 307
– molecules 23, 112
collision complex 109
collision frequency 30
collision impaction 391
collision number 24, 28
collisional cross section 28
colloid 15
column ozone 367
comet 310
common air 21
compensation point 392
competitive chemical pathways 146
complex ion 95
complex reaction 108
concentration, definition 15
conglomerate 49
coordination number 96
copper
– reaction with H2O2 170
– reaction with RS–NO 199
– reaction with superoxide 168
corresponding acids and bases 90
corrosion 138
Coulomb interaction 72
covalent bond 101
Criegee radical 257
crude oil, chemical composition 238
crust
– continental 309
– degassing 311
– depth and mass 309
– early earth 308
– main elements 309
– oceanic 309
– primordial 309
– volcanism 325

cryoscopic constant 46
crystalline solids 49
current 120
cyanate 204
cyanides 96, 204

DAC *see* direct air capture
dalton 17
Dalton's law 27
Debye–Hückel equation 73
decay serie 397
deep sea, carbon transport 342
deforestation 340
denitrification 177, 344
density 7, 16, 25, 27
deoxyribonucleic acid 319
detergent of the atmosphere 366
deuterium 303
dew
– bleaching property 169
– chemistry 173, 184
dew point 48
dialkyl sulphate 55
diamond 312
diatoms 297, 327
dicarbonyl compounds 260
dichlorine radical 281
diethylamine 201
diffusion 32
diffusion coefficient 32
diffusion coefficient for gases 33
diimine 181
dimerisation
– of amides 203
– of HNO 195, 344
– of HNO2 194
– of HO2 163
dimethyl formamide
– in air 201
– reaction with OH 203
dimethyl sulphide
– aqueous-phase chemistry 213
– emission 211
– oceanic emission 350
– reaction with OH 213
dimethylamine 201
dimethylsulphiopropionate 350
dimethylsulphone 213
dimethylsulphoxide 213

dinitrogen monoxide
– anhydride 197
– reaction with hydrated electron 125, 175
– stratospheric chemistry 184
– tautomer 195
– thermal formation 180
dinitrogen pentoxide
– gas-phase chemistry 188
– reaction with NaCl 188
– stratospheric chemistry 373
dinitrogen tetroxide, NOx equilibrium 187
dinitrogen trioxide
– aqueous-phase chemistry 191
– nitrolysation agent 198
– NOx equilibrium 187
direct air capture 406
dismutase 166, 170
dissimilatory sulphate reduction 349
dissipated work 66
dissociation constant 86
dissociation degree 89
dissolution 81
dissolution enthalpy 81
dissolved inorganic carbon 230
dithionate 223
Dobson spectrophotometer 367
Dobson unit 367
dolomite 291
double bond 103
doublet 132
drag 30
drinking water, ozonation 171
driving force 9, 70, 332
droplet, gas-liquid equilibrium 76
dry deposition 393
dry deposition velocity 393
dust 52
– air pollution 227
– alkaline 94, 230, 264
– chemical composition 52
– coal, explosion 226
– definition 51
– deposition 391
– fog and smoke 15
– from power plants 289
– from soils 52
– industrial 52, 94
– interstellar 304
– metals 287

– optical impact 128
– Saharan 264
dynamic equilibrium 74, 97

earthquake 325
ecosphere 330
ecosystem 330
effective Henry's law constant 80
electric charge 120
electric potential 120
electrical energy 117
electrical work 71
electrochemical potential 119
electrochemical reaction 117
electrochemical series 136
electrochemistry 117
electrolyt 86
electrolytic dissociation 73, 86
electromotive force 120
electron acceptor 118
electron donator 118
electron transfer
– onto CO 235
– onto CO2 234
– onto NO 197
– onto NO2 191
– onto NO3 190
– onto O2 166
– onto O3 173
electron transfer process 124
electroneutrality condition 359
electronic configuration 102
electronic excitation 131
electronic shell 100
element 13
– abundance 304
– crustal 309
– cycling 330
– formation by fusion 303
– in gaseous compounds 149
– major biological 330
– radioactive 304
– standard reduction potential 136
elemental carbon 226
elementary reaction 11, 106
endothermic 61, 64
energy, definition 24
enthalpy 63
entropy 66

environment, definition 1
enzymatic reduction 166
enzyme 116
equilibrium 73, 85, 97
equilibrium condition 68
equilibrium constant 85
equivalent 19
ethane
– emission 237
– gas-phase chemistry 250
– in air 238
– in natural gas 241
ethanediol 251
ethanol
– from fermentation 320
– reaction with OH 249
ethene
– from plants 254
– gas-phase chemistry 250
– hybrid orbitals 103
ethylamine 201
ethyne
– lifetime 256
– reaction with OH 256
– triple bond 105
Euler's number 110
evaporation enthalpy 75
excess chloride 273
excited states 130
exothermic 61, 64
extensive quantity 16
extinction coefficient 129
extinction module 129
Eyring plot 114

fallout 397
Faraday constant 93, 120
faujasite 43
faulting 325
feldspar 51, 307
Fenton chemistry 170
Fenton reaction 170
Fenton-like chemistry 170
fermentation 320
ferric/ferrous pair 292
first law of thermodynamics 60, 61
first-order reaction 109
Fischer–Hepp rearrangement 203
Fischer–Tropsch synthesis 412

fixed air 21
flue gas
– decarbonisation 404
– desulphurisation 211, 215
– mercury chemistry 294
fluorine
– in drugs 271
– in minerals 271
– volcanic 271
flux
– definition 8
– general equation 32
fog
– acid 354
– air pollution 357
– atmospheric water 11
– chemical composition 359
– impaction 391
– pollution in London 351
force, definition 24
forest
– damage 169, 351, 359
– new-particle formation 261
formaldehyde
– aqueous-phase chemistry 245, 248
– conversion to HCOOH 355
– from CH3OH oxidation 245
– from CH4 oxidation 244
– from isoprene oxidation 257
– H2 source 151
– hydrate 218
– interstellar 305
– photodisociation 154
– photolysis 244
– photolysis rate 135
formamide 203
formate
– aqueous-phase chemistry 174
– as buffer 355
– reaction with OH 234
formic acid 245
– as buffer 355
– from HCHO 247
– reaction with OH 174
formyl radical 245, 247
fossil fuel
– CO2 release 326
– energy and air pollution 227, 357
– origin 311, 313

fractions 18
free energy 68
free enthalpy 68
free radicals 141
freezing point depression 45, 46
Freundlich and Langmuir isotherm 83
friction 30
fugacity 72
Fukushima nuclear disaster 398
fullarenes 225
functional groups 44, 236, 258
fusion 303

galaxy 302
gamma radiation 396
gas law 18, 23
gas-liquid equilibrium 75
gas-to-particle conversion 55, 183, 215, 284
Gay-Lussac's law 25
general gas equation 18
geochemistry 6
geoengineering 116
ghostly lights 287
giant molecular clouds 304
giant polymers 318
Gibbs energy 68
Gibbs–Helmholtz equation 69
glass 14
global warming 225, 300, 402
global warming potential 151
glycolaldehyde 249, 250, 255
glycolic acid sulfate 252
glyoxal
– aqueous chemistry 247, 250
– dihydrate 252
– from aroimatic oxidation 261
– global formation 252
glyoxal sulfate 252
granite 314
granular material 49
graphite 225
graphitic carbon 54, 226
gravitational attraction 306
Great Oxidation Event 322
green chemistry 401
greenhouse gas 151, 276, 300, 346
Greenland ice core 290
gross primary production 338
groups 142

gypsum 291

Haber–Bosch process 178, 182
habitat 1
haematite 291
haems 198, 259
half-cell
– reactions 120
– standard potential 119
half-life 110
half-metal 13, 287
halides 263
halogenic acids, acidity 272
halogenide 263
halogens
– atoms 149
– compounds 263
– cycling 265
– from rocks 314
– organic, photodissociation 276
– stratosphere 371, 372
Hammett acidity function 92
Hartley band 160
heat 24, 61
heat capacity 63
Heisenberg uncertainty principle 102
helium
– interstellar 302
– isotopes 308
Helmholtz energy 68
Henderson–Hasselbalch equation 93
Henry's law constant 77–79
Herzberg continuum 368
Hess's law 64
heterocyclic compounds 236
heterogeneous catalysis 116
heterogeneous chemistry 139
heterotrophic 320
hexafluorosilicic acid 272
hole-electron pair 136
homogeneous catalysis 116
HOx cycle 370
Huggins band 160
humic substance 352
humic-like substance 173, 228
humidity 48
humus 51, 340
Hund's rule 131
hybrid orbitals 103, 105

hydrate 34
hydrated electron
– aqueous chemistry 121, 124
– formation 336
– history 122
– photochemical formation 136
hydrazine 181, 202
hydride 148, 152, 155, 317
hydrocarbons
– aqueous-phase oxidation 174
– biogenic emission 255
– deep crust 312
– from biomass burning 238
– functional groups 235
– gas-phase oxidation 239
– ozone precursor 377
– source of energy 237
– thermal dissociation 150, 312
hydrochloric acid
– from coal combustion 274
– from sea salt 264
– global emission 266
– sources 272
– stratospheric chemistry 373
hydrocyanic acid 204
hydrofluoric acid
– pK value 272
– sources 271
hydrogen
– aqueous chemistry 155
– atmospheric concentration 150
– atom 122, 124, 150, 154, 371
– clean production 411
– formation 154
– fuel 151
– interstellar 302
– lithosphere 312
– loss from Earth 316
– occurrence 150
– production 151
– reaction with OH 155
– residence time 150
– standard reduction potentials 121
hydrogen bond 40, 42, 44, 163
hydrogen cyanide
– chemistry 204
– formation in combustion 181
– interstellar occurrence 305
hydrogen dioxide 168

hydrogen economy 151
hydrogen ion 87, 90, 122, 352
hydrogen peroxide
– acid 89, 168
– aqueous-phase chemistry 168
– dry deposition velocity 393
– forest damage 169
– from Criegee radical 257
– gas-phase chemistry 162
– in autoxidation 138
– in plants 169
– in thunderstorm 169
– multiphase chemistry 171
– photodissociation 163
– reaction with amine 204
– reaction with carbonate radical 234
– reaction with DMS 213
– reaction with HCl 278
– reaction with hypochlorite 282
– reaction with sulphite 219, 222
– standard reduction potentials 121
hydrogen sulphide
– aqueous-phase chemistry 214
– reaction with OH 212
hydrological cycle 34
hydrolysis 93
hydrometeors 11
hydronium ion 43, 90
hydroperoxo radical see hydroperoxyl radical
– from CH2OH + O2 245
– from CH3O + O2 244
– from HCO + O2 245
hydroperoxyl radical
– from Criegee radical 257
– from H + O2 153
– from H2O2 photodissociation 163
– OH recycling 162, 217
– protolysis 167
– reaction with carbonate radical 234
– reaction with O3 162
– reaction with sulphite 222
– standard reductions potentials 121
hydroxide ion 43
hydroxyl radical
– air cleanser 366
– aqueous-phase chemistry 174
– atmospheric detergent 161
– dissociation 175
– formation from HONO photolysis 195

– from Criegee radical 257
– from dichlorine radical in water 281
– from Fenton reaction 170
– from O(1D) + H2O 161
– from O3 photocatalysis 195
– HO2 recycling 162
– lifetime 98
– reaction with alkanes 240
– reaction with alkenes 255
– reaction with CH3CHO, aqueous phase 249
– reaction with CH3OH, aqueous phase 245
– reaction with chloride 174
– reaction with CO 229
– reaction with CS2 211
– reaction with ethanol, aqueus phase 249
– reaction with HCHO 245
– reaction with HO2 162
– reaction with NH3 182
– reaction with nitrite 191
– reaction with NO2 188
– reaction with SO2 55, 217
– reaction with sulphite 222
– standard reduction potentials 121
– steady state 98
hydroxylamine 195
– biological intermediate 346
– derivates 202
hydroxymethanesulphonate 218
hydroxymethanesulphone acid 218
hypobromite 278
hypobromous acid 279
hypochlorite
– aqueous-phase chemistry 279
– discovery 278
– reaction with H2O2 166
hypochlorous acid 282
hypoiodous acid 279
hyponitrous acid 193, 197, 344

ice nucleation 46
ice nuclei 47
ice water content 34
ideal gas 23
ideal gas law 25
imine 202
impaction, particle 391
Industrial Revolution 401
inhibition 116
intensive quantity 60

interfacial chemistry 139, 169
interhalogens 383
intermolecular distance 48, 72, 133
intermolecular forces 71
internal energy 61, 64
interstellar chemistry 305, 306
interstellar cloud 304
interstellar dust 304
iodate 266, 269
iodide
– reaction with O3 386
iodine
– discovery 269
– in ocean 269
– in soils 269
– multiphase chemistry 390
– ozone depletion 270
– physiology 262
– reaction with OH 383
iodine monoxide, chemistry 383
ion 14
ionic bond 101
ionic product of water 90
iron
– biochemical importance 292
– biological chemistry 198
– complexes 291
– Earth core 308, 314
– meteorite 285
– NO complexes 198
– ocurrance 291
– oxalate 167
– oxidation 292
– oxides 291
– reaction with H2O2 171
– reaction with HO2 168
– redox role 119, 170
– rusting 138
– sediment 321
– standard electrode potential 136
irreversible 67
isobaric 63
isolated system 60, 68
isoprene 257
isopropylamine 201
isotopes 13

j-, diurnal variation 135
j-NO2, diurnal variation 135

Junge layer 211

Kelvin equation 77
kerogen 311
ketones 258
kinetic energy 26
kinetic theory of gases 23
Kirchhoff's law 65
krypton
– discovery 21
– evolution 308
krypton-85 398

Lambert–Beer law 129, 134
lapse rate 62
late heavy bombardment 309
law of mass action 85
lead
– standard electrode potential 136
– use 296
Lewis acid-base theory 88
lifetime, definition 395
ligands 96, 142
lightning 176, 180, 200, 285
limestone 94, 291
liquid water content 11, 20, 34, 418
litmus paper 269
litter 51, 330, 340
London smog 351
long-lived isotope 396
Los Angeles smog 374
Loschmidt constant 24, 26
Lyman-α line 369

magma 14, 309, 325
magmatic gas 325
magnesium
– in minerals 291
– occurrance 287
– standard electrode potential 136
magnesium chloride 274
magnetite 291
manganese
– occurrance 291
– reaction with superoxide 168
– standard electrode potential 136
mantle
– O2 production 314
– redox state 314

– seawater subduction 325
mass accommodation 140
mass action law 85
Mauna Loa record 340
Maxwell distribution 29
mean-free path 28, 30
meniscus 84
mercaptan *see* thiol
mercury
– atmospheric chemistry 294
– occurrance 292
– power-plant chemistry 293
– volcanic emission 288
mercury chloride 294
metabolisation 334
metallic bond 101
metalloid *see* half-metal
meteorite
– ammonia and methane 312
– carbonaceous 306
– chemical composition 145
– FeCl2 314
– organic matter 310
– phosphide 285
methanation 412, 413
methane
– hybrid orbitals 103
– reaction with OH 244
– residence time 98
– stratospheric chemistry 371
methane hydrate 307
methanediol 218, 245
methanesulphonic acid 55, 213
methanethiol 212
methanol
– biological formation 336
– emission 237
– reaction with OH 245
methanol economy 412
methoxy radical
– reaction with NO2 205
methyl chloride 267, 276
methyl iodide
– from wetlands 264
– oceanic emission 264
methyl radical 244
– reaction with NO 205
methylamine 201
methylglyoxal 261

methylhydroperoxide 241
methylperoxo radical 244
Mie scattering 128
Miller–Urey experiment 312, 318
mineral
– formation, early Earth 307
– hydrogen content 150
– water content 43
mineraloid 49
mixing ratio, definition 17
mixture, definition 14
molar heat capacity 62
molar mass 17
mole, definition 15
molecular abundance, interstellar 305
molecular orbital theory 102
molecule 13
molecule speed 29
monoethanolamine 404
monomethylamine 201
Montsouris Observatory ozone measurements
    365
multiphase chemistry 56
multiphase system 11
multiplicity 132

nanoparticles 20, 53, 206, 226
natural gas 241, 311
natural products 236
natural radiation 397
neon 21, 308
Nernst equation 120
Nernst heat theorem 67
net ecosystem production 340
net primary production 51, 339
neutron 13, 100, 302, 396
Newton's constant 425
nickel
– standard electrode potential 136
– volcanic emission 288
nitramine 202
nitrate
– aerosol 52
– aqueous-phase chemistry 190, 191
– in dust 54
– photodissociation 191
nitrate radical
– aqueous-phase chemistry 190
– health effect 187

– photodissociation 187
– photolysis rate 135
– reaction with NOx 187
– reaction with RH 188
nitrenium ion 183
nitric acid
– acid rain 355
– aqueous-phase chemistry 190
– dry deposition velocity 393
– formation from NO2 191
– gas-phase chemistry 189
– nitrogen cycle 345
– particle formation 55
– photolysis rate 135
– stratospheric chemistry 372
nitrification 343
nitrifying bacteria 344
nitriles 204
nitrite
– aqueous-phase chemistry 191
– in rainwater 194
– reaction with H2O2 194
– reaction with O3 194
nitrogen
– abundance 175
– cycle 343
– discovery 21
– fertiliser production 346
– residence time 176
– river-runoff 347
– thermolysis 180
nitrogen dioxide
– dry deposition velocity 393
– gas-phase chemistry 186
– HNO2 formation 192
– photocatalysis 194
– photolysis rate 135
– radical 140
– reaction with carbonate radical 234
– reaction with OH 189
– reaction with RO 205
– soil emission 344
nitrogen fixation 179, 181, 182
nitrogen monoxide
– aqueous-phase chemistry 195
– biological chemistry 198
– dry deposition velocity 393
– gas-phase chemistry 185, 186
– global emission 179

– lightning formation 343
– photosensitised conversion 197
– radical 140
– reaction with RO 205
– reaction with ROx 241
– thermal formation 180
nitrogen oxides
– biological chemistry 198
– gas-phase chemistry 185
– role in O3 formation 378
nitrogen trioxide see nitrate radical
nitroglycerine 205
nitroperoxo carbonate 197
nitrosamine 196, 202
nitrosation 198, 203
nitrosonium 196
nitrosoperoxocarboxylate 233
nitrosothiol 199
nitrosyl chloride 196
nitrous acid
– aqueous-phase chemistry 190
– as base 196
– formation from alkyl nitrite 206
– from NO2 + NO2 192
– gas-liquid equilibrium 361
– gas-phase chemistry 189
– photodissociation 189
– photolysis rate 135
– photosensitised formation 193
nitroxy organosulphate 55
nitroxyl chloride 275
nitroxyl radical 193, 196, 344
nitryl chloride 275
noble gas 311
non-metal 13, 287, 288
noosphere 329
normality 19
NOx-NOy
– gas-phase chemistry 189
– multiphase chemistry 199
nuclear bomb tests 397
nucleation
– heterogeneous 57, 385
– homogeneous 51, 55, 148
nucleophile 167, 282
nucleophilic attack 173, 221
nucleophilic substitution 203, 269
number, definition 16

O(1D) 159, 161
O(3P) 159
ocean
– carbon dioxide 232
– chemical composition 36
odd oxygen 199, 377, 379
olivine 307
open system 60
orbitals 100, 102
organic acids
– acid rain 355
– as buffer 95
– electrolytes 87
– from ozonolysis 257
– rainwater 254
organic chemistry 4, 235
organic matter
– and soil chlorine 267
– as electron donor 173
– burial 322
– concentration in air 239
– emission 237
– in meteorite 311
– in rock 311
– in soil 51, 340
– photosynthetic formation 337
– self-organising 320
organic nitrate 205
organic nitrite 205
organic peroxide 243, 419
organic radical chemistry scheme 243
organic radicals 239
organic reactive oxygen species 239
organic sulphides 208
organic sulphur compounds 209, 212
organic thiols 209
organohalogens
– gas-phase chemistry 263
– in soils 265
– stratospheric chemistry 372
organophosphines 286
orthosilicic acid 315, 327
Ostwald's solubility 78
overall reaction 108
oxalic acid 248, 251–253
oxidant 118, 119
oxidation 117
oxidation number 102
oxidation potential 119

oxidation state 117
oxidation states of elements 118
oxidative stress 169, 322
oxo or oxy 156
oxoacids 156
oxyfuel technology 415
oxygen
– abundance 158
– atmospheric concentration 318
– dissociation energy 157
– early atmosphere 322
– ground state 157
– photodissociation 158
– standard reduction potentials 121
oxygen anion radical 175
oxygen atom transfer 173
oxygenated hydrocarbons 158, 238
oxygenic photosynthesis 321
oxyhydrogen reaction 153
ozonation 171
ozone
– aqueous-phase chemistry 172
– background concentration 381
– budget 382
– decay in aqueous phase 171
– depletion in cloud 380
– diurnal variation 379
– dry deposition velocity 393
– evolution 322
– formation reaction 159
– history 365
– in situ formation rate 379
– in-cloud concentration 381
– monitoring 375
– Mt. Brocken statistics 381
– multiphase chemistry 380
– net formation rate 379
– oxidation of cyanide 204
– photodissociation 159
– photolysis rate 135
– reaction with alkenes 256
– reaction with halogens 277, 383
– seasonal variation 375
– standard redcuction potentials 121
– stratosphere 366
– stratospheric chemistry 369
– trend, Antarctica 368
– trend, troposphere 375
ozone acid 172

ozone formation cycle 377
ozone hole 367
ozone layer 366, 370
ozone precursors 377
ozone titration 379
ozone-depleting substances 371
ozonide anion 172
ozonolysis 256

PAN 206
panspermia 320
particulate matter
– acidity 363
– climate impact 146
– definition 20
– phosphorous 285
– soluble ions 364
– sulphate 215
Pauli principle 131
perchlorate 266, 277
perchloric acid 277
peridotite 315
peroxo or peroxy 157
peroxoacetic acid 252
peroxoacyl radical 206
peroxoacylnitrate 206
peroxodisulphate 221, 223
peroxohypochlorous acid 282
peroxomonosulphate 220
peroxonitrite 197
peroxonitrous acid 198
peroxosulphate radical 219, 220
peroxy radical 241
peroxynitrate, organic 205
petroleum 311
pH value 92
phase equilibrium 74
phenols 236, 260, 353
phosgene 267
phosphane 284
phosphate
– buffer 95
– chemistry 284
– in environment 158
– in rainwater 283
phosphide 285
phosphine *see* phosphane
phosphonium ion 284
phosphorous acid 283

phosphorus 282
photocatalysis 135
photocatalytic ozonation 173
photochemical smog chemistry 229
photochemistry 126
photodissociation 126, 132
photolysis 159
photolysis rate coefficient 133
photosensitiser 137, 166, 175, 193, 213
photosensitising 137
photosphere 127
photosynthesis 332, 333, 336
Pitzer theory 73
Planck constant 425
Planck–Einstein relation 425
Planck's constant 113
Planck's quantum 133
plasma 12, 14
PM1, PM2.5 and PM10 52
polar stratospheric cloud 368, 372
polar vortex 373
pollutant 10, 22
pollution chemistry 299
polonium 397
polycyclic aromatic hydrocarbons 53, 259
polyhalides 261
polywater 43
postaccrecationary period 307
potash industry 274
potassium arsenite 365
potassium iodide 365
potential temperature 62
powder material 49
power, definition 24
precipitate 81
precipitation process 56
pressure, definition 24
pressure-volume work 25, 61, 76
primitive atmosphere 316
prokaryote 181
propagating reaction 143
proton 13, 40, 90, 150, 302, 396, 425
proton-driven disproportionation 167
protoplanet 308
protosun 307
pseudo-first-order 110
pyrite 291

quantum numbers 130

quantum yield 133, 134
quarks 302
quartz 14, 51, 158, 285
quenching 132, 133

radiative transfer equation 134
radicals 140
radioactive decay 395
radioactive decay series 396
radioactive isotopes 397
radionuclides 396
radon 397
rain drop
– number concentration 37
– size 37
rainwater
– acidity 364
– chemical composition site 39
– CO2 equilibrium 363
– deposition 395
– organic acid acidity 254
– sampling 20
– trace elements 289
Raoult's law 46, 77
rate 7
rate law 109
rate-determining step 108
Rayleigh scattering 128
reaction order 109
reaction quotient 109
reaction rate 107, 111
reaction rate constant 109
reactive halogen compounds 265
reactive oxygen species 119, 137, 148, 164, 209,
    336
real mixtures 72
rearrangement reaction 192, 197, 203, 246, 256
red giant star 304
redox potential 119
redox process 117
redox state 314, 316
reductant 118
reduction 117
reduction potential 119
reduction–oxidation reaction 117
reflection, radiation 129
relative density 27
relative humidity 47
renewable resources 399

reservoir distribution 146
residence time 99, 110
residual layer 380
resistance model 394
respiration 322
reversibility 66
reversible heat 67
rhodanide 204
ribonucleic acid 319
river water, chemical composition 36
RNA world 319
rock
– degassing 312
– early earth 309
– silicate 315
– volatile gases 314
rock salt 290
root mean square velocity 26, 28
rust 138
rusting 138

sal ammoniac 182
salt cycle 264
sand 50
saturation pressure 47
saturation ratio 77
scattering, radiation 128
schreibersite 285
Schumann–Runge band 369
Schumann–Runge continuum 369
sea salt
– aerosol 52
– emission 265
– formation 264
– HCl degassing 200, 263, 272
– heterogeneous chemistry 385, 390
– reaction with NOy 188
– size range 53, 264
seawater
– carbonate chemistry 232
– chemical composition 317
– CO2 dissolution 233
– Na/Cl ratio 273
– pH 231
– SiO2 solubility 327
– subduction 325
second law of thermodynamics 65
second-order reaction 109
secondary atmosphere 316

secondary organic aerosol 53
sedimentation 391
selective catalytic reduction 186
semiconductor 136
serpentine 307, 315
silica 309
silicates 146, 287, 291, 304, 309, 315
silicon 101
– earth 224
– occurrance 297
silicon dioxide
– in cells 297
– in equilibrium with H2SiO4 315
– in PM 297
– in rocks 309
– water solubility 327
– weathering 327
silicon tetrafluoride 271
silicosis 297
singlet 131
singlet dioxygen 137, 279, 282
singlet oxygen 159, 162
size fraction
– diesel soot 226
– PM 52
– sea salt 265
size fraction, PM 20
smoke 210
smoke plague 227, 357
sodium
– element reaction with water 125
– enrichment due to HCl degassing 273
– in PM 54, 275
– occurrance 290
sodium chloride 101
sodium hypochlorite 281
soil dust
– alkaline 38
– chemical composition 52
– emission 52
– phosphorous 283
– PM contribution 52
– resuspension 53
– source of trace metals 288
soil parent mineral 51
soils 49
solar constant 127
solar nebula 306
solar radiation 127

solar spectrum 127
Solar System 306, 308
solar wind 306
solid-aqueous equilibrium 81
solubility coefficients 78
solubility equilibrium 81
solubility product 82
solution, definition 14
solvation 81
soot 54, 210, 226
spectral quantities 133
speed 7
spin 130
spontaneous freezing 46
spontaneous processes 66, 67, 69
spudomen 290
standard electrode potential 136
standard enthalpy 64
standard redox potentials 120
standard reduction potential 121
standard state function 64
star formation 304
starch-iodide paper 269, 387
state functions 60
state of matter 12
stationary 98
steady state 8, 97, 332
steady-state approximation 97
Stefan–Boltzmann constant 425
stepwise reaction 106
stratosphere 128
stratospheric multiphase chemistry 373
strong electrolyte 87
sub-cloud scavenging 363
subduction 325, 327
sublimate 294
substance 14
sulphate
– aerosol climate impact 351
– aqueous-phase chemistry 219, 223
– formation rates 224
– from DMS 213, 350
– from SO2 oxidation 215
– multiphase chemistry 217
– organic 55
sulphate radical 220
sulphate reduction 210
sulphite
– aldehyde adducts 218

– aqueous-phase chemistry 217
– oxidation, pH dependency 222
– oxidation scheme 220
– reaction with H2O2 219
sulphite radical 220
sulphonic acid 55
sulphur
– alchemy 207
– atmospheric chemistry 208
– cycle 348
– elemental 215
– elemental, in early atmosphere 322
– elemental, reaction with O2 212
– global turnover 350
– in coal 211
– in Odyssey 207
– natural global emission 208
– occurrence 210
– organic compounds 207
– oxides 215
– oxoacids 215
– radicals 211, 214, 221
– sources 209
sulphur dioxide
– air pollution 207, 351, 357
– aqueous-phase chemistry 219
– atmospheric concentration 210
– emission 210, 351
– emission trend 210
– from DMS oxidation 350
– gas-liquid equlilibrium 361
– multiphase chemistry 217
– reaction with OH 216
– residence time 217
– rock degassing 314
– SOA formation 55
– volcanic 210, 325
– volcanic emission 326
sulphur hexafluoride 262
sulphuric acid 217
– acid rain 355
– from S(IV) oxidation 219
– from SO2 and Criegee radical 257
– from SO2 oxidation 215
– particle formation 55
– stratospheric 373
sulphurous acid 217
superoxide anion
– aqueous-phase chemistry 167

– biological chemistry 196
– conjugated base to HO2 167
– formation from O2 167
– from O3 decay 171
– from oxalate photodissociation 167
– reaction with carbonate radical 234
– reaction with NO2 192
supersaturation 77
surface tension 44
surface work 71
surface-active substance 45
sustainable chemistry 399
sustainable society 300
sylvine 290

termination reaction 143
terpene 257
terrestrial radiation 130
tetraethyl lead 296
The Great Stink 356
Theia 308
thermochemistry 64
thermodynamics, definition 59
thiol 207
third law of thermodynamics 67
third-order reaction 109
three-way catalytic converters 187, 420
threshold wavelength 134
thunderstorm 169
titanium, standard electrode potential 136
titanium dioxide
– in aerosol 193
– in soil dust 278
– semiconductor 136
toluene
– in air 239
– reaction with OH 260
total column ozone 367
total suspended matter 20
town fog 357
trace metals
– in PM 290
– in rainwater 289
– toxicity 288
transfer complex 109
transition metal ions 168, 288
transition state 114
tremolite 307, 315
trimethylamine 201

triple bond 105
triplet 131
troilite 307
troposphere 10
tunneling 124

Umkehr measurement 367
urban pollution 356
– organic acid 254
– ozone 379, 380
– sulphur dioxide 210
UV radiation
– absorption through O3 370
– atmospheric penetration 369
– early atmosphere 312
– O2 level early Earth 322
– O3 depletion 146
– role in life evolution 312

valence bond theory 102
valence hybrids 105
valency 102
van't Hoff's reaction isobar 86
vapour pressure lowering 45
vehicle emission control 187
velocity
– dry deposition 391
– molecule 25, 28
vibrational-rotational 132
vinoxy radical 256
viscosity 30, 32
vital elements 288
vital force 4
volatile organic compound
– aqueous chemistry 250
– biomass burning 238
– contribution to O3 382
– gas-phase chemistry 248
– natural emission 340
– SOA formation 54
volcanic eruption 325
volcanoes
– emission of metals 287
– emissions 325
– eruptions 325
volume, definition 16
voluntary 66, 69

washout 397
wastewater treatment 139, 169

water
– activity 90
– and carbon cycle 327
– and life 319
– bleaching properties 169
– chemistry 123
– chlorination 281
– cryoscopic constant 46
– dissociation 90
– early atmosphere 316
– freezing 46
– from comets 310
– from rock degassing 314
– hydrosphere 33
– in clouds 11
– in rocks 314, 315
– ion product 90
– ligand 95
– pH 361
– photolysis 315
– properties 40, 43
– reaction with Criegee-radical 257
– reaction with O(1D) 161
– solvent 43
– stratospheric 368
– surface tension 45
– UV protection 322
– volcanic 325
water cycle 35
water radical cation 122
water splitting 195, 334
water structure 42, 43
water treatment 171, 281
water vapour in air 22, 47
water-gas shift 415
weak electrolyte 87
weathering 327
wet deposition 394
wetlands 208, 264, 286, 349
winter smog 227
work 24, 25, 61, 68, 423
work of mixing 71

xenon
– evolution 308

Zeldovich mechanism 181
zero-order reactions 115
zinc
– occurrence 295
– standard electrode potential 136

Periodic Table

a) New IUPAC System
b) Chemical Abstract System
Blue: radioactive elements

Period	1	2	3	4	5	6	7	8	9	10	11	12	13	14	15	16	17	18
a)	1	2	3	4	5	6	7	8	9	10	11	12	13	14	15	16	17	18
b)	Ia	IIa	IIIb	IVb	Vb	VIb	VIIb	VIIIb	VIIIb	VIIIb	Ib	IIb	IIIa	IVa	Va	VIa	VIIa	VIIIa
1	1 H																	2 He
2	3 Li	4 Be											5 B	6 C	7 N	8 O	9 F	10 Ne
3	11 Na	12 Mg											13 Al	14 Si	15 P	16 S	17 Cl	18 Ar
4	19 K	20 Ca	21 Sc	22 Ti	23 V	24 Cr	25 Mn	26 Fe	27 Co	28 Ni	29 Cu	30 Zn	31 Ga	32 Ge	33 As	34 Se	35 Br	36 Kr
5	37 Rb	38 Sr	39 Y	40 Zr	41 Nb	42 Mo	43 Tc	44 Ru	45 Rh	46 Pd	47 Ag	48 Cd	49 In	50 Sn	51 Sb	52 Te	53 I	54 Xe
6	55 Cs	56 Ba	57 La	72 Hf	73 Ta	74 W	75 Re	76 Os	77 Ir	78 Pt	79 Au	80 Hg	81 Tl	82 Pb	83 Bi	84 Po	85 At	86 Rn
7	87 Fr	88 Ra	89 Ac	104 Rf	105 Db	106 Sg	107 Bh	108 Hs	109 Mt	110 Ds	111 Rg	112 Eka-Hg	113 Eka-Tl	114 Eka-Pb	115 Eka-Bi	116 Eka-Po	117 Eka-At	118 Eka-Rn

lanthanides

58 Ce	59 Pr	60 Nd	61 Pm	62 Sm	63 Eu	64 Gd	65 Tb	66 Dy	67 Ho	68 Er	69 Tm	70 Yb	71 Lu

actinides

90 Th	91 Pa	92 U	93 Np	94 Pu	95 Am	96 Cm	97 Bk	98 Cf	99 Es	100 Fm	101 Md	102 No	103 Lr

www.ingramcontent.com/pod-product-compliance
Lightning Source LLC
Chambersburg PA
CBHW080127220326
41598CB00032B/4979